Compendium of Plant Genomes

Whole-genome sequencing is at the cutting edge of life sciences in the new millennium. Since the first genome sequencing of the model plant Arabidopsis thaliana in 2000, whole genomes of about 70 plant species have been sequenced and genome sequences of several other plants are in the pipeline. Research publications on these genome initiatives are scattered on dedicated web sites and in journals with all too brief descriptions. The individual volumes elucidate the background history of the national and international genome initiatives; public and private partners involved; strategies and genomic resources and tools utilized; enumeration on the sequences and their assembly; repetitive sequences; gene annotation and genome duplication. In addition, synteny with other sequences, comparison of gene families and most importantly potential of the genome sequence information for gene pool characterization and genetic improvement of crop plants are described.

Interested in editing a volume on a crop or model plant?

Please contact Dr. Kole, Series Editor, at ckole2012@gmail.com

More information about this series at http://www.springer.com/series/11805

Philipp Simon · Massimo Iorizzo ·
Dariusz Grzebelus · Rafal Baranski
Editors

The Carrot Genome

Springer

Editors
Philipp Simon
Vegetable Crops Research Unit
USDA-ARS
Madison, WI, USA

Massimo Iorizzo
Plants for Human Health Institute
North Carolina State University
Kannapolis, NC, USA

Dariusz Grzebelus
University of Agriculture in Krakow
Kraków, Poland

Rafal Baranski
Faculty of Biotechnology and Horticulture
University of Agriculture in Krakow
Kraków, Poland

ISSN 2199-4781 ISSN 2199-479X (electronic)
Compendium of Plant Genomes
ISBN 978-3-030-03388-0 ISBN 978-3-030-03389-7 (eBook)
https://doi.org/10.1007/978-3-030-03389-7

Library of Congress Control Number: 2019934354

This Springer imprint is published by the registered company Springer Nature Switzerland AG
The registered company address is: Gewerbestrasse 11, 6330 Cham, Switzerland

Contents

Contributors

Maria Soledad Alessandro Estación Experimental Agropecuaria La Consulta, Instituto Nacional de Tecnología Agropecuaria, Mendoza, Argentina

Charlotte Allender School of Life Sciences, University of Warwick, Warwick, UK

Rafal Baranski Faculty of Biotechnology and Horticulture, Institute of Plant Biology and Biotechnology, University of Agriculture in Krakow, Krakow, Poland

Hamed Bostan Plants for Human Health Institute, North Carolina State University, Kannapolis, NC, USA

Mathilde Briard IRHS, Agrocampus Ouest, INRA, Université d'Angers, SFR QuaSaV, Angers, France

Johanne Brunet USDA-Agricultural Research Service, Vegetable Crops Research Unit, Department of Entomology, University of Wisconsin, Madison, WI, USA

Pablo F. Cavagnaro National Scientific and Technical Research Council (CONICET), National Institute of Agricultural Technology (INTA) E.E.A. La Consulta, San Carlos, Mendoza, Argentina;
Faculty of Agricultural Sciences, National University of Cuyo, Mendoza, Argentina;
National Scientific and Technical Research Council (CONICET), Buenos Aires, Argentina;
La Consulta Experimental Station, National Institute of Agricultural Technology (INTA), La Consulta, Mendoza, Argentina;
Horticulture Institute, Faculty of Agricultural Sciences, National University of Cuyo, Lujan de Cuyo, Mendoza, Argentina

Valérie Le Clerc IRHS, Agrocampus Ouest, INRA, Université d'Angers, SFR QuaSaV, Angers, France

Shelby Ellison USDA-Agricultural Research Service, Vegetable Crops Research Unit, Department of Horticulture, University of Wisconsin-Madison, Madison, WI, USA

Claudio R. Galmarini Estación Experimental Agropecuaria La Consulta, Instituto Nacional de Tecnología Agropecuaria, Mendoza, Argentina;
Consejo Nacional de Investigaciones Científicas y Técnicas, Buenos Aires, Argentina;
Facultad de Ciencias Agrarias, Universidad Nacional de Cuyo, Chacras de Coria, Luján, Mendoza, Argentina

Emmanuel Geoffriau Agrocampus Ouest, Institut de Recherche en Horticulture et Semences – UMR 1345, Angers, France

Dariusz Grzebelus Faculty of Biotechnology and Horticulture, Institute of Plant Biology and Biotechnology, University of Agriculture in Krakow, Krakow, Poland

Ewa Grzebelus Faculty of Biotechnology and Horticulture, Institute of Plant Biology and Biotechnology, University of Agriculture in Krakow, Krakow, Poland

Mwafaq Ibdah Newe Ya'ar Research Center, Agriculture Research Organization, Ramat Yishay, Israel

Massimo Iorizzo Department of Horticultural Sciences, Plants for Human Health Institute, North Carolina State University, Kannapolis, NC, USA

Marina Iovene CNR, Institute of Biosciences and BioResources (CNR-IBBR), Portici, NA, Italy

Bettina Linke Department of Biology, Humboldt University, Berlin, Germany

Aneta Lukasiewicz Faculty of Biotechnology and Horticulture, Institute of Plant Biology and Biotechnology, University of Agriculture in Krakow, Krakow, Poland

Alicja Macko-Podgórni Faculty of Biotechnology and Horticulture, Institute of Plant Biology and Biotechnology, University of Agriculture in Krakow, Krakow, Poland

Jennifer R. Mandel Department of Biological Sciences, University of Memphis, Memphis, TN, USA

Andrew Muchlinski Department of Biological Sciences, Virginia Polytechnic Institute and State University, Blacksburg, VA, USA

Bhagwat Nawade Newe Ya'ar Research Center, Agriculture Research Organization, Ramat Yishay, Israel

Thomas Nothnagel Institute for Breeding Research on Horticultural Crops, Federal Research Centre for Cultivated Plants, Quedlinburg, Germany

Marti Pottorff Plants for Human Health Institute, North Carolina State University, Kannapolis, NC, USA

Douglas Senalik USDA-Agricultural Research Service, Vegetable Crops Research Unit, Department of Horticulture, University of Wisconsin-Madison, Madison, WI, USA

Philipp W. Simon USDA-Agricultural Research Service, Vegetable Crops Research Unit, Department of Horticulture, University of Wisconsin-Madison, Madison, WI, USA

David M. Spooner USDA-Agricultural Research Service, Vegetable Crops Research Unit, Department of Horticulture, University of Wisconsin-Madison, Madison, WI, USA

Dorothea Tholl Department of Biological Sciences, Virginia Polytechnic Institute and State University, Blacksburg, VA, USA

Lindsey J. du Toit Department of Plant Pathology, Washington State University Mount Vernon NWREC, Mount Vernon, WA, USA

Allen Van Deynze University of California, Davis, Seed Biotechnology Center and Plant Breeding Center, Davis, CA, USA

Mossab Yahyaa Newe Ya'ar Research Center, Agriculture Research Organization, Ramat Yishay, Israel

1 Economic and Academic Importance

Philipp W. Simon

Abstract

Carrot is a relatively recently domesticated vegetable crop that provides a significant source of dietary vitamin A to consumers. Earlier cultivar development for carrot was most extensive in temperate regions of Europe and Asia, but cultivars adapted to tropical and sub-tropical climates have contributed significantly to an increase in global carrot production in the last 50 years. Carrot germplasm includes a broad range of genotypic and phenotypic diversity that contributes to its wide adaptability. There has not been an extensive written historical record for carrot, where color and flavor were the most frequently noted attributes of the crop from its origins in Central Asia through its early development into the Middle East, North Africa, Europe, and Asia. Carotenoids and anthocyanins account for carrot colors and have been a major focus for carrot researchers, and the use of carrot in demonstrating biological totipotency and in providing the first evidence of plant transfer of mitochondrial DNA to the plastid genome has generated significant attention for carrot. The economic importance of carrot in agriculture and academic contributions attributable to carrot that are summarized in this chapter suggest an optimistic future for improved crop production and expanded basic research opportunities that are broadened with the availability of a carrot genome sequence.

1.1 Introduction

Carrot is a crop with a wide range of phenotypic variation utilized by breeders (Simon et al. 2008) and genotypic variation that is only beginning to be fully evaluated (Iorizzo et al. 2016). Carrots are among the top 10 vegetables, based on global production records of primary vegetables, after tomatoes, onions, cabbage, cucumbers, and eggplant (FAO 2017). Most of the 22 vegetables among those in that class of primary vegetables are members of the Amaryllidaceae, Brassicaceae, Compositae, Cucurbitaceae, Leguminosae, Poaceae, Solanaceae, where several major crops rank high in terms of global production for each of those families. In contrast, carrots are the only member of the Apiaceae in that class of primary vegetables, but several other vegetable crops, including celery, cilantro, fennel, and arracacha, and many spice crops are also significant Apiaceaous crops grown globally (Rubatzky et al. 1999). Carrot today is grown globally with extensive adaptation to temperate production areas in Europe, Asia, and the

P. W. Simon (✉)
USDA Agricultural Research Service, Vegetable Crops Research Unit, Department of Horticulture, University of Wisconsin, 1575 Linden Dr., 53706 Madison, WI, USA
e-mail: philipp.simon@ars.usda.gov

P. Simon et al. (eds.), *The Carrot Genome*, Compendium of Plant Genomes,
https://doi.org/10.1007/978-3-030-03389-7_1

Americas, but with more recent cultivar development for sub-tropical and even tropical climate (Simon et al. 2008). Orange carrots are rich in provitamin A and carotenoids (see Chap. 9), and with the expansion of carrot production in warmer climates, they can provide a sustainable, locally produced food to contribute to reducing the incidence of vitamin A deficiency, which continues to be particularly prevalent in those warmer climates (Tanumihardjo 2012). While categorized as a cool-season crop, the adaptation of carrots to warmer climates raises a positive indication for continued expanded carrot production into the future.

1.2 Global Production and Economic Value

Global carrot production has risen steadily in the last 50 years (FAO 2017), with a threefold increase in production area (Table 1.1; Fig. 1.1) and twofold increase in yield (Table 1.2) to result in a sixfold increase in total production (Table 1.3; Fig. 1.2). With these increases, the average global increase in per capita carrot production has risen 2.7-fold in the last 50 years (Table 1.4; Fig. 1.3). All of these increases in carrot production and availability have risen slightly ahead of the averages for the 22 primary vegetables, so that carrot today accounts for 5.5% of the per capita vegetable availability globally (Table 1.4). This increase was particularly steep in Asia. A rise in the economic value of the carrot crop follows a similar trend with a sixfold increase in global production value in the last 50 years (Table 1.5) and a twofold increase in value per hectare to the grower (Table 1.6). As with production trends, economic increases in the value of the carrot crop were particularly high in Asia (Tables 1.5, 1.6).

Unfortunately, FAO statistics combine turnip production with carrot, as they do for several other primary vegetable crops, like cauliflower and broccoli. Consequently, statistics presented in Tables 1.1, 1.2, 1.3, 1.4, 1.5, 1.6 and Figs. 1.1, 1.2, 1.3 include the combined values for carrot and turnip. Carrots account for most of the production values, based on crop-specific information available for the USA and Europe. Turnip production was less than 2% of carrot production in the USA in 1950 (USDA 1954), and publication of U.S. statistics for turnips was discontinued in 1963. Turnip production in Europe in the early 1990s was <1% that of carrot (Hinton 1991).

The portion of the carrot crop grown under organic production management practices has grown in recent decades in the more well-developed carrot markets of North America and Europe, accounting for 11% of the 2016 U.S. market (USDA 2017) and 25–30% of Danish and German markets (Willer and Lernoud 2016). Consumers place a high value on nutritional quality and flavor (Yiridoe et al. 2005), and the generally positive public impression of carrots as a nutritious food may account for increasing organically grown carrot consumption. The broad range of genetic diversity and new tools for improving carrot flavor available to breeding programs (see Chap. 16) provide promising prospects for flavor improvement, and while production of organic carrots is not without pest, disease, and weed challenges, progress has been made in managing them (Simon et al. 2017).

1.2.1 Historical Records

The first archeological record for carrot was seed found at Bronze Age campsites of around 4500 years ago in Switzerland and southern Germany (Neuweiler 1931), where it was speculated that seed was likely used as a spice or medicinal herb, as many other Apiaceous plants are used today (Rubatzky et al. 1999). Carrot tissue preparations were also found on a Roman shipwreck off Tuscany of around 2100 years ago where it was included in what is thought to be a medicinal preparation including several other plants (Smithsonian Insider 2010). Relatively little was written about carrot during its early history other than periodic references to its color and flavor (Banga 1957a, b, 1963). The 1963 work of Banga is the most extensive publication dedicated to carrot to date, where he reviewed and analyzed not only written historical records

Table 1.1 Global production area of 22 primary vegetables and of carrots + turnips comparing the average for the 1961–1965 period to the 2011–2015 period

Year	Region	Vegetables, primary (ha)	Carrots and turnips (ha)	Percent of vegetables[a]	Change[b]
1961–1965	World	13,057,559	383,965	2.9	
	Africa	1,028,571	36,782	3.6	
	Americas	1,949,102	57,351	2.9	
	Asia	5,782,030	88,985	1.5	
	Europe	4,215,052	197,915	4.7	
	Oceania	82,804	2932	3.5	
2011–2015	World	34,640,706	1,166,885	3.4	303%
	Africa	5,620,793	112,093	2.0	304%
	Americas	2,736,117	114,494	4.2	199%
	Asia	23,049,230	670,127	2.9	753%
	Europe	3,101,668	263,703	8.5	133%
	Oceania	132,899	6467	4.9	221%

Data from FAO (2017)
[a]Percent of vegetables is the carrot + turnip percentage of the primary vegetables
[b]Change is the carrot + turnip value for 2011–2015 relative to 1961–2015

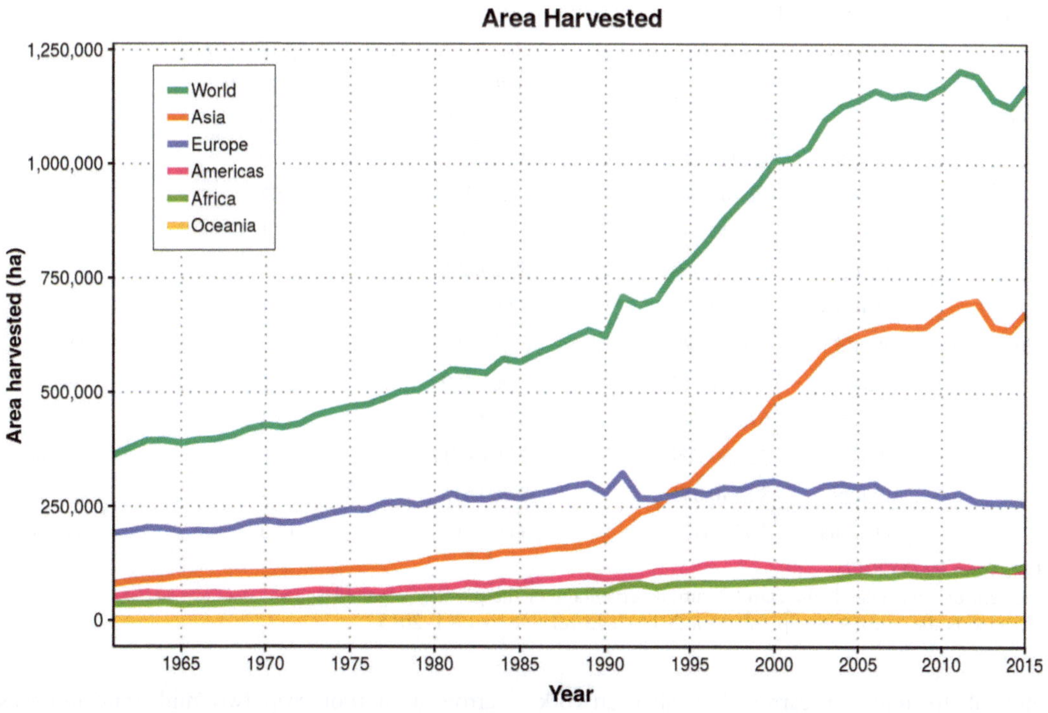

Fig. 1.1 Global and regional carrot + turnip production area 1961–2015

Table 1.2 Global yield of 22 primary vegetables and of carrots + turnips comparing the average for the 1961–1965 period to the 2011–2015 period

Year	Region	Vegetables, primary (hg/ha)	Carrots and turnips (hg/ha)	Percent of vegetables[a]	Change[b]
1961–1965	World	95,996	166,893	174	
	Africa	66,371	93,467	141	
	Americas	106,438	213,260	200	
	Asia	89,318	127,386	143	
	Europe	114,450	183,271	160	
	Oceania	111,474	280,789	252	
2011–2015	World	184,330	329,021	178	197%
	Africa	85,937	184,041	214	197%
	Americas	220,969	303,285	137	142%
	Asia	195,966	357,687	183	281%
	Europe	246,758	324,813	132	177%
	Oceania	191,741	526,737	275	188%

Data from FAO (2017)
[a]Percent of vegetables is the carrot + turnip percentage of the primary vegetables
[b]Change is the carrot + turnip value for 2011–2015 relative to 1961–2015

Table 1.3 Global crop production of 22 primary vegetables and of carrots + turnips comparing the average for the 1961–1965 period to the 2011–2015 period

Year	Region	Vegetables, primary (ton)	Carrots and turnips (ton)	Percent of vegetables[a]	Change[b]
1961–1965	World	133,903,539	6,413,270	4.8	
	Africa	7,705,073	343,979	4.5	
	Americas	22,366,990	1,222,371	5.5	
	Asia	55,344,331	1,134,616	2.1	
	Europe	47,578,050	3,629,651	7.6	
	Oceania	909,095	82,654	9.1	
2011–2015	World	724,328,890	38,352,663	5.3	598%
	Africa	51,501,312	2,061,469	4.0	599%
	Americas	67,561,800	3,469,613	5.1	284%
	Asia	520,393,097	23,919,717	4.6	2108%
	Europe	82,140,041	8,560,515	10.4	236%
	Oceania	2,732,641	341,350	12.5	413%

Data from FAO (2017)
[a]Percent of vegetables is the carrot + turnip percentage of the primary vegetables
[b]Change is the carrot + turnip value for 2011–2015 relative to 1961–2015

thought to refer to carrot, but also artwork thought to depict carrot, and early seed catalog illustrations and descriptions of carrots. The early written and illustrative evidence attributed to carrots as a root crop two millennia ago was disputed by Banga who was not convinced that carrot was the root crop described. He concluded that carrot was not developed as a root crop

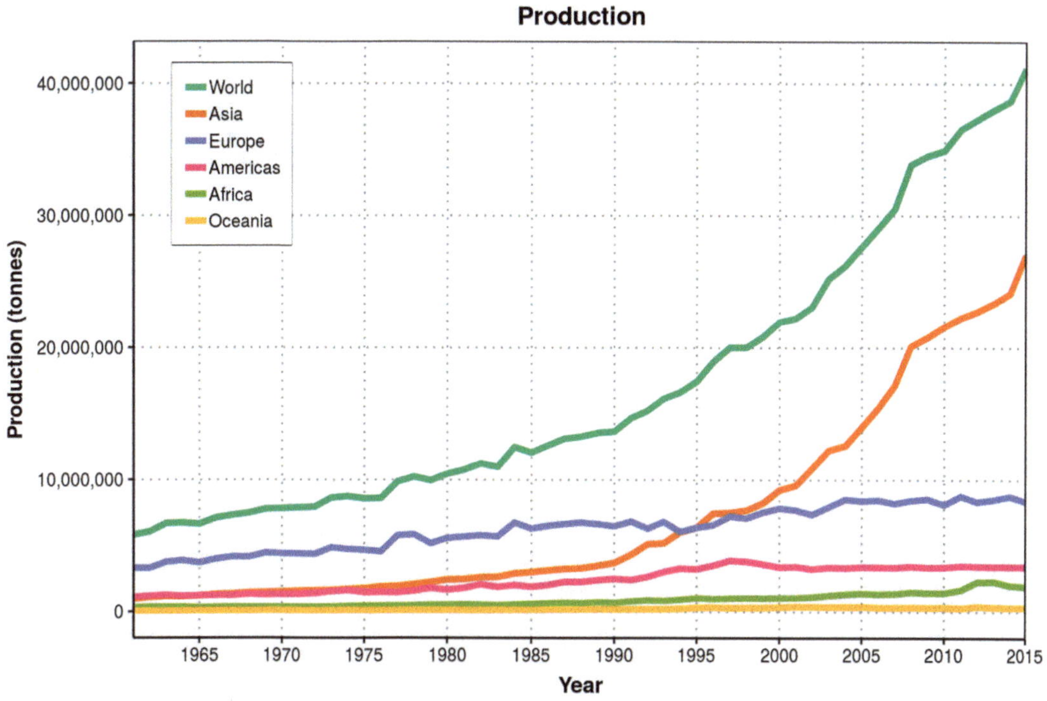

Fig. 1.2 Global and regional carrot + turnip total crop production 1961–2015

Table 1.4 Global production per capita of 22 primary vegetables and of carrots + turnips comparing the average for the 1961–1965 period to the 2011–2015 period

Year	Region	Vegetables, primary (kg)	Carrots and turnips (kg)	Percent of vegetables[a]	Change[b]
1961–1965	World	41.70	1.99	4.8	
	Africa	25.07	1.12	4.5	
	Americas	49.27	2.69	5.5	
	Asia	30.62	0.63	2.0	
	Europe	76.25	5.81	7.6	
	Oceania	53.91	4.89	9.1	
2011–2015	World	100.39	5.32	5.3	266%
	Africa	45.35	1.82	4.0	162%
	Americas	69.68	3.58	5.1	133%
	Asia	120.12	5.52	4.6	882%
	Europe	111.06	11.58	10.4	199%
	Oceania	71.19	8.89	12.5	182%

Data from FAO (2017)
[a]Percent of vegetables is the carrot + turnip percentage of the primary vegetables
[b]Change is the carrot + turnip value for 2011–2015 relative to 1961–2015

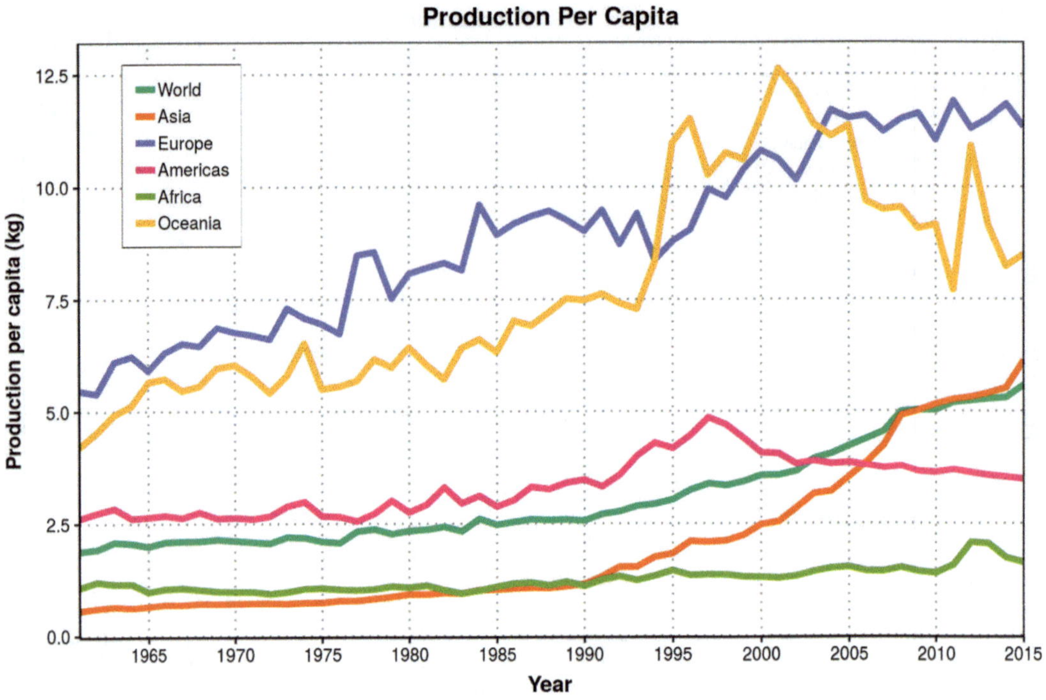

Fig. 1.3 Global and regional carrot + turnip production per capita 1961–2015

Table 1.5 Global gross production value of 22 primary vegetables and of carrots + turnips comparing the average for the 1961–1965 period to the 2011–2015 period

Year	Region	Vegetables, fresh (1000 Int. $)	Carrots and turnips (1000 Int. $)	Percent of vegetables[a]	Change[b]
1961–1965	World	12,120,283	1,600,105	13.2	
	Africa	991,392	85,822	8.7	
	Americas	826,997	304,980	36.9	
	Asia	7,449,353	283,086	3.8	
	Europe	2,809,668	905,594	32.2	
	Oceania	42,873	20,622	48.1	
2011–2015	World	52,135,962	9,568,951	18.4	598%
	Africa	3,527,670	514,334	14.6	599%
	Americas	1,413,772	865,665	61.2	284%
	Asia	44,987,006	5,967,945	13.3	2108%
	Europe	2,095,569	2,135,840	101.9	236%
	Oceania	111,945	85,166	76.1	413%

Data from FAO (2017)
[a]Percent of vegetables is the carrot + turnip percentage of the primary vegetables
[b]Change is the carrot + turnip value for 2011–2015 relative to 1961–2015

Table 1.6 Global value per hectare of production of 22 primary vegetables and of carrots + turnips comparing the average for the 1961–1965 period to the 2011–2015 period

Year	Region	Vegetables, fresh (1000 Int. $/ha)	Carrots and turnips (1000 Int. $/ha)	Percent of vegetables[a]	Change[b]
1961–1965	World	0.9282	4.1673	449	
	Africa	0.9639	2.3333	242	
	Americas	0.4243	5.3178	1253	
	Asia	1.2884	3.1813	247	
	Europe	0.6666	4.5757	686	
	Oceania	0.5178	7.0339	1359	
2011–2015	World	1.5050	8.2004	545	197%
	Africa	0.6276	4.5885	731	197%
	Americas	0.5167	7.5608	1463	142%
	Asia	1.9518	8.9057	456	280%
	Europe	0.6756	8.0994	1199	177%
	Oceania	0.8423	13.1690	1563	187%

Data from FAO (2017)
[a]Percent of vegetables is the carrot + turnip percentage of the primary vegetables
[b]Change is the carrot + turnip value for 2011–2015 relative to 1961–2015

during the Roman Empire, but rather about 1100 years ago, and was Central Asian in origin, as Vavilov suggested and molecular evidence supports (Iorizzo et al. 2013). More recently, Stolarczyk and Janick (2011) evaluated evidence for an earlier origin of carrots as a root crop in Turkey, Greece, and Italy, including support for orange storage and carrot color. As new archeological and artistic evidence for carrot arises, the early history of carrot will hopefully become clearer.

Carrot root color was a primary focus of early descriptions of the crop as noted above, and Vilmorin (1859) also wrote quite extensively about the origins of orange color in carrots where he evaluated intercrosses of wild and cultivated carrots. The genetics of carrot color due to carotenoids continues to be a major focus for carrot research today (also see Chap. 14), but it was the development of in vitro methods for plant propagation that brought carrot most widely into the basic scientific literature.

1.3 Totipotency and Future Directions

On the occasion of the 125th anniversary of Science magazine, the editors generated 125 questions that point to critical knowledge gaps addressing the question: What don't we know? In this broad-ranging sweep of questions characterized as "opportunities to be exploited" (Siegfried 2009), compelling scientific questions that could not be answered were raised, and most of those questions dealt with physics, mathematics, and human health. Only six dealt specifically with plant sciences, and the only one of those six to be included among the 25 top questions that were included as separate articles in that 125th anniversary issue of Science was "How does a single somatic cell become a whole plant" (Miller 2009). This article noted that nearly 50 years earlier "scientists learned they could coax carrot cells to undergo… embryogenesis in the lab", referring to the seminal work of Steward et al. (1958, 1964) that provided the foundation for the concept that became known as totipotency in plants and the inspiration for pluripotent stem cell research in humans. Totipotency has, in fact, been a focus of numerous research efforts, yet the biology of totipotency, having been observed in not only carrot but also in many other plants, remains a largely unanswered question.

Significant efforts have been made in carrot, advancing the basic scientific knowledge of totipotency, carotenoid accumulation, and a wide range of other research topics, and applied research has increased the productivity and improved the quality of the crop significantly in the last 50 years. Carrot genomic information has already contributed to our understanding of organelle evolution with first evidence of plant transfer of mitochondrial DNA to the plastid genome discovered in carrot, as highlighted in Chap. 12. The availability of the carrot genome sequence will provide future research efforts with an additional valuable tool to better understand and improve this important vegetable crop.

References

Banga O (1957a) Origin of the European cultivated carrot. Euphytica 6:54–63

Banga O (1957b) The development of the original European carrot material. Euphytica 6:64–76

Banga O (1963) Main types of the western carotene carrot and their origin. W.E.J. Tjeenk Willink, Zwolle, The Netherlands

FAO (2017). www.fao.org/statistics

Hinton L (1991) The European market for fruit and vegetables. Elsevier, London

Iorizzo M, Senalik DA, Ellison SL et al (2013) Genetic structure and domestication of carrot (*Daucus carota* L. subsp. *sativus* L.) (Apiaceae). Am J Bot 100:930–938

Iorizzo M, Ellison S, Senalik D, Zeng P et al (2016) A high-quality carrot genome assembly provides new insights into carotenoid accumulation and asterid genome evolution. Nat Genet 48:657–666

Miller G (2009) How does a single somatic cell become a whole plant? Science 209:86

Neuweiler E (1931) Die Pflanzenreste aus dem spätbronzezeitlichen Pfahlbau "sumpf" bei Zug. Vierteljahrschr Naturf Ges Zurich 76:776–732

Rubatzky VE, Quiros CF, Simon PW (1999) Carrots and related vegetable Umbelliferae. CABI, New York

Siegfried T (2009) Praise of hard questions. Science 309:75

Simon PW, Freeman RE, Vieira JV et al (2008) Carrot. In: Prohens J, Carena MJ, Nuez F (eds) Handbook of crop breeding, vol 1. Vegetable breeding. Springer, Heidelberg, pp 327–357

Simon PW, Zystro J, Roberts PA et al (2017) The CIOA (Carrot Improvement for Organic Agriculture) project: location, cropping system, and genetic background influence carrot performance including top height and flavor. Acta Hort 1153:1–8

Smithsonian Insider (2010) DNA sequencing reveals simple vegetables in ancient Roman medicines. https://insider.si.edu/2010/10/dna-sequencing-reveals-simple-vegetables-in-ancient-roman-medicines/

Steward FC, Mapes MO, Smith JO (1958) Growth and organized development of cultured cells. I. Growth and division of freely suspended cells. Am J Bot 45:693–703

Steward FC, Mapes MO, Kent AE, Holsten RD (1964) Growth and development of cultured plant cells. Science 743:20–27

Stolarczyk J, Janick J (2011) Carrot: history and iconography. Chron Hortic 51:13–18

Tanumihardjo S (ed) (2012) Carotenoids and human health. Springer, New York

Vilmorin M (1859) Notice sur I'amelioration de la carrotte sauvage. In: Vilmorin L (ed) Notice sur I'amelioration des plantes par Ie semis. Librairie Agricole, Paris, pp 5–29

USDA (2017). www.nass.usda.gov/Publications/Todays_Reports/reports/census17.pdf

USDA (1954). www.nass.usda.gov

Willer H, Lernoud J (2016) The world of organic agriculture—statistics and emerging trends. Research Institute of Organic Agriculture (FIBL) and IFOAM, Berlin

Yiridoe EK, Bonti-Ankomah S, Martin RC (2005) Comparison of consumer perceptions and preference toward organic versus conventionally produced foods: a review and update of the literature. Renew Agric Food Syst 20:193–205

Daucus: Taxonomy, Phylogeny, Distribution

2

David M. Spooner

Abstract

Cultivated carrot (*Daucus carota* subsp. *sativus*) is the most important member in the Apiaceae family in terms of economy and nutrition and is considered the second most popular vegetable in the world after potato. Despite its global importance, the systematics of *Daucus* remains under active revision at the species, genus, and subtribal levels. The phylogenetic relationships among the species of *Daucus* and close relatives in the Apioideae have been clarified recently by a series of molecular studies using DNA sequences of the plastid genes *rbcL* and *matK*; plastid introns rpl16, rps16, rpoC1; nuclear ribosomal DNA internal transcribed spacer (ITS) sequences; and plastid DNA restriction sites. Of these DNA markers, the ITS region consisting of ITS1, the intervening spacer, and ITS2 has served as the main marker used. Recently, next-generation DNA sequencing methodologies have been used. We review these techniques and how they are impacting the taxonomy of the genus *Daucus*.

D. M. Spooner (✉)
USDA-Agricultural Research Service, Vegetable Crops Research Unit, Department of Horticulture, University of Wisconsin–Madison, 1575 Linden Dr, 53706-1590 Madison, WI, USA
e-mail: David.Spooner@ars.usda.gov

2.1 Taxonomy of the Apiaceae (Umbelliferae)

The Apiaceae (Umbelliferae) family contains 466 genera and 3820 species (Plunkett et al. in press) and is one of the largest families of seed plants. It is nearly cosmopolitan in distribution, but most diverse in temperate regions of the northern hemisphere (Downie et al. 2000a, b, c; Heywood 1983). It is well supported as a monophyletic family, closely related to the families Araliaceae, Pittosporaceae, and Myodocarpaceae, and these, along with three smaller families, constitute the order Apiales, containing about 5400 species (Judd et al. 2016; Plunkett et al. 1996b).

The Apiaceae is well defined morphologically by a suite of characters, typically including herbs with compound leaves, stems usually hollow in the internodes and with secretory canals containing ethereal oils, resins, and other compounds; alternate compound leaves or simple and deeply divided or lobed leaves with sheathing petioles; determinate inflorescences containing simple to compound umbels often subtended by involucral bracts; small flowers with 5 sepals, 5 petals, 5 stamens, and 2 connate carpels with an inferior ovary; 2 small stigmas; with the fruit a schizocarp (dry fruits breaking into one-seeded segments) with each of the two mericarps attached to an entire and deeply divided forked central stalk (carpophone) (Judd et al. 2016).

P. Simon et al. (eds.), *The Carrot Genome*, Compendium of Plant Genomes,
https://doi.org/10.1007/978-3-030-03389-7_2

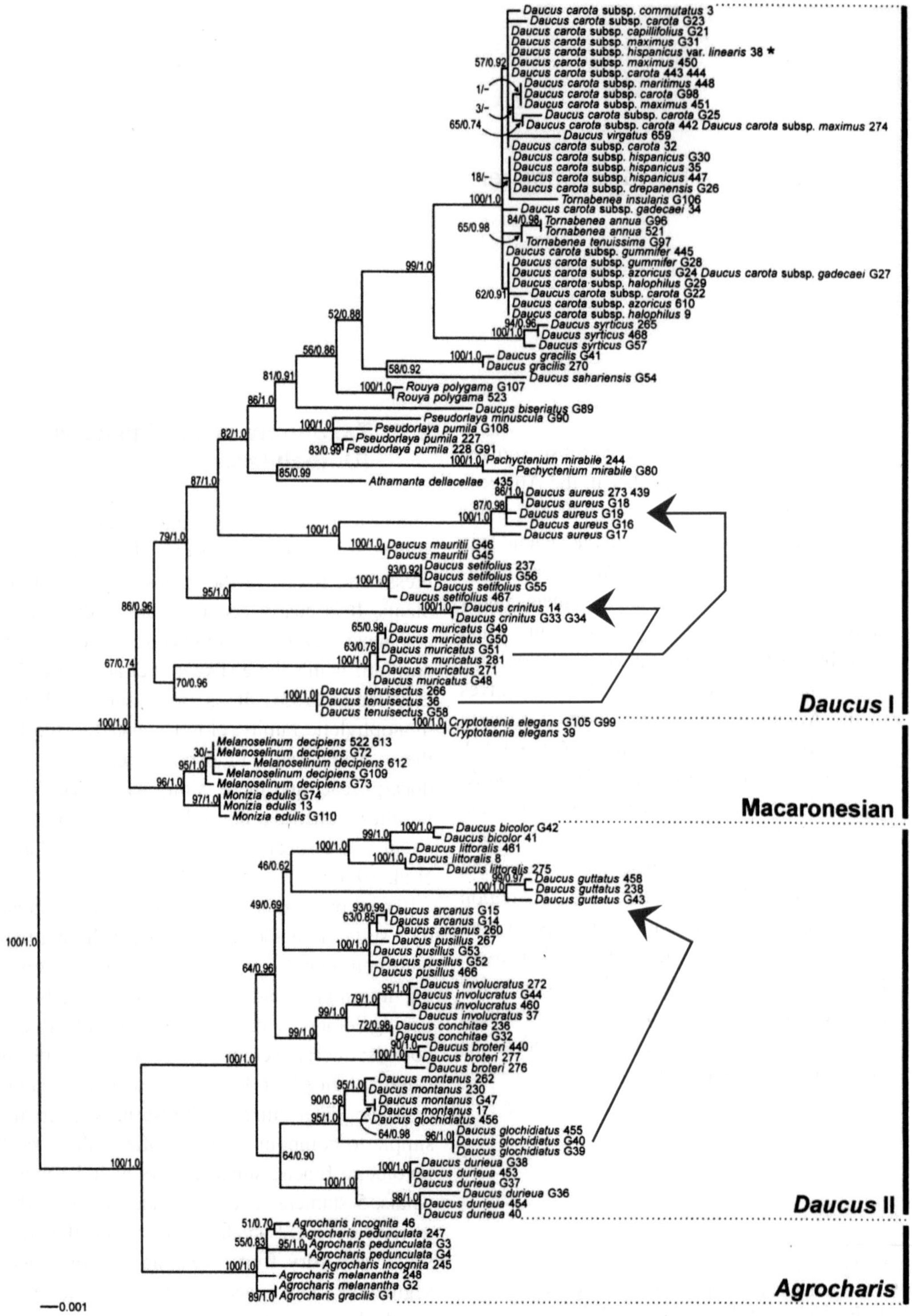
Daucus carota subsp. commutatus 3
Daucus carota subsp. carota G23
Daucus carota subsp. capillifolius G21
Daucus carota subsp. maximus G31
Daucus carota subsp. hispanicus var. linearis 38 *
Daucus carota subsp. maximus 450
Daucus carota subsp. carota 443 444
Daucus carota subsp. maritimus 448
Daucus carota subsp. carota G98
Daucus carota subsp. maximus 451
Daucus carota subsp. carota G25
Daucus carota subsp. carota 442 Daucus carota subsp. maximus 274
Daucus virgatus 659
Daucus carota subsp. carota 32
Daucus carota subsp. hispanicus G30
Daucus carota subsp. hispanicus 35
Daucus carota subsp. hispanicus 447
Daucus carota subsp. drepanensis G26
Tornabenea insularis G106
Daucus carota subsp. gadecaei 34
Tornabenea annua G96
Tornabenea annua 521
Tornabenea tenuissima G97
Daucus carota subsp. gummifer 445
Daucus carota subsp. gummifer G28
Daucus carota subsp. azoricus G24 Daucus carota subsp. gadecaei G27
Daucus carota subsp. halophilus G29
Daucus carota subsp. carota G22
Daucus carota subsp. azoricus 610
Daucus carota subsp. halophilus 9
Daucus syrticus 265
Daucus syrticus 468
Daucus syrticus G57
Daucus gracilis G41
Daucus gracilis 270
Daucus sahariensis G54
Rouya polygama G107
Rouya polygama 523
Daucus biseriatus G89
Pseudorlaya minuscula G90
Pseudorlaya pumila G108
Pseudorlaya pumila 227
Pseudorlaya pumila 228 G91
Pachyctenium mirabile 244
Pachyctenium mirabile G80
Athamanta dellacellae 435
Daucus aureus 273 439
Daucus aureus G18
Daucus aureus G19
Daucus aureus G16
Daucus aureus G17
Daucus mauritii G46
Daucus mauritii G45
Daucus setifolius 237
Daucus setifolius G56
Daucus setifolius G55
Daucus setifolius 467
Daucus crinitus 14
Daucus crinitus G33 G34
Daucus muricatus G49
Daucus muricatus G50
Daucus muricatus G51
Daucus muricatus 281
Daucus muricatus 271
Daucus muricatus G48
Daucus tenuisectus 266
Daucus tenuisectus 36
Daucus tenuisectus G58
Daucus I
Cryptotaenia elegans G105 G99
Cryptotaenia elegans 39
Melanoselinum decipiens 522 613
Melanoselinum decipiens G72
Melanoselinum decipiens 612
Melanoselinum decipiens G109
Melanoselinum decipiens G73
Monizia edulis G74
Monizia edulis 13
Monizia edulis G110
Macaronesian
Daucus bicolor G42
Daucus bicolor 41
Daucus littoralis 461
Daucus littoralis 8
Daucus littoralis 275
Daucus guttatus 458
Daucus guttatus 238
Daucus guttatus G43
Daucus arcanus G15
Daucus arcanus G14
Daucus arcanus 260
Daucus pusillus 267
Daucus pusillus G53
Daucus pusillus G52
Daucus pusillus 466
Daucus involucratus 272
Daucus involucratus G44
Daucus involucratus 460
Daucus involucratus 37
Daucus conchitae 236
Daucus conchitae G32
Daucus broteri 440
Daucus broteri 277
Daucus broteri 276
Daucus montanus 262
Daucus montanus 230
Daucus montanus G47
Daucus montanus 17
Daucus glochidiatus 456
Daucus glochidiatus 455
Daucus glochidiatus G40
Daucus glochidiatus G39
Daucus durieua G38
Daucus durieua 453
Daucus durieua G37
Daucus durieua G36
Daucus durieua 454
Daucus durieua 40
Daucus II
Agrocharis incognita 46
Agrocharis pedunculata 247
Agrocharis pedunculata G3
Agrocharis pedunculata G4
Agrocharis incognita 245
Agrocharis melanantha 248
Agrocharis melanantha G2
Agrocharis gracilis G1
Agrocharis
0.001

◀ **Fig. 2.1** Reproduction of the upper part of the *Daucus* maximum likelihood phylogeny of Banasiak et al. (2016), using combined nuclear internal transcribed spacer region of ribosomal DNA (ITS) and plastid (*rps16* intron, *rpoC1* intron, and *rpoB-trnC* intergenic spacer) data, with numbers above the branches representing bootstrap support and posterior probability values. The arrows show hard incongruence between Banasiak et al. (2016) and the nuclear ortholog phylogenies of Arbizu et al. (2014b, 2016b)

This large suite of distinctive characters makes the Apiaceae and its constituent species easily recognized to family, but divisions within the family have been the subject of long dispute including circumscription and relationships of the genus *Daucus* (Constance 1971; Plunkett and Downie 1999) Traditionally, the Apiaceae has been divided into three subfamilies, the Saniculoideae, Hydrocotyloideae, and Apioideae, with the Apioideae, containing the genus *Daucus*, by far the largest of these three traditional subfamilies. Drude (1898) recognized 8 tribes and 10 subtribes within the Apioideae. Molecular phylogenetic studies have confirmed the monophyly of the subfamily Apioideae but not many of its tribes and subtribes (Downie et al. 2001). Downie et al. (2001) recognized nine tribes in the Apiaceae subfamily Apioideae, and placed *Daucus*, and 12 other genera, in tribe *Scandiceae* Spreng., subtribe *Daucinae* Dumort. (the other 12 genera being *Agrocharis* Hochst., *Ammodaucus* Coss. and Durieu, *Cuminum* L., *Laser* Borkh. ex P. Gaertn., B. Mey. and Schreb., *Laserpitium* L., *Melanoselinum* Hoffm., *Monizia* Lowe, *Orlaya* Hoffm., *Pachyctenium* Maire and Maire and *Polemannia* Eckl. and Zeyh., *Polylophium* Boiss., *Pseudorlaya* (Murb.) Murb., and *Thapsia* L.).

A genus-level treatment of *Daucus* by Sáenz Laín (1981) used morphological and anatomical data and recognized 20 species. Rubatzky et al. (1999) later estimated 25 species of *Daucus*. The phylogenetic relationships among the species of genus *Daucus* and close relatives in the Apioideae have been clarified by a series of molecular studies using DNA sequences of the plastid genes *rbcL* and *matK*; plastid introns rpl16, rps16, rpoC1; nuclear ribosomal DNA internal transcribed spacer (ITS) sequences; and plastid DNA restriction sites (e.g., Arbizu et al. 2014b, 2016a, b; Banasiak et al. 2016; Downie and Katz-Downie 1996; Downie et al. 1996, 1998, 2000a, b, c, 2001, 2010; Katz-Downie et al. 1999; Lee 2002; Lee and Downie 1999, 2000, 2006; Plunkett et al. 1996a; Spalik and Downie 2007; Spalik et al. 2001a, b; Weitzel et al. 2014). Of these DNA markers, the ITS region consisting of ITS1, the intervening spacer, and ITS2 has served as the main marker. A recent study of ITS, and other DNA regions proposed as standard barcodes (*psbA-trnH*, *matK*, and *rbcL*) in 1957 species in 385 diverse genera in the Apiaceae have shown ITS to serve to identify species 73.3% of the time, higher than any of the other individual markers tested (Liu et al. 2014).

A study by Banasiak et al. (2016) using DNA sequences from nuclear ribosomal ITS and three plastid markers (*rps16* intron, *rpoC1* intron, and *rpoB-trnC* intergenic spacer) is the latest of a series of studies to investigate ingroup and outgroup relationships of *Daucus* (Fig. 2.1). This study redefined and expanded the genus *Daucus* to include the following genera and species into its synonymy: *Agrocharis* Hochst. (4 species), *Melanoselinum* Hoffm. (1 species), *Monizia* Lowe (1 species), *Pachyctenium* Maire and Pamp. (1 species), *Pseudorlaya* (Murb.) Murb. (2 species), *Rouya* Coincy (1 species), *Tornabenea* Parl. (6 species), *Athamanta dellacellae* E. A. Durand and Barratte, and *Cryptotaenia elegans* Webb ex Bolle (these latter two genera with only some of its members transferred to *Daucus*).

Banasiak et al. (2016) made the relevant nomenclatural transfers into *Daucus* (Table 2.1) and following this classification, the genus *Daucus* contains ca. 40 species and now includes winged and completely unadorned ("obsolete") fruits in addition to its traditionally recognized spiny fruits. As summarized in Banasiak et al. (2016) and presented in graphic form in Fig. 5 of this paper, winged versus spiny versus obsolete fruits presented major traditional taxonomic characters at higher levels in the Apiaceae (e.g.,

Table 2.1 Taxonomic circumscription of *Daucus* following Arbizu et al. (2014b, 2016b) and Banasiak et al. (2016), their cladistic relationships, and diploid chromosome numbers

Taxon	2*n*	Heywood (1978)—sections within *Daucus*	Sáenz Laín (1981)—sections within *Daucus*	Clade (Arbizu et al. 2014b, 2016b; Banasiak et al. 2016)	Banasiak et al. (2016)—sections within *Daucus*	Fruit type—morphology secondary ribs (Banasiak et al. 2016)	Countries of occurrence
Daucus arcanus García-Martín and Silvestre	22			Daucus II	*Anisactis*	Spiny	Spain
Daucus aureus Desf.	22	*Chrysodaucus* Thell.	*Chrysodaucus*	Daucus I	*Daucus*	Spiny	Spain (Canary Islands), Algeria, Egypt, Libya, Morocco, Tunisia, Cyprus, Israel, Lebanon, Syria, Italy
Daucus bicolor Sm.	–	*Pseudoplatyspermum* (Thell.)	*Platyspermum*	Daucus II	*Anisactis*	Spiny	Greece, Turkey
Daucus biseriatus Murb.	–			Daucus I	*Daucus*	Spiny	Algeria
Daucus carota subsp. *capillifolius* (Gilli) C. Arbizu	18	*Daucus*	*Daucus*	Daucus I	*Daucus*	Spiny	Libya, Tunisia
Daucus carota subsp. *carota* L.	18	*Daucus*	*Daucus*	Daucus I	*Daucus*	Spiny	Widely naturalized worldwide
Daucus carota subsp. *gummifer* (Syme) Hook. f.	18	*Daucus*	*Daucus*	Daucus I	*Daucus*	Spiny	Coastal Mediterranean, coastal Atlantic in UK, France, Tunisia, Italy
Daucus carota subsp. *maximus* (Desf.) Ball	18	*Daucus*	*Daucus*	Daucus I	*Daucus*	Spiny	Spain, Algeria, Morocco, Tunisia, Afghanistan, Cyprus, Iran, Israel, Jordan, Lebanon, Syria, Turkey, Pakistan, Greece, Italy (former) Yugoslavia, France, Portugal, Spain
Daucus carota subsp. *sativus* (Hoffm.) Arcang.	18	*Daucus*	*Daucus*	Daucus I	*Daucus*	Spiny	Cultivated worldwide
Daucus conchitae Greuter	–		Incertae sedis	Daucus II	*Anisactis*	Spiny	Greece
Daucus crinitus Desf.	22 (18 possibly an error)	*Meoides* Lange	*Daucus*	Daucus I	*Daucus*	Spiny	Algeria, Morocco, Tunisia, Portugal, Spain
Daucus durieua Lange	22	*Anisactis* DC.	*Anisactis*	Daucus II	*Anisactis*	Spiny	Spain, Algeria, Libya, Morocco, Tunisia, Cyprus, Israel, Lebanon, Syria, Portugal, Spain
Daucus glochidiatus (Labill.) Fisch. and C. A. Mey.	44	*Anisactis*	*Anisactis*	Daucus II	*Anisactis*	Spiny	Australia, New Zealand
Daucus gracilis Steinh.	–	*Daucus*	*Daucus*	Daucus I	*Daucus*	Spiny	Algeria, Tunisia
Daucus guttatus Sm.	20, 22	*Daucus*	*Daucus*	Daucus II	*Anisactis*	Spiny	Egypt, Libya, Cyprus, Iran, Iraq, Israel, Lebanon, Syria, Turkey, Albania, Bulgaria, Greece, Italy, Romania (former), Yugoslavia
Daucus hochstetteri A. Braun ex Drude	–	*Anisactis*		Daucus II		Spiny	Eritrea, Ethiopia
Daucus involucratus Sm.	20, 22	*Daucus*	*Daucus*	Daucus II	*Anisactis*	Spiny	Cyprus, Turkey, Greece

(continued)

Table 2.1 (continued)

Taxon	2*n*	Heywood (1978)—sections within *Daucus*	Sáenz Laín (1981)—sections within *Daucus*	Clade (Arbizu et al. 2014b, 2016b; Banasiak et al. 2016)	Banasiak et al. (2016)—sections within *Daucus*	Fruit type—morphology secondary ribs (Banasiak et al. 2016)	Countries of occurrence
Daucus jordanicus Post	–	*Daucus*	*Daucus*	–	Incertae sedis	Spiny	Libya, Israel, Jordan
Daucus littoralis Sm.	–	*Daucus*	*Platyspermum*	Daucus II	*Anisactis*	Spiny	Egypt, Libya, Cyprus, Iran, Israel, Jordan, Lebanon, Syria, Turkey
Daucus mauritii Sennen	–	–	–	–		Spiny	Morocco
Daucus montanus Humb. and Bonpl. ex Schult.	66	*Anisactis*	*Anisactis*	Daucus II	*Anisactis*	Spiny	Mexico, Costa Rica, El Salvador, Guatemala, Honduras, Venezuela, Bolivia, Colombia, Ecuador, Peru, Argentina, Chile
Daucus muricatus (L.) L.	22	*Platyspermum* (Hoffm.) DC.	*Platyspermum*	Daucus I	*Daucus*	Spiny	Portugal, Algeria, Libya, Morocco, Tunisia, Italy, France, Portugal, Spain
Daucus pusillus Michx. (= *D. montevidensis* Link ex Spreng.)	22	*Leptodaucus* Thell.	*Daucus*	Daucus II	*Anisactis*	Spiny	Canada, United States, Mexico, Brazil, Argentina, Chile, Uruguay
Daucus sahariensis Murb.	18	*Daucus*	*Daucus*	Daucus I	*Daucus*	Spiny	Algeria
Daucus setifolius Desf.	22	*Meoides*		Daucus I	*Daucus*	Spiny	Algeria, Morocco, Tunisia, Portugal, Spain
Daucus setulosus Guss. ex DC.	–	*Meoides*	*Meoides*	Daucus I		Spiny	Greece, Turkey
Daucus syrticus Murb.	–	*Daucus*	*Daucus*	Daucus I	*Daucus*	Spiny	Egypt, Libya, Tunisia
Daucus tenuisectus Coss. ex Batt.	–	*Daucus*	*Daucus*	Daucus I	*Daucus*	Spiny	Morocco
Daucus virgatus (Poir.) Maire	–			Daucus I	*Daucus*	Winged	Algeria, Tunisia
New *Daucus* species following the taxonomic expansion of Banasiak et al. (2016)							
Daucus annuus (Bég.) Wojew. et al. (≡ *Tornabenea annua* Bég.)	–			Daucus I	*Daucus*	Winged	Cape Verde
Daucus dellacellae (Asch. and Barbey ex E. A. Durand and Barratte) Spalik, Banasiak and Reduron ≡ *Athamanta dellacellae* Asch. and Barbey ex E. A. Durand and Barratte	–			Daucus I	*Daucus*	Obsolete	Libya
Daucus insularis (Parl. ex Webb) Spalik et al. (≡ *Tornabenea insularis* (Parl. ex Webb) Parl.)	–			Daucus I	*Daucus*	Winged	Cape Verde
Daucus tenuissimus (A. Chev.) Spalik et al. (≡ *Melanoselinum tenuissimum* A. Chev. (≡ *Tornabenea tenuissima* (A. Chev.) A. Hansen and Sunding)	–			Daucus I	*Daucus*	Winged	Madeira
Daucus rouyi Spalik and Reduron (≡ *Rouya polygama* (Desf.) Coincy)	20			Daucus I	*Daucus*	Winged	Algeria, Tunisia, Italy (Corsica, Sardinia)

(continued)

Table 2.1 (continued)

Taxon	2n	Heywood (1978)—sections within *Daucus*	Sáenz Laín (1981)—sections within *Daucus*	Clade (Arbizu et al. 2014b, 2016b; Banasiak et al. 2016)	Banasiak et al. (2016)—sections within *Daucus*	Fruit type—morphology secondary ribs (Banasiak et al. 2016)	Countries of occurrence
Daucus pumilus (L.) Hoffmanns. and Link (≡ *Pseudorlaya pumila* (L.) Grande)	26			Daucus I	*Daucus*	Spiny	Portugal, Spain, Morocco, France, Italy, Greece, Israel
Daucus minusculus Pau ex Font Quer (≡ *Pseudorlaya minuscula* (Pau ex Font Quer) Laínz)	16			Daucus I	*Daucus*	Spiny	Portugal, Spain, Morocco
Daucus mirabilis (Maire and Pamp.) Reduron et al. (≡ *Pachyctenium mirabile* Maire and Pamp.)	–			Daucus I	*Daucus*	Spiny winged proximally, naked dorsally	Libya
Daucus dellacellae (E. A. Durand and Barratte) Spalik et al. (≡ *Athamanta dellacellae* E. A. Durand and Barratte)	–			Daucus I	*Daucus*	Obsolete	Libya
Daucus elegans (Webb ex Bolle) Spalik et al. (≡ *Cryptotaenia elegans* Webb ex Bolle)	16			Macronesian	*Daucus*	Obsolete	Canary Islands
Daucus decipiens (Schrad. and J. C. Wendl.) Spalik et al. (≡ *Melanoselinum decipiens* Schrad. and J. C. Wendl. (≡ *Melanoselinium decipiens* (Schrad. and J. C. Wendl.) Hoffm.)	–			Macronesian	*Melanoselinum*	Winged	Madeira
Daucus edulis (Lowe) Wojew. et al. (≡ *Monizia edulis* Lowe)	22			Macronesian	*Melanoselinum*	Winged	Madeira
Daucus incognitus (C. Norman) Spalik, Reduron and Banasiak, comb. nov. ≡ *Caucalis incognita* C. Norman ≡ *Agrocharis incognita* (C. Norman) Heywood and Jury	44			Agrocharis	*Agrocharis*	Spiny	Tropical Africa
Daucus melananthos (Hochst.) Reduron, Spalik and Banasiak, comb. nov. ≡ *Agrocharis melanantha* Hochst.	–			Agrocharis	*Agrocharis*	Winged	Tropical Africa
Daucus pedunculatus (Baker f.) Banasiak, Spalik and Reduron, comb. nov. ≡ *Caucalis pedunculata Baker f.* ≡ *Agrocharis pedunculata* (Baker f.) Heywood and Jury in Launert	–			Agrocharis	*Agrocharis*	Winged	Tropical Africa
Possibly *Daucus* but not yet examined with molecular data							
Agrocharis gracilis Hook. f.	–			Unknown	Unknown	Spiny	Tropical Africa
Daucus dellacellae (Asch. and Barbey ex E. A. Durand and Barratte) Spalik, Banasiak and	22			Unknown	Unknown		Libya

(continued)

Table 2.1 (continued)

Taxon	2n	Heywood (1978)—sections within *Daucus*	Sáenz Laín (1981)—sections within *Daucus*	Clade (Arbizu et al. 2014b, 2016b; Banasiak et al. 2016)	Banasiak et al. (2016)—sections within *Daucus*	Fruit type—morphology secondary ribs (Banasiak et al. 2016)	Countries of occurrence
Reduron, comb. nov. ≡ *Athamanta dellacellae* Asch. and Barbey ex E. A. Durand and Barratte							
Daucus jordanicus Post	–	*Daucus*	*Daucus*	Unknown	Unknown	Spiny	Libya, Israel, Jordan
Daucus microscias Bornm. and Gauba	–			Unknown	Unknown	Spiny	Iran, Iraq
Daucus reboudii Coss.	–			Unknown	Unknown	Spiny	Algeria, Tunisia
Tornabenea bischoffii J. A. Schmidt	–			Unknown	Unknown	Winged	Cape Verde
Tornabenea humilis Lobin and K. H. Schmidt	–			Unknown	Unknown	Winged	Cape Verde
Tornabenea ribeirensis K. H. Schmidt and Lobin	–			Unknown	Unknown	Winged	Cape Verde

Drude 1897–1898). Winged fruits are considered to be adapted to wind dispersal (Jongejans and Telenius 2001; Theobald 1971), and spiny fruits to animal dispersal (Jury 1982; Spalik et al. 2001a; Williams 1994) and likely under strong selective pressure. The above phylogenetic analyses, however, show these fruit characters to be highly homoplastic and of limited value in delimiting monophyletic groups.

The above classification philosophy followed by Banasiak et al. (2016) in placing all members of a monophyletic clade into a single genus (here *Daucus*) is not universally accepted, and others may revise the circumscription of these genera. For example, a dissenting classification philosophy of relying solely on molecular data for classification is presented by Stuessy and Hörandl (2014), who recognize a "holophyletic" group as one that includes the immediate ancestor and all its descendants, independent of whatever divergence occurs within each of the derivative lineages (Ashlock 1971). A paraphyletic group, in contrast, is one that derives from a common ancestor but that does not contain all its descendants (Hennig 1966) and is an unacceptable taxon following cladistic conventions. Stuessy and Hörandl (2014) point out that adaptive radiation, common in oceanic islands, produces patterns where new populations continue to accrue reproductive isolation and speciation such that they produce quite distinctive new forms, often recognized as new genera, leaving parental populations intact. As examples in the Daucinae, Stuessy et al. (2014) cite the genus *Monizia* in the Madeira Islands, but other possibilities could be the genus *Tornabenea* or the species *Cryptotaenia elegans* on the Cape Verde Islands or the genus *Melanoselinum* on the Madeira Islands. Critical data bearing on this classification question rest in the distinctiveness and divergence of these new island forms. Because we have not studied these subsumed genera in detail, we currently take no position on these differences in classification, awaiting additional data and perspectives from others, such as Martínez-Flores (2016) and Plunkett et al. (in press) who maintain more traditional classifications of *Daucus*.

2.2 Distribution of *Daucus*

Phylogenetic analysis of ITS sequences supports southern Africa as the ancestral origin of the Apiaceae subfamily Apioideae (Banasiak et al. 2013). Phylogenetic analysis of ITS sequences supports an Old World Northern Hemisphere origin for *Daucus*, with one or two dispersals to the Southern Hemisphere (Spalik et al. 2010). The center of diversity of *Daucus* in its traditional sense is in the Mediterranean region (Sáenz Laín 1981). *Daucus* species also occur elsewhere, with one species (*D. glochidiatus*) in Australia, four species in the American continent (*D. carota*, *D. montanus*, *D. montevidensis*, *D. pusillus* Michx.). Following the expanded classification of *Daucus* by Banasiak et al. (2016), the now included genus *Agrocharis* extends the range of *Daucus* into tropical Africa (Townsend 1989).

2.3 New Taxonomic Approaches: Next-Generation Sequencing (NGS)

A major innovation in plant systematics is the development of high-throughput, "next-generation" DNA sequencing (NGS) to infer phylogenetic relationships (Egan et al. 2012; E. M. Lemmon and A. R. Lemmon 2013). NGS typically first involves large-scale sequencing of all components of the genome, with the Illumina platform currently the most commonly used. Some genomes, such as plastid and mitochondria, have much higher coverage than single- to low-copy nuclear DNA and can be factored out of the nuclear genome in NGS data by coverage statistics. The utility of NGS sequencing is markedly improved when a high-quality whole-genome "reference" sequence is available that serves as a heterologous template to guide mapping of sequences of related germplasm. Such whole-genome reference sequences are available in carrot for the plastid genome (Ruhlman et al. 2006) and for the plastid and nuclear genome (Iorizzo et al. 2016). As summarized below, recent phylogenetic studies in *Daucus* have used high-throughput DNA sequencing to infer phylogenetic relationships at the genus level using orthologous nuclear DNA sequences, also at the genus level using whole plastid DNA sequences, and at the species level using genotyping-by-sequencing (GBS).

2.3.1 Next-Generation DNA Phylogenetic Studies at the Genus Level Using Orthologous Nuclear DNA Sequences

In the past, there has been a paucity of validated nuclear orthologs for phylogenetic studies, and hence, most molecular taxonomic studies have relied heavily on a few plastid and/or ribosomal genes (Small et al. 2004). Phylogenies reconstructed with only one or a few independently inherited loci may result in unresolved or incongruent phylogenies due to data sampling (Graybeal 1998), horizontal gene transfer, or differential selection and lineage sorting at individual loci (Maddison 1995). Following a phylogenetic study by Spooner et al. (2013) where eight nuclear orthologs were used in *Daucus* but designed without NGS techniques, Arbizu et al. (2014b) identified 94 nuclear orthologs in *Daucus*, constructed a phylogeny with these, and determined 10 of them to provide essentially the same phylogeny as all 94, paving the way for additional and most cost-effective nuclear ortholog phylogenetic studies in carrot. The 94 (and 10) nuclear ortholog phylogeny was highly resolved, with 100% bootstrap support for most of the external and many of the internal clades. They resolved multiple accessions of many different species as monophyletic with strong support, but failed to support other species. This phylogeny had many points of agreement with Banasiak et al. (2016), including resolving two major clades (Daucus I and II in their study, labeled clade A and B in Arbizu et al. 2014b), with a clade A' containing all examined $2n = 18$ chromosome species (*D. carota* all subspecies, *D. capillifolius*, *D. syrticus*), with the other clade A species being and *D. aureus* and *D. muricatus* (as sister taxa), and *D. tenuisectus*. Two non-*Daucus* species (*Rouya polygama* and

Pseudorlaya pumila) resolved sister to Daucus clade A'. Clade B (Daucus II in Banasiak et al. 2016) contained six wild *Daucus* species *D. glochidiatus*, *D. guttatus*, *D. involucratus*, *D. littoralis*, and *D. pusillus*, but *D. guttatus* was not monophyletic within this clade.

2.3.2 *An Expansion of the Above Study—The Daucus Guttatus* Complex

As mentioned above, the nuclear ortholog study of Arbizu et al. (2014b) resolved a monophyletic group (clade B) of six wild *Daucus* species *D. glochidiatus*, *D. guttatus*, *D. involucratus*, *D. littoralis*, and *D. pusillus*. Some of these species are morphologically similar and difficult to distinguish, causing frequent misidentifications. Arbizu et al. (2016b) used the group of ten nuclear orthologs mentioned above in the study of Arbizu et al. (2014b), and morphological data (Arbizu et al. 2014a), and a greatly expanded subset of accessions of these species, to refine phylogenetic structure of the group. The nuclear ortholog data resolved four well-supported clades (Fig. 2.2), that in concert with morphological data, and nomenclatural data from a study of type specimens (Martínez-Flores et al. 2016) served to identify four phenetically most similar species *D. bicolor*, *D. conchitae*, *D. guttatus*, and *D. setulosus*. Internested among these four similar species were phenetically more distinctive species *D. glochidiatus*, *D. involucratus*, *D. littoralis*, and *D. pusillus*. They presented a key to better distinguish all of these eight species. In summary, their research clarified species variation in the *D. guttatus* complex, resolved interspecific relationships, provided the proper names for the species, and discovered morphological characters allowing proper identification and key construction of members of the *D. guttatus* complex and related species.

2.3.3 Next-Generation DNA Phylogenetic Studies at the Genus Level Using Whole Plastid DNA Sequences

The plastid genome has many features that make it useful for plant phylogenetic studies, including its small size (generally 120–160 kbp), high copy number (as many as 1000 per cell), generally conservative nature (Wolfe et al. 1987), and varying rates of change in different regions of the genome, allowing studies at different phylogenetic levels (Raubeson and Jansen 2005). Hence, earlier sequence-based plant phylogenetic studies used genes or gene regions from the plastid. Relative to the Apioideae, the subfamily of the Apiaceae including *Daucus,* systematic studies have used plastid restriction site data; DNA sequence data from plastid genes; from plastid introns; from plastid intergenic spacer regions. Using NGS sequencing approaches, Downie and Jansen (2015) sequenced five complete plastid genomes in the Apiales (Apiaceae + Araliaceae): *Anthriscus cerefolium* (L.) Hoffm., *Crithmum maritimum* L., *Hydrocotyle verticillata* Thunb., *Petroselinum crispum* (Mill.) Fuss, and *Tiedemannia filiformis* (Walter) Feist and S. R. Downie subsp. *greenmanii* (Mathias and Constance) Feist and S. R. Downie, and compared the results obtained to previously published plastomes of *Daucus carota* subsp. *sativus* and *Panax schin-seng* T. Nees. They discovered the *rpl32-trnL*, *trnE-trnT*, *ndhF-rpl32*, *5'rps16-trnQ*, and *trnT-psbD* intergenic spacers to be among the most fast-evolving loci, with the *trnD-trnY-trnE-trnT* combined region presenting the greatest number of potentially informative characters overall that may possess ideal phylogenetic markers in these families.

Spooner et al. (2017) explored the phylogenetic utility of entire plastid DNA sequences in *Daucus*, using Illumina sequencing, and

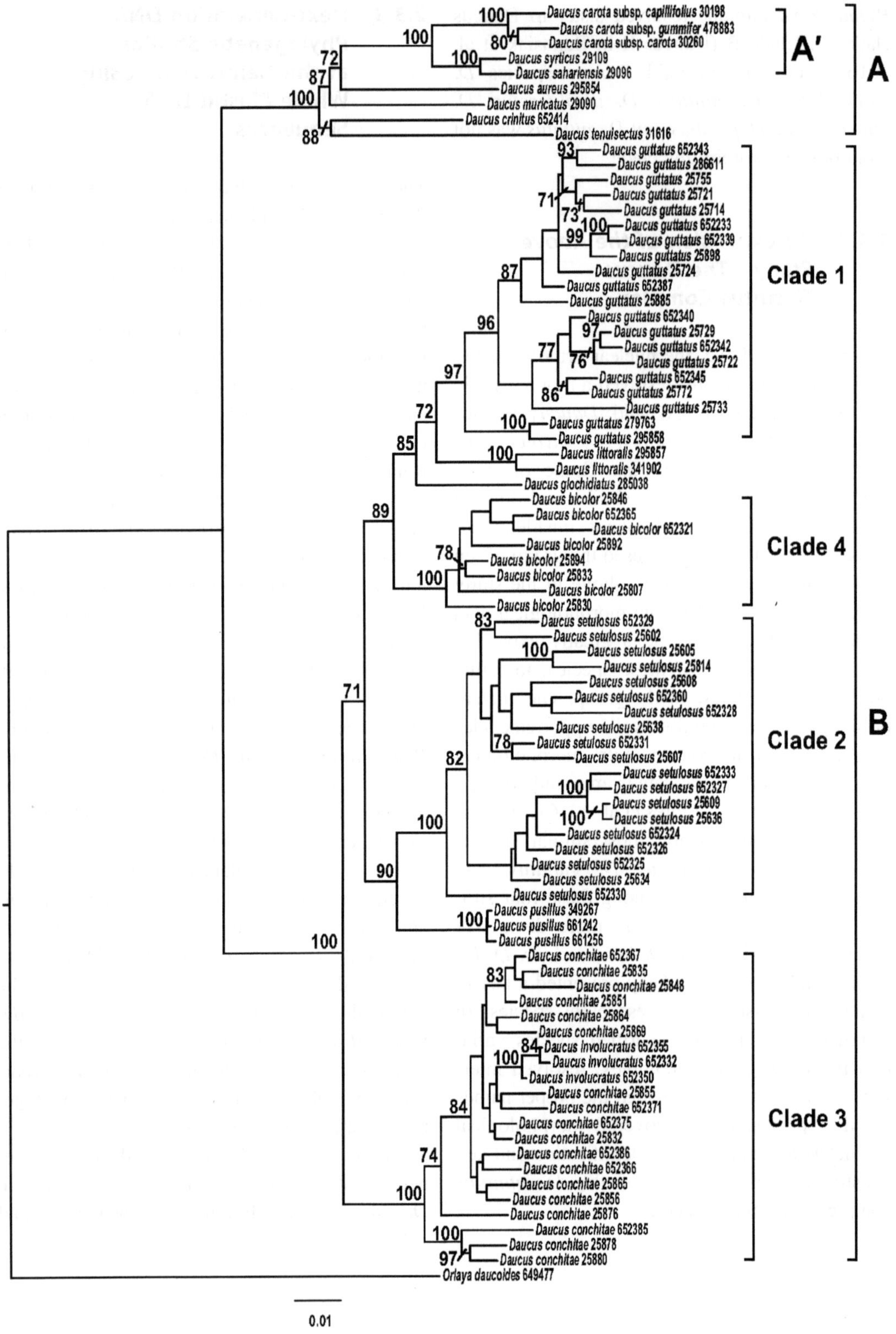
Daucus carota subsp. capillifolius 30198
Daucus carota subsp. gummifer 478883
Daucus carota subsp. carota 30260
Daucus syrticus 29109
Daucus sahariensis 29096
Daucus aureus 295854
Daucus muricatus 29090
Daucus crinitus 652414
Daucus tenuisectus 31616
Daucus guttatus 652343
Daucus guttatus 286611
Daucus guttatus 25755
Daucus guttatus 25721
Daucus guttatus 25714
Daucus guttatus 652233
Daucus guttatus 652339
Daucus guttatus 25898
Daucus guttatus 25724
Daucus guttatus 652387
Daucus guttatus 25885
Daucus guttatus 652340
Daucus guttatus 25729
Daucus guttatus 652342
Daucus guttatus 25722
Daucus guttatus 652345
Daucus guttatus 25772
Daucus guttatus 25733
Daucus guttatus 279763
Daucus guttatus 295858
Daucus littoralis 295857
Daucus littoralis 341902
Daucus glochidiatus 285038
Daucus bicolor 25846
Daucus bicolor 652365
Daucus bicolor 652321
Daucus bicolor 25892
Daucus bicolor 25894
Daucus bicolor 25833
Daucus bicolor 25807
Daucus bicolor 25830
Daucus setulosus 652329
Daucus setulosus 25602
Daucus setulosus 25605
Daucus setulosus 25814
Daucus setulosus 25608
Daucus setulosus 652360
Daucus setulosus 652328
Daucus setulosus 25638
Daucus setulosus 652331
Daucus setulosus 25607
Daucus setulosus 652333
Daucus setulosus 652327
Daucus setulosus 25609
Daucus setulosus 25636
Daucus setulosus 652324
Daucus setulosus 652326
Daucus setulosus 652325
Daucus setulosus 25634
Daucus setulosus 652330
Daucus pusillus 349267
Daucus pusillus 661242
Daucus pusillus 661256
Daucus conchitae 652367
Daucus conchitae 25835
Daucus conchitae 25848
Daucus conchitae 25851
Daucus conchitae 25864
Daucus conchitae 25869
Daucus involucratus 652355
Daucus involucratus 652332
Daucus involucratus 652350
Daucus conchitae 25855
Daucus conchitae 652371
Daucus conchitae 652375
Daucus conchitae 25832
Daucus conchitae 652386
Daucus conchitae 652366
Daucus conchitae 25865
Daucus conchitae 25856
Daucus conchitae 25876
Daucus conchitae 652385
Daucus conchitae 25878
Daucus conchitae 25880
Orlaya daucoides 649477
A′
A
Clade 1
Clade 4
Clade 2
B
Clade 3
0.01

◀ **Fig. 2.2** Maximum parsimony phylogenetic reconstruction of the *Daucus guttatus* complex using 10 nuclear orthologs showing resolution of the species in the *Daucus guttatus* complex. Numbers above branches represent bootstrap values. Clades 1, 2, and 3 were identified in Arbizu et al. (2014b)

compared the results with prior phylogenetic results using plastid and nuclear DNA sequences. The phylogenetic tree of the entire data set (Fig. 2.3) was highly resolved, with 100% bootstrap support for most of the external and many of the internal clades. Subsets of the plastid data, such as *matK, ndhF*, or the putative maximally informative regions of the plastid genome outlined by Downie and Jansen (2015) are only partly successful in *Daucus*, resulting in polytomies and reduced levels of bootstrap support. Additionally, there are areas of hard incongruence (strongly supported character conflict because of differences in underlying evolutionary histories) with phylogenies using nuclear data (Fig. 2.1).

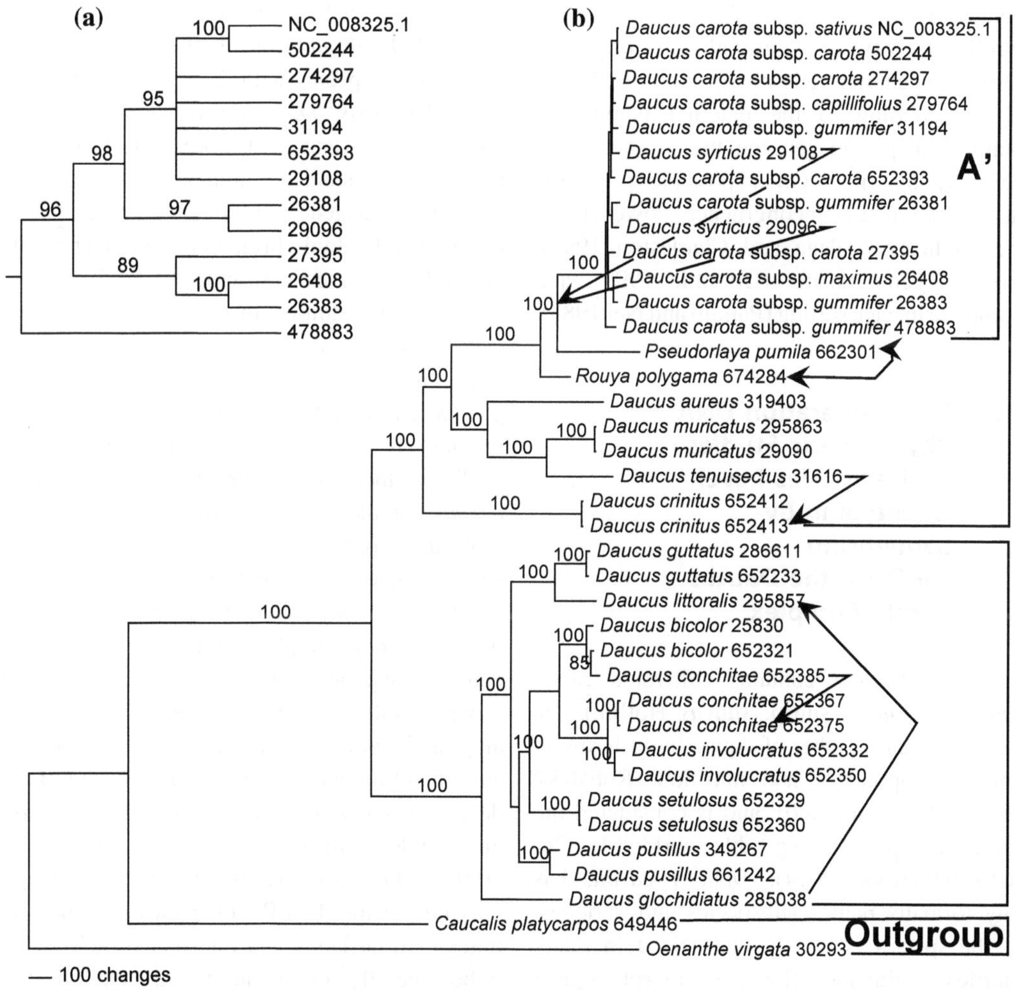

Fig. 2.3 Maximum likelihood cladogram of the entire plastid DNA sequences of Spooner et al. (2017), with the three main clades indicated, with arrows highlighting hard topological incongruence with the nuclear ortholog phylogenies of Arbizu et al. (2014b, 2016b); the two accessions of *Daucus syrticus* resolve as a sister group to all accessions of *D. carota*. **a** Represents expanded topological detail of the upper portion of the entire tree shown on **b**. The values above the branches are bootstrap support values

Incongruence between plastid and nuclear genes are not uncommon in phylogenetic studies in the Apiaceae (e.g., Lee and Downie 2006; Yi et al. 2015; Zhou et al. 2009), indeed throughout many angiosperms (Wendel and Doyle 1998). These incongruent results showed the value of resequencing data to produce a well-resolved plastid phylogeny of *Daucus*, and highlighted caution to combine plastid and nuclear data, if at all. The value of generating phylogenies from both nuclear and plastid sequences is that hard incongruence can be quite informative, suggesting such evolutionary processes as "plastid capture" where incongruence can be caused by a history of hybridization between plants with differing plastid and nuclear genomes (Rieseberg and Soltis 1991), and backcrossing to the paternal parent but retaining the plastid genome that is (typically) maternally inherited. Other possible processes that can lead to such incongruence, however, are gene duplication (Page and Charleston 1997), horizontal gene transfer (Doolittle 1999), and incomplete lineage sorting (Pamilo and Nei 1988).

2.3.4 Next-Generation DNA Phylogenetic Studies at the Species Level—Genotyping-by-Sequencing (GBS) for the *Daucus Carota* Complex

The genus *Daucus* contains cultivated carrot (*Daucus carota* L. subsp. *sativus* Hoffm.), the most important member of Apiaceae in terms of economic importance and nutrition (Rubatzky et al. 1999; Simon 2000), and is considered the second most popular vegetable worldwide after potato (Heywood 2014). *Daucus carota* has many formally named subspecies and varieties, and the species is widely naturalized in many countries worldwide. The great morphological variation in *D. carota* has resulted in more than 60 infraspecific taxa, making *D. carota* the most problematic species group in the Apiaceae (Heywood 1968a, b; Small 1978; Thellung 1926). Cultivated carrots and closely related wild carrots (other subspecies and varieties of *D. carota* sensu lato) belong to the *Daucus carota* complex. Its constituent taxa all possess $2n = 18$ chromosomes and have weak biological barriers to interbreeding. *D. carota* undergoes widespread hybridization experimentally and spontaneously with commercial varieties of carrot and the wild subspecies of *D. carota* (e.g., Ellis et al. 1993; Hauser 2002; Hauser and Bjørn 2001; Krickl 1961; McCollum 1975, 1977; Nothnagel et al. 2000; Rong et al. 2010; Sáenz de Rivas and Heywood 1974; Steinborn et al. 1995; St. Pierre and Bayer 1991; St. Pierre et al. 1990; Umiel et al. 1975; Vivek and Simon 1999; Wijnheijmer et al. 1989). In addition, there are other closely related wild species with $2n = 18$ chromosomes (*D. sahariensis, D. syrticus)* based on shared karyotypes (Iovene et al. 2008), the genus-level phylogenetic studies summarized above, and they represent gene pool 1 species to cultivated carrot. The haploid chromosome number for the genus *Daucus* (sensu stricto) ranges from $n = 8$ to $n = 11$. In addition to the $n = 8$ diploid species, diploid chromosome numbers in *Daucus* range from $2n = 16$ to 22, and a tetraploid (*D. glochidiatus*) and a hexaploid (*D. montanus*) species have been reported (Table 2.1).

To put the taxonomic problem of the *Daucus carota* complex into historical context, several molecular approaches have examined its diversity and genetic relationships. St. Pierre et al. (1990) used isozymes to study 168 accessions of the *D. carota* complex from 32 countries and could not separate named subspecies into distinct groups. Nakajima et al. (1998) used random amplified polymorphic DNA (RAPD) and amplified fragment length polymorphism (AFLP) data and showed all accessions of *D. carota* group into a major clade. Vivek and Simon (1998, 1999) used restriction fragment length polymorphisms (RFLPs) of nuclear, plastid, and mitochondrial DNA and interpreted their results to be generally concordant with the classification proposed by Sáenz Laín (1981), but studied just one additional subspecies (subsp. *drepanensis*). Using AFLPs, Shim and Jørgensen (2000) showed wild and cultivated carrot clustered separately. Bradeen et al. (2002) used AFLPs and intersimple sequence repeats (ISSR) and concluded wild carrots had no substructure. Rong et al. (2014) obtained a *Daucus* phylogeny using SNPs and found the subspecies of *D. carota* to be intermixed with each other. Lee and Park (2014) proposed *D. sahariensis*, *D. syrticus*, and *D. gracilis* to be the likely closest relatives to *D. carota*. In an attempt to characterize the populations of *D. carota* present in São Miguel Island (Azores, Portugal), Matias Vaz (2014) used one nuclear ortholog, nuclear ribosomal DNA ITS, and morphological descriptors and concluded that the classification of *D. carota* remained problematic. Other morphological studies (Arbizu et al. 2014a; Mezghani et al. 2014; Small 1978; Spooner et al. 2014; Tavares et al. 2014) likewise not distinguish the subspecies of *D. carota*. However, Iorizzo et al. (2013) used 3326 single nucleotide polymorphisms (SNPs) to study the genetic structure and domestication of carrot and found a clear separation between wild (subsp. *carota*) and cultivated (subsp. *sativus*) accessions of *D. carota*.

These taxonomic problems have practical considerations for germplasm curators and taxonomists who have relied on local floras for identifying these taxa such as floras from Algeria (Quézel and Santa 1963), the Azores (Schäfer 2005), Europe (Heywood 1968b), the Iberian Peninsula and Balearic Islands (Pujadas Salvà 2003), Libya (Jafri and El-Gadi 1985), Morocco (Jury 2002), Palestine (Zohary 1972), Portugal (Franco 1971), Syria (Mouterde 1966), Tunisia (Le Floc'h et al. 2010; Pottier-Alapetite 1979), and Turkey and the East Aegean Islands (Cullen 1972). Unfortunately, the keys and descriptions in these floras lack consensus about both the number of infraspecific taxa and characters best distinguishing them. For instance, 11 wild subspecies were recognized by Heywood (1968a, b), five by Sáenz Laín (1981: subsp. *carota*, subsp. *gummifer*, subsp. *hispanicus*, subsp. *maritimus*, and subsp. *maximus*), five by Arenas and García--Martin (1993), and Pujadas Salvà (2002) proposed nine subspecies for the Iberian Peninsula plus Balearic Islands (subsp. *carota*, subsp. *cantabricus*, subsp. *commutatus*, subsp. *gummifer*, subsp. *halophilus*, subsp. *hispanicus*, subsp. *majoricus*, subsp. *maximus*, and subsp. *sativus*).

Molecular investigations are trying to resolve the natural taxa in *D. carota*. "Reduced-representation" methods obtain partial DNA polymorphisms throughout the genome and have been shown to be very useful at the species level. Genotyping-by-sequencing (GBS) is one such reduced-representation method that generates sequence variants or single nucleotide polymorphisms (SNPs) (Elshire et al. 2011). GBS provides a powerful and cost-effective molecular approach for phylogeny reconstruction, producing abundant large-scale genomic data to infer phylogenetic relationships among recently diverged species or populations (e.g., Balfourier et al. 2007; Escudero et al. 2014; Good 2011; Wong et al. 2015). It captures both neutral genetic diversity and loci that affect quantitative traits of interest, because of the full-genome coverage of the GBS markers. It shows little to no ascertainment bias because markers are developed directly on the population being genotyped. Genetic relatedness among genotypes calculated using GBS markers is based on patterns of neutral and functional genetic variation across the genome.

Arbizu et al. (2016a) used GBS to examine the subspecies of *D. carota*. They obtained SNPs covering all nine *D. carota* chromosomes from 162 accessions of *Daucus* and related genera. They scored a total of 10,814 or 38,920 SNPs with a maximum of 10 or 30% missing data, respectively. Consistent with prior results, the phylogenetic tree separated species with $2n = 18$ chromosome from all other species in a single clade. Most interestingly, there was a strong geographic component to this phylogeny, with the wild members of *D. carota* from central Asia in a clade with eastern members of subsp. *sativus*. The other subspecies of *D. carota* were in four clades associated with geographic groups, suggesting that the subspecies are not natural groups. In summary, the wide range of morphological and molecular studies summarized above documents poor substructure of either morphologically or phylogenetically stable groups in *D. carota*. These results were concordant with results from recent morphological studies that led Spooner et al. (2014) to question whether many wild subspecies recognized within *D. carota* are valid taxa.

2.4 Conclusions

In summary, the taxonomy of *Daucus* at both the genus and species levels has been improved markedly in the last years by a series of morphological and molecular studies. Earlier studies using limited sets of plastid and nuclear markers have shown nuclear ribosomal ITS to be the most useful marker. Next-generation sequencing techniques are corroborating many of these studies, but adding details, especially cautioning combining nuclear and plastid data in combined data approaches. The phylogenetic study of Banasiak et al. (2016) has clarified ingroup and outgroup relationships and has resulted in an expanded concept of the genus. Continuing studies at the species and genus levels with NGS data and with additional collections are helping to refine our understanding of *Daucus* and should eventually lead to a much needed formal taxonomic revision taking into account phylogeny, keys, descriptions, illustrations, typifications, distributions, and maps.

References

Arbizu C, Ellison SL, Senalik D, Simon PW, Spooner DM (2016a) Genotyping-by-sequencing provides the discriminating power to investigate the subspecies of *Daucus carota* (Apiaceae). BMC Evol Biol 16:234

Arbizu C, Reitsma KR, Simon PW, Spooner DM (2014a) Morphometrics of *Daucus* (Apiaceae): a counterpart to a phylogenomic study. Amer J Bot 101:2005–2016

Arbizu C, Ruess H, Senalik D, Simon PW, Spooner DM (2014b) Phylogenomics of the carrot genus (*Daucus*, Apiaceae). Amer J Bot 101:1666–1685

Arbizu C, Simon PW, Martínez-Flores F, Ruess H, Crespo MB, Spooner DM (2016b) Integrated molecular and morphological studies of the *Daucus guttatus* complex (Apiaceae). Syst Bot 41:479–492

Arenas JA, García-Martin F (1993) Atlas carpológico y corológico de la subfamilia Apioideae Drude (Umbelliferae) en España peninsular y Baleares. Ruizia 12:222–234

Ashlock PD (1971) Monophyly and associated terms. Syst Zool 20:63–69

Balfourier F, Roussel V, Strelchenko P, Exbrayat-Vinson F, Sourdille P, Boutet G, Koenig J, Ravel C, Mitrofanova O, Beckert M, Charmet G (2007) A worldwide bread wheat core collection arrayed in a 384-well plate. Theor Appl Genet 114:1265–1275

Banasiak Ł, Piwczyński M, Uliński T, Downie SR, Watson MF, Shakya B, Spalik K (2013) Dispersal patterns in space and time: a case study of Apiaceae subfamily Apioideae. J Biogeor 40:1324–1335

Banasiak Ł, Wojewódzka A, Baczyński J, Reduron J-P, Piwczyński M, Kurzyna-Młynik R, Gutaker R, Czarnocka-Cieciura A, Kosmala-Grzechnik S, Spalik K (2016) Phylogeny of Apiaceae subtribe Daucinae and the taxonomic delineation of its genera. Taxon 65:563–585

Bradeen JM, Bach IC, Briard M, le Clerc V, Grzebelus D, Senalik DA, Simon PW (2002) Molecular diversity analysis of cultivated carrot (*Daucus carota* L.) and wild *Daucus* populations reveals a genetically nonstructured composition. J Amer Soc Hort Sci 127:383–391

Constance L (1971) History and classification of the Umbelliferae (Apiaceae). In: Heywood VH (ed) The biology and chemistry of the Umbelliferae. Academic Press, London, pp 1–11

Cullen J (1972) *Daucus*. In: Davis PH (ed) Flora of Turkey and the East Aegean Islands. Edinburgh University Press, Edinburgh, pp 531–536

Doolittle WF (1999) Lateral genomics. Trends Cell Biol 9:M5–M8

Downie SR, Jansen RK (2015) A comparative analysis of whole plastid genomes from the Apiales: expansion and contraction of the inverted repeat, mitochondrial to plastid transfer of DNA, and identification of highly divergent noncoding regions. Syst Bot 40:336–351

Downie SR, Katz-Downie DS (1996) A molecular phylogeny of Apiaceae subfamily Apioideae: evidence from nuclear ribosomal DNA internal transcribed spacer sequences. Amer J Bot 83:234–251

Downie SR, Katz-Downie DS, Cho KJ (1996) Phylogenetic analysis of Apiaceae subfamily Apioideae using nucleotide sequences from the chloroplast rpoC1 intron. Mol Phylo Evol 6:1–18

Downie SR, Katz-Downie DS, Spalik K (2000a) A phylogeny of Apiaceae tribe Scandiceae: evidence from nuclear ribosomal DNA internal transcribed spacer sequences. Amer J Bot 87:76–95

Downie SR, Katz-Downie DS, Watson MF (2000b) A phylogeny of the flowering plant family Apiaceae based on chloroplast DNA rpl16 and rpoC1 intron sequences: towards a suprageneric classification of subfamily Apioideae. Amer J Bot 87:273–292

Downie SR, Plunket GM, Watson MF, Spalik K, Katz-Downie DS, Valiejo-Roman CM, Terentieva EI, Troitsky AV, Lee BY, Lahham J, El-Oqlah A (2001) Tribes and clades within Apiaceae subfamily Apioideae: the contribution of molecular data. Edinb J Bot 58:301–330

Downie SR, Ramanath S, Katz-Downie DS, Llanas E (1998) Molecular systematics of Apiaceae subfamily Apioideae: phylogenetic analyses of nuclear ribosomal DNA internal transcribed spacer and plastid *rpoC1* intron sequences. Amer J Bot 85:563–591

Downie SR, Spalik K, Katz-Downie DS, Reduron JP (2010) Major clades within Apiaceae subfamily Apioideae as inferred by phylogenetic analysis of nrDNA ITS sequences. Plant Diver Evol 128:111–136

Downie SR, Watson MF, Spalik K, Katz-Downie DS (2000c) Molecular systematics of Old World Apioideae (Apiaceae): relationships among some members of tribe Peucedaneae sensu lato, the placement of several island endemic species, and resolution within the apioid superclade. Canad J Bot 78:506–528

Drude CGO (1897–1898) Umbelliferae. In: Engler A, Prantl K (eds) Die natürlichen Pflanzenfamilien, vol. 3, no. 8. Engelmann, Leipzig, p 63–250

Egan A, Schlueter J, Spooner DM (2012) Applications of next-generation sequencing in plant biology. Amer J Bot 99:175–185

Ellis P, Hardman J, Crowther T, Saw P (1993) Exploitation of the resistance to carrot fly in the wild carrot species *Daucus capillifolius*. Ann Appl Biol 122:79–91

Elshire RJ, Glaubitz JC, Sun Q, Poland JA, Kawamoto K, Buckler ES, Mitchell SE (2011) A robust, simple genotyping-by-sequencing (GBS) approach for high diversity species. PLoS ONE 6:e19379

Escudero M, Eaton DAR, Hahn M, Hipp AL (2014) Genotyping-by-sequencing as a tool to infer phylogeny and ancestral hybridization: a case study in *Carex* (Cyperaceae). Mol Phylo Evol 79:359–367

Franco JA (1971) Nova flora de Portugal (Continente e Açores). Sociedade Astória, Lisbon

Good JM (2011) Reduced representation methods for subgenomic enrichment and next-generation sequencing. In: Orgogozo V, Rockman MV (eds) Molecular methods for evolutionary genetics, methods in molecular biology. Springer, New York, pp 85–103

Graybeal A (1998) Is it better to add taxa or characters to a difficult phylogenetic problem? Syst Biol 47:9–17

Hauser TP (2002) Frost sensitivity of hybrids between wild and cultivated carrots. Conserv Genet 3:73–76

Hauser TP, Bjørn GK (2001) Hybrids between wild and cultivated carrots in Danish carrot fields. Genet Res Crop Evol 48:499–506

Hennig W (1966) Phylogenetic systematics (trans: Davis DD & Zangerl R). University of Illinois Press, Urbana

Heywood VH (1968a) The *Daucus carota-gingidium* complex. Feddes Rep 79:66–68

Heywood VH (1968b) *Daucus*. In: Tutin TG, Heywood VH, Burges NA, Valentine DH, Walters SM, Webb DA (eds) Flora Europaea. Cambridge University Press, Cambridge, pp 373–375

Heywood VH (1978) Multivariate synthesis of the tribe Caucalideae. In: Cauwet-Marc A-M, Carbonnier J (eds) Actes du 2ème Symposium International sur les Ombellifères, CNRS, Perpignan, p 727–736

Heywood VH (1983) Relationships and evolution in the *Daucus carota* complex. Israel J Bot 32:51–65

Heywood VH (2014) The socio-economic importance of the Apiales. J Fac Pharm Istanbul 44:113–130

Iorizzo M, Ellison S, Senalik D, Zeng P, Satapoomin P, Bowman M, Iovene M, Sanseverino W, Cavagnaro P, Yildiz M, Macko-Podgórni A, Moranska E, Grzebelus E, Grzebelus D, Ashrafi H, Zheng Z, Cheng S, Spooner D, Van Deynze A, Simon P (2016) A high-quality carrot genome assembly reveals new insights into carotenoid accumulation and asterid genome evolution. Nat Genet 48:657–666

Iorizzo M, Senalik DA, Ellison SL, Grzebelus D, Cavagnaro PF, Allender C, Brunet J, Van Deynze A (2013) Genetic structure and domestication of carrot (*Daucus carota* subsp. *sativus*) (Apiaceae). Amer J Bot 100:930–938

Iovene M, Grzebelus E, Carputo D, Jiang J, Simon PW (2008) Major cytogenetic landmarks and karyotype analysis in *Daucus carota* and other Apiaceae. Amer J Bot 95:793–804

Jafri SMH, El-Gadi A (1985) *Daucus*. In: Jafri SMH, El-Gadi A (eds) Flora of Libya. Al-Faateh University, Faculty of Science, Department of Botany Tripoli, p 130–144

Jongejans E, Telenius A (2001) Field experiments on seed dispersal by wind in ten umbelliferous species (Apiaceae). Pl Ecol 152:67–78

Judd WS, Campbell CS, Kellogg EA, Stevens PF, Donoghue MJ (2016) Plant systematics: a phylogenetic approach. Sinauer Associates Inc, Sunderland, Massachusetts

Jury SL (1982) Tuberculate fruits in the Umbelliferae. In: Cauwet-Marc AM, Carbonnier J (eds) Actes du 2ème symposium international sur les ombellifères "contributions pluridisciplinaires à la systématique". Braun-Brumfield, Ann Arbor, pp 149–159

Jury SL (2002) *Daucus*. In: Valdés B, Rejdali M, Achhal El, Kadmiri A, Jury JL, Montserrat JM (eds) Catalogue des plantes vasculaires du Nord de Maroc, incluant des clés d'identification. Consejo Superior de Investigaciones Científicas, Madrid, p 467–469

Katz-Downie DS, Valiejo-Roman CM, Terentieva EI, Troitsky AV, Pimenov MG, Lee B, Downie SR (1999) Towards a molecular phylogeny of Apiaceae subfamily Apioideae: additional information from nuclear ribosomal DNA ITS sequences. Plant Syst Evol 216:167–195

Krickl M (1961) Karotten: Zur Frage der Verkreuzung mit der wilden Karotte. Saatgut-Wirtschaft 13:135–136

Le Floc'h É, Boulos L, Véla E (2010) Catalogue synonymique commenté de la flore de Tunisie. République Tunisienne, Ministère de l'Environnement Et Du Développment Durable Banque Nationale de Gènes, Tunis

Lee BY (2002) Taxonomic review on the African Umbelliferous genus *Agrocharis*: inferences based on molecular data. Israel J Plant Sci 50:211–216

Lee BY, Downie SR (1999) A molecular phylogeny of Apiaceae tribe Caucalideae and related taxa: inferences based on ITS sequence data. Syst Bot 24:461–479

Lee BY, Downie SR (2000) Phylogenetic analysis of cpDNA restriction sites and rps16 intron sequences reveals relationships among Apiaceae tribes Caucalideae, Scandiceae and related taxa. Plant Syst Evol 221:35–60

Lee C-S, Downie SR (2006) Phylogenetic relationships within *Cicuta* (Apiaceae tribe Oenantheae) inferred from nuclear rDNA ITS and cpDNA sequence data. Canad J Bot 84:453–468

Lee BY, Park C-W (2014) Molecular phylogeny of *Daucus* (Apiaceae): evidence from nuclear ribosomal DNA ITS sequences. J Spec Res 3:39–52

Lemmon EM, Lemmon AR (2013) High-throughput genomic data in systematics and phylogenetics. Annu Rev Ecol Evol Syst 44:99–121

Liu J, Shi L, Han J, Li G, Lu H, Hou J, Zhou X, Meng F, Downie SR (2014) Identification of species in the angiosperm family Apiaceae using DNA barcodes. Mol Ecol Resour 14:1231–1238

Maddison WP (1995) Phylogenetic histories within and among species. In: Hoch PC, Stevenson AG (eds) Experimental and molecular approaches to plant biosystematics Monographs in systematics. Botanical Garden, St. Louis, Missouri, pp 273–287

Martínez-Flores F (2016) Sistemática del género *Daucus* L. (Apiaceae): implicaciones taxonómicas y filogenéticas. Doctoral thesis, Universidad de Alicante, Spain

Martínez-Flores F, Arbizu CI, Reitsma K, Juan A, Simon PW, Spooner DM, Crespo MB (2016) Lectotype designation for seven species names in the *Daucus guttatus* complex (Apiaceae) from the central and eastern Mediterranean basin. Syst Bot 41:464–478

Matias Vaz AM (2014) Estudo morfológico e filogenético das subespécies *Daucus carota* ssp. *azoricus* e *Daucus carota* ssp. *maritimus* na ilha de S. Miguel. M.Sc. thesis, Universidade dos Açores, Ponta delgada

McCollum GD (1975) Interspecific hybrid *Daucus carota* *D. capillifolius*. Bot Gaz 136:201–206

McCollum GD (1977) Hybrids of *Daucus gingidium* with cultivated carrots (*D. carota* subsp. *sativus*) and *D. capillifolius*. Bot Gaz 138:56–63

Mezghani N, Zaouali I, Amri WB, Rouz S, Simon PW, Hannachi C, Ghrab Z, Neffati M, Bouzbida B, Spooner DM (2014) Fruit morphological descriptors as a tool for discrimination of *Daucus* L. germplasm. Genet Res Crop Evol 61:499–510

Mouterde P (1966) Nouvelle Flore de Liban et de la Syrie. Dar El-Machreq, Beirut

Nakajima Y, Oeda K, Yamamoto T (1998) Characterization of genetic diversity of nuclear and mitochondrial genomes in *Daucus* varieties by RAPD and AFLP. Plant Cell Rep 17:848–853

Nothnagel T, Straka P, Linke B (2000) Male sterility in populations of *Daucus* and the development of alloplasmic male-sterile lines of carrot. Plant Breed 119:145–152

Page RD, Charleston MA (1997) From gene to organismal phylogeny: reconciled trees and the gene tree/species tree problem. Mol Phylo Evol 7:231–240

Pamilo P, Nei M (1988) Relationships between gene trees and species trees. Mol Biol Evol 5:568–583

Plunkett GM, Downie SR (1999) Major lineages within Apiaceae subfamily Apioideae: a comparison of chloroplast restriction site and DNA sequence data. Amer J Bot 86:1014–1026

Plunkett GM, Pimenov MG, Reduron J-P, Kljuykov EV, van Wyk B-E, Ostroumova TA, Henwood MJ, Tilney PM, Spalik K, Watson MF, Lee B-Y, Pu F-D, Webb CJ, Hart JM, Mitchell AD, Muckensturm B (in press) Apiaceae. In: Kadereit JW, Bittrich V (eds) The families and genera of vascular plants, vol. 15. Springer, Berlin, Germany

Plunkett GM, Soltis DE, Soltis PS (1996a) Evolutionary patterns in Apiaceae: inferences based on *matK* sequence data. Syst Bot 21:477–495

Plunkett GM, Soltis DE, Soltis PS (1996b) Higher level relationships of Apiales (Apiaceae and Araliaceae) based on phylogenetic analysis of rbcL sequences. Amer J Bot 83:499–515

Pottier-Alapetite G (1979) *Daucus*. In: Pottier-Alapetite G (ed) Flore de la Tunisie: angiospermes-Dicotylédones (Apetales—Dialypetales). Ministère de l'Enseignement Supérieur et de la Recherche Scientifique et le Ministère de l'Agriculture, Tunis, pp 615–621

Pujadas Salvà AJ (2002) El complejo de *Daucus carota* L. (Apiaceae) en la flora Ibérica. Anales Jard Bot Madrid 59:368–375

Pujadas Salvà AJ (2003) *Daucus*. In: Nieto-Feliner G, Jury SL, Herrero A (eds) Flora Iberica: plantas vasculares de la Península Ibérica e islas Baleares. Anales Jard Bot Madrid CSIC 97–125

Quézel P, Santa S (1963) *Daucus*. In: Quézel P, Santa S (eds) Nouvelle Flore de L'Algérie et des Régions Désertiques Méridionales. Éditions du Centre National de la Recherche Scientifique, Paris, p 659–663

Raubeson LA, Jansen RK (2005) Chloroplast genomes of plants. In: Henry RJ (ed) Plant diversity and evolution: phenotypic variation in higher plants. CABI Publishing, Wallingford, UK, p 45–68

Rieseberg LH, Soltis DE (1991) Phylogenetic consequences of cytoplasmic gene flow in plants. Evol Trends Plant 5:65–84

Rong J, Janson S, Umehara M, Ono M, Vrieling K (2010) Historical and contemporary gene dispersal in wild carrot (*Daucus carota* ssp. *carota*) populations. Ann Bot 106:285–296

Rong J, Lammers Y, Strasburg JL, Schidlo NS, Ariyurek Y, de Jong TJ, Klinkhame PGL,

Smulders MJM, Vrieling K (2014) New insights into domestication of carrot from root transcriptome analyses. BMC Genom 15:895

Rubatzky VE, Quiros CF, Simon PW (1999) Carrots and related vegetable Umbelliferae. CABI, New York, NY, USA

Ruhlman T, Lee S-B, Jansen RK, Hostetler JB, Tallon LJ, Town CD, Daniell H (2006) Complete plastid genome sequence of *Daucus carota*: implications for biotechnology and phylogeny of angiosperms. BMC Genom 7:222

Sáenz de Rivas C, Heywood VH (1974) Preliminary study on the *Daucus* species of peninsular Spain. An Jard Bot AJ Canavilles 31:97–118

Sáenz Laín C (1981) Research on *Daucus* L. (Umbelliferae). An Jard Bot Madrid 37:481–534

Schäfer H (2005) Flora of the Azores. Margraf Publishers, Weikersheim

Shim SI, Jørgensen RB (2000) Genetic structure in cultivated and wild carrots (*Daucus carota* L.) revealed by AFLP analysis. Theor Appl Genet 101:227–233

Simon PW (2000) Domestication, historical development, and modern breeding of carrot. Plant Breed Rev 19:157–190

Small E (1978) A numerical taxonomic analysis of the *Daucus carota* complex. Canad J Bot 56:248–276

Small RL, Cronn RC, Wendel JF (2004) Use of nuclear genes for phylogeny reconstruction in plants. Aust Syst Bot 17:145–170

Spalik K, Downie SR (2007) Intercontinental disjunctions in *Cryptotaenia* (Apiaceae, Oenantheae): an appraisal using molecular data. J Biogeogr 34:2039–2054

Spalik K, Piwczyński M, Danderson CA, Kurzyna-Młynik R, Bone TS, Downie SR (2010) Amphitropic amphiantarctic disjunctions in Apiaceae subfamily Apioideae. J Biogeogr 37:1977–1994

Spalik K, Wojewódzka A, Downie SR (2001a) The evolution of fruit in Scandiceae subtribe Scandicinae (Apiaceae). Canad J Bot 79:1358–1374

Spalik K, Wojewódzka A, Downie SR (2001b) Delimitation of genera in Apiaceae with examples from Scandiceae subtribe Scandicinae. Edinb J Bot 58:331–346

Spooner DM, Rojas P, Bonierbale M, Mueller LA, Srivastav M, Senalik D, Simon P (2013) Molecular phylogeny of *Daucus* (Apiaceae). Syst Bot 38:850–857

Spooner DM, Ruess H, Iorizzo M, Senalik D, Simon P (2017) Entire plastid phylogeny of the carrot genus (*Daucus*, Apiaceae): concordance with nuclear data and mitochondrial and nuclear DNA insertions to the plastid. Amer J Bot 104:296–312

Spooner DM, Widrlechner MP, Reitsma KR, Palmquist DE, Rouz S, Ghrabi-Gammar Z, Neffati M, Bouzbida B, Ouabbou H, El Koudrim M, Simon PW (2014) Reassessment of practical subspecies identifications of the USDA *Daucus carota* germplasm collection: morphological data. Crop Sci 54:706–718

Steinborn R, Linke B, Nothnagel T, Börner T (1995) Inheritance of chloroplast and mitochondrial DNA in alloplasmic forms of the genus *Daucus*. Theor Appl Genet 91:632–638

St. Pierre MD, Bayer RJ (1991) The impact of domestication on the genetic variability in the orange carrot, cultivated *Daucus carota* ssp. *sativus* and the genetic homogeneity of various cultivars. Theor Appl Genet 82:249–253

St. Pierre MD, Bayer RJ, Weiss IM (1990) An isozyme-based assessment of the genetic variability within the *Daucus carota* complex (Apiaceae: Caucalideae). Canad J Bot 68:2449–2457

Stuessy TF, Hörandl E (2014) Evolutionary systematics and paraphyly: introduction. Ann Missouri Bot Gard 100:2–5

Stuessy TF, König C, López Sepúlveda P (2014) Paraphyly and endemic genera of oceanic islands: implications for conservation. Ann Missouri Bot Gard 100:50–78

Tavares AC, Loureiro J, Castro S, Coutinho AP, Paiva J, Cavaleiro C, Salgueiro L, Canhoto JM (2014) Assessment of *Daucus carota* L. (Apiaceae) subspecies by chemotaxonomic and DNA content analyses. Biochem Syst Ecol 55:222–230

Thellung A (1926) *Daucus*. In: Hegi G (ed) Illustrierte Flora von Mitteleuropa. J. F. Lehmanns Verlag, München, pp 1501–1526

Theobald WL (1971) Comparative anatomical and developmental studies in the Umbelliferae. In: Heywood VH (ed) The biology and chemistry of the Umbelliferae. Academic Press, London, pp 177–197

Townsend CC (1989) Flora of tropical East Africa: Umbelliferae. Balkema, Rotterdam

Umiel N, Jacobsen R, Globerson D (1975) Pollination of the cultivated carrot (*Daucus carota* L.) by the wild carrot (*D. carota* var. *maximus*), and its implications on commercial seed production. Hassadeh 56:478–480. (In Hebrew, English summary)

Vivek BS, Simon PW (1998) Genetic relationships and diversity in carrot and other *Daucus* taxa based on nuclear restriction fragment length polymorphisms. J Amer Soc Hort Sci 123:1053–1057

Vivek BS, Simon PW (1999) Phylogeny and relationships in *Daucus* based on restriction fragment length polymorphisms (RFLPs) of the chloroplast and mitochondrial genomes. Euphytica 105:183–189

Weitzel C, Rønsted N, Spalik K, Simonsen HT (2014) Resurrecting deadly carrots: towards a revision of *Thapsia* (Apiaceae) based on phylogenetic analysis of nrITS sequences and chemical profiles. Bot J Linn Soc 174:620–636

Wendel JF, Doyle JJ (1998) Phylogenetic incongruence: window into genome history and molecular evolution. In: Soltis DE, Soltis PS, Doyle JJ (eds) Molecular systematics of plants II: DNA sequencing. Chapman and Hall, Kluwer, Boston, pp 265–296

Wijnheijmer EHM, Brandenburg WA, Ter Borg SJ (1989) Interactions between wild and cultivated carrots (*Daucus carota* L.). Euphytica 40:147–154

Williams CF (1994) Genetic consequences of seed dispersal in three sympatric forest herbs. II. Microspatial genetic structure within populations. Evolution 48:1959–1972

Wolfe KH, Li WH, Sharp PM (1987) Rates of nucleotide substitution vary greatly among plant mitochondrial, chloroplast, and nuclear DNAs. Proc Natl Acad Sci USA 84:9054–9058

Wong MML, Gujaria-Verma N, Ramsay L, Yuan HY, Caron C, Diapari M, Vandenberg A, Bett KE (2015) Classification and characterization of species within the genus *Lens* using genotyping-by-sequencing (GBS). PLoS ONE 10:e122025

Yi T-S, Jin G-H, Wen J (2015) Chloroplast capture and intra- and inter-continental biogeographic diversification in the Asian—New World disjunct plant genus *Osmorhiza* (Apiaceae). Mol Phylo Evol 85:10–21

Zhou J, Gong X, Downie SR, Peng H (2009) Towards a more robust molecular phylogeny of Chinese Apiaceae subfamily Apioideae: additional evidence from nrDNA ITS and cpDNA intron (rpl16 and rps16) sequences. Mol Phylo Evol 53:56–68

Zohary M (1972) Flora Palaestina. Israel Academy of Sciences and Humanities, Jerusalem

3 Carrot Floral Development and Reproductive Biology

Bettina Linke, Maria Soledad Alessandro, Claudio R. Galmarini and Thomas Nothnagel

Abstract

The defining characteristic of the botanical family of Apiaceae (former Umbelliferae) is the inflorescence. The flowers aggregate in terminal umbels that may be commonly compound, often umbelliform cymes. Likewise, flowers of the carrot are clustered in flat, dense umbels, partially with zygomorphic petals at the edges. Carrot producers and consumers mainly consider the vegetative phase, namely the storage root as a vegetable. Nevertheless, the reproductive phase is an important topic for genetic research, for breeding new cultivars and for seed production. Hence, improved knowledge on the genetic control mechanisms of reproduction such as flowering time, flower development and architecture, pollen fertility and male sterility, as well as seed set is of essential importance. The chapter reviews key steps on carrot floral development and reproductive biology, especially under consideration of the comprehensive genomic data set recently obtained from carrot.

3.1 Key Steps of Reproductive Biology of Carrot

The reproductive phase comprises different ontogenetic stages, starting with the induction of the shoot meristem for stalk elongation, the transition into an inflorescence meristem, and the complex process of flower development. Flower development includes the induction of flower meristems and the specification and subsequent formation of flower organs. After pollination, fertilization, and seed development, the reproductive cycle ends with the senescence and decline of the plant (Fig. 3.1).

The transition from the vegetative phase to the generative phase is controlled by external factors and by internal cues of the plant (compare to 3.2.1). The onset of flowering is initiated by a period of low temperature and includes several physiological and morphological changes

B. Linke
Department of Biology, Humboldt University, Berlin, Germany

M. S. Alessandro · C. R. Galmarini
Estación Experimental Agropecuaria La Consulta, Instituto Nacional de Tecnología Agropecuaria, La Consulta, Mendoza, Argentina

C. R. Galmarini
Consejo Nacional de Investigaciones Científicas y Técnicas, Buenos Aires, Argentina

C. R. Galmarini
Facultad de Ciencias Agrarias, Universidad Nacional de Cuyo, Chacras de Coria, Luján, Mendoza, Argentina

T. Nothnagel (✉)
Institute for Breeding Research on Horticultural Crops, Federal Research Centre for Cultivated Plants, Quedlinburg, Germany
e-mail: thomas.nothnagel@julius-kuehn.de

P. Simon et al. (eds.), *The Carrot Genome*, Compendium of Plant Genomes,
https://doi.org/10.1007/978-3-030-03389-7_3

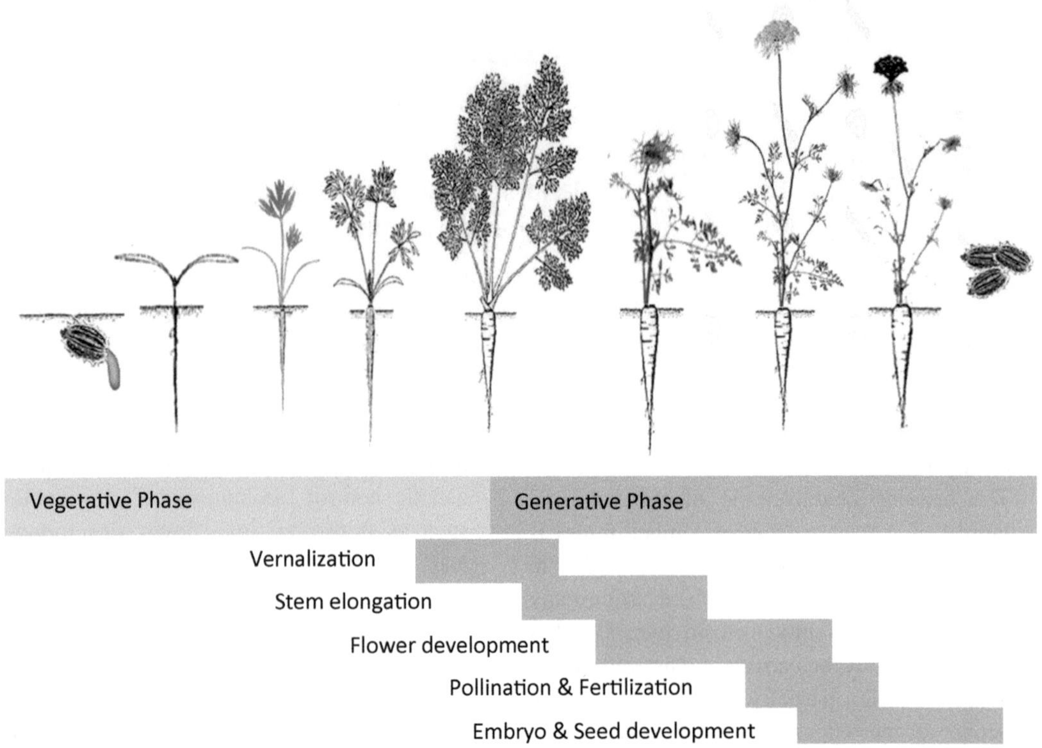

Fig. 3.1 Vegetative and generative phase of the carrot. Main characteristics of the generative phase that are mentioned in this chapter are shaded in gray

(Le Dily et al. 1991). Roots become fibrous and inedible even before the first visible inflorescences appear. The leaf-producing vegetative apical meristem changes into an uplifted conical reproductive meristem. The differentiation into a reproductive meristem also includes the elongation of the stem (shoot apex) and its transformation into an inflorescence apex (Borthwick et al. 1931).

The first floral axis grows slightly upward and elongates to more than one meter until (first) branching occurs. The main shoot terminates into an inflorescence structure, which is designated as a 'primary umbel' (P; Fig. 3.2b). Lateral shoots develop 'higher-order umbels' (S, T; Fig. 3.2b). Each individual shoot bears umbels of third, fourth, or even higher order. The inflorescence is a compound umbel comprising of several subunits, the so-called umbellets.

Depending on the genotype, the primary umbel may contain more than 50 umbellets, each about 50 individual flowers. The number of lateral shoots and the number of developing umbels are influenced by genetic factors, as well as by environmental conditions or seed plant spacing as was already shown a long time ago (Austin and Longden 1967; Gray and Steckel 1983a, b; Gray et al. 1983; Harrington 1951; Hawthorn 1952). Carrots normally develop epigynous, hermaphrodite flowers with a five-lobed calyx, each with five petals and stamens, and a two-celled, inferior ovary, with each locule bearing a single functional ovule. The upper surface of the carpels covers a nectar cell containing disk, the stylopodium, which is important to increase the attractiveness for insects (Broussard et al. 2017; Mas et al. 2018). Wild and cultivated carrots tend to exhibit andromonoecy

Fig. 3.2 **a** Flowering carrot plant. **b** Scheme of the carrot inflorescence indicating the stalk with the primary stem and secondary branches (modified according to Rubatzky et al. 1999; abbreviations: P = primary, S = secondary, T = tertiary order). **c** The flowering structure is designated as a compound umbel. **d** Carrot umbel with a typical dark central flower. **e** Hermaphroditic flower in the protandric stage. **f** Floret during the receptive phase of the stigma. **g** Ovule-less, staminate flowers before opening. **h** Flower during pollen flow

and have hermaphroditic and staminate flowers on the same inflorescence (Fig. 3.2e–h). The ratio of both flower types is varying in umbels of different order and tends to produce staminate flowers in later orders. On the primary umbel, approximately 95% of the flowers are hermaphroditic, whereas staminate flowers are found more toward the center of the umbel (Lamborn and Ollerton 2000).

The floral sex ratio varies with regard to the appropriate genotype of the plant. Several authors hypothesized that the sex differentiation is determined late within the flower development, allowing the plant a quick response to environmental changes (Koul et al. 1989; Reuther and Claßen-Bockhoff 2013; Rubashevskaia 1931). In numerous carrot accessions and wild relatives, the central umbellet can be reduced to one or a few white or dark-red flowers (Fig. 3.2d). A previous hypothesis that the dark central flower has an attracting function for insect pollinators was experimentally disproven (Lamborn and Ollerton 2000).

Floral development is centripetal and protandrous, with dehiscence of anthers before the stigma becomes receptive (Fig. 3.2e–h). The receptivity of the stigma is visually associated with a separation of the paired styles. Opening of the flower initially occurs at umbellets of the periphery and at the primary umbel and takes approximately a week for the entire umbel. In weekly succession, flower opening of the higher-order umbels follows gradually. The flowering period of an individual umbel may vary between 7 and 10 days and of the whole plant between 30 and 60 days (Rubatzky et al. 1999). In wild carrots, protandrous dichogamy has been suggested as a strategy to reduce geitonogamy (Koul et al. 1989). Secondary umbels are only produced after the primary umbel has been pollinated (Westmoreland and Muntan 1996).

Once the flowers are fertilized and the seed set has been initiated, the umbels are closing by developing a nest-like shape, possibly to protect the developing fruits. Later they re-expand to release the seeds. Fruits contain two seeds that are enclosed in a spiny pericarp, which probably aids their dispersal on animal fur (Lamborn and Ollerton 2000). The fruits are dry schizocarps consisting of two ribbed or winged mericarps that can separate upon maturity, of which each is an individual seed. Mericarps are small, longer than they are wide, and form the longitudinal hemisphere of the fruit. Seeds have secretion ducts containing essential oils (Mockute and Nivinskiene 2004; Staniszewska et al. 2005; Yahyaa et al. 2017). Genotype and environmental conditions are responsible for the seed quality characterized by traits as germability, vigor, dormancy, or the disease contamination.

In the following sections, key steps of carrot reproductive biology are emphasized, namely the onset of flowering after vernalization and the formation of the flower architecture (compare to Fig. 3.1). This includes the specification of floral organs according to the ABC(DE) model of flower formation that has been well characterized in model plants like Arabidopsis or Antirrhinum (Bowman et al. 1989; Coen and Meyerowitz 1991; Theissen and Saedler 2001). The specification of the 'floral organ identity' by several classes of homeotic genes (mainly MADS-box genes) will be accentuated with a further aim to connect the process of flower development with the subject of CMS (cytoplasmic male sterility) as an important tool for breeder's application.

3.2 Genetic Control of the Different Pathways of Reproductive Biology

3.2.1 Vernalization and Stalk Elongation

The molecular basis of flowering has been thoroughly studied in *Arabidopsis thaliana*, which has been used as a model plant for flowering studies among dicots. Flowering induction is dependent on a complex gene network regulated by endogenous factors and environment. Light and temperature (acting through photoperiod and vernalization pathways) are the most important environmental factors regulating flowering time (Amasino and Michaels 2010). *CONSTANS* (*CO*) is a photoperiod-dependent gene in which cis-regulatory variations are responsible for variations in flowering time (Rosas et al. 2014). Vernalization and autonomous pathways promote flowering by reducing the transcription level of *FLOWERING LOCUS C* (*FLC*), thereby enhancing the expression of floral integrator genes *FLOWERING LOCUS T* (*FT*) and *SUPPRESSION OF OVEREXPRESSION OF CONSTANS 1* (*SOC1*) (Helliwell et al. 2006; Michaels and Amasino 1999). *VERNALIZATION INSENSITIVE 3* (*VIN3*) gene family members repress different subsets of the *FLC* gene family, whose members are also differentially regulated during the course of vernalization via epigenetic changes (Kim and Sung 2013).

FLC-like genes have been identified in other taxa like Brassica and Raphanus, both within the Brassicaceae family (Bagget and Kean 1989; Ferreira et al. 1995; Lan and Paterson 2000; Kole et al. 2001; Osborn et al. 1997) and in Beta (Reeves et al. 2007). Among monocots, flowering in winter cereals has been extensively studied and a repressor that inhibits flowering in the fall season has been found, similar to the one seen in Arabidopsis (Yan et al. 2004). *VRN1* represses another genetically defined gene, *VRN2*, which blocks the transition to flowering before vernalization (Yan et al. 2004). This repression enables the activation of the FT ortholog *VRN3* and leads to subsequent induction of flowering (Yan et al. 2006).

Although bolting and flowering are important constraints to carrot production, only a few genetic or molecular studies have focused on this process. The stage of growth when carrots seedlings are not responsive to low-temperature vernalization is known as juvenility. That condition usually ends when carrot plants have initiated 8–12 leaves, and storage roots are greater than 4–8 mm in diameter (Atherton et al. 1990; Galmarini and Della Gaspera 1996; Galmarini et al. 1992; Fig. 3.1). After a vernalization period, with temperatures between 0 and 10 °C at long days, floral stem elongation and flowering are induced (Atherton and Basher 1984; Dickson and Peterson 1960; Sakr and Thompson 1942). Carrot roots quickly become much lignified after vernalization, even before the floral stalk elongates, so that the initiation of flowering also results in a complete loss of commercial value (Rubatzky et al. 1999).

In almost all crop species known as biennials, early flowering or annual plants usually are found. In carrot, the level of response to cold treatments is cultivar-dependent. Late-flowering cultivars, from seed-to-seed crops, require approximately 11–12 weeks at 5 °C to be permanently vernalized (Atherton et al. 1990; Hiller and Kelly 1979). Early flowering cultivars require shorter vernalization periods, 1–4 weeks (Alessandro and Galmarini 2007; Dias-Tagliacozzo and Valio 1994).

In late-flowering cultivars, the response to vernalization has been more extensively studied. Chilling treatments to carrot plants (cv. 'Chantenay Red Cored') maintained in darkness or photoperiods of less than 12 h resulted in more rapid flowering than chilling under longer photoperiods (Craigon et al. 1990). The temperatures used during chilling treatment also influence the rate of flower bud appearance and rate of stem internode extension, as temperatures increased from −1 °C to an optimum of 6 °C the rate of elongation increased, but the rate decreased as temperatures were increased to a maximum of about 16 °C (Atherton et al. 1990). After vernalization, flowering could be suppressed with continuous low-light, short-day photoperiods or by a few days of high temperature (28–35 °C) if stem elongation had not yet occurred (Fisher 1956; Ou et al. 2017). Long days following the vernalization stimulated flowering (Atherton and Basher 1984).

Physiological changes occur during vernalization before any morphological changes are evident. Endogenous gibberellic acid (GA) levels rise as a response to cold treatment which stimulates the flowering process (Nieuwhof 1984; Schwab and Neumann 1975). Exogenous application of gibberellins successfully induces flowering in carrot, although this technique is not widely used (Galmarini et al. 1995).

In carrot the annual habit is clearly dominant, and the observed segregation ratios in F_2 and BC_1 families, derived from the cross between an early flowering (annual) cultivar 'Criolla INTA' and two late-flowering (biennial) petaloid male sterile lines, adjust to the model of one single dominant gene conditioning the annual habit (Alessandro and Galmarini 2007). The gene controlling early flowering habit in carrot was named *Vrn1* and mapped using an F_2 progeny. On a map of 355 markers covering all 9 chromosomes with an average marker distance of 1.88 cM, *Vrn1* mapped to chromosome 2 (Fig. 3.3) with flanking markers at 0.70 and 0.46 cM (Alessandro et al. 2013). Recent studies with new segregating families done by Wohlfeiler et al. (2019) suggest that two genes are controlling the annual habit. Furthermore, data of Villeneuve and Latour (2017) indicated that epigenetic factors also influence flower initiation. Using transcriptome analysis, Ou et al. (2017) identified 45 flowering time-related unigenes in carrot that were classified into five categories including photoperiod, vernalization, autonomous and gibberellin pathway, and floral integrators.

Several homologs of *LATE ELONGATED HYPOCOTYL* (*LHY*), *CONSTANS-LIKE 2* (*COL2*), *SUPPRESSION OF OVER-EXPRESSION OF CONSTANS 1* (*SOC1*), *FLOWERING LOCUS C* (*FLC*), and *GIBBERELLIC ACID INSENSITIVE* (*GAI*) exhibited differential expression between the two carrot libraries analyzed, based on digital gene expression analysis, and their expression was significantly correlated with that of other flowering

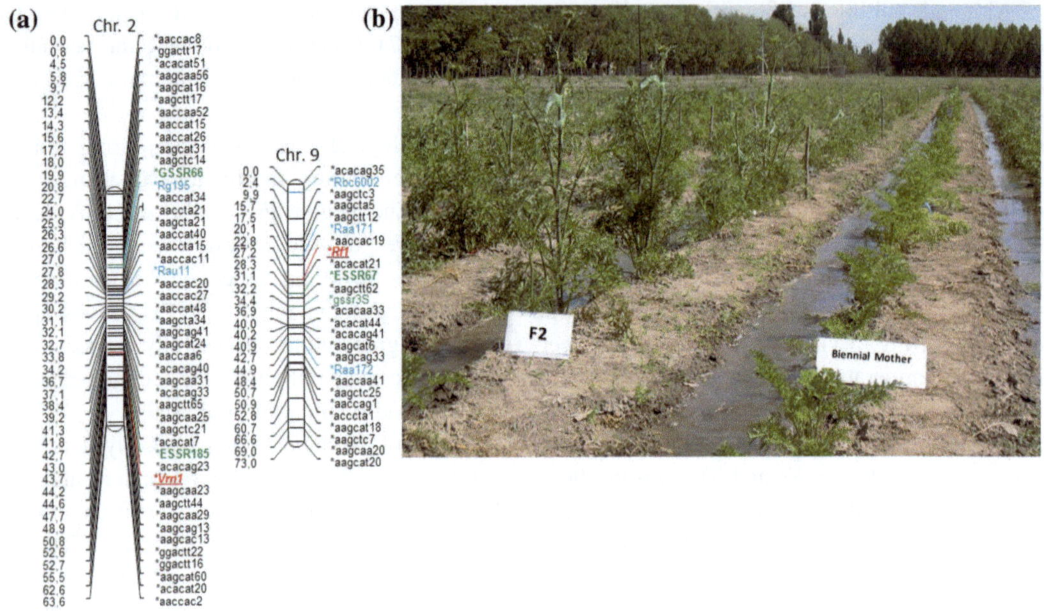

Fig. 3.3 **a** Part of a carrot linkage map according to Alessandro et al. (2013) indicating chromosomes 2 and 9 with the assigned loci *Vrn1* and *Rf1* (loci in red). Microsatellites mapped in this work are denoted in green. RAPDs are denoted in blue. *Vrn1*, vernalization gene (compare to 3.2.1); *Rf1*, male fertility restorer gene (compare to 3.2.4.2). **b** Bolting behavior of a carrot F_2 population used for developing the map (75% bolting plants) and its biennial mother (no bolting plants) in the field in the spring. Both genotypes were sown in the autumn

time-related unigenes (Ou et al. 2017). In addition, a member of the *FT* (*FlOWERING LOCUS T*)*-like* genes with potential roles in the promotion of flowering has been initially characterized (Zhan et al. 2017). It remains to be shown whether the large genomic data sets available for carrot (Iorizzo et al. 2016) will improve knowledge on these and other gene candidates and their regulatory circuits during the control of flowering time.

3.2.2 Flower Development

During the switch to the flowering program, the apical shoot meristem shifts from vegetative growth to the development of floral apices. While approaching the reproductive state, the vegetative shoot apex changes its globular shape into a rather conical/longitudinal structure. After transitioning into an inflorescence apex, the meristem structure becomes more flattened and starts to produce involucral bracts and umbel primordia. Compound umbels in a young state bear different developmental stages, with older stages at the margins and younger stages in the center of the apex. Each umbel represents a single developmental unit and is covered by bracts. Within one umbel, single umbellets begin to develop from the meristem. Within a single umbellet, primordia of single florets develop in a similar sequential arrangement as is described for the major umbel (Fig. 3.2).

The formation of carrot single flowers is a temporally and spatially tightly regulated process. It is dependent on the initiation of organ primordia in a correct position and on a correct identity to ensure that sepals, petals, stamens, and carpels are arranged in a proper composition to form the final flower architecture. Flower morphogenesis during different developmental stages has been comprehensively analyzed regarding botany, taxonomy and compared to other species like model plants as Arabidopsis (e.g., Ajani et al. 2016; Borthwick et al. 1931; Erbar and Leins 2004; Kitagawa et al. 1994; Leins and Erbar 1997; Linke et al. 1999, 2003; Nothnagel et al. 2000).

Among members of the Apiaceae-Apioideae, Ajani et al. (2016) described three patterns of floral organ development, especially emphasizing the meristem conditions underlying a divergent organ initiation and a characteristic stamen promotion. Main differences among the taxa considered the direction and timing of floral organ initiation (Ajani et al. 2016; Erbar and Leins 2004; Leins and Erbar 1997). In carrot, it has been shown that primordia initiation followed a 'grouped' pattern arranged in a spiral sequence of organ formation. A gradual formation of large protuberances was observed, followed by a successive splitting into one petal, one stamen, and one sepal primordium, respectively (Ajani et al. 2016). If compared to other members of the Apiaceae, carrot flowers revealed a flatter meristem and the stamen primordia developed earlier and grow equally fast or even faster than the petals. The sequential versus simultaneous and centripetal versus divergent primordia initiation in the Apiaceae-Apioideae was discussed as a consequence of meristem size and spatial constraints. In spite of a grouped primordia initiation, the organ position remained usually constant suggesting a different regulation from that of 'organ sequence' and that of 'organ identity' (Ajani et al. 2016).

The placing of correct organs in right positions within a flower requires a proper determination of their organ identity (compare to 3.2.3). To analyze the specification of organ identity, organogenesis of single flowers has been briefly studied and compared to those of model plants like Arabidopsis or Antirrhinum (Linke et al. 1999, 2003; and references therein). Flower formation has been subdivided into seven stages. This included the formation of single floret primordia from the inflorescence meristem (stages S0–S3), the establishment of organ primordia within single flowers (until stage S5), and the beginning of organ differentiation during stages S6–S7. During stages S0–S3, single flowers are formed within an umbellet which is indicated by a beginning separation from the flower bottom. Whole-flower primordia are distinguished by their characteristic shapes. At stage S5, the organ primordia of sepals, petals, and stamens have been already initiated in the outer three flower whorls and their differentiation occurs. Primordia of the perianth organs (future sepals and petals) become distinguishable by their different orientation relative to the floral apex. Petals and stamen filaments continue to extend and are curved toward the center of the apex, which was also reported by Kitagawa et al. (1994).

3.2.2.1 Development of the Male Sporophyte

The basal-distal differentiation of the stamens starts from the early floral stage 6 onwards (Linke et al. 1999, 2003). Indentations at the base of the globular-stamen primordia give rise to the formation of filaments, whereas the distal organ region acquires an oval shape to form out the anthers. A furcation initiates the development of a bi-lobed structure. The subsequent invagination partitions the anther into a paired structure finally resulting in the typical tetra-lobed anther symmetry. Locules or pollen sacs are arranged in a pair-wise structure, each pair belongs to a separate unit, the theca. In each anther, four laterally symmetrical locules develop by forming two identical adaxial locules and two abaxial locules. Anther and pollen development are complex processes finally leading to the release of viable pollen. Within the locules of the anther, organized sporophytic layers are subsequently established by forming epidermis, endothecium, middle layer, and the tapetum, which is important for nutrition and development of the subsequent pollen grains. The microspore mother cells (MMCs) undergo meiosis to produce haploid microspores that progress to develop in the center of the locule (see Fig. 3.6). After two mitotic divisions, trinucleic pollen is formed. During these stages, the tapetum and the middle layer start to degenerate. At the mature stage, the anther undergoes dehiscence to allow the release of mature adherent, tricolporate and mostly barrel-shaped up to long-oval pollen grains (Linke and Börner 2005; Struckmeyer and Simon 1986; Zenkteler 1962).

3.2.2.2 Development of the Female Sporophyte

The development of the female organ in the center of the flower initiates rather late, if compared to the other organs. The formation of the carpels starts by forming a groove in the center of the floral meristem where two C-shaped carpel primordia appear. From stage 6 onwards, the first evidence of carpel formation reveals a slight rise at two points on opposite sides of the circular area of the flower meristem. The two regions continue to elevate and acquire a crescent shape. Then, the primordial carpels slope to their respective margins, continue to elevate, and form out a cavity in the ovary that is gradually divided into two locules. Later, the inturned margins of the two carpels are formed, as the first indication of its double structure. From the inturned margins of each carpel, two ovule primordia are formed one slightly above the other. Only the lower ovule in each locule becomes functional (Borthwick et al. 1931). Further Borthwick (1931) described the stages of megasporogenesis in carrot starting with a single archesporial cell directly as a macrospore mother cell up to the embryo sac formation containing synergids, antipodals, the egg cell, and the polar nucleus as well as the nucleus of the primary endosperm (compare to Sect. 3.2.6). More recent investigations supported these observations by histological studies and by scanning electron microscopy (SEM) (Ajani et al. 2016; Linke et al. 1999; Struckmeyer and Simon 1986).

3.2.3 Genes Involved in Flower Formation

Flower development is depending on a complex gene regulatory network (Immink et al. 2010; Liu and Mara 2010). Analyses of 'homeotic' flower mutants of these model plants have led to the prediction of the basic ABC model of flower development (Bowman et al. 1989; Coen and Meyerowitz 1991; Schwarz-Sommer et al. 1990). The model has been later expanded to include class D genes, which promote ovule development and class E, or *SEPALLATA* (*SEP*) genes, which act as cofactors of A, B, C, and D class genes (e.g., reviewed by Theissen 2001; Theissen and Saedler 2001).

The model explains that different gene classes (ABCDE) act in combination with each other to determine the identity of the flower organs in the four flower whorls consisting of perianth (sepals, petals) and reproductive floral organs (stamens, carpels). In the first (outer) flower whorl, sepals are specified by class A and E genes. In the second flower whorl, petals are specified by the action of class A, B, and E function. In the third whorl, stamens (male) organs are specified by class B and C and E function. The female organ in the center of the flower is specified by the action of class C and class E and D gene function (compare to Fig. 3.4).

Most of the participating genes encode for transcription factors. A vast majority of the pivotal regulating genes belong to the MADS-box family (e.g., reviewed by Smaczniak et al. 2012).

3.2.3.1 The MADS-Box Gene Family of Transcription Factors

According to their conserved structure, MADS-box genes can be subdivided into defined gene groups or 'clades' with subfamily-specific functions in flower development. Thirteen different paralogous MADS-box gene subfamilies have previously been analyzed and defined by phylogeny reconstructions. According to an unambiguous system for the nomenclature of these subfamilies, they are named after the first clade member that has been identified members of the early identified and well-characterized subfamilies, such as *AG* (*AGAMOUS*)-, *DEF* (*DEFICIENS*)-, *GLO* (*GLOBOSA*)-, and *SQUA* (*SQUAMOSA*)-*like* genes, typically share similar expression patterns and highly related functions (Doyle 1994; Purugganan et al. 1995; Smaczniak et al. 2012; Theissen 2001; Theissen and Saedler 1995). Genes belonging to each of these groups

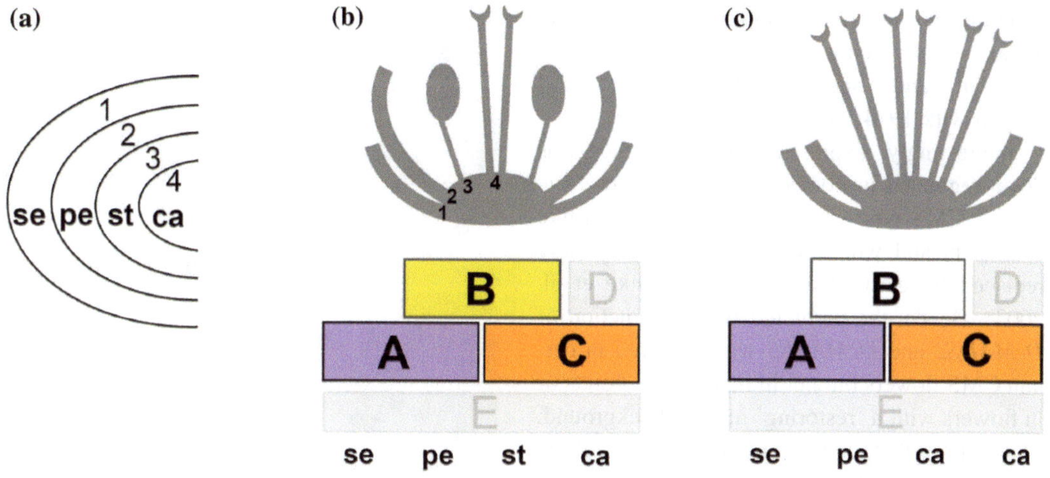

Fig. 3.4 The ABC(DE) model of flower formation focused on the (basic) A, B, and C classes. Schemes of a wild-type flower and a class B mutant are exemplarily shown. The basic ABC model predicts that three classes of genes act in overlapping fields to specify the identity of the four types of floral organs. Class A alone specifies sepal identity, classes A and B together specify petal identity, classes B and C together determine the identity of stamens, and class C alone specifies carpel identity. Class D genes specify ovule identity, and class E genes act as cofactors of the A, B, C, and D class genes. The activity fields of the class D and the class E genes are shaded in gray but are not highlighted here (see text). **a** The four flower whorls from which sepals (se), petals (pe), stamens (st), and carpels (ca) arise are numbered. **b** A wild-type flower with a correct organ identity is indicated by a scheme and with a focus on the activity fields of the gene classes A, B, and C. **c** The scheme shows the phenotype of a class B mutant, where a lack of activity of the B class transcription factors (indicated by a white box) results in florets with sepals in whorl 2 and carpels in whorl 3

play a critical role in the determination of the identity of flower organs according to the ABCDE model as was mentioned above.

In carrot, the firstly identified MADS-box sequences, *DcMADS1–DcMADS5*, were identified from a flower-specific cDNA library that has been prepared using young florets of different developmental stages (Linke et al. 2003). Based upon sequence similarities to certain members of the MADS-box gene subfamilies, *DcMADS1* has been previously assigned to the *SQUA* group and *DcMADS2* and *DcMADS3* to the *GLO* group and the *DEF* group, respectively. *DcMADS4* shared most similarities to the AG group and *DcMADS5* mostly matched to members of *SEP* (former *AGL2*) group (Hernández-Hernández et al. 2007; Linke et al. 2003; Zahn et al. 2005). In a F_2-linkage map of the carrot, two of the five MADS-box genes, *DcMADS3* and *DcMADS5*, have been assigned to LG-5 and LG-7, respectively (Budahn et al. 2014).

A detailed temporal and spatial expression analysis of *DcMADS1–5* by in situ hybridization suggested an assignment of these genes to the group of organ identity genes (Linke et al. 2003). Expression of *DcMADS1* was observed in organ primordia arising in whorl one and two during the whole differentiation of perianth organs. Expression was comparable to that of *SQUAMOSA* (Antirrhinum) or *AP1* (Arabidopsis), and both genes have been predicted to be involved in the specification of sepal and petals. Expression of *DcMADS1* was further observed in pedicels of single flowers and during the earlier stages when inflorescence meristems started to develop. *DcMADS2* and *DcMADS3* reveal striking sequence similarities to the group of B class MADS-box genes, which specify the identity of stamens and, hence, the development of anthers and pollen. Both genes were expressed in petals and stamens throughout development and likely play a similar role in carrot as shown for

Arabidopsis and Antirrhinum (e.g., Bowman et al. 1989; Schwarz-Sommer et al. 1990; Theissen and Saedler 2001; Weigel and Meyerowitz 1994; and references therein).

Interestingly, expression of both genes was down-regulated in homeotic flowers of the 'carpeloid' CMS type where petals reveal sepal-like characters and where stamens were completely replaced by carpel-like structures (Linke et al. 2003). Hence, the predicted B class function of *DcMADS2* and *DcMADS3* was impaired in certain CMS flowers but not in male fertile florets or in flowers with a 'restoring' nuclear background. This implies that the role of MADS factors of the B class was disturbed in CMS plants (see Figs. 3.4, 3.6, and Sect. 3.2.4.6).

DcMADS4 structurally belongs to the *AG* clade indicating high-sequence similarities to the Arabidopsis *AGAMOUS* gene, a major determinant in the specification of stamens and carpels throughout development (Ito et al. 2007; Yanofsky et al. 1990). Regarding the fact that different members of *AG*-like genes exist in numerous plant species and also in the carrot, a final assignment to an ultimate function is not yet possible. However, expression patterns are congruent to the function of *AGAMOUS* as a C class gene and were observed in stamens and carpels throughout organ development (Linke et al. 2003; Fig. 3.5).

The *DcMADS5* gene has a significant sequence similarity to members of the *SEP* group (Linke et al. 2003; Table 3.1). Genes of this subfamily are essential for proper advanced B and C function in Arabidopsis (Pelaz et al. 2000) but revealed a strong heterogeneity in most of the yet-analyzed plant species. Expression of *DcMADS5* was observed in single floral primordia, before any organ primordia appeared, but was absent in the center of the inflorescence meristem.

A more precise classification within the *SEP* clade requires further analyses. Hence, the combination of both structural data and expression patterns predicts roles for *DcMADS2* and *DcMADS3* as B class genes and for *DcMADS4* as a C class gene during specification of the organ identity of stamens and carpels (Linke et al. 2003).

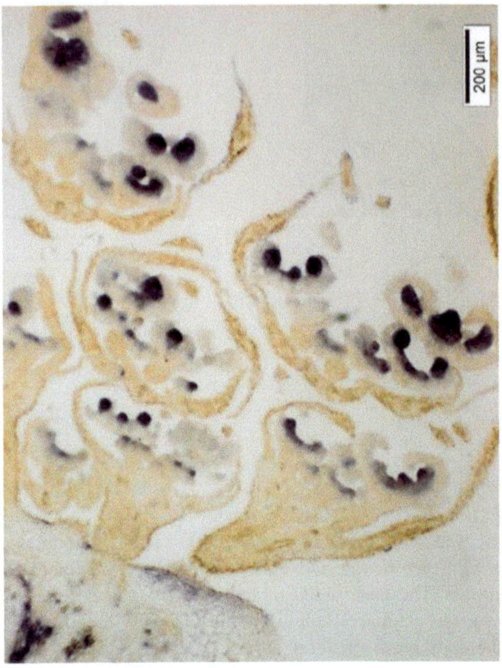

Fig. 3.5 Temporal and spatial expression patterns of *DcMADS4* by *in situ* hybridization of mRNA. Expression patterns are shown on tissue sections. The compound carrot inflorescence indicates several umbellets that are covered by bracts and bear several single florets. Expression of *DcMADS4* can be observed in the primordia of stamens and carpels (whorls three and four) but not in the perianth organs (sepals, petals) of the two outer flower whorls. Bar 0.2 mm

Using *DcMADS1–5* as queries in BLAST searches against the carrot genome, database analyses revealed their assignment to chromosomes 1, 2, 3, 4 and 9, respectively. Exon numbers deduced from exon-intron structures were obtained from the genome data (Table 3.2).

Hence, the current analysis of the genuine GenBank accessions for *DcMADS1–5* confirmed the predicted subgroup classification regarding the comprehensive genome data sets (Iorizzo et al. 2016). Several members of the MADS-box gene family are involved in other developmental processes such as flowering time control. Major roles have been assigned for *FLOWERING LOCUS C* (*FLC*) or *SUPPRESSOR OF OVEREXPRESSION*

Table 3.1 Mapping of *DcMADS1–5* to chromosomes of the carrot genome

Gene name	ORF (bp)	Reference (GenBank)	Subgroup	Genome annotation	Locus tag	Chr.	Exon count
DcMADS1	975	AJ271147.1/CAC81068.1	SQUA group (AP1/FUL)	DCAR_030052*	LOC108200812	9	9
DcMADS2	842	AJ271148.1/CAC81069.1	GLO group (PI)	DCAR_014369	LOC108217347	4	14
DcMADS3	887	AJ271149.1/CAC81070.1	DEF group (AP3)	DCAR_009949	LOC108211719	3	7
DcMADS4	1128	AJ271150.1/CAC81071.1	AG group	DCAR_003963	LOC108214703	1	10
DcMADS5	909	AJ271151.1/CAC81072.1	SEP group/ (former AGL2)	DCAR_007203*	LOC108206005	2	7

In silico mapping of *DcMADS1–5* to the carrot genome. Columns indicate the name of the sequences (*mads1–mads5* according to the GenBank format), the ORF (open reading frames) lengths of the genuine cDNA clones, the accession numbers of the nucleotide and of the deduced protein sequence (GenBank), as well as the classification into specific clades/subgroups of the MADS-box gene family. An alternative nomenclature using abbreviations basing on the terminology of leading members of Arabidopsis genes is mentioned in brackets. Abbreviations for representative genes assigning the clades/subgroups are as follows: *SQUA*, *SQUAMOSA* (Antirrhinum); *AP1*, *APETALA1* (Arabidopsis); *FUL*, *FRUITFUL* (Arabidopsis); *GLO*, *GLOBOSA* (Antirrhinum); *DEF*, *DEFICIENS* (Antirrhinum); *AP3*, *APETALA3* (Arabidopsis); *AG*, *AGAMOUS* (Arabidopsis); *SEP*, *SEPALLATA* (Arabidopsis); *AGL2*, *AGAMOUS-like2* (Arabidopsis). The assignments to certain annotations of the carrot genome (Iorizzo et al. 2016) are shown; annotations with a model RefSeq state are marked by stars (*). Locus tags and the appropriate chromosomes (Chr.) are indicated. Numbers of exon counts deduced from the genome regions are shown

Table 3.2 Inheritance models explaining CMS in carrot in context to cytoplasmic and nuclear-genetic factors

CMS type	Cytoplasm origin	Loci in the nuclear genome	Models of inheritance	Reference
Brown anther	Sa *D.c. sativus*	*Ms*	Dominant *Ms*. allele controls male sterility	Welch and Grimball (1947)
	Sa cv. 'Tendersweet'	*Ms1*, *Ms2*, *Ms3*	Dominant allele(s) at any of three duplicated genes are necessary to maintain sterility (postulated for both cytoplasms), dominant alleles at one or more epistatic loci restore fertility	Thompson (1961)
	Sa	*Ms4*, *Ms5*	Male sterility when recessive *Ms5* or dominant *Ms4*	Hanschke and Gabelmann (1963)
	Line 551324	*aa*, *B.*, *D.*, *E.*	Consistent with Hanschke and Gabelmann (1963), two additional, complementary dominant loci can restore the fertility	Banga et al. (1964)
	Sa cv. 'Nantes-4,' 'Moskovskaya-zimnaya A515'	*msms*	Male sterility is conditioned by a recessive gene	Kononkov and Mokhov (1972), Zhidkova et al. (1991)
	Sa		Consistent with Banga et al. (1964)	Morelock (1974)
	Sa cv. 'Selecta'	*ms*	Recessive allele *ms* would be sterile, gene action is influenced by temperature, at constant high temperature the penetrance and expressivity of the ms gene is reduced 50%	Michalik (1974)
	Sa cv. 'Marktgärtner'	*aa*, *B.*, *D.*, *E.*	Validation of the model of Banga et al. (1964)	Weit (1979), Dame et al. (1988)

(continued)

Table 3.2 (continued)

CMS type	Cytoplasm origin	Loci in the nuclear genome	Models of inheritance	Reference
Petaloid	Sp *D.c. carota*	*Ms1*, *Ms2*, *Ms3*	Dominant allele at any of three duplicated genes are necessary to maintain sterility	Thompson (1961), Wolyn and Chahal (1998)
	Sp US sources	*M. ll tt*	Three independent genes, one dominant *M*, two recessive genes *ll* and *tt*. heterozygous *Mm* plants can be restored at high temperature	Mehring-Lemper (1987)
	Sp n.d. American sources	*Ms3*, *Ms4*, *Ms5*	Validation of the model of Thompson (1961) in a nuclear background of a Russian variety	Timin and Vasilevsky (1997)
	Sp US sources	*Rf1*	Consistent with Hanschke and Gabelmann (1963), and Banga et al. (1964), one dominant loci can restore the fertility	Alessandro et al. (2013)
	D.c. gummifer	*Gum1* (*Gum2*)	One or two homozygous recessive alleles *gum1*, *gum2* seem to be responsible for the male sterile phenotype	Nothnagel et al. (2000)
Petaloid like	*D.c. maritimus*	*Mar1* (*Mar2*)	One or two dominant alleles lead to the male sterile phenotype	Nothnagel et al. (2000)
	D.c. gadecaei	*Gad1* (*Gad2*)	One or two dominant alleles lead to the male sterile phenotype	Nothnagel et al. (2000)

All listed inheritance models based on the assumption that a male sterility-inducing cytoplasm interact with nuclear components. Cytoplasm sources: Sa, male sterility-inducing cytoplasm associated with the 'brown anther' CMS; Sp, male sterility-inducing cytoplasm associated with the 'petaloid' CMS. Loci abbreviations are given as mentioned in appropriate publications

OF CONSTANS (*SOC*) in Arabidopsis and several other crops (compare to 3.2.1). It remains to be shown whether the yet-identified carrot sequences with structural similarities to these and several other MADS-box genes (Ou et al. 2017) reveal similar or divergent roles in carrot.

A brief characterization of their temporal and spatial expression patterns by histological analyses throughout appropriate developmental stages of most of the structurally explored sequences is yet unknown. Hence, despite the fact that large transcriptional and/or genomic data sets are now available for carrot (Iorizzo et al. 2016; Ou et al. 2017) a well-designed analysis of appropriate candidate genes could be a next step to on the way to a successive functional assignment.

3.2.4 Cytoplasmic Male Sterility (CMS)

Nowadays, it is known that CMS is based on a complex interplay between maternally inherited (mitochondrial) and biparental (nuclear) genetic information (reviewed, e.g., by Chen and Liu 2014; Hanson and Bentolila 2004; Linke and Börner 2005; Schnable and Wise 1998). As in other crops, the trait of cytoplasmic male sterility (CMS) is a prerequisite to enable hybrid breeding in carrot. Regarding the historical breeding context, this particular phenomenon of male flower organ development has been described nearly 130 years ago, even though the genetic background was not yet resolved. Male sterility, the dysfunction of stamens resulting in the lack of development of functional pollen, was reported for carrot by Beiyernick (1885), Staes (1889) and Warenstorf (1896). Since not only pollen production, but also sporophyte formation itself can be affected, the definition of this phenomenon has been extended as a partial or a complete stamen degeneration that causes in unrolled filaments and indehiscent anthers (Knuth 1898). Male sterile flowers enable directed, insect-assisted pollination of breeding lines on a commercial level. Along with many crops, the research on carrot was accelerated with the discovery of heterosis and the subsequent research on hybrid breeding at the first decades of the twentieth century. Yet it took a lot of research to get from the establishment of a practicable hybrid breeding system to the admission of the first hybrid in the 1960s (Simon et al. 2008 and book Chap. 9). Despite the intensive research up to the recent time, the evolutionary, developmental, and molecular-genetic background of the highly complex phenomenon of cytoplasmic male sterility (CMS) used in hybrid breeding is only partially understood. Nowadays hybrid breeding is the preferred method for the commercial carrot breeding, considering the two basic forms of cytoplasmic male sterility 'brown anther' and 'petaloid' (further summarized by Simon et al. 2008).

3.2.4.1 Phenotypic Characteristics of Male Sterility in Carrot

The 'brown anther' type of male sterility is characterized by forming at first stamens that appear phenotypically normal (Fig. 3.6). In combination with a disturbed microsporogenesis and pollen production, these stamens subsequently persist in a rudimentary state and later anthers turn brown. The 'brown anther' CMS type was first discovered in the cultivar 'Tendersweet' (Welch and Grimball 1947). In the following years, the 'brown anther' sterility was selected in several other cultivars worldwide (Banga et al. 1964; Braak and Kho 1958; Dame et al. 1988: Michalik 1971; Kononkov and Mokhov 1972; Litvinova 1973; Litvinova et al. 1980) and also in wild relatives (McCollum 1966; Nothnagel et al. 2000; Rubashevskaia 1931).

Investigation of the microsporogenesis and the development of the tapetum (Zenkteler 1962) revealed only subtle morphological differences between the 'brown anther' CMS flowers and their fertile counterparts prior to the tetrad formation. A complete microspore abortion was observed during advanced stages. This observation has been discussed as a consequence of a periplasmodial tapetum structure and a deterioration of the anther wall. First irregularities were observed during the microspore pachytene stage. Zenkteler (1962) showed abnormal meiosis, an abnormal tapetum development with enlarged cells forming plasmodial structures and a complete pollen abortion. During the microspore separation, the tapetal nuclei and the tapetal cells increased to a twice of that of the male fertile counterpart. Later, the nuclei of the tapetal plasmodium and the microspores decreased rapidly. It has been observed that during advancing development, an expanded plasmodium infiltrated the locule and became aggregated with clumps of microspores finally leading to the collapse of the anther. In other male sterile materials, it has been shown that a persisting tapetum starved the developing microspores to death. Finally, anthers became shrunken and revealed dark-brown structures (Struckmeyer and Simon 1986; Zenkteler 1962; Fig. 3.6e).

The 'petaloid' male sterility was first discovered in 1953 by Munger in a North American wild carrot (*D. carota* subsp. *carota*) and was later termed 'Cornell-CMS' (Thompson 1961; reviewed by Peterson and Simon 1986).

McCollum (1966) detected petaloid structures, staminodes, and sterile stamens in a wild carrot population received from Sweden. Petaloidy has been also found in other North American—'Wisconsin-CMS' (Morelock et al. 1996) and Canadian wild carrots—'Guelph-CMS' (Wolyn and Chahal 1998; Fig. 3.2).

Petaloidy resembles specific 'homeotic' mutations characterized by a replacement of the stamen by petals or petal-like structures (compare to Fig. 3.6g, h). The additional petaloid structures and partially the originated petals can also reveal different shapes. Besides a complete transformation into petal-shaped structures, also incomplete organ transformations have been described that include different basal-distal transitions ranging from filamentous to spoon-like to three-lobed protrusions. Furthermore, the coloring of the florets can range from white, yellowish, white-green, green, to purple flowers (Dyki et al. 2010; Eisa and Wallace 1969a, b; Struckmeyer and Simon 1986; Wolyn and Chahal 1998). In some cases, flowers have pistils with multiple stigmata instead of stamens that

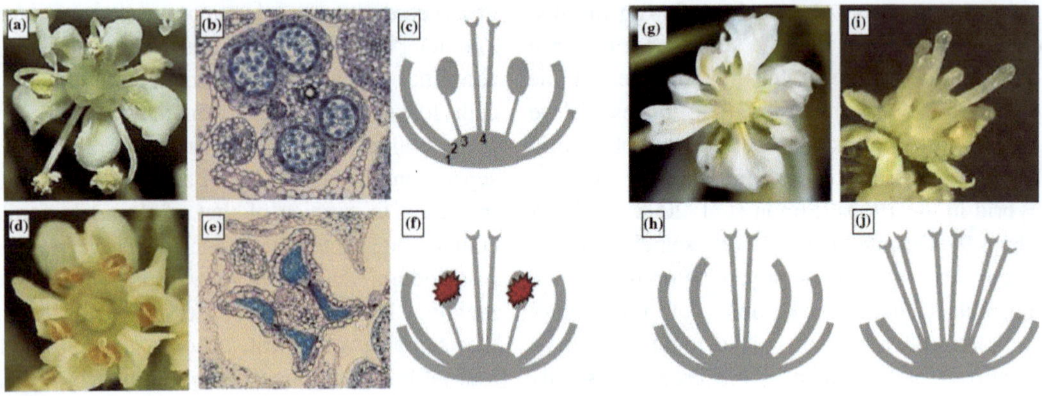

Fig. 3.6 Homeotic and non-homeotic carrot CMS flowers. **a–c** Male fertile flower. **b** Transverse anther section indicating four locules, each of which contains developing microspores. The internal sporophytic layer, the tapetum, is marked by a dark-blue staining. **c** Flower architecture of male fertile flower consisting of sepals, petals, stamens and the bipartite carpel, flower whorls are numbered. **d–e** Brown anther (b.a.) CMS flower; male organs are brownish and shriveled. **e** Transverse anther section of a b.a. CMS flower at a comparable developmental stage as shown in **b**. Locules are collapsed, microspores are compressed to a dead mass, and tapetum cells are not visible. **f** Scheme of the 'non-homeotic' b.a. CMS type indicating a principally unmodified flower architecture; advanced anther defects are marked. **g–j** Homeotic CMS flowers. **g** 'Petaloid' CMS flower indicating sepaloid petals with green midribs and petal-like structures instead of stamens. **h** Scheme of a petaloid CMS flower. **i** 'Carpeloid' CMS flower indicating sepal-like structures in whorl 2 and carpel-like structures instead of stamens in whorl 3. **j** Scheme of a 'carpeloid' CMS flower with an impaired organ identity that resembles B class mutants according to the ABC(DE) model, where MADS-box gene activity of the B class is impaired (compare to Fig. 3.4)

normally occupy the position in the third flower whorl. These florets are termed 'carpeloid' (Dyki et al. 2010; Kitagawa et al. 1994; Struckmeyer and Simon 1986; Wolyn and Chahal 1998; Fig. 3.2i).

In the last decade of the twentieth century, three new CMS sources were selected in the wild relatives *D. carota* subsp. *gummifer*, *D. carota* subsp. *maritimus*, and *D. carota* subsp. *gadecaei*. The CMS-GUM type is characterized by a nearly complete loss of petals and stamen in an early stage of organ development. The CMS-MAR type is comparable to the common petaloid CMS flower types. Flowers of the CMS-GAD type have only short filament-like stamen rudiments. Further analyses for the application of these novel CMS types for breeding are in progress (Linke et al. 1999; Nothnagel et al. 1997, 2000; Fig. 3.7).

The majority of published data reported that the sterility-type 'brown anther' was found in a lot of cultivars as well as in wild relatives. In contrast, male sterility based on 'petaloidy' was only identified in wild relatives and has been subsequently introduced into the nuclear-genetic background of the cultivated carrot (compare to Table 3.2).

3.2.4.2 Genetic Analysis of CMS and Fertility Restoration

Genetic studies have early hypothesized a cytoplasmic-nuclear inheritance for most of the identified male sterile plants of both classes of phenotypes. This was due to the fact that besides complete male sterile crossing progenies (suggesting maternal cytoplasmic inheritance), crosses with some pollinating lines segregated into male sterile and male fertile plants or led to progenies which are completely restored to fertility. Although Jones (1950) and Lamprecht (1951) considered that male sterility in carrot has a cytoplasmic origin, first experimental data were presented by Gabelman (1956) who suggested the existence of some nuclear genes interacting with a male sterility-inducing (S) cytoplasm.

Thompson (1961) assumed a common inheritance for both CMS systems. Three duplicated

Fig. 3.7 Umbel sections and single flowers of the GUM, MAR, and GAD-CMS type (**a–c**) identified in the carrot wild relatives *Daucs carota* ssp. *gummifer, D.c. maritimus,* and *D.c. gadecaei*, respectively. **a** The flower phenotype is characterized by the absence of petals and anthers. **b** The anthers are transformed into petaloid structures. **c** The development of anthers is early interrupted, and only filament-like rudiments are expressed

dominant genes *Ms1, Ms2, Ms3* necessary to maintain male sterility and an epistatic locus to restore fertility in both CMS systems. However, most other authors favored separate inheritance models for the 'brown anther' and 'petaloid' CMS. Banga et al. (1964) suggested that two duplicate genes, one recessive *(aa)* and one dominant *(B.)*, led to the expression of male sterility, while dominant alleles of either of two complementary genes (*D*, *E*) can restore the fertility.

Similar or identical hypotheses have been reported by Michalik (1974), Morelock (1974), and Weit (1979). For the 'petaloid' CMS, data of Timin and Vasilevsky (1997) and Wolyn and Chahal (1998) supported the hypothesis reported by Thompson (1961). Comprehensive studies of Mehring-Lemper (1987) using 'petaloid' CMS lines of American origin led to a model of three independent genes, one dominant gene *M* and two recessive genes, *l* and *t*. Plants heterozygous for the *M* locus can show partial restoration to fertility at high temperature (Table 3.2).

The broad application of CMS in commercial hybrid breeding (compare to Chap. 6) supports the published models of inheritance, in particular with regards to major genes (Dame et al. 1988; Kozik et al. 2012; Michalik 1978; Weit 1979). The development of maintainer lines which guarantees a 100% expression of male sterile plants in the maternal hybrid line is very expensive and time-consuming due to the management of the relatively complicated inheritance, the appearance of inbreeding depression, and the failure or limited availability of double-haploid (DH) lines (Elen 1970; Roth 1981; Stein et al. 1985). Molecular markers would be a helpful tool to facilitate this problem for commercial breeders. On the base of an F_2 segregation population, a single dominant nuclear gene determining restoration of petaloid cytoplasmic male sterility (*Rf1*) was identified and mapped to chromosome 9 (Alessandro et al. 2013) (compare to Fig. 3.3). Efforts are being done to map and clone the postulated *Rf1* gene. The *Rf1* locus supports an inheritance model where a recessive gene is responsible for the male sterility such as was published by Banga et al. (1964) and Hanschke and Gabelmann (1963) for the 'brown anther' CMS, as well as for the 'petaloid' CMS postulated by Mehring-Lemper (1987).

3.2.4.3 Molecular Research on CMS

CMS has been well studied in many crop plants. To obtain direct evidence for a causal role of certain mitochondrial genes in CMS remains a problem as long as it is not feasible to genetically manipulate plant mitochondria. In many cases, CMS is caused by rearrangements of the mitochondrial DNA leading to new open reading frames (ORFs). These ORFs have a chimerical structure since they are composed of fragments derived from other genes and/or non-coding sequences (Chen and Liu 2014; Hanson and Bentolila 2004; Linke and Börner 2005; Schnable and Wise 1998). It should be emphasized that other chimeric mitochondrial genes have been discovered that are clearly not associated with CMS or any other phenotype (e.g., Marienfeld et al. 1997). Alternatively, mitochondrial gene/genome rearrangements may alter the expression of common mitochondrial genes coding for proteins involved in respiration/ATP synthesis, e.g., because of co-transcription with a new flanking gene (Linke and Börner 2005; and references therein). Chen and Liu (2014) have summarized 28 types of CMS from 13 crop species. At least 10 essential mitochondrial genes, most belonging to the mitochondrial electron transfer chain (mtETC) pathways, have been found to be involved in the formation of CMS genes. Among them, *cox1*, *atp8*, and *atp6* are frequently involved in the origination of CMS genes in different plant species. In addition, most CMS genes encode transmembrane proteins (Chen and Liu 2014).

In carrot, the application of molecular tools fulfilled several expectations to identify causative nuclear and organellar (mitochondrial) genes leading to CMS but also to improve practical aspects regarding the development of molecular markers. The contribution of extra-chromosomal genetic information to the expression of CMS has been early demonstrated by genetic analyses (see above). Molecular analyses have also shown that the mitochondrial genome is associated with the CMS trait. A maternal mode of inheritance of the mitochondrial (mt)DNA has been observed in carrot CMS plants by several authors (Börner et al. 1995; Nothnagel et al. 2000; Scheike et al. 1992; Steinborn et al. 1995). The application of molecular-genetic tools to CMS research further revealed a large variability of the mitochondrial genome within the carrot (Steinborn et al. 1992). Detailed analyses indicated relatively large genetic distances between carrot cultivars and wild relatives. This included a high degree of heteroplasmy and intra-individual SNP variations in several mitochondrial genes (Mandel et al. 2012; Mandel and McCauley 2015; see Chap. 12). Considering the fact that molecular markers can support breeding research, discrimination of different 'mitotypes' was a strong aim of research also beyond identification of certain CMS-associated genes. Initially, CMS-associated 'mitotypes' have been distinguished by restriction fragment analyses of mitochondrial DNA sequences/genes, and later PCR-based makers were applied. Restriction fragment analysis of mitochondrial (and chloroplast) DNAs from a 'brown anther' and a 'petaloid' cytoplasmic male sterile (CMS) line revealed unique patterns for each CMS line distinct from those of male fertile cytoplasms (Börner et al. 1995; Scheike et al. 1992). In addition, expression analyses of several mitochondrial genes indicated alterations on the RNA and protein level in fertile and CMS cytoplasms (Börner et al. 1995; Scheike et al. 1992).

Several years later, mitochondria-specific sequence-tagged site (STS) primer pairs deduced from randomly amplified polymorphic DNA (RAPD) markers were reported to distinguish SpC (Cornell-CMS) and SpW (Wisconsin-CMS) cytoplasms from a collection of three male fertile inbred carrot lines and five open-pollinated cultivars (Nakajima et al. 1999). Two of these primer pairs amplified fragments that were associated with either 'petaloid' (Sp) cytoplasms or male fertile (N-) cytoplasms. Bach et al. (2002) applied PCR-based markers to distinguish the mitochondrial genomes of 'petaloid' and male fertile carrot. The authors have developed fourteen primer pairs that amplify marker fragments from either the Sp or the N cytoplasms and three primer pairs that amplify fragments with length polymorphisms. The markers target the *nad6*, *cob*, *atp1*, *atp6*, *atp8* (former *orfB*), and *atp9* loci from the mitochondrial genomes of a diverse collection of male fertile and 'petaloid' carrots.

Hence, it was clearly possible to distinguish different mitotypes of the carrot (Bach et al. 2002; Börner et al. 1995; Kanzaki et al. 1991; Nakajima et al. 1999; Nothnagel et al. 2000; Scheike et al. 1992; Steinborn et al. 1995).

In addition, a maternal inheritance in combination with unique 'mitotype' features was frequently associated with specific CMS flower types. This was also the case for three novel CMS sources with unique flower phenotypes (Linke et al. 1999, 2003; Nothnagel et al. 2000). As was mentioned above, attempts to identify a causative association of certain mitochondrial genes with the CMS trait require concomitant DNA, RNA, and protein analyses in CMS plants and in corresponding male fertile (maintainer) and restored (Rf) genotypes (Börner et al. 1995; Scheike et al. 1992).

Rearrangements in the vicinity of certain mitochondrial genes have been identified by different working groups (Table 3.3). A C-terminal extension of the *cox1* reading frame was identified by Robison and Wolyn (2006a); irrespective of this difference, Western blot analyses of mitochondrial proteins revealed comparable products of the same size in male fertile and 'petaloid' CMS plants. The authors discussed that the products of the *cox1* gene are probably unaffected in structure and function in the different mitotypes and not involved in petaloid CMS. Rurek et al. (2001) have shown that 'petaloid' and male fertile carrots differed in the nucleotide sequence and editing of mRNA of the *nad3* gene. However, a causative association to expression of CMS was not observed. Rearrangements of the *atp8* (former *orfB*) and *atp9*-loci of the 'petaloid' CMS cytoplasms have been described in detail (Bach et al. 2002; Szklarczyk et al. 2000). Variations in markers specific for the 3′-primed configurations of the 'petaloid' Sp cytoplasm for *atp8* revealed that the duplicated *atp8* genes have a rearranged structure since these cytoplasms were combined with the nuclear backgrounds of cultivated carrot (Bach et al. 2002; Szklarczyk et al. 2000). Earlier analyses of the *atp8* gene suggested an involvement in CMS (Nakajima et al. 1999, 2001). However, Robison and Wolyn (2006b) argued that presently there is no evidence to support a role for the *atp8* gene(s) of the carrot in the 'petaloid' CMS type. The authors have shown that there was no change in the quantity or size of transcripts or translated products from these reading frames in flowers that have been restored to fertility under the action of nuclear *Ms* or *Rf* alleles.

Molecular analysis of the mitochondrial DNA, mRNA, and protein of the *atp9* gene revealed differences between fertile and the male

Table 3.3 Mitochondrial analyses in male fertile and in CMS plants of the carrot

Mitochondrial genes/gene portions	Genotype	Analyses of DNA, RNA or protein	Comment	Reference
coxI, coxII, coxIII, atp1 (former *atpA*), *rrn26*	Sp, N, GUM	DNA, RNA	RFLP analyses using different 'composite' probes for Southern and Northern hybridization; maternal inheritance of mitochondrial DNA shown; parents and progenies of several intraspecific crosses included	Steinborn et al. (1995)
cob, atp4 (former *orf25*) *atp6*	Sp, N	DNA	RFLP fragment containing parts of *cob, atp4, atp6* sequenced; mitochondrial DNA of carrot CMS suspension cultures analyzed	Kanzaki et al. (1991)
nad2, nad3, cox1, coxII, coxII, atp1 (*former atpA*), *atp6*	Sp, N, Rf; Sa, N, Rf	DNA, RNA, protein	RFLP analyses with heterologous DNA probes of *Oenothera*; DNA (Southern hybridization), RNA (Northern hybridization), and protein analyses (*in organello* translation); 'brown anther' and 'petaloid' CMS plants compared to corresponding male fertile and restored (Rf) plants	Scheike et al. (1992), Börner et al. (1995)

(continued)

Table 3.3 (continued)

Mitochondrial genes/gene portions	Genotype	Analyses of DNA, RNA or protein	Comment	Reference
coxI, *cob*, *atp1*	Sp, Sa	DNA	Random amplified polymorphic DNA (RAPD) analyses, Southern hybridization using heterologous probes (beet, pea, wheat); discrimination of cybrids	Yamamoto et al. (2000)
nad6, *cob*, *atp1*, *atp6*, *atp9*, *atp8*	Sp, N	DNA	14 PCR-primer pairs that amplify marker fragments from either 'petaloid' or male fertile cytoplasms; length polymorphisms shown by three primer pairs	Bach et al. (2002)
cob, *cox2*, *cox3*	GUM, GAD, MAR	DNA	RFLP analyses/Southern hybridization; maternal inheritance and discrimination of the mitotypes of the GUM, GAD, MAR cytoplasms shown	Nothnagel et al. (2000)
nad3	Sp, N	DNA, RNA	RNA editing of *nad3*; 'petaloid' and male fertile 'maintainer' plants differed in the nucleotide sequence and in RNA editing of the *nad*3 gene	Rurek et al. (2001)
coxI	Sp, N	DNA, RNA, protein	Co-expression with *rps7* shown; C-terminal transcript extensions identified; identical sizes of the cox1 protein detected by Western immunoblotting revealed no causal association with the 'petaloid' CMS type	Robison and Wolyn (2006a)
atp6	Sp, N	DNA, RNA	Different lengths, copies, sequences, and expression levels of *atp6*; discrimination of male fertile ('Kuroda') and 'petaloid' ('Wuye-BY') CMS lines	Tan et al. (2018)
atp8 (former *orfB*)	Sp, N Sp, N, Rf	DNA, RNA DNA, RNA protein	Three types of *orfB*-related genes, *orfB-F1*, *orfB-F2*, and *orfB-CMS* identified; *orfB-CMS* suggested as a novel chimeric *orfB*-related gene associated with 'petaloid' CMS; thirteen varieties of the carrot including seven CMS lines investigated; later analyses revealed no alterations of *atp8* proteins between male fertile, 'petaloid' and restored (Rf) genotypes, indicating that different structure is not associated with 'petaloid' CMS	Nakajima et al. (1999, 2001), Robison and Wolyn (2006b)
atp9	Sp, N	DNA, RNA, protein	Quantitative aspects of *atp9*-organization and expression; partial RNA editing and multiple 5′-termini of certain *atp9*-transcripts, elevated protein of *atp9* in petaloid flowers; heteroplasmic conditions of *atp9* in different CMS cytoplasms suggested	Szklarczyk et al. (2000, 2014)
rpo, *dpo*	Sp, N	DNA, RNA	Plasmid- or plasmid-like RNA and DNA polymerases identified; expression patterns analyzed by Northern hybridization and RT-PCR in male fertile and 'petaloid' CMS plants; potential association with CMS not mentioned	Robison and Wolyn (2005)

Different analyses of mitochondrial sequences, genes, or gene portions in carrot are summarized. Due to the application of different molecular methods, a separated listing of single genes was not always possible. Genotypes of the analyzed cytoplasms or CMS types are as follows: Sp, 'petaloid' CMS; Sa, 'brown anther' CMS; GUM, Gummifer-CMS; GAD, Gadeacei-CMS; MAR, Maritimus-CMS; N, male fertile plants (including 'maintainer' genotypes of CMS); Rf, plants restored to male fertility

sterile flower phenotypes, including rearrangements of genes as a potential cause for the CMS phenotypes (Börner et al. 1995; Nakajima et al. 2001; Scheike et al. 1992; Szklarczyk et al. 2000, 2014). However, a clear causal association to the CMS trait was not yet shown. Differences in lengths, copies, and expression profiles of the *atp6* gene in fertile and male sterile lines of carrot were observed, too, but a certain association to CMS is yet unclear (Bach et al. 2002; Kanzaki et al. 1991; Scheike et al. 1992; Tan et al. 2018). Mitochondrial DNA- and RNA-directed polymerases (*dpo* and *rpo*), encoded by mitochondrial plasmid-like structures, have been initially characterized, but yet without any causative association to CMS (Robison and Wolyn 2005). In summary, despite several variations of mitochondrial genes in CMS-inducing cytoplasms, further investigations are required to state if any of these modifications are directly involved in the cause of CMS.

3.2.4.4 Sequence of the Mitochondrial Genome

Previous sequence analyses basing on restriction digestion mapping demonstrated a complex structural organization of the mitochondrial genome of the petaloid CMS cytoplasm (Robison and Wolyn 2002). In male fertile plants, a de novo assembly of the carrot mitochondrial genome has been generated using next-generation sequencing (Iorizzo et al. 2012, 2016). Analyses of structure and gene content have confirmed earlier results that a large amount of genetic variation exists at the organelle genome level even between samples sharing a very close genetic relationship (Iorizzo et al. 2016; see Chaps. 11 and 12). Hence, the natural plasticity of the mitochondrial genome makes it difficult to identify CMS-associated gene regions. Additional data on gene expression in mitochondria of sterile CMS plants in comparison with restored fertile plants having identical mitochondrial genomes are required. The availability of complete genome data can support further research on the CMS trait.

3.2.4.5 'Restorer of Fertility' Genes

'Restorer of fertility' (*Rf*) genes are encoded by the nuclear genome and 'counteract' or suppress the mitochondrial-associated defects leading to CMS. Restorer genes have been cloned in several plant species including crops, and more than half of the identified *Rf* genes encode PPR proteins (e.g., reviewed by Chen and Liu 2014). PPR proteins belong to a group of RNA-binding proteins, mostly acting in organellar post-transcriptional mRNA processing, such as editing, splicing, cleavage, degradation, and translation. However, besides PPR proteins, restorer genes can indicate protein properties as an aldehyde dehydrogenase (maize), as a glycine-rich protein (rice), as an amino acid mitochondrial sorting protein with an acyl-carrier protein synthase-like domain (rice), or as a putative peptidase (sugar beet). A generation of PPR and diverse other types of restorer (*Rf*) genes indicates that plants have evolved complex pathways to counteract the effects of CMS (reviewed by Chen and Liu 2014). Despite a brief genetic characterization (see above), gene candidates involved in the restoration to fertility have not yet been identified in carrot. However, segregation studies revealed a single dominant nuclear gene (*Rf1*) responsible for fertility restoration of the 'petaloid' CMS and could be mapped to chromosome 9 (Alessandro et al. 2013; Fig. 3.2). It remains to be shown, whether PPR proteins or other 'restoring' gene functions of the nuclear genome can be identified through additive informative contents deduced from the novel genome data available in carrot (Iorizzo et al. 2016).

3.2.4.6 Nuclear-Encoded Genes Influenced by the Expression of CMS

As was mentioned above, carrot CMS flowers of the 'brown anther' type are an example of the appearance of a CMS phenotype where the general flower architecture is not altered (see Fig. 3.6). This contrasts with another group of CMS plants in which early steps of flower formation are impaired. This type of CMS has been studied in tobacco cybrids (plants regenerated from fused protoplasts with the nuclear genome of tobacco, *Nicotiana tabacum*, and the cytoplasm including mitochondria from another member of the Solanaceae, *Hyoscyamus niger*), wheat, and carrot (Kofer et al. 1991; Linke et al. 2003; Murai et al. 2002; Zubko et al. 2001; reviewed by Carlsson et al. 2008; Linke and Börner 2005). Such CMS plants develop 'homeotic' flowers, in which male organs are replaced by another flower organ, e.g., by petals or even carpels, the female flower organs (Fig. 3.6). The homeotic type of CMS flowers resembles nuclear mutants with defective nuclear genes involved in the specification of the identity of flower organs. Studies on the 'carpeloid' type of carrot CMS flowers for the first time demonstrated a mitochondrial effect on the expression of MADS-box genes of the B class, which specify the identity of petals and stamens (Linke et al. 2003; Figs. 3.4 and 3.6). In several homeotic CMS flowers of other plants, a reduced transcript accumulation of genes for MADS-box proteins with B function was identified (Geddy et al. 2005; Hama et al. 2004) as was summarized earlier (Carlsson et al. 2008; Linke and Börner 2005). Hence, first candidates of nuclear 'target genes' were identified, the expression of which is affected by the CMS state. They encode MADS-box transcription factors specifying organ identity in flower development. Those target genes are supposed to contribute to the defective formation of male flower organs and pollen, respectively.

3.2.5 Pollination and Fertilization

The protandry of carrot flowers leads to anther dehiscence (and stamens fall) before the stigma becomes receptive. An individual flower completes anthesis within 2 days. Stigma receptivity starts 4 days after anthesis, when styles are separated, and appear to last more than a week. In contrast to anther dehiscence, which is staggered, stigmas of all flowers in an umbel become receptive around the same time, after dehiscence of the anthers of all flowers in the umbel had been completed (Koul et al. 1989). Hence, strong protandry is quite effective in promoting outcrossing which lies at 95% (Becker 1943; Rong et al. 2010; Thompson 1961; Webb 1981). Pollination success is influenced by environmental conditions, such as adequate and various pollinators (Abrol 1997; Ahmad and Aslam 2002; Bell 1971; Flemion and Henrickson 1949; Hawthorn et al. 1960; Pérez-Bañón et al. 2007), as well as by the exposure of nectar to enhance quick pollen flow during the approximately 10 days of stigma receptivity period (Broussard et al. 2017). Especially the volatile production and the sugar content of the nectar affect the attraction for pollinators, e.g., in cases of honeybee attraction. Mas et al. (2018) showed a strong negative correlation between nectar aldehydes like nonanal and decanal compounds on the seed yield. Comprehensive analyses of natural compounds might strengthen knowledge in this area (Keilwagen et al. 2017; Yahyaa et al. 2017).

An adequate pollen deposition on the stigma assumes that the pollen tube growth until fertilization follows the general pathway known for angiosperms. The structure and path of the pollen tubes were investigated in detail. The tubes were found to grow intercellularly down through the conducting tissue of the style to its base and then superficially along a groove leading to a canal communicating with each locule. Tubes growing down one style may enter the locule immediately below or grow through the transverse canal and into the other locule (Borthwick 1931). After the pollen tube enters the gametophyte, the two sperm cells are released. One of the two sperm cells fertilize the egg cell forming the diploid zygote. At this point, the fertilization actually takes place. The other sperm cell is combined with two polar nuclei of the central cell to form the primary endosperm (double fertilization).

3.2.6 Embryogenesis and Seed Development

Seed development is initiated by the process of double fertilization, which leads to the development of the embryo and the endosperm. Division of the endosperm nucleus takes place, and the number of nuclei becomes noticeable before the zygote divides. The endosperm becomes cellular at about the time the embryo is at the two-cell stage (Borthwick 1931). Typically for Apiaceae as well as for carrot is the fact that after the first division of the fertilized zygote, the second cell division forms a linear four-celled structure and further a linear eight-celled structure. The embryo proper entirely derives from the distal cell of the 4-celled stage (Borthwick 1931). The subsequent development of the suspensor and the embryo proper up to the globular and heart stage as well as the following steps to the early cotyledon stage have been briefly described by Lackie and Yeung (1996).

A graduated time line from anthesis (days after anthesis—DAA) to seed maturity was described for carrot by Gray et al. (1983). Using this scale, Becu and Broască (2012) have shown that the individual layers of the pericarp-epicarp, mesocarp, and endocarp, all joined to the seed, are visible approximately 14 DAA. During this time, the integument consists of a single cell layer. The endosperm passes a rapid cell division and expansion phase reaching a maximum at approximately 28–25 DAA. Between 14 and 21 DAA, starch grain depositions in the endosperm are detectable. Further, a lignin deposition surrounding the cells walls of the endocarp begins to grow out at 21 DAA. Protein and lipid bodies in the cells of the endosperm can be observed at 28 DAA (Corner 1976; Graham 2008; Miranda et al. 2017). Seeds reach their physiological maturity at 35–56 DAA depending on the cultivar and environmental conditions when the endosperm occupies the whole seed volume (Gray et al. 1984; Miranda et al. 2017; Nascimento et al. 2003). An association between the physiological maturity, the germination, and vigor of seeds was shown (Miranda et al. 2017, see 3.3). The seed maturation phase initiates the decline of the plant (Fig. 3.1). At 63 DAA, the seed dry content increases to a maximum. During these stages, the pericarp is partially collapsed, and the lignified endocarp is the resistance layer of the carrot seed. The embryo at maturity reaches a volume which is equivalent of 2–3% of the endosperm volume.

Developmental and environmental factors of seed dormancy have been described in carrot, especially in wild relatives (Borkrid et al. 1988; Dale 1974; Dale and Harrison 1966; Sylwester 1960). The preceding late stages of embryogenesis are characterized by extensive physiological changes to introduce maturation and subsequent post-abscission, followed by pre-desiccation and the desiccation phases. In general, these developmental stages include metabolic changes like lipid deposition, deposition of storage proteins, and finally the dehydration steps that introduce the desiccation state (reviewed by Holdsworth et al. 1999).

3.2.6.1 Molecular Data on Embryogenesis and Seed Development

The model state of carrot for research on 'somatic embryogenesis' since the 1960s (Reinert 1958; Steward et al. 1958) supported the study of several developmental aspects similar to zygotic embryogenesis, which have been evaluated in detail. Several genes have been identified that show a similar expression during the zygotic embryogenesis (Borkrid et al. 1988; Thomas and Wilde 1985, 1987). Several representative gene candidates involved in the processes from embryogenesis to seed maturation have been also identified in the carrot. The *SERK* (*somatic embryogenesis receptor kinase*) gene was firstly isolated from carrot embryogenic cells and has been found to be expressed in somatic and zygotic embryos up to the globular state but in no other plant tissues. *SERK* encodes a receptor-like kinase protein containing five leucine-rich repeats (Schmidt et al. 1997). In different plant species, diverse roles of SERK proteins have been shown in different signaling pathways also beyond plant development (Ikeda et al. 2006). During the lipid deposition state, cuticular material is already accumulated on epidermal layers during the development of the embryo (Meijer et al. 1993). Lackie and Yeung (1996) stated that the cuticular material is not deposited on the embryo proper until after the protoderm has formed and is then found in all subsequent developmental stages. The main function of the embryo cuticle may be inferred from sections of carrot embryos surrounded by a partially liquefied endosperm. To render the nutrients that are stored there accessible to the growing embryo, the cellular endosperm is dissolved by hydrolytic enzymes. To protect the embryo itself from these enzymes, the formation of a water-repellent coating would clearly be beneficial (Sterk et al. 1991). A cDNA encoding an extracellular protein (EP2) with homology to plant lipid transfer proteins has been identified and characterized by in situ hybridization in carrot (Sterk et al. 1991). Expression of the *EP2* gene was observed in protoderm cells of zygotic embryos, in epidermal cells of the cotyledons as well as in the epidermis of the pericarp and in the region where both mericaps started to separate. Expression of *EP2* has been further observed in inflorescences, where it transiently marks epidermal cells of all flower organs. Expression ceased upon maturation of sepals, petals, and stamens, but remained apparent in epidermal cells of the integuments of the ovary and re-appeared when both the epidermal cells of the inner and outer integument were combined to the seed coat. Hence, expression of *EP2* revealed a strong value as an epidermal histological marker throughout the reproductive phase of carrot (Sterk et al. 1991). Shiota et al. (1998) have isolated the *ABI-3* (*abscisic acid insensitive3*) gene of the carrot (*C-ABI3*); expression was specifically observed in developing seeds during mid- to late embryogenesis (from the heart stage onwards) prior to the increase in levels of endogenous ABA that was followed by desiccation of seeds. Expression patterns during both zygotic and somatic embryogenesis were comparable in Arabidopsis and in carrot. A regulation of ABA-induced gene expression and a repression from the phase transition to germination has been shown (Ikeda-Iwai et al. 2002, 2003). Expression of the transcription factor *LEAFY COTYLEDON 1* (*LEC1*) was observed in developing seeds in the heart stage of the Arabidopsis embryo (Meinke et al. 1994). During somatic embryogenesis *LEC1* and *LEC1*-homologs revealed similar expression patterns in Arabidopsis, maize, and carrot (Ikeda-Iwai et al. 2002; Yazawa et al. 2004; Zhang et al. 2002), indicating that *LEC1* has a common and important role in both zygotic and somatic embryogenesis. In situ hybridization analyses in carrot showed expression of *C*(carrot)-*LEC1* in the peripheral region of the embryos but not in the endosperm (Yazawa et al. 2004). Due to its mutant phenotype, where cotyledons acquire leaf-like structures, *LEC1* has been described as one of the primary factors that regulate the transition from embryogenesis to germinative growth (Yazawa et al. 2004; Yazawa and Kamada 2007). Interaction of *LEC1* with *ABI3* appeared to potentiate ABA responses (Parcy et al. 1994, 1997). Late-embryogenesis-abundant (LEA) proteins are stored in seeds. Its genes are usually expressed in the late stage of embryogenesis (e.g., EMB-1; Wurtele et al. 1993). Several of these genes are expressed in both embryonic cultures and immature seeds of carrot and have been found to be induced in somatic and zygotic embryos, when they are treated with abscisic acid (ABA). The LEAs are critical proteins for zygotic embryos to acquire desiccation tolerance and seed dormancy (summarized by Ikeda et al. 2006). Hence, during the last phases of development, the zygotic embryos finally exhibit desiccation tolerance, and dormancy occurs at the final stage of seed development. All these steps are indicative for important adaptive traits that enable the seeds of many species to remain quiescent until conditions become favorable for germination (Gubler et al. 2005).

3.3 Practical Aspects of Seed Production

Regarding carrot seed production, two main technologies are used: the root-to-seed system and the seed-to-seed system. The first one follows the biennial habit of the species; in the first cycle roots are produced, and after selection roots are usually placed into refrigerated storage (1–5 °C) for 1–7 months. The roots are replanted in a second season for seed production. In general, it takes around 14–17 months (Gaviola 2013). The seed-to-seed method reduces the time required for seed production; nevertheless, it does not allow root selection, and it cannot be used to produce the categorized basic seed. Under the seed-to-seed method, carrot seeds are sown during summer in areas with cold winters, that allow plant vernalization, flowering occurs in spring and in the following summer seeds are harvested; the cycle requires 12 months (Gaviola 2013).

Seed production of hybrid cultivars occurs in a similar manner, but some factors complicate the process. First, an adapted plant row design of the parental lines is necessary. Frequently, a proportion of 8 plant rows of the CMS line (maternal parent) and 4 plant rows of the male fertile pollinator line (paternal parent) are used in the field. Second, the synchronization of flowering to guarantee an adequate quality of pollen production of the paternal line as well as a high seed set potential of the maternal line are prerequisites. Pollination in both open-pollinated and hybrid cultivars is mostly accomplished by providing honeybee hives, supporting the naturally occurring insect pollinators (Abrol 1997; Hawthorn et al. 1960; Sinha and Chakrabacti 1992; Thompson 1961).

Cultivar integrity is ensured by geographic isolation between two seed lots by at least few kilometers including carrot wild relatives. The unlimited outcrossing character of both cultivated and wild carrot descending from the *Daucus carota* complex may be a serious problem for the commercial seed production in some regions (Rong et al. 2010).

The genotype influences seed yield and seed size/weight. Whereas the seed yield is mainly dependent on the plant density per m^2, the seed size/weight is rather influenced by environmental conditions during anthesis (Gray and Steckel 1983a, b). At present, the expected seed yield of open-pollinated cultivars in the temperate regions is reaching between 600 and 1000 kg/ha (Duczmal and Tylkowska 1997; Gaviola 2013). In contrast, seed production in the tropical regions is usually lower despite using higher altitudes to achieve satisfactory vernalization and figures of about 300 kg/ha (Pereira et al. 2008). The Asiatic types only produce about 250 kg/ha when seeded in the tropics (Nagarajan and Pandita 2001). The yield of hybrid seed production varies between 400 and 700 kg/ha, but the costs are much higher if compared to seed production in open-pollinated cultivars (Gaviola 2013). The thousand seed weight (TSW) can range from less than 0.5 to more than 3.0 g (Bonnet 1991; Lesprit 1991). It should be mentioned that the seed quality is further influenced by various factors such as physiology or abiotic and biotic stress during the seed developing and ripening process.

The physiological maturity has been defined as the moment when the seeds reach maximum dry matter accumulation, showing that the translocation of assimilates from the plant to the seed has ceased (Demir and Ellis 1992; Harrington 1972). At this point, seed deterioration is minimal and may or may not coincide with the maximum physiological quality, i.e., maximum germination and vigor (see 3.2.6). As an example, carrot seeds of the cv. 'Brasília,' under the conditions Brasília, DF, Brazil, had their endosperm completely developed at 28 DAA, and from 28 to 35 DAA, anatomical changes hardly occur in the seeds. Physiological maturity, represented by maximum dry matter, occurred at 35 DAA, when the seed moisture content was about 56% and the color of the pericarp is yellowish green. Maximum germination and vigor were reached at about 30 DAA and continued until 63 DAA (Miranda et al. 2017).

It has been shown that seed maturity influences the viability of the seed (Gray and Steckel 1983a, b; Hawthorn et al. 1960, 1962; Miranda et al. 2017) and the variability of embryo length (Gray et al. 1984), indicating that the timing of harvesting of seed crops is likely a major factor for seed quality (Sandin 1980).

Embryo size at seed harvest and the variability of the embryo size among seeds were directly correlated with variation in seedling size and root size at harvest (Austin et al. 1969; Gray and Steckel 1983a, b). Variability in seedling weight and the spread of seedling emergence times were closely related to the variation in embryo length, but not to the coefficient of variation of seed weight (Gray and Steckel 1983a, b; Gray et al. 1984; Nagarajan and Pandita 2001).

Seed maturation does not occur uniformly and is dependent on the positions of flowers on the maternal plant, from which the seeds are originated. Flowers and seeds in different orders of umbels (see Fig. 3.2) exhibit various levels of maturity. Joyce et al. (1989) evaluating seed production in two different cultivars in 2 years have reported that the maximum seed dry weight occurred approximately 40–45 days after flowering (DAF) in both cultivars. Maximum germination (International Seed Testing Association 14-day count) occurred 40 and 55 DAF in cvs 'Chantenay' and 'Amsterdam', respectively, but the maximum 7-day count and the minimum coefficient of variation of embryo length did not occur until 60 DAF in cv. 'Chantenay' and 55–65 DAF in cv. 'Amsterdam'. Percentage germination was negatively and linearly related to seed moisture content, chlorophyll content in the seed coat, and seed distortion, the relationships accounting for 77, 71, and 64% of the variance in the 7-day germination count, respectively. The corresponding values for the 14-day count were 63, 61, and 50% (Joyce et al. 1989).

Abiotic and biotic stress factors such as rainfall, soil structure and pH, nutrient deficiency, as well as infections by pathogenic and saprophytic bacteria and fungi (e.g., *Alternaria radicina*, *Xanthomonas campestris*) and attacks by several insects and arthropods further affect the seed development and quality. Physiological disturbances as well as pathogens can cause necrotic changes in embryo and endosperm (Habdas et al. 1997). Pathogen infection during flowering and seed development may promote seed-borne diseases (Bereśniewicz and Duczmal 1994; Bulajić et al. 2009; Duczmal and Tylkowska 1997; Kuan et al. 1985; Pryor and Gilbertson 2001; Pryor et al. 1994; Strandberg 1988; Trivedi et al. 2010; Umesh et al. 1998). Loss of seed yield and seed viability by *Lygus campestris* attack was reported and can contribute to embryoless seeds (Arnott 1956; Flemion and Henrickson 1949).

The commercial carrot seed production is guided by specialized companies and is widely mechanized. The complex process includes the seed production on field and covers seed harvest as well as the post-harvest processing. The latter comprises techniques of seed reprocessing, purity, and quality control, as well as seed storage, coating, priming, and packaging performances.

3.4 Conclusions and Future Directions

Current knowledge on carrot reproductive biology is limited if compared to that of model plants and inadequate despite powerful genetic research on carrot and their wild relatives over the last decades. A possible reason for this is the prioritization of research on economically relevant traits such as yield, quality, or resistance. A focus on reproductive biology is missing but important for breeders and seed producers. The current knowledge helps to introduce characters from late-flowering cultivars to early flowering germplasm in breeding programs with more accuracy. Nevertheless, in future the knowledge on the reproductive biology of carrot should be improved and will clearly benefit from the availability of comprehensive carrot genome data. So far, highlights include the application of male sterility for hybrid breeding and the first genetic and molecular identification of genes involved in vernalization, flower architecture, and fertility restoration. Currently, the genetic background of seed development, dormancy, and senescence is completely unknown. New molecular tools such as GBS/GWAS or the CRISPr/Cas-technique (Klimek-Chodacka et al. 2018) are now available to investigate more details and should be complemented by in-depth experiments and an excellent phenotyping/chemotyping (metabolome) analysis.

References

Abrol DP (1997) Impact of insect pollination on carrot seed production. Insect Environ 3:61

Ahmad M, Aslam M (2002) Pollinators visiting carrot (*Daucus carota* L.) seed crop. J Res Sci 13:31–35

Ajani Y, Bull-Herenu K, Claßen-Bockhoff R (2016) Pattern of flower development in Apiaceae-Apioideae. Flora 221:38–45

Alessandro MS, Galmarini CR (2007) Inheritance of vernalization requirement in carrot. J Am Soc Hort Sci 132:525–529

Alessandro MS, Galmarini CR, Iorizzo M, Simon PW (2013) Molecular mapping of vernalization requirement and fertility restoration genes in carrot. Theor Appl Genet 126:415–423

Amasino RM, Michaels SD (2010) The timing of flowering. Plant Physiol 154:516–520

Arnott DA (1956) Some factors reducing carrot seed yields in British Columbia. Proc Entomol Sci BC 52:27–30

Atherton JG, Basher EA (1984) The effects of photoperiod on flowering in carrot. J Hort Sci 59:213–215

Atherton JG, Craigon J, Basher EA (1990) Flowering and bolting in carrot. I. Juvenility, cardinal temperatures and thermal times for vernalization. J Hort Sci 65:423–429

Austin RB, Longden PC (1967) Some effects on seed size and maturity on the yield of carrot crops. J Hort Sci 42:339–353

Austin RB, Longden PC, Hutchinson J (1969) Some effects of 'hardening' carrot seed. Ann Bot 33:883–895

Bach I, Olesen A, Simon P (2002) PCR-based markers to differentiate the mitochondrial genomes of petaloid and male fertile carrot (*Daucus carota* L.). Euphytica 127:353–365

Bagget JR, Kean D (1989) Inheritance of annual flowering in *Brassica olereacea*. HortScience 24:662–664

Banga O, Petiet J, Van Bennekom JL (1964) Genetical analysis of male-sterility in carrots, *Daucus carota* L. Euphytica 13:75–93

Becker T (1943) Blütenbiologische Studien an Zwiebeln, Möhren, Sellerie und Petersilie. Kühn Archiv 60:466–492

Becu R, Broascã L (2012) Comparative histoanatomical aspects of the fruit of some Apiaceae lindl. fruit used for therapeutic purposes. Ann Soc Nat Biol Cell 17:265–270

Beiyernick MW (1885) Gynodioecie bei *Daucus carota* L. Neder Kruidk Arch Ser 4:345–355

Bell CR (1971) Breeding systems and floral biology of the Umbelliferae, or evidence for specialization in unspecialized flowers. In: Heywood VH (ed) The biology and chemistry of the Umbelliferae. Academic Press, London, pp 93–107

Bereśniewicz M, Duczmal KW (1994) The effect of environmental conditions on carrot seed health. Plant Var Seeds 7:151–160

Bonnet A (1991) Production and quality of hybrid carrot seeds. In: Eucarpia carrot 91—proceedings of the IVth meeting on breeding of carrots, pp 137–145

Borkrid C, Choi JH, Jin ZH, Franz G, Hatzopoulos P, Chorneau R, Bonas U, Pelegri F, Sung ZR (1988) Developmental regulation of embryonic genes in plants. Proc Natl Acad Sci USA 85:6399–6403

Börner T, Linke B, Nothnagel T, Scheike R, Schulz B, Steinborn R, Brennicke A, Stein M, Wricke G (1995) Inheritance of nuclear and cytoplasmic factors affecting male sterility in *Daucus carota*. In: Kück U, Wricke G (eds) Genetic mechanisms for hybrid breeding. Advances in plant breeding, vol 18. Blackwell Science, Berlin, pp 111–122

Borthwick HA (1931) Development of the macrogametophyte and embryo of *Daucus carota*. Bot Gaz 92:23–44

Borthwick HA, Phillips M, Robbins WW (1931) Floral development in *Daucus carota*. Am J Bot 18:784–786

Bowman JL, Smyth DR, Meyerowitz EM (1989) Genes directing flower development in *Arabidopsis*. Plant Cell 1:37–52

Braak JP, Kho YO (1958) Some observations on the floral biology of the carrot (*Daucus carota* L.). Euphytica 92:23–44

Broussard MA, Mas F, Howlett B, Pattemore D, Tylianakis JM (2017) Possible mechanisms of pollination failure in hybrid carrot seed and implications for industry in a changing climate. PLoS One 12:1–23

Budahn H, Baranski R, Grzebelus D, Kielkowska A, Straka P, Metge K, Linke B, Nothnagel T (2014) Mapping genes governing flower architecture and pollen development in a double mutant population of carrot. Front Plant Sci 5:1–10

Bulajić A, Djekić I, Lakić N, Krstić B (2009) The presence of *Alternaria* spp. on the seed of Apiaceae plants and their influence on seed emergence. Arch Biol Sci 61:871–881

Carlsson J, Leino M, Sohlberg J, Sundström JF, Glimelius K (2008) Mitochondrial regulation of flower development. Mitochondrion 8:74–86

Chen L, Liu YG (2014) Male sterility and fertility restoration in crops. Ann Rev Plant Biol 65:579–606

Coen ES, Meyerowitz EM (1991) The war of the whorls: genetic interactions controlling flower development. Nature 353:31–37

Corner EJH (1976) The seeds of dicotyledons, vol 2. Cambridge University Press, p 311

Craigon J, Atherton JG, Basher EA (1990) Flowering and bolting in carrot. II. Prediction in growth room, glasshouse and field environments. J Hort Sci 65:547–554

Dale HM (1974) The biology of canadian weeds. 5. *Daucus carota*. Can J Plant Sci 54:673–685

Dale HM, Harrison PJ (1966) Wild carrot seeds, germination and dormancy. Weeds 14:201–204

Dame A, Bielau M, Stein M, Weit E (1988) Zur Entwicklung von männlich sterile Linien bei der Speisemöhre (*Daucus carota* L. ssp. *sativus* (Hoffm.) Arcang.). Arch Gartenbau Berlin 36:345–352

Demir I, Ellis RH (1992) Changes in seed quality during seed development and maturation in tomato. Seed Sci Res 2:81–87

Dias-Tagliacozzo GM, Valio IF (1994) Effect of vernalization on flowering of *Daucus carota* (Cvs Nantes and Brasilia). Rev Bras Fisiol Veg 6:71–73

Dickson MH, Peterson CE (1960) The influence of gibberellin on the flowering of carrots. Can J Plant Sci 40:468–473

Doyle JJ (1994) Evolution of a plant homeotic multigene family: towards connecting molecular systematics and molecular developmental genetics. Syst Biol 43:307–328

Duczmal K, Tylkowska K (1997) Carrot seed market and prospects for carrot seed production in Poland. J Appl Genet 38A:5–12

Dyki B, Nowak R, Stepowska A (2010) The influence of flower structures on the seeds productivity of the carrot breeding lines. Veget Crops Res Bull 72:5–13

Eisa HM, Wallace DH (1969a) Morphological and anatomical aspects of petaloidy in the carrot (*Daucus carota* L.). J Am Soc Hort Sci 94:545–548

Eisa HM, Wallace DH (1969b) Factors influencing petaloidy expression in the carrot, *Daucus carota* L. J Am Soc Hort Sci 94:647–649

Elen AJJ (1970) A model for identifying "general" B-lines to brown anther type male-sterile carrots. Eucarpia Sect Hort Versailles 41–46

Erbar C, Leins P (2004) Sympetaly in Apiales (Apiaceae, Araliaceae, Pittosporaceae). South Afr J Bot 70:458–467

Ferreira ME, Satagopan J, Yandell BS, Williams PH, Osborn TC (1995) Mapping loci controlling vernalization requirement and flowering time in *Brassica napus*. Theor Appl Genet 90:727–732

Fisher JE (1956) Studies on the photoperiodic and thermal control of flowering in carrots. Plant Physiol 31:36

Flemion F, Henrickson ET (1949) Further studies on the occurrence of embryoless seeds and immature embryos in the Umbelliferae. Contrib Boyce Thompson Inst 15:291–297

Gabelman WH (1956) Male sterility in vegetable breeding. Genetics in plant breeding. In: Brookhaven symposium in biology: genetics in plant breeding, no 9, pp 113–122

Galmarini CR, Della Gaspera P (1996) Determinación de requerimientos de pre-vernalizaciónenzanahorias (*Daucus carota* L.) anuales. In: Actas de la XXI Reunión Argentina de Fisiología Vegetal, Mendoza, Argentina, p 82

Galmarini CR, Borgo R, Tizio R (1992) Determination of a pre-vernalization phase in carrot (*Daucus carota* L.) cv. Flakkee. Turrialba 42:140–142

Galmarini C, Borgo R, Gaviola JC, Tizio R (1995) Effect of gibberellic acid on seed production of carrot (*Daucus carota* L.): effect of different concentrations and application dates on vegetative cycle length and seed yield and quality. Horticultura Argentina 14:47–86

Gaviola JC (2013) Effects of sowing and transplant date and planting density of carrot stecklings on seed production. Horticultura Argentina 32:5–13

Geddy R, Mahé L, Brown GG (2005) Cell-specific regulation of a *Brassica napus* CMS-associated gene by a nuclear restorer with related effects on a floral homeotic gene promoter. Plant J 41:333–345

Graham IA (2008) Seed storage oil mobilization. Ann Rev Plant Biol 59:115–142

Gray D, Steckel JRA (1983a) Some effects of umbel order and harvest date on carrot seed variability and seedling performance. J Hort Sci 58:73–82

Gray D, Steckel JRA (1983b) Seed quality in carrots: the effects of seed crop plant density, harvest date and seed grading on seed and seedling variability. J Hort Sci 58:393–401

Gray D, Steckel JRA, Ward JA (1983) Studies on carrot seed production: effects of plant density on yield and components of yield. J Hort Sci 58:83–90

Gray D, Ward JA, Steckel JRA (1984) Endosperm and embryo development in *Daucuc carota* L. J Exp Bot 35:459–465

Gubler F, Millar AA, Jacobsen JV (2005) Dormancy release, ABA and pre-harvest sprouting. Curr Opin Plant Biol 8:183–187

Habdas H, Staniaszek M, Szafirowska A (1997) Cytological aspects of low carrot seeds quality. J Appl Genet 38A:193–195

Hama E, Takumi S, Ogihara Y, Murai K (2004) Pistillody is caused by alterations to the class-B MADS-box gene expression pattern in alloplasmic wheats. Planta 218:712–720

Hanschke PE, Gabelmann WH (1963) Digenic control of male sterility in carrots, *Daucus carota* L. Crop Sci 3:383–386

Hanson MR, Bentolila S (2004) Interactions of mitochondrial and nuclear genes that affect male gametophyte development. Plant Cell 4(16):154–169

Harrington JF (1951) Effect of spacing and size of root on carrot seed yield and germination. Proc Am Soc Hort Sci 58:165–167

Harrington JF (1972) Seed storage longevity. In: Kozlowsky TT (ed) Seed biology, vol 3. Academic Press, New York, pp 145–245

Hawthorn RL (1952) Interrelations of soil moisture, nitrogen, and spacing in carrot seed production. Proc Am Soc Hort Sci 60:321–326

Hawthorn RL, Bohart GE, Toole EH, Nye WP, Levin MD (1960) Carrot seed production as affected by insect pollination. Bull Utah Agric Exp Station 422:18

Hawthorn RL, Toole EH, Toole VK (1962) Yield and viability of carrot seeds as affected by position of umbel and time of harvest. Proc Am Soc Hort Sci 80:401–407

Helliwell CA, Wood CC, Robertson M, James Peacock W, Dennis ES (2006) The Arabidopsis FLC protein interacts directly in vivo with SOC1 and FT chromatin and is part of a high-molecular-weight protein complex. Plant J 46:183–192

Hernández-Hernández T, Martínez-Castilla LP, Alvarez-Buylla ER (2007) Functional diversification of B MADS-box homeotic regulators of flower development: adaptive evolution in protein-protein interaction domains after major gene duplication events. Mol Biol Evol 24:465–481

Hiller LK, Kelly WC (1979) Post-vernalization temperature effects on seedstalk elongation and flowering in carrots, *Daucus carota* L. J Am Soc Hort Sci 104:253–257

Holdsworth M, Kurup S, McKibbin R (1999) Molecular and genetic mechanisms regulating the transition from embryo development to germination. Trends Plant Sci 4:275–280

Ikeda M, Umehara M, Kamada H (2006) Embryogenesis-related genes; its expression and roles during somatic and zygotic embryogenesis in carrot and Arabidopsis. Plant Biotechnol 23:153–161

Ikeda-Iwai M, Satoh S, Kamada H (2002) Establishment of a reproducible tissue culture system for the induction of Arabidopsis somatic embryos. J Exp Bot 53:1575–1580

Ikeda-Iwai M, Umehara M, Satoh S, Kamada H (2003) Stress induced somatic embryogenesis in vegetative tissues of *Arabidopsis thaliana*. Plant J 34:107–114

Immink RG, Kaufmann K, Angenent GC (2010) The 'ABC' of MADS domain protein behaviour and interactions. Semin Cell Dev Biol 21:87–93

Iorizzo M, Senalik D, Szklarczyk M, Grzebelus D, Spooner D, Simon P (2012) De novo assembly of the carrot mitochondrial genome using next generation sequencing of whole genomic DNA provides first evidence of DNA transfer into an angiosperm plastid genome. BMC Plant Biol 12:61

Iorizzo M, Ellison S, Senalik D, Zeng P, Satapoomin P, Huang J, Bowman M, Iovene M, Sanseverino W, Cavagnaro P, Yildiz M, Macko-Podgórni A, Moranska E, Grzebelus E, Grzebelus D, Ashrafi H, Zheng Z, Cheng S, Spooner D, Van Deynze A, Simon P (2016) A high-quality carrot genome assembly provides new insights into carotenoid accumulation and asterid genome evolution. Nat Genet 48:657–666

Ito T, Ng KH, Lim TS, Yu H, Meyerowitz EM (2007) The homeotic protein AGAMOUS controls late stamen development by regulating a jasmonate biosynthetic gene in Arabidopsis. Plant Cell 19:3516–3529

Jones DF (1950) The interrelation of plasmagenes and chromogens in pollen production in maize. Genetics 35:507–512

Joyce R, Steckel A, Gray D, Rowse HR (1989) Relationships between indices of seed maturity and carrot seed quality. Ann Appl Biol 114:177–183

Kanzaki H, Takeda M, Kameya T (1991) Sequence analysis of a mitochondrial DNA fragment isolated from cultured cells of carrot cytoplasmic male-sterile strain. Jpn J Genet 66:719–724

Keilwagen J, Lehnert H, Berner T, Budahn H, Nothnagel T, Ulrich D, Dunemann F (2017) The terpene synthase gene family of carrot (*Daucus carota* L.): identification of QTLs and candidate genes associated with terpenoid volatile compounds. Front Plant Sci 8:1930–1948

Kim DH, Sung S (2013) Coordination of the vernalization response through a VIN3 and FLC gene family regulatory network in *Arabidopsis*. Plant Cell 25:454–469

Kitagawa J, Posluszny U, Gerrath JM, Wolyn DJ (1994) Developmental and morphological analyses of homeotic cytoplasmic male sterile and fertile carrot flowers. Sex Plant Reprod 7:41–50

Klimek-Chodacka M, Oleszkiewicz T, Lowder LG, Qi Y, Baranski R (2018) Efficient CRISPR/Cas9-based genome editing in carrot cells. Plant Cell Rep 37:575–586

Knuth P (1898) Handbuch der Blütenbiologie I-III. Engelmann, Leipzig

Kofer W, Glimelius K, Bonnett HT (1991) Modifications of mitochondrial DNA cause changes in floral development in homeotic-like mutants of tobacco. Plant Cell 3:759–769

Kole C, Quijada P, Michaels SD, Amasino RM, Osborn TC (2001) Evidence for homology of flowering-time genes VFR2 from *Brassica rapa* and FLC from *Arabidopsis thaliana*. Theor Appl Genet 102:425–430

Kononkov PF, Mokhov AI (1972) Comparison of the biology of flowering and fertilization in sterile and fertile forms of carrot. Nauch Dokl Vyssh Shikoly Biol 12:73

Koul P, Koul AK, Hamal IA (1989) Reproductive biology of wild and cultivated carrot (*Daucus carota* L.). New Phytol 112:437–443

Kozik E, Nowak R, Nowakowska M, Dyki B (2012) Level of sterility and morphological flower differentiation of petaloid male-sterile plants of carrot. J Agric Sci 4:187–194

Kuan TL, Minsavage GV, Gabrielson RL (1985) Detection of *Xanthomonas campestris* pv. *carotae* in carrot seed. Plant Dis 69:758–760

Lackie S, Yeung EC (1996) Zygotic embryo development in *Daucus carota*. Can J Bot 74:990–998

Lamborn E, Ollerton J (2000) Experimental assessment of the functional morphology of inflorescences of *Daucus carota* (Apiaceae): testing the 'fly catcher effect'. Funct Ecol 14:445–454

Lamprecht H (1951) Über partielle und Semisterilität, insbesondere bei *Pisum sativum*. Z Pflanzenzücht 35:422–433

Lan TH, Paterson AH (2000) Comparative mapping of quantitative trait loci sculpting the curd of *Brassica oleracea*. Genetics 155:1927–1954

Le Dily F, Duyme M, Villeneuve F (1991) Carbohydrate composition of carrot (*Daucus carota* L.) during field development and cold moisture storage. In: Eucarpia carrot 91—proceedings of the IVth meeting on breeding of carrots, pp 113–115

Leins P, Erbar C (1997) Floral developmental studies: some old and new questions. Int J Plant Sci 158:3–12

Lesprit E (1991) Genetic evaluation of 10 male-sterile lines in carrot (*Daucus carota* L.). 1. Seed yield potential. In: Eucarpia carrot 91—proceedings of the IVth meeting on breeding of carrots, pp 55–67

Linke B, Börner T (2005) Mitochondrial effects on flower and pollen development. Mitochondrion 5:389–402

Linke B, Nothnagel T, Börner T (1999) Morphological characterization of modified flower morphology of three novel alloplasmic male sterile carrot sources. Plant Breeding 118:543–548

Linke B, Nothnagel T, Börner T (2003) Flower development in carrot CMS plants: mitochondria affect the expression of MADS box genes homologous to GLOBOSA and DEFICIENS. Plant J 34:27–37

Litvinova MK (1973) Cytoembryological investigations of the male sterility in carrot. Nauch Tr Vononezh USSR 56:14–18

Litvinova MK, Klyuchinkova ZA, Ermokova NA, Kuznetsov YUP (1980) Study of sterile families of carrot used in breeding for heterosis. Genet Selekt Semenovod Saratov USSR 89:97–100

Liu Z, Mara C (2010) Regulatory mechanisms for floral homeotic expression. Semin Cell Dev Biol 21:80–86

Mandel JR, McCauley DE (2015) Pervasive mitochondrial sequence heteroplasmy in natural populations of wild carrot, *Daucus carota* spp. *carota* L. PLoS One. https://doi.org/10.1371/journal.pone.0136303

Mandel JR, McAssey EV, Roland KM, McCauley DE (2012) Mitochondrial gene diversity associated with the atp9 stop codon in natural populations of wild carrot (*Daucus carota* ssp. *carota*). J Hered 103:418–425

Marienfeld JR, Unseld M, Brandt P, Brennicke A (1997) Mosaic open reading frames in the *Arabidopsis thaliana* mitochondrial genome. Biol Chem 378:859–862

Mas F, Harper A, Horner R, Welsh T, Jaksons P, Suckling DM (2018) The importance of key floral bioactive compounds to honey bees for the detection and attraction of hybrid vegetable crops and increased seed yield. J Sci Food Agric 98:4445–4453. https://doi.org/10.1002/jsfa.8967

McCollum GD (1966) Occurrence of petaloid stamens in wild carrot (*Daucus carota* L.) from Sweden. Swed Econ Bot 2:361–367

Mehring-Lemper M (1987) Genetisch-züchterische Untersuchungen zur Schaffung von Hybridsorten bei Möhren (*Daucucs carota* L.). Dissertation, University of Hannover

Meijer EA, De Vries SC, Sterk P, Gardella DWJ, Wirtz KWA, Hendriks T (1993) Characterization of the non-specific lipid transfer protein EP2 from carrot (*Daucus carota* L.). Mol Cell Biochem 123:159–166

Meinke DW, Franzmann LH, Nickle TC, Yeung EC (1994) Leafy cotyledon mutants of Arabidopsis. Plant Cell 6:1049–1064

Michaels SD, Amasino RM (1999) FLOWERING LOCUS C encodes a novel MADS domain protein that acts as a repressor of flowering. Plant Cell 11:949–956

Michalik B (1971) Studia genetyczne nad meska bezplodnoscia u marchwi (*Daucus carota* L.). Plant Breed Acclimatization Seed Prod 15:446–474

Michalik B (1974) Partial male sterility in carrots. In: Proceedings of the 19th international horticultural congresss I. Section VII, vegetables, pp 721–761

Michalik B (1978) Stability of male sterility in carrot under different growth conditions. Bull Acad Sci Ser Biol 26:827–832

Miranda RM, dos Santos Dias DCF, de Toledo Picoli EA, Pereira da Silva P, Nascimento WM (2017) Physiological quality, anatomy and histochemistry during the development of carrot seeds (*Daucus carota* L.). Ciênc Agrotec 41:169–180

Mockute D, Nivinskiene O (2004) The sabinene chemotype of essential oil of seeds of *Daucus carota* L. ssp. *carota* growing wild in Lithuania. J Essent Oil Res 16:227–281

Morelock TE (1974) Influence of cytoplasmic sources of expression of male sterility in carrot *D. carota* L. Dissertation, University of Wiscosin

Morelock TE, Simon PW, Peterson CE (1996) Wisconsin wild: another petaloid male-sterile cytoplasm for carrot. HortScience 31:887–888

Murai K, Takumi S, Koga H, Ogihara Y (2002) Pistillody, homeotic transformation of stamens into pistil-like structures, caused by nuclear–cytoplasm interaction in wheat. Plant J 29:169–181

Nagarajan S, Pandita VK (2001) Effect of umbel shape on root characters and subsequent seed yield in Asiatic carrot (*Daucus carota*). Ind J Agric Sci 71:98–101

Nakajima Y, Yamamoto T, Muranaka T, Oeda K (1999) Genetic variation of petaloid male-sterile cytoplasm of carrots revealed by sequence-tagged sites (STSs). Theor Appl Genet 99:837–843

Nakajima Y, Yamamoto T, Muranaka T, Oeda K (2001) A novel *orfB*-related gene of carrot mitochondrial genomes that is associated with homeotic cytoplasmic male sterility (CMS). Plant Mol Biol 46:99–107

Nascimento WM, Vieira JV, Alvares MC (2003) Physiological maturity of carrot seeds cv. Alvorada under tropical conditions. Acta Hort 607:49–51

Nieuwhof M (1984) Effect of gibberellic acid on bolting and flowering of carrot (*Daucus carota* L.). Sci Hortic 24:211–219

Nothnagel T, Straka P, Budahn H (1997) Development of new cms-systems for carrot breeding. J Appl Genet 38A:172–177

Nothnagel T, Straka P, Linke B (2000) Male sterility in populations of *Daucus* and the development of alloplasmic male-sterile carrot lines. Plant Breeding 119:145–152

Osborn TC, Kole C, Parkin IA, Sharpe AG, Kuiper M, Lydiate DJ, Trick M (1997) Comparison of flowering time genes in *Brassica rapa*, *B. napus* and *Arabidopsis thaliana*. Genetics 146:1123–1129

Ou CG, Mao JH, Liu LJ, Li CJ, Ren HF, Zhao ZW, Zhuang FY (2017) Characterising genes associated with flowering time in carrot (*Daucus carota* L.) using transcriptome analysis. Plant Biol 19:286–297

Parcy F, Valon C, Raynal M, Gaubier-Comella P, Delseny M, Giraudat J (1994) Regulation of gene expression programs during Arabidopsis seed development: roles of the ABI3 locus and of endogenous abscisic acid. Plant Cell 6:1567–1582

Parcy F, Valon C, Kohara A, Misera S, Giraudat J (1997) The ABSCISIC ACID-INSENSITIVE3, FUSCA3, and LEAFY COTYLEDON1 loci act in concert to control multiple aspects of Arabidopsis seed development. Plant Cell 9:1265–1277

Pelaz S, Ditta GS, Baumann E, Wisman E, Yanofsky MF (2000) B and C floral organ identity functions require SEPALLATA MADS-box genes. Nature 405:200–203

Pereira RS, Nascimento WM, Vieira JV (2008) Carrot seed germination and vigor in response to temperature and umbel orders. Sci Agric (Piracicaba, Braz.) 65:145–150

Pérez-Bañón C, Petanidou T, Marcos-Garcia MA (2007) Pollination in small islands by occasional visitors: the case of Daucus carota subsp. commutatus (Apiaceae) in the Columbretes archipelago. Spain. Plant Ecology 192:133–151

Peterson CE, Simon PW (1986) Carrot breeding. In: Bassett MJ (ed) Breeding vegetable crops. AVI, Westport, CT, pp 321–356

Pryor BM, Gilbertson RL (2001) A PCR-based assay for detection of *Alternaria radicina* on carrot seed. Plant Dis 85:18–23

Pryor BM, Davis RM, Gilbertson RL (1994) Detection and eradication of *Alternaria radicina* on carrot seed. Plant Dis 78:452–456

Purugganan MD, Rounsley SD, Schmidt RJ, Yanofsky MF (1995) Molecular evolution of flower development: diversification of the plant MADS-box regulatory gene family. Genetics 140:345–356

Reeves PA, He Y, Schmitz RJ, Amasino RM, Panella LW, Richards CM (2007) Evolutionary conservation of the FLOWERING LOCUS C-mediated vernalization response: evidence from the sugar beet (*Beta vulgaris*). Genetics 176:295–307

Reinert J (1958) Untersuchungen über die Morphogenese an Gewebekulturen. Ber Dtsch Bot Ges 71:15

Reuther K, Claßen-Bockhoff R (2013) Andromonoecy and developmental plasticity in *Chaerophyllum bulbosum* (Apiaceae-Apioideae). Ann Bot 112:1495–1503

Robison MM, Wolyn DJ (2002) Complex organization of the mitochondrial genome of petaloid CMS carrot. Mol Genet Genomics 268:232–239

Robison MM, Wolyn DJ (2005) A mitochondrial plasmid and plasmid-like RNA and DNA polymerases encoded within the mitochondrial genome of carrot (*Daucus carota* L.). Curr Genet 47:57–66

Robison MM, Wolyn DJ (2006a) Petaloid-type cms in carrot is not associated with expression of *atp8* (*orfB*). Theor Appl Genet 112:1496–1502

Robison MM, Wolyn DJ (2006b) A 60 kDa COX1 protein in mitochondria of carrot irrespective of the presence of C-terminal extensions in the cox1 reading frames. Mol Gen Genomics 275:68–73

Rong J, Janson S, Umehara M, Ono M, Vrieling K (2010) Historical and contemporary gene dispersal in wild carrot (*Daucus carota* ssp. *carota*) populations. Ann Bot 106:285–296

Rosas U, Mei Y, Xie Q, Banta JA, Zhou RW, Seufferheld G, Gerard S, Chou L, Bhambhra N, Parks JD, Flowers JM, McClung CR, Hanzawa Y, Purugganan MD (2014) Variation in Arabidopsis flowering time associated with cis-regulatory variation in CONSTANS. Nat Commun 5:3651

Roth E (1981) Zur Problematik der Inzuchtwirkung bei Ausgangslinien der Speisemöhre (*Daucus carota* L.). Tag Ber Akad Landwirtsch Wiss DDR 191:105–116

Rubashevskaia MK (1931) Observation on the wild carrot in cultivation and under natural conditions. Bull Appl Bot Genet Plant Breeding 26:194–252

Rubatzky VE, Quiros CF, Simon PW (1999) Carrots and related vegetable umbelliferae. CABI Publishing, New York

Rurek M, Szklarczyk M, Adamczyk N, Michalik B, Augustyniak H (2001) Differences in editing of mitochondrial nad3 transcripts from CMS and fertile carrots. Acta Biochim Pol 48:711–717

Sakr ES, Thompson HC (1942) Effect of temperature and photoperiod on seedstalk development in carrots. Proc Am Soc Hort Sci 41:343–346

Sandin NH (1980) Optimum harvest time for *Daucus carota* (carrot) seed crops in Sweden. In: Hebblethwaite PD (ed) Seed production. Butterworths, London, pp 553–559

Scheike R, Gerold E, Brennicke A, Mehring-Lemper M, Wricke G (1992) Unique patterns of mitochondrial genes, transcripts and proteins in different male-sterile cytoplasms of *Daucus carota*. Theor Appl Genet 83:419–427

Schmidt EDL, Guzzo F, Toonen MAJ, de Vries SC (1997) A leucine-rich repeat containing receptor-like kinase marks somatic plant cells competent to form embryos. Development 124:2049–2062

Schnable PS, Wise RP (1998) The molecular basis of cytoplasmic male sterility and fertility restoration. Trends Plant Sci 3:175–180

Schwab B, Neumann KH (1975) Influence of gibberellic sprays on the flowering carrots. Z Pflanzenernähr Bodenk 1:13–18

Schwarz-Sommer Z, Huijser P, Nacken W, Saedler H, Sommer H (1990) Genetic control of flower development by homeotic genes in *Antirrhinum majus*. Science 250:931–936

Shiota H, Satoh R, Watabe K, Harada H, Kamada H (1998) *CABI3*, the carrot homologue of the *Arabidopsis ABI3*, is expressed during both zygotic and somatic embryogenesis and functions in the regulation of embryo-specific ABA-inducible genes. Plant Cell Physiol 39:1184–1193

Simon PW, Freeman RE, Vieira JV, Boiteux LS, Briard M, Nothnagel T, Michalik B, Kwon YS (2008) Carrot. In: Prohens J, Nuez F (eds) Handbook of plant breeding: vegetables II: Fabaceae, Liliaceae, Solanaceae, and Umbelliferae. Springer, New York, pp 327–357

Sinha SH, Chakrabacti AK (1992) Insect pollination in carrot seed crop. Seed Res 20:37–40

Smaczniak C, Immink RG, Muiño JM, Blanvillain R, Busscher M, Busscher-Lange J, Dinh QD, Liu S, Westphal AH, Boeren S, Parcy F, Xu L, Carles CC, Angenent GC, Kaufmann K (2012) Characterization of MADS-domain transcription factor complexes in Arabidopsis flower development. Proc Natl Acad Sci USA 109:1560–1565

Staes G (1889) Die Blumen von *Daucus carota*. Bot Jahrb 1:124

Staniszewska M, Kula J, Wieczorkiewicz M, Kusewicz D (2005) Essential oils of wild and cultivated carrots—the chemical composition and antimicrobial activity. J Essent Oil Res 17:579–583

Stein M, Weit E, Wolfram H (1985) Zuchtmethodische Aspekte der Nutzung der Pollensterilität bei Zwiebeln und Möhren. Arch Gartenbau Berlin 33:345–353

Steinborn R, Weihe A, Boerner T (1992) Mitochondrial genome diversity within a cultivar of *Daucus carota* ssp. *sativus* revealed by restriction fragment analysis of single plants. Plant Breeding 109:75–77

Steinborn R, Linke B, Nothnagel T, Boerner T (1995) Inheritance of chloroplast and mitochondrial DNA in alloplasmic forms of the genus Daucus. Theor Appl Genet 91:632–638

Sterk P, Booij H, Schellekens GA, Van Kammen A, De Vries SC (1991) Cell-specific expression of the carrot EP2 lipid transfer protein gene. Plant Cell 3:907–921

Steward FC, Mapes MO, Mears K (1958) Growth and organized development of cultures cells. II. Organization in cultures grown from freely suspended cells. Am J Bot 45:705–708

Strandberg JO (1988) Detection of *Alternaria dauci* on carrot seed. Plant Dis 72:531–534

Struckmeyer BE, Simon PW (1986) Anatomy of fertile and male-sterile carrot flowers from different genetic sources. J Am Soc Hort Sci 111:965–968

Sylwester EP (1960) Beware of wild carrot. Hoard's Dairyman 105:330–331

Szklarczyk M, Oczkowski M, Augustyniak H, Linke B, Börner T, Michalik B (2000) Organisation and expression of mitochondrial *atp9* genes from CMS and fertile carrots. Theor Appl Genet 100:263–270

Szklarczyk M, Szymanski M, Wójik-Jagla M, Simon PW, Weihe A, Börner T (2014) Mitochondrial atp9 genes from petaloid male-sterile and male-fertile carrots differ in their status of heteroplasmy recombination involvement, post-transcriptional processing as well as accumulation of RNA and protein product. Theor Appl Genet 127:1689–1701

Tan GF, Wang F, Zhang XY, Xiong AS (2018) Different lengths, copies and expression levels of the mitochondrial atp6 gene in male sterile and fertile lines of carrot (*Daucus carota* L.). Mitochondrial DNA Part A 29:446–454

Theissen G (2001) Development of floral organ identity: stories from the MADS house. Curr Opin Plant Biol 4:75–85

Theissen G, Saedler H (1995) MADS-box genes in plant ontogeny and phylogeny: Haeckel's 'biogenetic law' revisited. Curr Opin Genet Dev 5:628–639

Theissen G, Saedler H (2001) Plant biology. Floral quartets. Nature 409:469–471

Thomas TL, Wilde D (1985) Analysis of gene expression in carrot somatic embryos. In: Terzi M, Pitto L, Sung ZR (eds) Somatic embryogenesis. IPRA, Rome, pp 77–85

Thomas TL, Wilde D (1987) Analysis of carrot somatic embryo gene expression programs. In: Green CE, Folmers DA, Hackett WT, Biesboer DD (eds) Plant biology. Alan R. Liss, New York, pp 83–93

Thompson DS (1961) Studies on the inheritance of male-sterility in carrot, *Daucus carota* L. var. *sativa*. Proc Am Soc Hort Sci 78:332–338

Timin NI, Vasilevsky VA (1997) Genetic peculiarities of carrot (*Daucus carota* L.). J Appl Genet 38A:232–236

Trivedi RS, Hampton JG, Townshend JM, Jaspers MV, Ridgway HJ (2010) First report of *Alternaria carotiincultae* on carrot seed produced in New Zealand. Plant Dis 94(1168):3

Umesh KC, Davis RM, Gilbertson RL (1998) Seed contamination thresholds for development of carrot bacterial blight caused by *Xanthomonas campestris* pv. *carotae*. Plant Dis 82:1271–1275

Villeneuve F, Latour F (2017) Influence of sowing time and chilling exposure on flower induction in carrot (*Daucus carota* L.). Acta Hortic 1153:47–54

Warenstorf C (1896) Blütenbiologische Beobachtungen aus der Ruppiner Flora im Jahr 1895. Bot Ver Germany 38:15–63

Webb CJ (1981) Andromonoecism, protandry, and sexual selection in Umbelliferae. New Zeal J Bot 19:335–338

Weigel D, Meyerowitz EM (1994) The ABCs of floral homeotic genes. Cell 78:203–209

Weit E (1979) Investigations on line maintainance in carrot. Tag Ber Akad Landwirtsch Wiss DDR 168:421–430

Welch JE, Grimball EI (1947) Male sterility in carrot. Science 12:594

Westmoreland D, Muntan C (1996) The influence of dark central florets on insect attraction and fruit production in Queen Anne's Lace (*Daucus carota* L.). Am Midl Nat 135:122–129

Wohlfeiler J, Alessandro MS, Cavagnaro PF, Galmarini CR (2019) Multiallelic digenic control of vernalization requirement in carrot (Daucus carota L.). Euphytica 215:37–47. https://doi.org/10.1007/s10681-019-2360-2

Wolyn DJ, Chahal A (1998) Nuclear and cytoplasmic interactions for petaloid male sterile accessions of wild carrot (*Daucus carota* L). J Am Soc Hort Sci 123:849–853

Wurtele ES, Wang H, Durgerlan S, Nikolau BJ, Ulrich TH (1993) Characterization of a gene that is expressed early in somatic embryogenesis of *Daucus carota*. Plant Physiol 102:303–312

Yahyaa M, Ibdah M, Mazouk S, Ibdah M (2017) Profiling of the terpene metabolome in carrot fruits of wild (*Daucus carota* L. ssp. *carota*) accessions and characterization of a geraniol synthase. J Agric Food Chem 66:2378–2386

Yamamoto T, Nakajima Y, Oeda K (2000) Morphological changes in homeotic cytoplasmic male-sterile carrots combined with fertile cytoplasm by asymmetrical cell fusion. Plant Cell Rep 19:363–370

Yan L, Loukoianov A, Blechl A, Tranquilli G, Ramakrishna W, San Miguel P, Bennetzen JL, Echenique V, Dubcovsky J (2004) The wheat VRN2 gene is a flowering repressor down-regulated by vernalization. Science 303:640–1644

Yan L, Fu D, Li C, Blechl A, Tranquilli G, Bonafede M, Sanchez A, Valarik M, Yasuda S, Dubcovsky J (2006) The wheat and barley vernalization gene VRN3 is an orthologue of FT. Proc Natl Acad Sci USA 103:19581–19586

Yanofsky MF, Ma H, Bowman JL, Drews GN, Feldmann KA, Meyerowitz EM (1990) The protein encoded by the Arabidopsis homeotic gene agamous resembles transcription factors. Nature 346:35–39

Yazawa K, Kamada H (2007) Identification and characterization of carrot HAP factors that form a complex with the embryo-specific transcription factor C-LEC1. J Exp Bot 58:3819–3828

Yazawa K, Takahata K, Kamada H (2004) Isolation of the gene that encodes carrot leafy cotyledon 1 and expression analysis during somatic and zygotic embryogenesis. Plant Physiol Biochem 42:215–223

Zahn LM, Leebens-Mack J, DePamphilis CW, Ma H, Theissen G (2005) To B or Not to B a flower: the role of DEFICIENS and GLOBOSA orthologs in the evolution of the angiosperms. J Hered 96:225–240

Zenkteler M (1962) Microsporogensis and tapetal development in normal and male-sterile carrots (*Daucus carota*). Am J Bot 49:341–348

Zhan Z, Zhang C, Zhang H, Li X, Wen C, Liang Y (2017) Molecular cloning, expression analysis, and subcellular localization of *FLOWERING LOCUS T* (*FT*) in carrot (*Daucus carota* L.). Mol Breeding 37:149–158

Zhang S, Wong L, Meng L, Lemaux PG (2002) Similarity of expression patterns of knotted1 and ZmLEC1 during somatic and zygotic embryogenesis in maize (*Zea mays* L.). Planta 215:191–194

Zhidkova NI, Bunin MS, Kalinia LM, Romanyak AN, Tsvetkova MA, Asyakin BP (1991) The main directions and achievements in carrot breeding in the USSR. In: Bonnet A (ed) Eucarpia carrot 91, Avignon, Montfavet, France, pp 31–39

Zubko MK, Zubko EI, Ruban AV, Adler K, Mock HP, Misera S, Gleba YY, Grimm B (2001) Extensive developmental and metabolic alterations in cybrids *Nicotiana tabacum* (+ *Hyoscyamus niger*) are caused by complex nucleo-cytoplasmic incompatibility. Plant J 25:627–639

Gene Flow in Carrot

4

Jennifer R. Mandel and Johanne Brunet

Abstract

In this chapter, we first present characteristics of carrots that will affect gene flow and discuss dispersal via pollen by insect pollinators and via seeds by wind and animals. Although carrot is often referred to as a biennial, we introduce the various life history strategies observed in wild carrot populations as these can impact population growth and the range expansion of wild carrots over the landscape. We then review the studies of gene flow between crops, between crop and wild carrot and among wild carrot populations, concentrating on studies that used molecular markers. The consequences of these different types of gene flow (among cultivars, between crop and wild, and among wild) are then discussed. A major goal of biotechnology risk assessment for crops is to improve predictions of the fate of escaped genes either to other crop fields or to wild populations. We suggest as a priority for future studies to incorporate population dynamics with population genetics when modeling the fate of introduced genes. Improving our understanding of the factors that affect the spread of escaped genes will lead to the design of better management strategies to contain and limit their spread.

4.1 Introduction

4.1.1 Biology and Life History Strategies Influence Gene Flow

Carrot, *Daucus carota* L. is a diploid, highly outcrossed and insect-pollinated species in the family Apiaceae (see also Chap. 2). Plants are andromonoecious (both male and hermaphroditic flowers on a plant) and have protandrous hermaphroditic flowers (pollen shed before stigmas become receptive) (see also Chap. 3). Some plant populations are reported to exhibit gynodioecy as male sterile plants (functionally female plants) co-occur with hermaphroditic individuals (Ronfort et al. 1995). The hermaphroditic plants in gynodioecious populations are likely andromonoecious. The male sterility trait has been a very useful tool in the development of cultivated hybrid carrot. Small, white flowers are grouped into umbels that flower sequentially on a plant (Koul et al. 1989). The primary umbel, located at the tip of the flowering stalk, is the largest and first umbel to flower. Plants have one primary umbel and can have many secondary, tertiary, and higher order umbels. Carrots are

J. R. Mandel (✉)
Department of Biological Sciences,
University of Memphis, Memphis, TN, USA
e-mail: jmandel@memphis.edu

J. Brunet
USDA-Agricultural Research Service, Vegetable Crops Research Unit, Department of Entomology,
University of Wisconsin, Madison, WI, USA

P. Simon et al. (eds.), *The Carrot Genome*, Compendium of Plant Genomes,
https://doi.org/10.1007/978-3-030-03389-7_4

monocarpic or semelparous, i.e., they reproduce once and then die (Lacey 1982), and they do not reproduce vegetatively (Gross 1981). Carrots generally require vernalization (some period of exposure to cold following the seedling stage) to flower, though the degree of vernalization required varies greatly across accessions (lines) and may be minimal in accessions adapted to warmer climates (Alessandro and Galmarini 2007; Alessandro et al. 2013). Wild carrot or Queen Anne's lace (*D. carota* L. subsp. *carota*) seeds can remain dormant in the soil for over a year or two prior to germinating, creating a seed bank (Magnussen and Hauser 2007).

In wild carrot, substantial variation in life history strategies exists, and because carrots are monocarpic, the age at which a plant reproduces and therefore disperses its pollen and seeds will vary among individuals with different life history strategies. While carrots are typically thought of as biennial, a significant amount of variation can exist both within and among wild carrot populations, where biennials co-occur with monocarpic perennials, winter annuals and at times, summer annuals (Table 4.1).

Biennials reproduce in their second year: seeds produced the previous fall germinate in the spring, plants overwinter as rosettes and they flower and set seed during their second summer (de Jong et al. 2016; Lacey 1982; Lacey and Pace 1983). In contrast, monocarpic perennials flower and then die in their third or later summers and survive as rosettes over more than one winter (de Jong et al. 2016; Lacey and Pace 1983; Lacey 1986). However, some populations consist mostly of annuals (de Jong et al. 2016). Winter annuals germinate in the fall (typically from seed produced that summer), overwinter as rosettes, and flower in the summer (de Jong et al. 2016; Lacey 1982; Lacey and Pace 1983). Summer annuals, on the other hand, germinate in the spring (likely from seed produced in the fall) and flower that same growing season (de Jong et al. 2016; Harrison and Dale 1966).

The relative prevalence of the different life history strategies in a population is affected by both genetics and environment. For example, seeds set earlier during the flowering season are more likely to become annuals (Lacey and Pace 1983). In the USA, there is a latitudinal gradient in the mean age of reproduction with earlier reproduction (more annuals) in the South and later reproduction (more biennials and perennials) in the North (Lacey 1988). The latitudinal gradient in age of reproduction persists under common garden conditions suggesting a genetic basis (Lacey 1988). There is also a strong environmental component to age at reproduction where annuals are more common in resource-rich environments and biennials and perennials are abundant in nutrient-poor environments (Lacey 1986; Verkaar and Schenkeveld 1984). Moreover, a greater proportion of perennials are found in older successional fields (Gross 1981; Gross and Werner 1982; Holt 1972; Lacey 1982). Finally, age of reproduction is associated with the ratio of first-to-second-year survival: in the South, a greater first-to-second-year survival ratio is associated with early reproduction while in the North, where biennials and perennials are more common, second-year survival tends to be greater than first-year survival (Lacey 1988).

Life history strategies will affect the growth rate of carrot populations (see Sect. 4.4).

Table 4.1 Summary of the different life history strategies reported in wild carrots

Life history strategy	Seeds produced	Seed germination	Rosette	Flowering
Biennial	Fall	Spring	Survive one winter	Summer of second year
Monocarpic perennial	Fall	Spring	Survive two or more winters	Summer of third or next years
Winter annual	Fall	Fall	Survive winter	Summer of first year
Summer annual	Fall	Spring	Spring	Summer of first year

Populations comprised mostly of annuals will have greater reproductive rates relative to populations with a majority of biennial plants. This occurs because annual plants reproduce after one year instead of two years for biennials or three or more years for perennials. Biennials would have to produce twice as many seeds as annuals in order to attain a similar reproductive output. The proportion of annuals relative to biennials in a population will vary with the environment. Environments with low seed germination and low rosette survival favor perennials while annuals thrive in environments where seed germination and rosette survival are high (Van Etten and Brunet, unpublished data). Resource-rich environments are likely to have higher seed germination and rosette survival relative to nutrient-poor environments and this could explain the differences in the proportion of perennials versus annual plants observed in these two types of environments. The proportion of annuals, biennials and/or perennial individuals in a carrot population varies with latitude and with the resource quality of their habitat (Gross 1981; Holt 1972; Lacey 1986, 1988). Populations with a greater proportion of annual plants and a higher population growth rate will tend to expand and have greater gene flow. This occurs because more plants are likely to be flowering in these populations each year and therefore more pollen and seeds will be produced each year increasing the gene flow potential. We therefore expect, within a latitude, populations in resource-rich habitats to have greater gene flow relative to populations in resource-poor habitats. Range expansion will also be maximized when seeds land in resource-rich habitats.

4.1.2 Overlap of the Geographic Distributions of Cultivated and Wild Carrot and Opportunities for Gene Flow

Cultivated carrots (*Daucus carota* L. subsp. *sativus*) are grown worldwide (see also Chap. 2). The genetic evidence suggests domesticated carrot likely originated from wild carrot in Central Asia with the first domestication occurring approximately 1100 years ago (Iorizzo et al. 2013, 2016; see also Chap. 5). Cultivated carrots were grown on over 87,000 acres in the USA in 2015 (USDA, National Agricultural Statistics Service 2017). Carrots are typically grown for their roots throughout the USA with California producing over 85% of all carrots grown in the USA. Michigan and Texas are other important carrot-producing states (http://www.agmrc.org/commodities-products/vegetables/carrots). The majority of carrot seed production in the USA occurs in the Columbia Basin of Washington, the Madras area of Oregon, California, and in Idaho-in areas where wild carrot is not as common. Though wild carrot is quite common across the seed production area on the Olympic Peninsula in the Sequim-Dungeness Valley (Clallam County) in Washington State and has been classified as a noxious weed there due to the potential for contamination of commercial carrot crops (Fig. 4.1 and see Sect. 4.3.2). Seed and root production also occurs in other regions across the globe including Southern Europe, Chile, Australia, and Japan (Magnussen and Hauser 2007; Umehara et al. 2005).

The geographic distribution of wild carrot is also global occurring on all continents except Antarctica (Grzebelus et al. 2011), thereby providing ample opportunity for gene flow between the cultivated fields and wild populations (see Sect. 4.2.2). Wild carrot is common in temperate regions worldwide and is frequently found in full sun to partial shade in disturbed sites, along roads and in abandoned fields. In the USA, wild carrot plants flower in June–July in more southern populations (Tennessee, North Carolina, and Virginia) and in July–August in more northern populations (Michigan, Wisconsin) (Lacey 1984; Brunet, pers. obs.). The genetic evidence supports the introduction of wild carrot into North America as a weed from Europe (Iorizzo et al. 2013). Wild carrot is commonly found in the eastern, Midwestern and western parts of the USA but is less common in the Great Plains. It has been declared invasive in a number of states (http://www.invasiveplantatlas.org).

Fig. 4.1 Poster from wild carrot seed prevention campaign in Washington State. Credit to and permission from Clea Rome, Washington State University

4.1.3 Pollinators and Agents of Seed Dispersal as Facilitators of Gene Flow

In its native range across central Asia, wild carrot is reported as a pollination generalist, with the largest group of pollinators being Diptera (Ahmad and Aslam 2002; Westmoreland and Muntan 1996). In central England and in the USA, wild carrot flowers are also visited by a variety of generalist insects including 15 insect families: Andrenidae, Apidae, Calliphoridae, Crabronidae, Dolichopodidae, Empididae, Halictidae, Ichneumonidae, Lycaenidae, Muscidae, Nymphalidae, Sphecidae, Syrphidae, Tabaninae, and Vespidae (Lamborn and Ollerton 2000; Ramsey and Mandel, unpublished data). Cultivated carrots are also visited by a large variety of bees (Bohart and Nye 1960), and these include honey bees, leafcutting bees, and wild bees (Davidson et al. 2010; Howlett et al. 2015). Gene flow is expected to be high in carrots. In fact, wild carrot plants are highly outcrossed (96%) which indicates that the majority of the pollen-reaching stigmas and fertilizing ovules comes from other plants in the population (Rong et al. 2010). In addition, carrot pollen can remain viable for days although 50% viability was observed after 12 h (Umehara et al. 2005).

Seed dispersal in carrots can occur via wind or animals with wind likely the most frequent dispersal agent (Lacey 1981). Controlled air velocity studies indicate short-distance dispersal of seeds via wind of a scale of a few meters (Umehara et al. 2005). The presence of spines on carrot seeds suggests seed dispersal by animals via transportation outside of the body (epizoochory) (Lacey 1981; Umehara et al. 2005). Manzano and Malo (2006) demonstrated seed dispersal up to 400 km for carrot seeds attached to the fur of live sheep.

4.2 Gene Flow Studies with Molecular Markers

Gene flow homogenizes the genetic composition of populations and therefore limits genetic differentiation. Gene flow in plant populations can occur via pollen and via seeds, and in carrot insect pollinators move genes via pollen while seeds can be dispersed by wind or by animals. Depending on the sampling strategy and the program used to estimate gene flow, one can obtain contemporary or historical measures of gene flow. Contemporary measures may reflect gene flow over one ecological season, for example, using paternity analyses (Burczyk et al. 2002) or a Kindist approach (Robledo-Arnuncio et al. 2007). Contemporary measures may also quantify recent immigration over the last several generations, as is the case when using BayesAss (Wilson and Rannala 2003; illustrated in Mandel et al. 2016). In addition, when available, phenotypic and genetic markers such as transgenes can facilitate the process of detecting gene flow events (Greene et al. 2015). Historical measures of gene flow are typically based on measures of genetic differentiation among populations traditionally calculated using F_{ST} measures but more recently obtained with various Bayesian clustering methods (for example, STRUCTURE, Pritchard et al. 2000; illustrated in Brunet et al. 2012). Methods based on coalescence (MIGRATE) also provide historical gene flow estimates (Beerli 2006; Beerli and Felsenstein 1999). In addition, historical gene flow measures can be obtained from fine-scale spatial structure data (Vekemans and Hardy 2004). Measures of gene flow obtained from F_{ST} values assume isolation by distance and are calculated as the product of effective population size and migration as $Nm = (1 - F_{ST}/4\ F_{ST})$, e.g., where Nm is the number of migrants per generation and F_{ST} is the standardized measure of the genetic variance among populations (Wright 1951). The greater the level of genetic differentiation the lower the gene flow.

Gene flow in carrot can be studied to answer different questions. One may be interested in the transfer of genes among cultivated carrots with the goal of limiting gene flow to maintain cultivar purity. Gene flow from wild to cultivated carrot can also impact cultivar purity due to the presence of early bolters and less edible roots (Wijnheijmer et al. 1989). Gene flow from cultivars to wild populations has implications for the spread of cultivar genes and genetically modified genes, were genetically engineered crops to become available, into wild carrot populations. Gene flow among wild populations will also influence the spread of these genes. Below, we summarize the results of studies in carrot that examined these different aspects of gene flow using the various methods described above.

4.2.1 Crop-to-Crop Gene Flow

Given that carrot is highly outcrossing and generalist pollinated, maintaining cultivar purity is a high priority. Recommendations for minimum distances require maintaining at least 1000 m from another field with flowering plants and Grzebelus et al. (2011) report that carrot seed production in the USA typically maintains a 3–5 km minimum distance between different root and color types, respectively.

The carrot germplasm is a valuable resource for plant-breeding efforts including improvement to crop yield and protecting against pests and disease (see also Chap. 6). Therefore, understanding the genetic composition and structuring of diversity within the cultivated germplasm has been an important goal of carrot researchers (see also Chap. 5). Studies have been carried out with molecular markers using allozymes to whole genome sequencing with a goal to assess genetic relatedness among cultivar lines but also to understand and ensure the maintenance of cultivar purity, especially across agricultural landscapes. Chapter 5 provides substantial detail of the genetic diversity and structure of the germplasm; therefore, we will only highlight a few examples here as they pertain to gene flow among cultivars. Moreover, as less work has been carried out to evaluate the maintenance of cultivar purity in the landscape (but see Hauser and Bjørn 2001), we describe this as an area for future work in carrot (see Sect. 4.4).

Bradeen et al. (2002) used AFLPs and inter-simple sequence repeat (ISSR) markers to assay genetic variation in a diverse set of cultivar carrot accessions which included 73 open-pollinated lines from European, North American, and Asian primary cultivars. Genetic similarity coefficients ranged from 0.3 to 0.8 across the cultivated lines, and no strong genetic structuring was seen among cultivars (similar to their findings from the wild populations). The authors suggest that given carrot's outcrossing breeding history, and the lack of strict control over pollinations during seed production prior to the 1950s, gene flow among cultivar lines was probably extensive. The authors also suggest that gene flow among wild populations and cultivar lines during this time may have been widespread thus influencing the cultivated gene pool. Notably, the authors argue that human selection against hybrids (with maladapted phenotypes) during these breeding efforts likely played a role in preventing some gene flow among wilds and cultivars. However, later studies observed genetic differentiation between old and new breeding varieties (Shim and Jørgensen 2000) and between eastern (Asia) and western (Europe and America) cultivars (Baranski et al. 2012; Grzebelus et al. 2014; Iorizzo et al. 2013).

4.2.2 Crop-Wild and Wild-Crop Gene Flow

Wild and crop carrots belong to the same species, have similar flowering phenology and can easily hybridize (Grebenstein et al. 2013; Hauser and Shim 2007; Small 1984; Umehara et al. 2005). This implies that pollen reaching a plant's stigma has a high probability of setting a seed, which represents a gene flow event. In addition to the timing of flowering and genetic compatibility, physical proximity can strongly influence gene flow. In many regions of the world, populations of wild carrots are found in close proximity to cultivated carrot fields (Umehara et al. 2005; Magnussen and Hauser 2007; Mandel et al. 2016). In fact, in the USA, carrot seed production usually occurs in areas with little or no wild carrots in order to limit hybridization and maintain the purity of the crop.

Hybrids between cultivated and wild carrots have been detected in wild carrot populations (Hauser and Shim 2007; Magnussen and Hauser 2007). In addition, wild carrots have been found growing inside cultivated carrot fields (Wijnheijmer et al. 1989; Hauser and Bjørn 2001). Therefore, gene flow is bidirectional; it occurs both from cultivars to wild and from wild to cultivars. Gene flow from wild to cultivated carrots can affect cultivar purity. It is typically detected by the presence of early bolters in cultivated fields when hybrids between crop and wilds flower early. This occurs because cultivated carrot is biennial while many wild carrots, especially in Europe where many of these studies took place, are mainly annual (Wijnheijmer et al. 1989; Hauser and Bjørn 2001; Magnussen and Hauser 2007). Annual hybrids in cultivated fields are early bolters, and they can increase in frequency via seed dispersal as seeds are not removed from fields and can lead to pockets of early bolting plants in cultivated fields (Hauser and Bjørn 2001; Magnussen and Hauser 2007). In addition, early bolters can survive crop rotation due to their seed bank (seed dormancy) (Hauser and Bjørn 2001).

The presence of cultivated genes in wild carrot populations has been examined by comparing the genetic differentiation between cultivated and wild carrot populations for wild populations in close proximity to the cultivars and populations further away. Lower levels of genetic differentiation for wild populations in closer physical proximity to cultivated fields would suggest gene flow between cultivated and wild carrots. Magnussen and Hauser (2007) found that wild carrot populations located in closer proximity to the cultivar fields were more genetically similar to the cultivar lines relative to the wild populations located further away from the cultivar fields. A similar pattern was observed by Mandel et al. (2016). Using both nuclear and plastid DNA markers, Mandel et al. (2016) demonstrated gene flow between cultivated and wild carrots in both the eastern part of the USA where carrots are grown for their roots (Nantucket Island, MA) and in a region on the Olympic Peninsula where open-pollinated carrot seeds are produced (Sequim-Dungeness Valley, WA). In both regions, populations geographically closer to crop fields were genetically more similar to the crops than populations that were further away from sites where crops were grown. This study also found evidence that plastid DNA may move via pollen (paternal leakage). If this is common enough, plastid genes could be used as an extra marker for pollen movement in carrots, and not simply as a marker for seed movement. Although, it may be more difficult to separate these two processes.

Magnussen and Hauser (2007) also looked for the presence of hybrid individuals in wild carrot populations in Denmark as evidence of introgression of cultivar genes into wild carrot populations. Using amplified fragment length polymorphisms (AFLPs), the authors detected four hybrid individuals after testing 71 wild plants. The authors determined, based on the genetic data, that the hybrids were likely F2 or backcross individuals. These individuals Magnussen and Hauser (2007) detected could represent the second-generation hybrids in wild carrot populations, supporting the presence of introgression. However, the bidirectional gene flow in carrot, from wild to cultivated and cultivated to wild, and the presence of early bolters in cultivated fields introduce another possible explanation for the presence of backcross individuals in wild carrot populations. These backcross individuals found in wild carrot populations could result from bidirectional gene flow. For example, a gene flow event from wild to cultivated carrot could have produced hybrid bolters in the cultivated carrot fields. Subsequent gene flow from the early hybrid bolter in cultivated fields to wild carrot populations could have created the first-generation backcross individuals in wild carrot populations. While such a process is not the typical scenario invoked to explain the presence of first-generation backcrosses in wild populations, in carrot, it is a probable and interesting scenario and may represent a frequent route to production of backcross individuals. These potential routes to the formation of backcrosses in wild carrot populations require further investigation.

The presence of hybrids between cultivated and wild carrots, either in carrot fields or in wild carrot populations, is supported by several studies (Hauser and Bjørn 2001; Hauser et al. 2004; Magnussen and Hauser 2007). However, in wild populations, F1 hybrids show fitness fairly similar to wild carrots (Hauser and Shim 2007; Ghosh 2012; Umehara et al. 2005). However, Hauser (2002) reported that hybrids were less frost tolerant than wild plants indicating a selective disadvantage to hybrids. Though only preliminary data has been published for first-generation backcrosses to wild carrots (BC1), the data suggests a survival probability to flowering and umbel size very similar to wild carrots (Ghosh 2012). Interestingly, similar fitness to wild carrot does not translate into a fitness advantage to hybrids and suggest hybrids would not increase in frequency in the population via selection. While gene flow may be high enough to produce F1 hybrids, and potentially backcross individuals, in the bidirectional gene flow scenario described above, the fitness differences could not explain the potential for introgression of cultivar genes into wild carrot populations. Moreover, as noted above, Hauser (2002) reported that hybrids were less frost tolerant than wild plants indicating a selective disadvantage to hybrids which would act as a barrier to the introgression of cultivar genes into wild carrot populations. More studies are needed both to compare the fitness of F1 and later-generation hybrids to wild carrots in wild populations and to quantify the extent of introgression of cultivar genes into wild carrot populations.

Although cultivated carrots were derived from wild carrots (Iorizzo et al. 2013), and bidirectional gene flow occurs between cultivated and wild carrot (Wijnheijmer et al. 1989; Hauser and Bjørn 2001; Magnussen and Hauser 2007), the two groups remain genetically differentiated (Grzebelus et al. 2014; Iorizzo et al. 2013; St. Pierre and Bayer 1991; Shim and Jørgensen 2000). Using AFLPs, Shim and Jørgensen (2000) detected clear clustering of wild and cultivated carrot populations based on unweighted pair group method with arithmetic mean (UPGMA) and on principal component analysis (PCA). The UPGMA dendrogram based on genetic distance separated wild from cultivar while the PCA identified three distinct groups, the wild, the old varieties, and the more recently bred varieties. The analysis of molecular variance (AMOVA) on these three groups indicated strong differentiation with an F_{ST} value of 0.398. Using DArT microarray-based genotyping and the program STRUCTURE, Grzebelus et al. (2014) identified three carrot groups in their samples, the wild, eastern (Asia) and western (Europe and America) cultivars. Such separation among the three groups was also found by Iorizzo et al. (2013) who used 4000 SNP to describe the genetic diversity of carrot. Baranski et al. (2012) observed the separation between eastern and western cultivars based on simple sequence repeats (SSRs or microsatellites) but did not examine wild carrot. Therefore, studies based on a variety of genetic markers have identified three groups in carrots, the wild carrots, the eastern

(Asia) cultivars, and the Western cultivars (Europe and America). The genetic differentiation between wild and cultivated carrots indicates that gene flow remains restricted between these two groups overall. However, this finding does not preclude the possibility of gene flow occurring between specific cultivar fields and surrounding wild populations.

Despite being genetically differentiated from one another, the groups of wild and cultivated carrots both maintain high and similar levels of genetic diversity. In other words, the alleles may differ between the two groups leading to genetic differentiation but the number of alleles and level of heterozygosity remain high in both groups. This finding was identified by St. Pierre and Bayer (1991) using allozymes where they observed only a slight decrease in genetic diversity in cultivated relative to wild carrot accessions. It was later confirmed by Iorizzo et al. (2013) who detected no differences in genetic diversity between wild and cultivated carrots when using 4000 single nucleotide polymorphisms (SNPs). The differentiation between wild and cultivated carrots should permit the identification of genetic markers to detect cultivar genes in wild populations although to date, few such markers have been identified (Umehara et al. 2005).

4.2.3 Wild-Wild Gene Flow

Below, we summarize the different studies used to measure gene flow among wild carrot populations. Gene flow estimates varied, some being contemporary and others historical and used many of the approaches described earlier to measure gene flow, such as F_{ST}, Kindist, and BayesAss.

In populations of wild carrot collected from five locations in Denmark, Shim and Jørgensen (2000) used ten AFLPs to study genetic variation and population structure. Populations were located in Zealand and Jutland with geographical distances ranging from a few km to more than 200 km. The degree of genetic differentiation among wild populations was moderate with $G_{ST} = 0.18$ (Shim and Jørgensen 2000). Populations that were separated by just a few km were the most genetically similar with the one population collected from more than 200 km away showing the greatest divergence (Shim and Jørgensen 2000). This pattern suggests isolation by distance as gene flow decreases with increasing geographic distances.

Using nuclear microsatellite markers, Rong et al. (2010) estimated contemporary and historical measures of gene flow in two populations of wild carrot in the Netherlands: Meijendel and Alkmaar. The authors obtained historical gene flow estimate based on small-scale spatial genetic structure (SGS) (Vekemans and Hardy 2004). Because historical and contemporary gene flow estimates often differ, they also estimated contemporary gene flow using Kindist (Robledo-Arnuncio et al. 2007). They detected weak but statistically significant SGS in both populations. Limited gene flow can create such non-random distribution of genotypes at a small scale. The most common method to obtain historical gene flow measures from SGS data assumes isolation by distance (Vekemans and Hardy 2004). Using such a method, Rong et al. (2010) estimated that roughly 95% of the historical gene dispersal occurred at a distance of 8–24 m in Meijendel and 20–62 m in Alkmaar. The method Kindist fits a dispersal kernel to pollen dispersal data in order to determine pollen dispersal distances. It indirectly estimates gene flow based on a normalized measure of correlated paternity among offspring of mother plants sampled at different spatial distances. Rong et al. (2010) determined that an exponential power function best fitted the pollen dispersal data and observed low differentiation of pollen pools among mother plants ($\varphi ft = 0.057$). The authors estimated 95% of contemporary pollen dispersal could potentially occur over distances up to 1.8 km and 99% of pollen dispersal within 4.2 km. The authors also estimated an outcrossing rate of 96% for wild carrot using progeny arrays and the MLTR program developed by Ritland (1996, 2002).

In another study by Rong et al. (2013), the authors used 11 nuclear microsatellite markers to estimate both contemporary and historical estimates of gene flow among a metapopulation comprising 12 patches separated by a few km. In this study, contemporary gene flow was quantified using BayesAss (Wilson and Rannala 2003) while historical estimates used MIGRATE (Beerli 2006). Rong et al. (2013) observed low-to-moderate measures of genetic differentiation among populations ($F_{ST} = 0.082$). They identified a pattern of isolation by distance, where gene flow decreased as the geographical distances increased. The assignment-based method (BayesAss: Wilson and Rannala 2003) provided fairly low migration rate (m) estimates ranging from $m = 0.0008$ to 0.0898 (overall mean $m = 0.0032$) between the 12 wild carrot patches. However, historical estimates, based on coalescence theory were five times lower than contemporary estimates and ranged from 0.0003 to 0.0012 with an overall mean of 0.006.

Reiker et al. (2015) studied the genetic diversity and level of genetic differentiation among nine indigenous and ten restored wild carrot populations in Germany, where non-local (non-indigenous) seeds were used for restoration and the ten restored sites spanned a 200×200 km^2 region. Using ten nuclear microsatellites (developed previously by Cavagnaro et al. 2011), the authors detected high genetic diversity in both indigenous and restored carrot populations with observed heterozygosity levels greater than 0.75 in each population. The level of genetic differentiation among indigenous populations was low with an $F_{ST} = 0.030$. The degree of population genetic structure among all sampled populations was also low ($F_{ST} = 0.044$) leading the authors to note that carrot populations from their study (both indigenous and restored) were essentially randomly mating with one another and indicated that gene flow was high. The authors claimed that the higher pattern of genetic differentiation observed in Denmark by Shim and Jørgensen (2000) ($F_{ST} = 0.18$) resulted from the fact that the Denmark populations were at the northern edge of the distribution of wild carrot.

Mandel et al. (2016) used 15 nuclear microsatellites (also developed previously by Cavagnaro et al. 2011) and one polymorphic plastid marker to assess patterns of genetic diversity and genetic differentiation and infer patterns of gene flow among wild carrot populations (and cultivated lines, see Sect. 4.2.2) located in the Northwestern and Northeastern USA. The Northwestern USA study site comprised seven wild carrot populations and was located on the Olympic Peninsula in the Sequim-Dungeness Valley (Clallam County) in Washington State. This site was chosen for study because it co-occurs with locations where cultivated carrot is grown for seed production. Pairwise population distances ranged from 1.91 to 6.49 km. Estimates of pairwise population genetic differentiation measured by calculating Wright's F_{ST} varied from 0.049 to 0.288 (Table 4.2). The Northeastern USA study site

Table 4.2 Pairwise F_{ST} values (below diagonal) and geographic distance in km (above diagonal) for Olympic Peninsula wild carrot populations

Site	Hemlock	Kendall	Medsker	Eberle	Fasola	Fencebird	Prince
Hemlock[a]	-	2.66	2.94	6.49	4.56	4.67	4.94
Kendall[a]	0.134	-	2.17	4.18	1.91	2.52	3.83
Medsker[a]	0.049	0.109	-	3.92	3.13	2.18	2.03
Eberle	0.106	0.138	0.072	-	2.76	1.82	3.06
Fasola	0.209	0.288	0.196	0.139	-	1.80	3.87
Fencebird	0.141	0.183	0.126	0.106	0.236	-	2.13
Prince	0.157	0.190	0.129	0.089	0.165	0.137	-

[a]Sites that were not in close proximity to root production crop fields

Table 4.3 Pairwise F_{ST} values (below diagonal) and geographic distance in km (above diagonal) for Nantucket Island wild carrot populations

Site	Cliff Rd	Polpis Rd	Tuckernuck	Bart Farm	Moors Farm
Cliff Rd	-	6.54	11.02	3.87	4.24
Polpis Rd	0.109	-	17.53	8.54	2.78
Tuckernuck	0.135	0.112	-	10.93	15.20
Bart Farm[a]	0.171	0.139	0.169	-	5.77
Moors Farm[a]	0.178	0.143	0.169	0.064	-

[a]Sites that were not in close proximity to seed production crop fields

comprised five wild carrot populations and was located on Nantucket Island (Nantucket County) in Massachusetts. Pairwise population distances ranged from 2.78 to 17.53 km in that region and the level of genetic differentiation based on Wright's F_{ST} varied from 0.064 to 0.178 (Table 4.3), indicating low-to-intermediate levels of genetic differentiation. Higher levels of genetic differentiation can indicate lower gene flow. Low migration rates were also reported using the program BayesAss that measures gene flow over the last several generations; values of migration rates varying between 0.0057 and 0.0405 with some values greater than 0.20 were obtained for the pairwise Olympic Peninsula populations while values between 0.0077 and 0.0434 with one estimate at 0.2113 were obtained for the Nantucket populations.

While the above studies used nuclear makers to assess levels of gene flow in carrot. Mandel et al. (2012), studied the mitochondrial gene *Atp9* in 24 populations collected from the Eastern USA. This study found a surprisingly low level of mitochondrial population structure in wild carrot populations. In fact, although the observed F_{ST} value indicated population structuring ($F_{ST} = 0.34$), it was quite low when compared to estimates derived from haploid, maternally inherited markers in other angiosperm species (mean for 124 angiosperm species: $F_{ST} = 0.637$; Petit et al. 2005). Note that the generally high levels of organellar F_{ST} in angiosperms (compared to nuclear estimates) are due to reduced effective population size (and thus more effective genetic drift) due to haploidy and uniparental inheritance (Blanchard and Lynch 2000), as well as generally lower dispersal of seeds versus pollen. This finding of lower than expected organellar F_{ST} suggests wild carrot seeds may move much more efficiently than those of other angiosperm species on average and/or that organellar DNA may occasionally be transmitted via pollen (see Sect. 4.3.3). A lower than expected geographic structuring of mitochondrial diversity was also reported in a study wild populations of carrot from France, Greece, the Mediterranean Basin, and Asia. Ronfort et al. (1995) found on average that 4.4 mitochondrial haplotypes were present per population and that populations tended to share haplotypes indicating a moderate amount of gene flow among them.

Taken together, these results indicate that gene flow is occurring among wild carrot populations. However, the data suggests that the majority of pollen dispersal occurs at fairly short distances (Rong et al. 2010). Dispersal by seeds can occur over long distances by attaching themselves to the fur of animals (Manzano and Malo 2006). Genetic differentiation among carrot populations tends to be low, although intermediate levels have also been detected (Mandel et al. 2016; Reiker et al. 2015; Rong et al. 2013; Shim and Jørgensen 2000). Low genetic differentiation suggests high gene flow although rates of migration measured using the program BayesAss among wild carrot populations tended to be low (Rong et al. 2013; Mandel et al. 2016). Contemporary measures of dispersal were larger than historical gene flow measures (Rong et al. 2010, 2013). Genetic differentiation among wild carrot populations may increase as a result of their proximity to cultivated carrot fields (Mandel et al. 2016). More studies of gene flow among wild carrot populations are needed to generalize the patterns of gene flow occurring in wild carrot populations.

4.3 Consequences of Gene Flow

4.3.1 Gene Flow Among Crop Fields and Between Crop and Wild

Gene flow among crop fields has implications for maintaining cultivar purity. The flow of genetic material between fields of cultivated carrot for root production is less of a concern. However, for the production of carrot seed, stricter guidelines are generally followed to prevent unwanted gene flow. In the USA, commercial growers typically plant orange cultivar varieties with different root shapes a minimum of 3 km apart and fields with different root colors are kept a minimum of 5 km apart (Grzebelus et al. 2011).

Gene flow from wild to cultivated carrot can also negatively impact cultivar purity. Grzebelus et al. (2011) describe that carrot foundation seed is kept at minimum of 1 km distance to wild carrot sites to maintain purity and quality of the seed. In general, carrot seed production in the USA occurs in areas where wild carrot is uncommon; however, at least one carrot seed production site on the Olympic Peninsula struggles with potential contaminants from wild carrot. In fact, there is a campaign on the Peninsula by Washington State University and the Clallam County Extension office to prevent wild carrot from "going to seed." In Washington State, it is a noxious weed and is considered to threaten commercial seed production. An active public education campaign (including "Wanted" posters) to remove Queen Anne's Lace in the county was launched (Fig. 4.1). Proximity of wild carrot to seed production areas is also an issue in Europe and can result in the presence of bolters (flowering in their first year) in carrot production areas (Rong et al. 2010).

Gene flow from crop to wild carrot populations has implications for the introduction (gene flow and hybridization) and spread (introgression) of cultivar genes into wild populations. Moreover, carrot can be used as a model for the spread of transgenes or other genetically modified genes into wild populations (see Sect. 4.3.3). Gene flow from the crop fields into wild populations does occur and can be substantial as demonstrated earlier (see Sect. 4.2.1). Gene flow from cultivated to wild carrots produces F1 hybrids. However, the spread of the cultivar genes within and among wild carrot populations, i.e., the formation of first- and later-generation backcrosses and F2, indicating that the genes are introgressing into wild populations, requires either that the F1 hybrids have a selective advantage over wild plants, i.e., have greater seed set or survival relative to the wild plants, else selection is not occurring and the spread must result from neutral processes, a balance between gene flow and genetic drift. The evidence accumulated to date (see Sect. 4.2.1) does not suggest a selective advantage to F1 hybrids (Ghosh 2012; Hauser et al. 2004; Hauser and Shim 2007; Umehara et al. 2005) and one study (Hauser 2002) demonstrated a selective disadvantage to F1 hybrids. From that study, F1 hybrids survived frost less than wild carrots and only slightly better than the cultivars indicating that frost could limit their survival in the wild (Hauser 2002).

The current evidence indicates that selection would not favor the spread of cultivar genes within wild carrot populations. Future studies of wild carrot populations are therefore needed to explain why wild carrot populations closer to cultivated carrots are less genetically differentiated from cultivated carrots relative to wild populations located at a greater physical distance from these fields. In other words, is gene flow between cultivated and wild carrot sufficient to maintain such differences or are the cultivar genes introgressing into wild carrot populations? If cultivar genes are introgressing into wild carrot populations, we expect movement of cultivar genes among wild carrot populations and the presence of later-generation hybrids (i.e., backcrosses) in wild carrot populations. However, as noted earlier, the presence of first-generation backcrosses in wild populations could also result from gene flow of early bolter plants from cultivated to wild populations and therefore be the result of the bidirectional gene flow that occurs between cultivated and wild carrot. More studies are needed to determine how widespread cultivar genes are in wild carrot populations and to understand the mechanisms that contribute to the establishment of cultivar genes in wild carrot populations and to their spread over the wild carrot landscape.

4.3.2 Gene Flow Among Wild Carrot Populations

Gene flow among wild carrot populations has been reported from both wild carrot populations in the USA and across Europe where studies have been conducted (e.g. Mandel et al. 2016; Reiker et al. 2015; Ronfort et al. 1995; Rong et al. 2010, 2013). Gene flow among wild carrot populations will facilitate the spread of cultivar genes over the wild carrot landscape. Gene flow via seeds will facilitate establishment into new areas and may have contributed to the successful establishment of wild carrot across much of the globe (see Chap. 2). By homogenizing the genetic composition of wild carrot populations, gene flow can hinder processes of local adaptation as selection must be strong to counteract the effects of gene flow. Studies have reported local adaptation for life history of carrot populations with generally shorter generation times (becoming more annual-like) in lower latitudes as compared to more northern (Lacey 1988) suggesting some degree of local adaptation. In the Netherlands over a smaller latitudinal scale, de Jong et al. (2016) demonstrated only small differences in life history strategy and other fitness measures across six different populations of wild carrot and suggested that genetic differences among populations were minimal. Wild carrot populations in the Netherlands have a great majority of annuals (de Jong et al. 2016) while in the USA annuals are more common in the southern latitudes and biennial and perennials are more dominant at northern latitudes (Lacey 1988). A greater frequency of annuals could increase population growth rate relative to biennials (Van Etten and Brunet 2017, unpublished data). Therefore, a number of questions remain unanswered with regard to gene flow in wild carrot. We need more estimates of gene flow among wild carrot populations in different areas. The population dynamics of wild carrot populations is an understudied area and research in this area could yield insights into the establishment of weedy species to novel environments especially under a changing climate. The variation in life history strategies observed in wild carrot makes it a great system to examine the relationships between life history strategies and population growth rate and range expansion and to study the role of population dynamics in the spread of cultivar genes in wild carrot populations (see Sect. 4.4).

4.3.3 Implications for Transgene Escape

While genetically engineered (GE) crops provide many agricultural and consumer benefits, there have been growing concerns over the use of GE crops. One concern relates to the potential for the GE crop itself to become weedy and/or invasive (Craig et al. 2008). Another major issue relates to the risk of escape of GE genes from an agricultural setting to the wild by gene flow or introgressive hybridization. Given that crop plants and their sexually compatible wild relatives often overlap in terms of geographic proximity and phenology (Ellstrand 2003), the likelihood of gene escape can be quite high. Assuming that crop-wild gene flow does occur, the chief concern is that a GE gene escape could result in the production of an increasingly weedy or invasive wild plant populations (Burke 2004; Craig et al. 2008; Raybould and Gray 1994).

A number of strategies have been suggested for minimizing the risks associated with GE gene escape. One approach is the insertion of GE genes into an organellar genome (i.e., either the plastid or mitochondrial genome), as these cytoplasmic genomes are typically maternally inherited in angiosperms (Bilang and Potrykus 1998; Birky 2001; Daniell and Edwards 2011; Daniell et al. 1998; Gressel 1999; Grevich and Daniell 2005; Ijaz 2010; Verma and Daniell 2007). The resulting lack of transmission through pollen would presumably reduce the probability of "escape" to related weedy species (Bilang and Potrykus 1998; Daniell et al. 1998; Gressel 1999; Grevich and Daniell 2005; Ijaz 2010; Verma and Daniell 2007), as pollen would be unable to serve as a vector for gene transfer from crop plants to their sexually compatible wild relatives. Cytoplasmic inheritance is, however, far from universal (Corriveau and Coleman 1988; Ellis et al. 2008; McCauley et al. 2007; Reboud and Zeyl 1994; Röhr et al. 1998; Sears 1980; Zhang et al. 2003), and mathematical models have suggested that even low levels of transmission may be sufficient for the establishment and spread of advantageous transgenes in the wild (Haygood et al. 2004). Moreover, the paternal leakage of cytoplasmic inheritance observed in carrots makes this approach less reliable (Mandel et al. 2016).

Paternal leakage of cytoplasmic genomes via transmission through pollen can result in heteroplasmy (having a mixture of different plastid or mitochondrial genomes within the same individual) (reviewed in McCauley 2013; illustrated in Mandel et al. 2016). In wild carrot, substantial levels of heteroplasmy in the mitochondrial and plastid genomes of wild carrot have been reported, potentially caused by some degree of paternal leakage (Mandel et al. 2012, 2016; Mandel and McCauley 2015). Furthermore, Mandel et al. (unpublished data) have shown that heteroplasmy can be inherited from mother to offspring and maintained in the offspring. Thus, if a transgene escapes via pollen into wild populations, it may persist in the wild being maintained in the heteroplasmic state for at least one generation following the leakage event. No transgenic carrots have been released, so the concern in this system is not high; however, carrot serves as a good model for studying heteroplasmy and paternal leakage, and the results have implications for other crop systems where organellar placement of transgenes has been proposed.

An alternative strategy with similar outcomes would be to insert the GE gene(s) in area(s) of the carrot genome shown not to introgress into wild carrot populations. More research is needed in this area, and study of introgression using genomic approaches should help identify areas of the carrot genome that are not prone to introgression. With gene editing methods, these regions would be helpful to reduce unwanted introduction of these edited genes into wild populations if the edited genes were located on a "non-introgressive" section of a chromosome.

4.4 Future Approaches and the Need to Incorporate Population Dynamics into Studies of Introgression

The study of gene escape and spread is complicated: many ecological, demographic, and population genetic processes affect the spread and establishment of genes within and among populations including dispersal rates, life history traits, population growth rates, fitness, and the environment. With the growing concern that escaped GE genes could have significant detrimental effects on natural communities (Chapman and Burke 2006; Ellstrand et al. 2013; Rieger et al. 2002), a major goal in biotechnology risk assessment is to develop methods that allow the prediction of the fate of an escaped gene (both the rate of spread and the potential for establishment). A critical gap in this area of study and therefore in our understanding of the introgression process is the interplay between population genetics and population dynamics in leading to the successful spread of an escaped gene.

Population dynamics can provide data on traits or characteristics most influential to wild carrot population growth and provide information on which life stages to target to prevent unwanted spread of wild carrot. For example, using a stage structure model for a biennial life cycle with a reproductive and a non-reproductive stage, Van Etten and Brunet (2017) obtained increasing population growth (i.e., lambda > 1.0) for carrot populations with a growth rate of 1.9 when germination was low and 6.1 with high germination. The model was parameterized using values for reproduction, germination rate, over-winter survival and flowering rate, obtained from wild carrot populations in Wisconsin. The average seed production per plant was 4402 ± 484 seeds per plant (mean ± s.e.), germination rate of 0.027 (low) or 0.309 (high), overwintering survival of 0.409 and a flowering rate of 0.063. The model demonstrated that reproduction from a single plant could increase the population by 382 individuals within three years. Sensitivity analyses highlighted the transition from non-reproductive to reproductive, which included overwinter survival and flowering rate, as having the greatest impact on population growth. The proportion of non-reproductives that remained non-reproductive had the least impact on population growth.

This population dynamics approach can also be used to examine whether annuals or biennials are more likely to persist in different types of environment, environments that affect reproduction and survival. Using matrix modeling, Van Etten and Brunet (unpublished data) showed that annuals will be more common in favorable environments and remain at low frequency in harsher environments which favor biennials. Interestingly, in wild carrot populations, there is a strong environmental component to life history strategies with more annuals observed in richer environments and more biennials in environments with poor resources (Lacey 1986; Verkaar and Schenkeveld 1984). The matrix modeling approach can also help examine whether and how hybrids may influence population growth and under what conditions they can be maintained. While these questions examine population dynamics, besides the population growing, a GE or cultivar gene introduced into a wild carrot population via gene flow (F1) has a trajectory and, depending on its fitness relative to wild carrots, it can increase or decrease in frequency in the population or remain neutral (exhibits no fitness advantage or disadvantage). This trajectory will occur irrespective of whether a population is increasing or decreasing or remaining stable. Population genetics approaches have typically been used to study the spread of GE or cultivar genes in wild populations; however, combining population dynamics with population genetics, although not an easy task, would provide a clearer picture of the fate of GE genes introduced into wild populations. We therefore recommend that efforts be placed into this area of study because predicting the fate of introduced genes and understanding the factors that affect their spread will help design management strategies that best contain them and limit their spread.

4.5 Conclusions

In many carrot-producing regions throughout the world, wild carrot populations can be found growing in close proximity to cultivated carrot fields, and there are also reports of wild carrot growing within cultivated carrot fields. Cultivated and wild carrot are fully inter-fertile, often overlap in flowering time, and hybrids between cultivated and wild carrot may sometimes have high fertility and viability. Gene flow between cultivated and wild carrot is bidirectional and this bidirectionality may even contribute to the maintenance of first-generation backcross individuals in wild carrot populations. While gene flow between crops and wild carrots appears extensive, more work is needed to generalize about gene flow among wild carrot populations. The greater level of genetic differentiation among wild carrot populations at different geographic distances from cultivated carrots could simply reflect high gene flow from crop to wild carrot and not be an indicator of limited gene flow among wild carrot populations. The spread of wild carrots over the landscape suggests a good ability to disperse genes, and seed dispersal at least has been documented to happen over long distances.

It is not clear whether cultivar genes have introgressed into wild carrot populations. Gene flow does occur and F1 hybrids are formed and some first-generation backcrosses may occur but their spread beyond the populations closest to cultivar fields has not to date been demonstrated. Although wild populations in proximity to cultivated carrots are more similar genetically to cultivated carrots relative to wild populations further away, this could simply result from high gene flow between crop and wild carrot. In fact, there is little evidence that the F1 hybrid individuals have a selective advantage over wild carrot and they may actually be at a disadvantage. Without a selective advantage to the F1 hybrid, selection will not help cultivar genes increase in frequency within wild carrot populations and the frequency of cultivar genes would depend on the balance between gene flow and genetic drift. Because wild carrot populations tend to be large, at least in the USA, the effect of genetic drift is expected to be small and gene flow would dominate the process.

New methodologies such as gene editing increase the chances that modified genes will be released in carrot and in many other crops in the future. To prepare for this eventuality and to prevent unwanted gene escape and gene spread into wild populations of close relatives, it is important to have a good understanding of the process. Knowledge of gene flow from crop to wild and among wild populations are very important parameters in this process. An understanding of the impact of population dynamics and population genetics on the spread and introgression of cultivars or modified genes will help us better understand and prevent the potential consequences of such actions. Carrot is a good model system to study these processes to help develop and test models of the spread of modified genes into wild populations.

The wide variation in life history characteristics in carrots will influence the rate of growth of populations and can influence range expansion. These characteristics will influence gene flow among wild carrot populations, and this is an area of study where more work is needed to understand the influence of life history traits on gene flow. Moreover, the life history characteristics will affect how quickly cultivar genes can spread within wild carrot populations. There is also a need for combining population genetics and population dynamics approaches in the study of the spread of cultivar genes and of any GE genes in carrot and in other crop/wild systems. Carrot can serve as a great model to develop such approaches as the variation in life history strategies can facilitate the testing of different hypotheses for the mechanism of spread of cultivar and/or GE genes in wild populations.

References

Ahmad M, Aslam M (2002) Pollinators visiting carrot (*Daucus carota* L.) seed crop. J Res (Sci) 13:31–35

Alessandro MS, Galmarini CR (2007) Inheritance of vernalization requirement in carrot. J Am Soc Hort Sci 132:525–529

Alessandro MS, Galmarini CR, Iorizzo M, Simon PW (2013) Molecular mapping of vernalization requirement and fertility restoration genes in carrot. Theor Appl Genet 126:415–423

Baranski R, Maksylewicz-Kaul A, Nothnagel T, Cavagnaro PF, Simon PW, Grzebelus D (2012) Genetic diversity of carrot (*Daucus carota* L.) cultivars revealed by analysis of SSR loci. Genet Resour Crop Evol 59(2):163–170

Beerli P (2006) Comparison of Bayesian and maximum-likelihood inference of population genetic parameters. Bioinformatics 22:341–345

Beerli P, Felsenstein J (1999) Maximum likelihood estimation of migration rates and effective population numbers in two populations. Genetics 152:763–773

Bilang R, Potrykus I (1998) Containing excitement over transplastomic plants. Nat Biotechnol 16:333–334

Birky CW Jr (2001) The inheritance of genes in mitochondria and chloroplasts: laws, mechanisms, and models. Annu Rev Genet 35:125–148

Blanchard JL, Lynch M (2000) Organellar genes—why do they end up in the nucleus? Trends Genet 16:315–320

Bohart GE, Nye WP (1960) Insect pollinators of carrots in Utah. Utah Aer Res Bull 419:16p

Bradeen JM, Bach IC, Briard M, Le Clerc V, Grzebelus D, Senalik DA, Simon PW (2002) Molecular diversity analysis of cultivated carrot (*Daucus carota* L.) and wild *Daucus* populations reveals a genetically nonstructured composition. J Am Soc Hort Sci 127 (3):383–391

Brunet J, Larson-Rabin Z, Stewart CM (2012) The distribution of genetic diversity within and among populations of the Rocky Mountain columbine: the impact of gene flow, pollinators and mating system. Int J Plant Sci 173:484–494

Burczyk J, Adams WT, Moran GF, Griffin AR (2002) Complex patterns of mating revealed in a *Eucalyptus regnans* seed orchard using allozyme markers and the neighbourhood model. Mol Ecol 11(11):2379–2391

Burke JM (2004) When good plants go bad.... Evolution 58:1637–1638

Cavagnaro PF, Chung SM, Manin S, Yildiz M, Ali A, Alessandro MS, Iorizzo M, Senalik DA, Simon PW (2011) Microsatellite isolation and marker development in carrot—genomic distribution, linkage mapping, genetic diversity analysis and marker transferability across Apiaceae. BMC Genom 12 (1):386

Chapman MA, Burke JM (2006) Letting the gene out of the bottle: the population genetics of GM crops. New Phytol 170:429–443

Corriveau JL, Coleman AW (1988) Rapid screening method to detect potential biparental inheritance of plastid DNA and results for over 200 angiosperm species. Am J Bot 75:1443–1458

Craig W, Tepfer M, Degrassil G, Ripandelli D (2008) An overview of general features of risk assessments of genetically modified crops. Euphytica 164:853–880

Daniell H, Edwards KJ (2011) Chloroplast biotechnololgy. Plant Biotechnol J 9:525–526

Daniell H, Datta R, Varma S, Gray S (1998) Containment of herbicide resistance through genetic engineering of the chloroplast genome. Nat Biotechnol 16:345–348

Davidson MM, Butler RC, Howlett BG (2010) *Apis mellifera* and *Megachile rotundata*: a comparison of behaviour and seed yield in a hybrid carrot seed crop. New Zeal J Crop Hort Sci 38:113–117

de Jong TC, Grebenstein C, Tamis WLM (2016) Demography and life history of *Daucus carota* L. populations in the Netherlands. Flora 224:154–158

Ellis JR, Bentley KE, McCauley DE (2008) Detection of rare paternal leakage in controlled crosses of the endangered sunflower *Helianthus verticillatus*. Heredity 100:574–580

Ellstrand NC (2003) Current knowledge of gene flow in plants: implications for transgene flow. Philos Trans R Soc Lond B 358:1163–1170

Ellstrand NC, Meirmans P, Rong J, Bartsch D, Ghosh A, De Jong TJ, Haccou P, Lu BR, Snow AA, Neal Stewart C Jr, Strasburg JL (2013) Introgression of crop alleles into wild or weedy populations. Ann Rev Ecol Evol Syst 44:325–345

Ghosh A (2012) Calculating hazard rates of introgression with branching processes. Doctoral thesis, Leiden University, Leiden

Grebenstein C, Kos SP, de Jong TJ, Tamis WLM, de Snoo GR (2013) Morphological markers for the detection of introgression from cultivated into wild carrot (*Daucus carota* L.) reveal dominant domestication traits. Plant Biol 15:531–540

Greene SL, Kesoju SR, Martin RC, Kramer M (2015) Occurrence of transgenic feral alfalfa (*Medicago sativa* subsp. *sativa* L.) in alfalfa seed production areas in the United States. PLoS One. https://doi.org/10.1371/journal.pone.0143296

Gressel J (1999) Tandem constructs: preventing the rise of superweeds. Trends Biotechnol 17:361–366

Grevich JJ, Daniell H (2005) Chloroplast genetic engineering: recent advances and future perspectives. Crit Rev Plant Sci 24:83–107

Gross KL (1981) Predictions of fate from rosette size in four biennial plants species: *Verbascum thapsus*, *Oenothera biennis*, *Daucus carota*, and *Tragoponon dubius*. Oecologia 48:209–213

Gross KL, Werner PA (1982) Colonizing abilities of biennial plant-species in relation to ground cover—implications for their distributions in a successional sere. Ecology 63:921–931

Grzebelus D, Baranski R, Spalik K, Allender C, Simon PW (2011) *Daucus*. In: Wild crop relatives: genomic and breeding resources. Springer, Berlin, Heidelberg, pp 91–113

Grzebelus D, Iorizzo M, Senalik D, Ellison S, Cavagnaro P, Macko-Podgorni A, Heller-Uszynska K, Kilian A, Nothnagel T, Allender C, Simon PW (2014) Diversity, genetic mapping, and signatures of domestication in the carrot (*Daucus carota* L.) genome, as revealed by Diversity Arrays Technology (DArT) markers. Mol Breeding 33(3):625–637

Harrison P, Dale H (1966) The effect of grazing and clipping on the control of wild carrot. Weeds 14:285–288

Hauser TP (2002) Frost sensitivity of hybrids between wild and cultivated carrots. Cons Genet 3:75–78

Hauser TP, Bjørn GK (2001) Hybrids between wild and cultivated carrots in Danish carrot fields. Genet Resour Crop Evol 48:499–506

Hauser TP, Shim SI (2007) Survival and flowering of hybrids between cultivated and wild carrots (*Daucus carota*) in Danish grasslands. Environ Biosaf Res 237–247

Hauser TP, Bjorn GK, Magnussen L, Shim SI (2004) Hybrids between cultivated and wild carrot: a life history. In: den Nijs HCM, Bartsch D, Sweet J (eds) Introgression from genetically modified plants into wild relatives. CABI Publishing, Wallingford, UK

Haygood R, Ives AR, Andow DA (2004) Population genetics of transgene containment. Ecol Lett 7:213–220

Holt BR (1972) Effect of arrival time on recruitment, mortality, and reproduction in successional plant populations. Ecology 53(4):668–673

Howlett BG, Lankin-Vega GO, Pattermore DE (2015) Native and introduced bee abundances on carrot seed crops in New Zealand. New Zeal Plant Prot 68:373–379

Ijaz S (2010) Plant mitochondrial genome: "a sweet and safe home" for transgene. Afr J Biotechnol 9:9196–9199

Iorizzo M, Senalik DA, Ellison SL, Grzebelus D, Cavagnaro PF, Allender C, Brunet J, Spooner DM, Van Deynze A, Simon PW (2013) Genetic structure and domestication of carrot (*Daucus carota* subsp. *sativus*) (Apiaceae). Am J Bot 100:930–938

Iorizzo M, Ellison S, Senalik D, Zeng P, Satapoomin P, Huang J, Bowman M, Iovene M, Sanseverino W, Cavagnaro P, Yildiz M (2016) A high-quality carrot genome assembly provides new insights into carotenoid accumulation and asterid genome evolution. Nat Genet 48(6):657

Koul P, Koul AK, Hamal IA (1989) Reproductive biology of wild and cultivated carrot (*Daucus carota* L.). New Phytol 112(3):437–443

Lacey EP (1981) Seed dispersal in wild carrots. Mich Bot 20:15–20

Lacey EP (1982) Timing of seed dispersal in *Daucus carota*. Oikos 39:83–91

Lacey EP (1984) Seed mortality in *Daucus carota* populations: latitudinal effects. Am J Bot 71:1175–1182

Lacey EP (1986) The genetic and environmental control of reproductive timing in a short-lived monocarpic species *Daucus carota* (Umbelliferae). J Ecol 74:73–86

Lacey EP (1988) Latitudinal variation in reproductive timing of a short-lived monocarp, *Daucus carota* (Apiaceae). Ecology 69:220–232

Lacey EP, Pace R (1983) Effect of parental flowering and dispersal times on offspring fate in *Daucus carota*. Oecologia 60:274–278

Lamborn E, Ollerton J (2000) Experimental assessment of the functional morphology of inflorescences of *Daucus carota* (Apiaceae): testing the 'fly catcher effect'. Funct Ecol 14:445–454

Magnussen LS, Hauser TP (2007) Hybrids between cultivated and wild carrots in natural populations in Denmark. Heredity 99:185–192

Mandel JR, McCauley DE (2015) Pervasive mitochondrial sequence heteroplasmy in natural populations of wild carrot, *Daucus carota* spp. *carota* L. PLoS One 10(8):e0136303

Mandel JR, McAssey EV, Roland KM, McCauley DE (2012) Mitochondrial gene diversity associated with the atp9 stop codon in natural populations of wild carrot (*Daucus carota* ssp. *carota*). J Hered 103:418–425

Mandel JR, Ramsey AJ, Iorizzo M, Simon PW (2016) Patterns of gene flow between crop and wild carrot, *Daucus carota* (Apiaceae) in the United States. PLoS One 11(9):e0161971

Manzano P, Malo JE (2006) Extreme long-distance seed dispersal via sheep. Front Ecol Environ 4(5):244–248

McCauley DE, Sundby AK, Bailey MF, Welch ME (2007) Inheritance of chloroplast DNA is not strictly maternal in *Silene vulgaris* (Caryophyllaceae): evidence from experimental crosses and natural populations. Am J Bot 94:1333–1337

McCauley DE (2013) Paternal leakage, heteroplasmy, and the evolution of plant mitochondrial genomes. New Phytol 200(4), 966–977

Petit RJ, Duminil J, Fineschi S, Hampe A, Salvini D, Vendramin GG (2005) Comparative organization of chloroplast, mitochondrial and nuclear diversity in plant populations. Mol Ecol 14:689–701

Pritchard JK, Stephens M, Donnelly P (2000) Inference of population structure using multilocus genotype data. Genetics 155(2):945–959

Raybould AF, Gray AJ (1994) Will hybrids of genetically modified crops invade natural communities? Trends Ecol Evol 9:85–89

Reboud X, Zeyl C (1994) Organelle inheritance in plants. Heredity 72:132–140

Reiker J, Schulz B, Wissemann V, Gemeinholzer B (2015) Does origin always matter? Evaluating the influence of nonlocal seed provenances for ecological restoration purposes in a widespread and outcrossing plant species. Ecol Evol 5(23):5642–5651

Rieger MA, Lamond M, Preston C, Powles SB, Roush RT (2002) Pollen-mediated movement of herbicide resistance between commercial canola fields. Science 296 (5577):2386–2388

Ritland K (1996) Estimators for pairwise relatedness and individual inbreeding coefficients. Genet Res 67:175–185

Ritland K (2002) Extensions of models for the estimation of mating systems using n independent loci. Heredity 88:221–228

Robledo-Arnuncio JJ, Austerlitz F, Smouse PE (2007) POLDISP: a software package for indirect estimation of contemporary pollen dispersal. Mol Ecol Notes 7:763–766

Röhr H, Kues U, Stahl U (1998) Organelle DNA of plants and fungi: inheritance and recombination. Prog Bot 60:39–87

Ronfort J, Saumitou-Laprade P, Cugen J, Couvet D (1995) Mitochondrial DNA diversity and male sterility in natural populations of *Daucus carota* ssp. *carota*. Theor Appl Genet 91:150–159

Rong J, Janson S, Umehara M, Ono M, Vrieling K (2010) Historical and contemporary gene dispersal in wild carrot (*Daucus carota* spp. *carota*) populations. Ann Bot 106:285–296

Rong J, Xu S, Meirmans PG, Vrieling K (2013) Dissimilarity of contemporary and historical gene flow in wild carrot (*Daucus carota*) metapopulation under contrasting levels of human disturbance: implications for risk assessment and management of transgene introgression. Ann Bot 112:1361–1370

Sears BB (1980) Elimination of plastids during spermatogenesis and fertilization in the plant kingdom. Plasmid 4:233–255

Shim SI, Jørgensen RB (2000) Genetic structure in cultivated and wild carrots (*Daucus carota* L.) revealed by AFLP analysis. Theor Appl Genet 101 (1–2):227–233

Small E (1984) Hybridization in the domesticated-weed-wild complex. In: Grant WF (ed) Plant biosystematics. Academic Press, Toronto, Ontario, Canada, pp 195–210

St. Pierre MS, Bayer RJ (1991) The impact of domestication on the genetic variability in the orange carrot, cultivated *Daucus carota* ssp. *sativus* and the genetic homogeneity of various cultivars. Theor Appl Genet 82(2):249–253

Umehara M, Eguchi I, Kaneko D, Ono M, Kamada H (2005) Evaluation of gene flow and its environmental effects in the field. Plant Biotechnol J 22:497–504

USDA, National Agricultural Statistics Service 2017. United States Government Printing Office, Washington

Van Etten ML, Brunet J (2017) Using population matrix models to reduce the spread of wild carrot. Acta Hortic 1153:273–278

Vekemans X, Hardy OJ (2004) New insights from fine-scale spatial genetic structure analyses in plant populations. Mol Ecol 13:921–935

Verkaar HJ, Schenkeveld AJ (1984) On the ecology of short-lived forbs in chalk grasslands—life-history characteristics. New Phytol 98:659–672

Verma D, Daniell H (2007) Update on plastid transformation vectors chloroplast vector systems for biotechnology applications. Plant Physiol 145:1129–1143

Westmoreland D, Muntan C (1996) The influence of dark central florets on insect attraction and fruit production in Queen Anne's Lace (*Daucus carota* L.). Am Midl Nat 1:122–129

Wijnheijmer EHM, Brandenburg WA, Terborg SJ (1989) Interactions between wild and cultivated carrots (*Daucus carota* L.) in the Netherlands. Euphytica 40:147–154

Wilson GA, Rannala G (2003) Bayesian inference of recent migration rates using multilocus genotypes. Genetics 163:1177–1191

Wright S (1951) The genetical structure of populations. Ann Eugen 15:323–354

Zhang Q, Liu Y, Sodmergen (2003) Examination of the cytoplasmic DNA in male reproductive cells to determine the potential for cytoplasmic inheritance in 295 angiosperm species. Plant Cell Physiol 44:941–951

Carrot Domestication

5

Shelby Ellison

Abstract

The domestication syndrome of carrot (*Daucus carota* subsp. *sativus*) includes increased carotenoid, anthocyanin, and sugar content, loss of lateral root branching, biennial growth habit, and increased size and variation of root shape. Recent advances in high-throughput sequencing and computational techniques have facilitated new ways to study the genetic and genomic changes that accompany plant domestication. While most genetic studies now support a central Asian center of domestication for carrot much remains unknown regarding the genetic mechanisms that contribute to phenotypic changes associated with domestication. Most research to study the genetics of plant domestication uses a top-down approach, which begins with a phenotype of interest and then identifies causative genomic regions via genetic analyses such as quantitative trait locus (QTL) and linkage disequilibrium (LD) mapping. An alternative approach is to start by identifying genes or genomic regions with signatures of selection and then make use of genetic tools to identify the phenotypes to which these genes contribute, also referred to as a bottom-up approach. In this chapter, we present a thorough review of genetic and genomic studies that have used both top-down and bottom-up approaches to study the domestication syndrome of carrot.

5.1 Domestication Introduction and Overview

Darwin (1868) was first to describe how human selection altered plants to meet human food, fiber, shelter, medicinal, and aesthetic needs (Gepts 2004). The process of domestication follows a similar path in most species where the plant is first cultivated and then conscious and unconscious selection occurs to modify plant characteristics to meet human needs, typically resulting in a plant that has lost its ability to survive without human intervention (Harlan 1992). Early studies of where plant domestication first occurred were dominated by the centers of origin concept. This hypothesis, initially proposed by Candolle (1884) and later refined by Vavilov (1926), posits that domestication occurred in a few discrete geographies or "centers". Recent genomic and archeological data suggest that the concept of discrete centers of origin or diversity may oversimplify the actual histories of cultivated species. In many cases,

S. Ellison (✉)
USDA-Agricultural Research Service, Vegetable Crops Research Unit, 1575 Linden Dr., 53706 Madison, WI, USA
e-mail: Shelby.Ellison@ars.usda.gov

P. Simon et al. (eds.), *The Carrot Genome*, Compendium of Plant Genomes,
https://doi.org/10.1007/978-3-030-03389-7_5

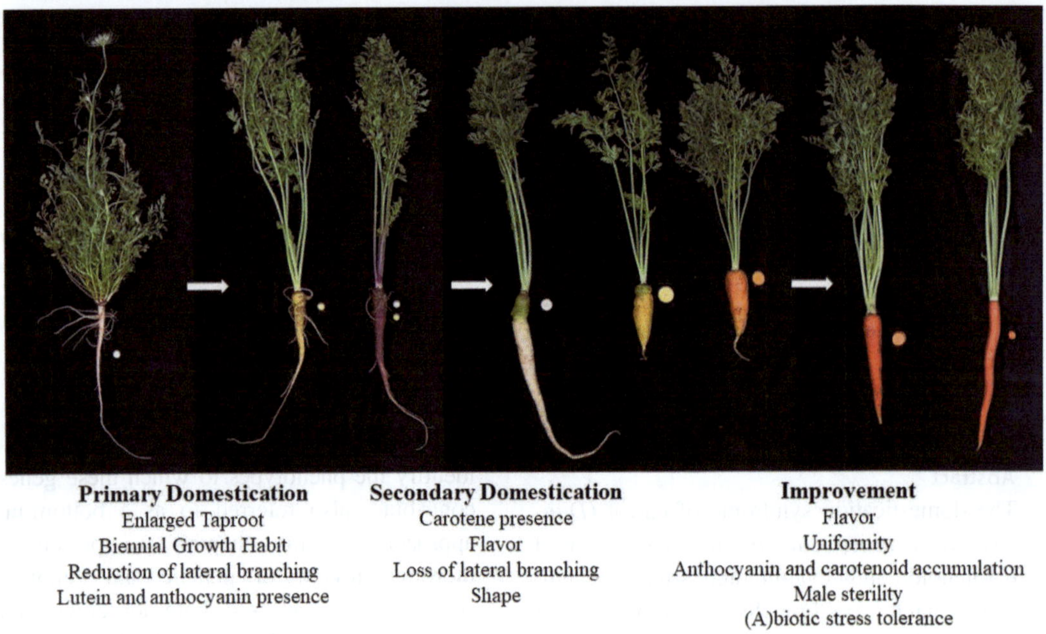

Fig. 5.1 Domestication syndrome in carrot during primary domestication, secondary domestication, and carrot improvement. White arrows (left to right) align with phenotypic shifts between wild and primary domesticates and secondary domesticates and secondary domesticates and improved varieties. Domestication syndrome traits associated with each stage are listed below the white arrows

the evolution of crop plants has been a more complex and continuous process (Harlan 1971; Meyer and Purugganan 2013). Although limited in universality, these concepts remain useful frameworks for finding and preserving valuable variation for plant breeding and determining the extent of genotypic and phenotypic evolution in crops (Ross-Ibarra et al. 2007).

The domestication syndrome is a set of phenotypic characteristics that are common across crop plants and include grain retention by loss of shattering (rice, barley, wheat, and soybean), reduction of lateral branching (maize and sunflower), increase in organ size (tomato, potato, and bean), and flowering-time modification (small grains, sunflower, maize, and soybean) (Harlan 1971; Meyer and Purugganan 2013; Zohary and Hopf 2000). After primary traits have been selected and fixed, the process of domestication often has directed more attention to quality traits such as color, shape, and flavor, and physiological traits contributing to uniformity (Doebley et al. 2006). The domestication syndrome of carrot (*Daucus carota* subsp. *sativus*) includes increased carotenoid, anthocyanin, sugar content, loss of lateral root branching, biennial growth habit, and increased size and variation of root shape (Fig. 5.1). After domestication, carrot improvement traits have included better flavor, nutrition, uniformity, (a)biotic stress tolerance, and male sterility for hybrid cultivar development (Fig. 5.1).

5.2 Wild Carrot Distribution

Wild carrot (*D. carota* subsp. *carota*), also known as Queen Anne's lace, is native to temperate regions of Europe and Western Asia, and has been introduced into America, New Zealand, Australia, and Japan (Bradeen et al. 2002; Iorizzo et al. 2013; Rong et al. 2010). It is speculated that the seed was the first part of the carrot plant used by humans, as observed by the presence of carrot seed at prehistoric human habitations in Switzerland and Southern Germany,

4000–5000 years ago (Neuweiler 1931). Wild carrot seed was likely used medicinally or as a spice (Andrews 1949; D. Brothwell and P. Brothwell 1969). The Romans used carrot seed as an aphrodisiac and to render the body safe from poison (Stolarczyk and Janick 2011). In fact, Roman-made pills containing carrot seed were recovered from a shipwreck that occurred approximately 130 BCE (Fleischer et al. 2010).

5.3 Carrot Domestication and Dispersal

The first evidence of carrot used as a storage root crop is in the Iranian Plateau (Afghanistan, Pakistan, and Iran) and the Persian Empire (modern day Turkey) in the tenth century AD (D. Brothwell and P. Brothwell 1969; Laufer 1919). The Iranian Plateau was described as the primary center of greatest carrot diversity (Heywood 1983; Mackevic 1929; Vavilov 1951) with Turkey being proposed as a secondary center of diversity (Banga 1963a, b; Clement-Mullet 1866; Vavilov 1951). Several recent studies support a Central Asian center of domestication by showing that domesticated carrots from Central Asia are more genetically similar to wild samples from the same region, as compared to wild samples from Turkey (Arbizu et al. 2016; Ellison et al. 2018; Iorizzo et al. 2013; Rong et al. 2014). Wild carrot roots lack carotenoid and anthocyanin pigments and are therefore white. The first descriptions of domesticated carrot roots included purple and yellow types and therefore these traits were likely some of the first human selections in carrot. Purple and yellow carrots spread west to Syria, North Africa, the Mediterranean region, and Southern Europe during the eleventh to fourteenth centuries (Fig. 5.2) (Simon 2000; Smartt and Simmonds 1976). Carrot arrived in Europe in the early middle ages after the Arab conquest and the revival of horticulture under Charlemagne. Documents from Muslim Spain and Christian Europe describe carrots as purple and yellow without any mention of orange (Banga 1957b, 1963b). Carrot spread eastward to China, India, and Japan during the thirteenth to seventeenth centuries (Banga 1957a, b, 1963a, b; Shinohara 1984) (Fig. 5.2). Purple root color was apparently popular in eastern regions, yellow more popular in the west although the red (likely purple) carrots in twelfth century Spain were described as more "juicy and tasty" than the "more coarse" yellow types (Banga 1957a; Clement-Mullet 1866). The Asiatic carrot was

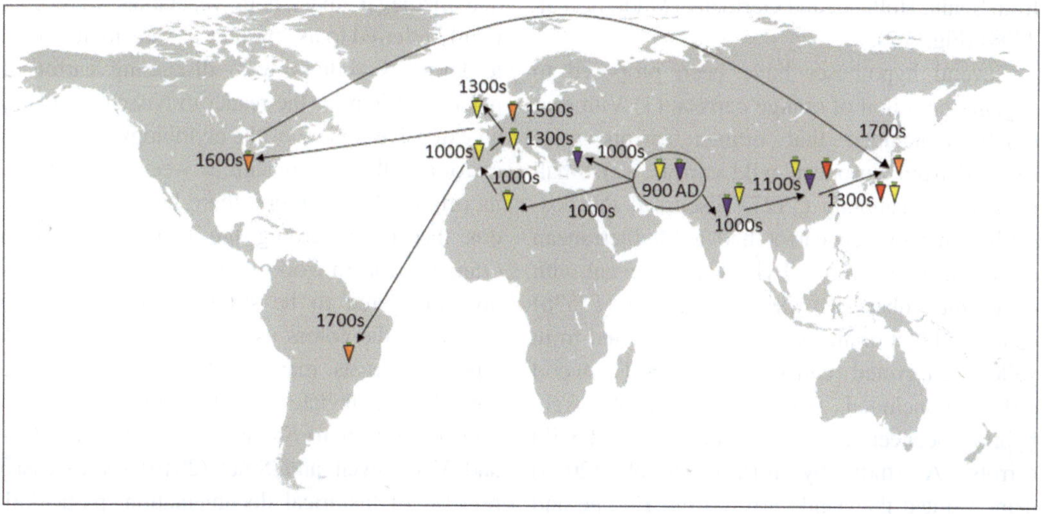

Fig. 5.2 Origin and spread of carrot throughout the world. Dates shown indicate the first known appearance of domesticated carrot within a region. Colors within the cartoon carrots indicate the most prevalent carrot pigment class found in that region at the time of first arrival

developed from the Afghan type and a true red type appeared in China and India around the 1700s (Laufer 1919).

In Europe, genetic improvement led to a wide variety of cultivars. White and orange-colored carrots were first described in Western Europe in the early 1600s (Banga 1963a, b). It is unknown why carrot breeders shifted their preference to orange types, but this preference has had a significant effect in providing a rich source of vitamin A, from α- and β-carotene, to carrot consumers ever since (Simon 2000). The modern orange carrot was stabilized by Dutch growers in the sixteenth and seventeenth centuries, supported from variety names and contemporary works of art (Banga 1957b; Stolarczyk and Janick 2011) (Fig. 5.2). Soon after orange carrots became popular, the first named carrot cultivars came to be described in terms of shape, size, color, and flavor, and the first commercially sold carrot seed became available (Banga 1957b; Simon 2000; Simon et al. 2008). Orange carrots first arrived in North America during the early seventeenth century (Rubatzky et al. 1999) (Fig. 5.2). After the eighteenth century, with the discovery of gold, there was a strong and systematic immigration from the Açores islands to the South of Brazil. Immigrants brought along many varieties of vegetables including white, yellow, purple, and red carrots from Spain, Holland, and Germany (Madeira et al. 2008) (Fig. 5.2).

Several hypotheses have been proposed to explain the origin of orange carrots: (1) Vilmorin (1859) concluded that orange carrots were selected from European wild carrots; (2) Small (1978) and Thellung (1927) discussed the possibility that orange carrot had a Mediterranean origin, resulting from a hybridization event with *D. carota* subsp. *maximus* (3) Banga (1957b) concluded that orange carrots were selected from yellow cultivated carrots; and (4) Heywood (1983) concluded that orange carrots were hybrids between European cultivated and wild carrots. A study by Iorizzo et al. (2013) demonstrated that wild carrots from Europe and samples of *D. maximus*, grouped into two separate clades that are phylogenetically distinct from all cultivated carrot, contrary to the hypotheses of Vilmorin (1859), Thellung (1927), Small (1978), and Heywood (1983). Additionally, Iorizzo et al. (2013) found orange carrots formed a sister clade with all other cultivated carrots (yellow, red, and purple) supporting the idea that orange carrot was selected from cultivated carrot. Their work provides support for Banga's hypothesis that orange root color was selected out of yellow, domesticated carrots (Banga 1957b). In fact, there now appears to be three genetic loci (Y, Y_2, Or) that must be fixed for the "domestication allele" to maximize carotenoid accumulation in carrot (Ellison et al. 2017, 2018; Iorizzo et al. 2016).

5.4 Historic Evidence of Carrot Domestication

The wide distribution of wild carrot, the absence of carrot remains in archeological excavations, and lack of historical documentation make it challenging to determine precisely where and when carrot domestication was initiated. A particular challenge is that of carrot and parsnip nomenclature. Carrot and parsnip have often been confused in historical references and in many cases were discussed interchangeably. In classical and medieval texts, both vegetables were commonly referred to as "pastinaca" making it difficult to know if authors were discussing carrots or parsnips. We refer the reader to Nissan (2014) for an extensive review of the etymology of the word carrot and the relationship between the word and its origins. Furthermore, there have been numerous theories regarding the first occurrence of orange carrots in works of art. Art works alone are not considered to be strong evidence for crop origins as the colors used are not always true to type and artists may use "artistic freedom" to embellish or deviate from the subject matter. We refer the reader to Stolarczyk and Janick (2011) and Vergauwen and Smet (2016) for extensive reviews of historical documentation, particularly artwork, as related to the origin of various pigmentation classes of carrot.

5.5 Genetics and Genomics of Carrot Domestication

Recent advances in high-throughput sequencing and computational techniques have facilitated new ways to study the genomic changes that accompany plant domestication. There is substantial interest in discovering the genes and genetic mechanisms that contribute to phenotypic changes associated with domestication, because their identification may facilitate trait manipulation during breeding. Most research to study the genetics of plant domestication uses a top-down approach, which begins with a phenotype of interest and then identifies causative genomic regions via genetic analyses such as quantitative trait locus (QTL) and linkage disequilibrium (LD) mapping. An alternative approach is to start by identifying genes or genomic regions with signatures of selection and then make use of genetic tools to identify the phenotypes to which these genes contribute, also referred to as a bottom-up approach (Ross-Ibarra et al. 2007).

5.6 Population Structure and Genetic Diversity

Many studies have analyzed population structure and genetic relatedness in carrot. Population structure can cause spurious correlations between the genetic background and traits of interest in association studies and therefore must be accounted for by using a mixed model approach such as proposed by Zhang et al. (2010). Furthermore, population structure and genetic relatedness can shed light on where domestication may have occurred and if gene flow is continuous between wild and domesticated populations. Finally, understanding the genetic diversity within breeding resources is important for developing carrot varieties with new beneficial alleles.

Strong population structure is commonly observed between wild and domesticated carrots (Bradeen et al. 2002; Rong et al. 2014; Shim and Jorgensen 2000) and between eastern (Central and Eastern Asia) and western (American and European) geographies (Baranski et al. 2012; Clotault et al. 2010; Ellison et al. 2018; Grzebelus et al. 2014; Iorizzo et al. 2013, 2016; Maksylewicz and Baranski 2013; Soufflet-Freslon et al. 2013). However, there is evidence of continuous gene flow where populations overlap geographically, such as in Europe and the USA where wild accessions are present in areas where domesticated carrot is grown. There is significant overlap in structure between wild and domesticated samples from the eastern group. This may be attributed to either recent admixture or to domesticated carrots sharing many of the same alleles as wild carrots from the region.

Further geographic substructure has been observed by Arbizu et al. (2016) including the Balkan Peninsula and the Middle East, North Africa exclusive of Morocco, and the Iberian Peninsula and Morocco with the two latter groups confirmed by Ellison et al. (2018). Interestingly, domesticated carrot germplasm in the USA, representing many market types, formed an unstructured population with only some evidence of structure within the hybrid imperator market class (Ellison et al. 2018; Luby et al. 2016). Ma et al. (2016) used 119 carrot accessions to investigate the relationship between Chinese carrots and western orange varieties. Their results indicated that western orange samples were clearly separated from Chinese carrots. They concluded that Chinese orange carrots were derived from Chinese red carrots according to the mixed distribution of red and orange accessions in the observed phylogeny, suggesting that Chinese orange carrots may have undergone a specific, independent process different from that of western orange.

Although a reduction of allelic diversity caused by a genetic bottleneck is a hallmark of domestication, cultivated carrot does not appear to have gone through a severe bottleneck. As compared to domesticated western carrot, there is slightly higher genetic diversity in wild and eastern germplasm with advanced breeding materials from the west containing the least amount of observed diversity. Barański et al. (2012) assessed 30 SSRs in a collection of 88 carrot accessions comprised of cultivars and landraces mainly from Asia, Europe, and North America and found genetic diversity of the Asian gene pool was higher than that of the western gene pool. Iorizzo et al. (2013) used 3326 SNPs to genotype 84 geographically well-distributed wild and domesticated carrots samples and observed no reduction of genetic diversity. Rong et al. (2014) used 622 SNPs to genotype 115 domesticated carrots, wild carrots, and other wild *D. carota* subspecies, and found genetic diversity was significantly reduced in western cultivars; however, a high proportion (85%) of genetic diversity found in wild carrot was retained in western cultivars. Consistent with these previous findings, Iorizzo et al. (2016) found nucleotide diversity estimates in wild carrots have a slightly higher level of genetic diversity than domesticated carrots as well as a clear reduction in genetic diversity in inbred breeding lines. Maksylewicz and Barański (2013) studied intra-population variation of 18 cultivated carrot populations of diverse origins using 27 SSRs and found accessions originating from continental Asia and Europe had more allelic variants and higher diversity than those from Japan and USA. Also, allelic richness and variability in landraces was higher than in F_1 hybrids and open-pollinated cultivars. Finally, Ellison et al. (2018) found little reduction in genetic diversity in 520 domesticated carrot compared to 154 wild carrot after analyzing over 30,000 SNPs.

5.7 Linkage Disequilibrium

Linkage disequilibrium (LD) between two loci decays gradually in proportion to the recombination rate and time as measured in numbers of generations. When mutations are under positive selection, the LD surrounding the mutations is maintained because of the hitchhiking effect which produces longer haplotypes at high frequencies within the population. Extended blocks of LD found in domesticated populations as compared to their wild counterparts can inform researchers of potential regions of the genome under selection. Additionally, LD decay rates are important for the design of powerful association studies as they inform necessary marker density and casual mutation discovery. Few studies to date have assessed LD in carrot. In 2010, Clotault et al. found carotenoid biosynthesis genes did not exhibit LD decay (mean $r^2 = 0.635$) within the 700–1000 bp analyzed. Soufflet-Freslon et al. (2013) observed the absence of LD decay across 4234 bp in the *CRTISO* sequence. Similarly large blocks of LD were found around the *Y* and *cult* candidate domestication genes (Iorizzo et al. 2016; Macko-Podgórni et al. 2017). The first estimates of genome-wide LD found very rapid decay in wild carrot and moderate decay in domesticated accessions. Furthermore, decay was uneven across the nine chromosomes and large blocks of LD were found to correlate with observed signatures of selection (Ellison et al. 2018). The observed rapid LD decay in carrot suggests genome-wide association studies (GWAS) should be very useful for identifying candidate genes as long as SNP density and coverage is comprehensive.

5.8 Top-Down Approach: QTL and LD Mapping

5.8.1 Anthocyanins

From a historical viewpoint, the appearance of purple-colored carrot coincided with that of the yellow carrot at the beginning of the domestication from white wild carrots in central Asia, 1100 years ago (Barański et al. 2016). In Southeastern Europe and Asia, purple carrot became an important crop during the early Middle Ages (Simon 2000; Stolarczyk and Janick 2011). Purple carrots accumulate abundant cyanidin-based anthocyanins in taproots. Discovery of candidate genes relating to anthocyanin production will be quite helpful when tracing the origins of carrot domestication.

Several studies have mapped purple pigmentation and anthocyanin content as well as analyzed transcriptional differences between purple and non-purple carrot taproots. Yildiz et al. (2013)

quantified the gene expression of six anthocyanin biosynthetic genes, *phenylalanine ammonia-lyase* (*PAL3*), *chalcone synthase* (*CHS1*), *flavanone 3-hydroxylase* (*F3H*), *dihydroflavonol 4-reductase* (*DFR1*), *leucoanthocyanidin dioxygenase* (*LDOX2*), and *UDP-glucose:flavonoid 3-O-glucosyltransferase* (*UFGT*), in three carrot inbreds with contrasting root color. Transcripts for five of these genes (*CHS1, DFR1, F3H, LDOX2, and PAL3*) accumulated at high levels in solid purple carrots, less in purple–orange carrot, and low or no transcript in orange carrots. In addition, they mapped the *P1* locus that conditions purple root color, to chromosome 3 near the anthocyanin biosynthetic genes, *F3H* and *FLS1*. In 2014, Cavagnaro et al. identified a total of 15 significant QTL, mapped to six chromosomes, for all anthocyanin pigments and purple epidermis pigmentation. Eight of the QTL with the largest phenotypic effects mapped to two regions of chromosome 3. Additionally, a single dominant gene conditioning anthocyanin acylation was identified and mapped. In 2016, Chen et al. cloned the *DcUSAGT1* gene from "Deep Purple" carrot taproots. *UDP-glucose: sinapic acid glucosyltransferase* (*USAGT*) helps stabilize the accumulation of anthocyanins. Expression profiles of *DcUSAGT1* showed high expression levels in the taproots of all three purple carrot cultivars tested but low expression levels in non-purple carrots. In many species, *R2R3-MYB* transcription factors form "MBW" complexes with other proteins and bind to the promoters of target genes to directly activate the transcription of structural genes in the anthocyanin pathway (Baudry et al. 2004). In 2017, Xu et al. observed that the expression pattern of *DcMYB6* was correlated with anthocyanin production. *DcMYB6* transcripts were detected at high levels in three purple carrot cultivars but at much lower levels in six non-purple carrot cultivars. Overexpression of *DcMYB6* in *Arabidopsis* led to enhanced anthocyanin accumulation in both vegetative and reproductive tissues and upregulated transcript levels of all seven tested anthocyanin-related structural genes.

5.8.2 Carotenoids

The presence and accumulation of carotenoids in carrot taproot is the hallmark of carrot domestication. Certainly, it is the most studied domestication trait and provides a clear phenotypic divide between wild and domesticated carrot. Although the historical record has several different accounts of when orange carrots first occurred, the majority of researchers and historians believe orange carrots rose in popularity in Europe hundreds of years after the first yellow and purple cultivars were observed in Central Asia. As more candidate genes relating to carotenoid accumulation are discovered, the origin of pigmentation in carrot will become elucidated.

Initial efforts to understand the phenotypic variation among white, yellow, and orange carrot storage roots identified two major loci, *Y* and Y_2 (Buishand and Gabelman 1979; Laferriere and Gabelman 1968). A digenic segregation pattern was observed in the F_2 when white roots were crossed to orange, with some evidence that a third gene, Y_1, was segregating. Bradeen and Simon (1998) used bulked segregant analysis and found AFLP markers flanking the Y_2 locus at a distance of 3.8 and 15.8 cM. Later, a SCAR marker for Y_2 was developed to facilitate marker-assisted selection for β-carotene (Bradeen and Simon 1998). Just et al. (2007) mapped twenty-two carotenoid biosynthetic pathway genes on a carrot genetic linkage map developed from a cross between orange-rooted and white-rooted carrot. The two major interacting loci *Y* and Y_2 were mapped to chromosomes 5 and 7, respectively, near carotenoid biosynthetic genes *zeaxanthin epoxidase*, *carotene hydroxylase,* and *carotenoid dioxygenase* (Cavagnaro et al. 2011; Just et al. 2009; Santos and Simon 2002). In 2016, Iorizzo et al. identified a candidate gene, *DCAR_032551*, for the *Y* locus on chromosome 5. This gene conditions carotenoid accumulation in carrot taproot and is a homolog of the Arabidopsis *PSEUDO-ETIOLATION IN LIGHT* (*PEL*) protein. *PEL* presumably acts as a repressor of photomorphogenesis. Only carrot varieties with a loss-of-function allele of the *PEL* gene accumulate carotenoids in the root, suggesting that their high pigment contents might result from a derepressed development of carotenoid-accumulating plastids (i.e., chloroplasts in the light but chromoplasts in the dark). Ellison et al. (2017) identified a single large effect QTL on the distal arm of chromosome 7 which overlapped with the previously identified β-carotene accumulation QTL, Y_2. Fine mapping efforts reduced the genomic region of interest to 650 kb including 72 genes. Transcriptome analysis within this fine-mapped region identified 17 differentially expressed genes included transcription factors and genes involved in light signaling and carotenoid flux, including a member of the *Di19* gene family involved in Arabidopsis photomorphogenesis, and a homolog of the *bHLH36* transcription factor involved in maize carotenoid metabolism.

Many carrot carotenoid studies have focused on a candidate gene approach utilizing known carotenoid biosynthetic genes, with particular interest in *phytoene synthase* (*PSY*), the proposed rate limiting enzyme in the carotenoid pathway (Santos et al. 2005). Maass et al. (2009) overexpressed *crtB*, a bacterial *PSY* gene, in white carrots, to increase PSY protein amounts. This resulted in increased carotenoids deposited in crystals, similar to carotenoid amounts and sequestration mechanisms found in Arabidopsis when *AtPSY* is overexpressed in green and non-green cells. Wang et al. (2014) utilized three backcross inbred lines (BC2S4) with different colored roots derived from a cross between an orange inbred line and related wild species to investigate the role of the duplicated *DcPSY* genes in root carotenogenesis. Expression levels of *DcPSY1* and *DcPSY2* were generally positively correlated with carotenoid content during root development. There were higher quantities of *DcPSY1* transcripts in carrot leaves compared with roots suggesting that *DcPSY1* seems to be more important in carotenoid accumulation in photosynthetic tissues. Similarly, Bowman et al. (2014) found increased *phytoene synthase 1* (*PSY1*) and *phytoene synthase 2* (*PSY2*) expression in orange carrot roots compared with yellow and white carrots.

Clotault et al. (2012) analyzed partial sequence from carotenoid biosynthetic pathway genes *IPI, PDS, CRTISO, LCYB, LCYE, CHXE,* and *ZEP* in 46 individuals representing a wide diversity of cultivated carrots. An excess of intermediate frequency polymorphisms, high nucleotide diversity, and/or high differentiation (F_{ST}) was found in cultivated *CRTISO, LCYB1,* and *LCYE* suggesting balancing selection may have targeted genes acting centrally in the carotenoid biosynthetic pathway. Rong et al. (2014) sequenced the root transcriptomes of cultivated and wild carrots and looked for expression patterns that differed radically between them. They found elevated expression of carotenoid-binding-protein genes in cultivars which could be related to the high carotenoid accumulation in roots. In 2014, Arango et al. found overexpression of *CYP97A3* in orange carrots strongly reduced α-carotene and total root carotenoids in the root and correlated with reduced PSY protein levels while *PSY* expression was unchanged. Furthermore, they identified a deficient *CYP97A3* allele containing a frame-shift insertion in orange carrots. Association mapping analysis using a large carrot population revealed a significant association of this polymorphism with both α-carotene content and the α-/β-carotene ratio and explained a large proportion of the observed variation in carrots. Jourdan et al. (2015) developed an unstructured population of 380 samples and genotyped 109 SNPs located in 17 carotenoid biosynthesis genes to test their association with carotenoid contents and color components. Total carotenoids and β-carotene contents were significantly associated with genes *zeaxanthin epoxydase* (*ZEP*), *phytoene desaturase* (*PDS*), and *carotenoid isomerase* (*CRTISO*) while α-carotene was associated with *CRTISO* and *plastid terminal oxidase* (*PTOX*) genes. Ma et al. (2017) looked at six different carrot cultivars to simultaneously analyze carotenoid contents by high-performance liquid chromatography and quantify the expression levels of genes involved in carotenoid biosynthesis of carrot by quantitative PCR. They found that genes involved in xanthophyll formation were expressed at high levels in yellow carrot cultivars. However, these genes were expressed at low levels in orange carrot cultivars.

Most recently, Ellison et al. (2018) used a diverse collection of domesticated varieties and wild carrot accessions to conduct an association analysis for orange pigmentation and revealed a significant genomic region that contained the *Or* gene. In other species, the *Or* gene differentiates non-colored plastids into chromoplasts, which provide the deposition sink for carotenoid accumulation (Lu et al. 2006). Analysis of sequence variation at the carrot *Or* locus revealed a non-synonymous mutation co-segregating with carotenoid content. This mutation was absent in all wild carrot samples and nearly fixed in all orange domesticated samples. The *Or* domestication allele appears to have been selected after the initial domestication of yellow carrots in the east, near the proposed center of domestication in Central Asia.

5.8.3 Flavor

Free sugars (sucrose, glucose, and fructose) are the major reserves in mature carrot roots (Alabran and Mabrouk 1973). Total sugar content is moderately heritable in carrot ($h^2 = 0.40$) and has a large effect on flavor which is highly correlated with sweetness ($R^2 = 0.95$) (Simon 2000). The type of sugar accumulated in carrot roots is conditioned by a single dominant gene, *Rs* (Freeman and Simon 1983). Carrots with the *Rs/-* genotype predominantly accumulate the reducing sugars glucose and fructose, while *rs/rs* carrots accumulate sucrose. The *Rs* allele occurs in nearly all wild carrots with only rare incidence of the *rs* allele (Freeman and Simon 1983). Yau et al. (2003, 2005) found an *rs/rs* inbred line that harbored a naturally occurring 2.5 kb insertion in the first intron of *acid soluble invertase isozyme II*. Co-dominant, PCR-based markers for *acid soluble invertase isozyme II* allowed genotyping of the *Rs* locus in 1-week-old carrot seedlings whereas mature carrot roots were needed to make this evaluation previously. More recently, Liu et al. (2018) surveyed the contents of soluble sugar and sucrose in four carrot cultivars at five different developmental stages. Three *DcSus* genes (*DcSus1*, *DcSus2*, and *DcSus3*), were identified and cloned in carrot. They found that during carrot root development, the soluble sugar content and sucrose content showed increasing trends, while *DcSus* activities had persisting declinations, which may be due to the decreasing expression levels of genes encoding sucrose synthase.

In carrots, terpenes are an important group of secondary metabolites that are important for taste and flavor but are also known to influence bitterness and harshness (Kramer et al. 2012). Indeed harsh flavor is highly correlated ($R^2 = 0.93$) with total volatile terpenoid content (Simon 2000). Keilwagen et al. (2017) used metabolite profiling to identify 31 terpenoid volatile compounds in carrot leaves and roots in a panel of 85 carrot accessions and genotyping-by-sequencing (GBS) was used to provide dense genome-wide marker coverage (>168,000 SNPs). A total of 30 QTL were identified for 15 terpenoid volatiles. Most QTL were detected for the monoterpene compounds ocimene, sabinene, β-pinene, borneol, and bornyl acetate. In total, 27 genomic regions across the nine carrot chromosomes associated with distinct mono- and sesquiterpene substances and terpene synthase candidate genes.

5.8.4 Flowering

Wild carrot is mostly biennial, but both annual and short-lived perennial forms often occur. As a biennial species, carrot plants develop leaves and storage roots during the first year of growth, and flowering is induced after a long vernalization period. In some cases, wild carrot and landraces adapted to warmer climates require less vernalization and can be classified as early flowering or annuals (Alessandro and Galmarini 2007). Eastern carrots have a greater tendency toward early flowering than western carrots, likely due to the somewhat warmer climates over the eastern production range. Once flowering occurs, the xylem quickly becomes lignified before the floral stalk elongates and the taproot becomes woody and inedible. During carrot domestication and improvement, there has been strong selection against premature flowering as it results in a complete loss of the commercial value of the crop. Despite its economic importance only a few studies have looked into the genetic control of flowering in carrot.

In 2013, Alessandro et al. created an F_2 population, derived from the intercross between the annual cultivar "Criolla INTA" and a petaloid male sterile biennial carrot. They evaluated early flowering habit, named Vrn1, which was found to be a dominant trait conditioned by a single gene. Vrn1 mapped to chromosome 2 with flanking markers at 0.70 and 0.46 cM. Ou et al. (2016) used RNA-seq in a wild carrot species sensitive to flower induction by vernalization and photoperiod and an orange cultivar to identify flowering-time genes and use digital gene expression (DGE) analysis to examine their expression levels. Homologs of *COL2*, *CONSTANS-LIKE 5* (*COL5*), *SUPPRESSION OF OVEREXPRESSION OF CONSTANS 1* (*SOC1*), *FLOWERING LOCUS C* (*FLC*), and *GIBBERELLIC ACID INSENSITIVE* (*GAI*) were differentially expressed between the early flowering wild carrots and domesticated carrots. Shen et al. (2018) used a set of backcross inbred lines developed by crossing a wild carrot with an orange cultivar to map days to initial flowering (DIF), main stalk length (MSL), and seed weight per plant (SWP). Two, four, and two QTLs associated with DIF, MSL, and SWP were discovered, respectively, with 14.6–23.8% phenotypic variance. The QTL for DIF mapped to chromosomes 1 and 5.

5.8.5 Root Shape

The ability to form a fleshy storage root, with reduced lateral branching, was undoubtedly one of the first selected domestication traits in carrot. Later, during carrot improvement, a vast array of carrot shapes and sizes become important for classifying market types grown in different regions of the world, many of which are still used today. Until recently, very few carrot root shape studies were conducted likely due to the difficulty of phenotyping root traits. Fortunately, modern advances in automated image analysis have started to help unravel complex traits such as root system architecture.

Macko-Podgórni et al. (2017) identified a candidate domestication syndrome gene, *DcAHLc1*, carrying three non-synonymous single nucleotide polymorphisms and one indel that systematically differentiates wild and cultivated accessions. This gene belongs to the *AT-hook motif nuclear localized* (*AHL*) family of plant regulatory genes which are involved in the regulation of organ development, including root tissue patterning. *AHL* genes work through direct interactions with other AHL family proteins and a range of other proteins that require intercellular protein movement. They speculate that *DcAHLc1* might be involved in the development of the carrot storage root, as the localization of the gene overlapped with one of the QTL for root thickening. Turner et al. (2018) developed an automated analysis platform that extracts size and shape components for carrot shoots and roots. This method reliably measures variation in shoot size and shape, petiole number, petiole length, and petiole width, root length, and biomass. They used the imaging pipeline to phenotype an F_2 mapping population consisting of 316 individuals which segregated for root and shoot morphologies and identified co-localization of quantitative trait loci for shoot and root characteristics on chromosomes 1, 2, and 7, suggesting these traits are controlled by genetic linkage and/or pleiotropy. Machaj et al. (2018) reported the first comparative transcriptome analysis between wild and cultivated carrot roots at multiple developmental stages. Comparisons of expression between cultivated and wild carrot found that transcription factors and genes encoding proteins involved in post-translational modifications were mostly upregulated, while those involved in redox signaling were mostly downregulated. Also, genes encoding proteins regulating cell cycle, involved in cell divisions, development of vascular tissue, water transport, and sugar metabolism were enriched in the upregulated clusters of cultivated carrot.

5.9 Bottom-up Approach: Signatures of Selection

One of the most observed trends during crop domestication is a reduction of genetic diversity caused by a genetic bottleneck (Doebley et al. 2006). Since selected genes experience more severe bottlenecks than unselected ones, the reduction of genetic diversity becomes uneven along chromosomes and creates distinct genetic characteristics that can be used to identify selective sweeps. Within domesticated populations, low genetic diversity (π), Tajima's D, and runs of extended haplotype homozygosity (EHH) can be used to find recent positive selection (Nielsen 2005). When the population data of both wild ancestors and modern domesticated accessions are available, the selective sweeps can be identified by comparing distinct genetic characteristics between two populations. A straightforward method is to scan the genome for regions with significant reduction of genetic diversity ($\pi_{wild}/\pi_{cultivar}$). Population differentiation statistics, such as F_{ST} which measures variation of allele frequency between two populations, can also be used to identify selection (Shi and Lai 2015). Additionally, a cross-population composite likelihood ratio (XP-CLR) approach jointly calculates multiple locus allele frequency differentiation to identify selective sweeps between two groups (Chen et al. 2010). With the newly sequenced carrot genome (Iorizzo et al. 2016) and the cost of genotyping rapidly declining, genome-wide scans for signatures of selection are now possible in carrot. Identified regions can be crossed referenced with genes found using traditional top-down approaches or scanned for potential candidates using the carrot genome.

Grzebelus et al. (2014) identified 27 DArT markers that showed signatures of selection and localized two of these markers to chromosomes 2 and 6. Macko-Podgórni et al. (2014) selected one of the DArT markers showing the strongest evidence for directional selection from the Grzebelus et al. (2014) work and converted it into a co-dominant cleaved amplified polymorphic site (CAPS) marker named *cult* which was used to differentiate wild and domesticated accessions. The *cult* marker was validated on 88 domesticated and wild carrot accessions. Macko-Podgórni et al. (2017) mapped *cult* to the distal portion of the long arm of carrot chromosome 2, where it overlapped with a plant regulatory gene (*DcAHLc1*) involved in the regulation of organ development, including root tissue patterning and confirmed that this gene had been selected, as reflected in both the lower nucleotide diversity in the cultivated gene pool, as compared to the wild, as well as high F_{ST}.

To identify genomic regions associated with domestication events, Iorizzo et al. (2016) analyzed genome-wide F_{ST} between wild and domesticated eastern accessions and found local differentiation signals on chromosomes 2, 5, 6, 7, and 8. Peaks on chromosomes 2, 5, and 7 overlapped with previously mapped domestication QTL for root thickness (*cult*) (Macko-Podgórni et al. 2017) and carotenoid content, (*Y* and Y_2) (Cavagnaro et al. 2011; Just et al. 2009). Ellison et al. (2017) found a drastic decrease in nucleotide diversity in the fine-mapped Y_2 region in orange cultivated accessions.

In the most extensive investigation of genome-wide selective sweeps to date, Ellison et al. (2018) surveyed F_{ST}, nucleotide diversity, and XP-CLR ratios in 500 kb genomic bins across the genomes of 520 cultivated and 154 wild carrot accessions. Twelve genomic regions were significant for all three methods of selective sweep detection. The candidate carotenoid gene, *Or*, was located in one of these 12 genomic locations. Interestingly, two DArT markers that showed signatures of selection in Grzebelus et al. (2014) overlapped with genomic regions on chromosome 2 and 6. Chromosome 2 was previously shown to carry the *Vrn1* trait (Alessandro et al. 2013) which was likely a target to favor biennial growth habit during the course of carrot domestication.

5.10 Concluding Remarks

The study of carrot domestication will continue to be an important area of focus in which the location, timing, and genes under selection will be under examination. There are new resources for domestication studies that were once limited to major crops but are now readily available for all crop species. Historical efforts to collect and preserve wild relatives, landraces, and cultivated varieties have strengthened the world's public genebanks, and the onset of low-cost sequencing and global interest in these genetic collections have initiated a transition from long-term storage facilities to active exploration. Within the next few years, over a thousand carrot wild relatives, landraces, and modern cultivars will be genotyped and available to the public scientific community. Although the focus of these efforts will be to enable plant breeders, these resources will be equally useful for studies of domestication, population genetics, genome evolution, and diversity in carrot.

References

Alabran DM, Mabrouk AF (1973) Carrot flavor, sugars, and free nitrogenous compounds in fresh carrots. J Agric Food Chem 21:205–208

Alessandro MS, Galmarini CR (2007) Inheritance of vernalization requirement in carrot. J Am Soc Hort Sci 132:525–529

Alessandro MS, Galmarini CR, Iorizzo M et al (2013) Molecular mapping of vernalization requirement and fertility restoration genes in carrot. Theor Appl Genet 126:415–423

Andrews AC (1949) The carrot as a food in the classical era. Class Philol 44:182–196

Arango J, Jourdan M, Geoffriau E et al (2014) Carotene hydroxylase activity determines the levels of both α-carotene and total carotenoids in orange carrots. Plant Cell 26:2223–2233

Arbizu CI, Ellison SL, Senalik D et al (2016) Genotyping-by-sequencing provides the discriminating power to investigate the subspecies of *Daucus carota* (Apiaceae). BMC Evol Biol 16:234

Banga O (1957a) Origin of the European cultivated carrot. Euphytica 6:54–63

Banga O (1957b) The development of the original European carrot material. Euphytica 7:64–76

Banga O (1963a) Main types of the western carotene carrot and their origin. WEJ Tjeenk, Willink, Zwolle, The Netherlands

Banga O (1963b) Origin and distribution of the western cultivated carrot. Genet Agrar 17:357–370

Barański R, Maksylewicz-Kaul A, Nothnagel T et al (2012) Genetic diversity of carrot (*Daucus carota* L.) cultivars revealed by analysis of SSR loci. Genet Res Crop Evol 59:163–170

Barański R, Goldman I, Nothnagel T et al (2016) Improving color sources by plant breeding and cultivation. In: Carle, Schweiggert R (ed) Handbook on natural pigments in food and beverages: industrial applications for improving food color. Springer, New York, pp 429–472

Baudry A, Heim MA, Dubreucq B et al (2004) TT2, TT8, and TTG1 synergistically specify the expression of BANYULS and proanthocyanidin biosynthesis in Arabidopsis thaliana. Plant J 39:366–380

Bowman MJ, Willis DK, Simon PW (2014) Transcript abundance of *phytoene synthase 1* and *phytoene synthase 2* is associated with natural variation of storage root carotenoid pigmentation in carrot. J Am Soc Hort Sci 139:63–68

Bradeen JM, Simon PW (1998) Conversion of an AFLP fragment linked to the carrot *Y2* locus to a simple, codominant, PCR-based marker form. Theor Appl Genet 97:960–967

Bradeen JM, Bach IC, Briard M et al (2002) Molecular diversity analysis of cultivated carrot (*Daucus carota* L.) and wild Daucus populations reveals a genetically nonstructured composition. J Am Soc Hort Sci 127:383–391

Brothwell D, Brothwell P (1969) Food in antiquity. Johns Hopkins University Press, Baltimore, pp 111–112

Buishand JG, Gabelman WH (1979) Investigations on the inheritance of colour and carotenoid content in phloem and xylem of carrot roots (*Daucus carota* L.). Euphytica 28:611–632

Cavagnaro PF, Chung SM, Manin S et al (2011) Microsatellite isolation and marker development in carrot—genomic distribution, linkage mapping, genetic diversity analysis and marker transferability across Apiaceae. BMC Genom 12:386

Cavagnaro PF, Iorizzo M Yildiz et al (2014) A gene-derived SNP-based high resolution linkage map of carrot including the location of QTL conditioning root and leaf anthocyanin pigmentation. BMC Genom 15:1118

Chen H, Patterson N, Reich D (2010) Population differentiation as a test for selective sweeps. Genome Res 20:393–402

Chen Y-Y, Xu Z-S, Xiong A-S et al (2016) Identification and characterization of DcUSAGT1, a UDP-glucose: sinapic acid glucosyltransferase from purple carrot taproots. Plos One 11:e154938

Clement-Mullet JJ (1866) Le livre de l'agriculture d'Ibn-al-Awam. A. Franck, Paris

Clotault J, Geoffriau E, Lionneton E et al (2010) Carotenoid biosynthesis genes provide evidence of geographical subdivision and extensive linkage disequilibrium in the carrot. Theor Appl Genet 121:659–672

Clotault J, Peltier D, Soufflet-Freslon V et al (2012) Differential selection on carotenoid biosynthesis genes as a function of gene position in the metabolic pathway: a study on the carrot and dicots. Plos One 7:e38724

Darwin CR (1868) The variation of animals and plants under domestication. Murray, London

De Candolle A (1884) Origin of cultivated plants. Kegan Paul, Trench & Co, London, UK

Doebley JF, Gaut BS, Smith BD (2006) The molecular genetics of crop domestication. Cell 127:1309–1321

Ellison S, Senalik D, Bostan H et al (2017) Fine mapping, transcriptome analysis, and marker development for *Y2*, the gene that conditions β-Carotene accumulation in carrot (*Daucus carota* L.). G3: Genes Genom Genet 8:2665–2675

Ellison S, Luby C, Corak K et al (2018) Carotenoid presence is associated with the Or gene in domesticated carrot. Genet. Accepted from https://doi.org/10.1534/genetics.118.301299

Fleischer RM, Touwaide A, Appetiti E et al (2010) Composition of pharmaceuticals from a 1st century BC/AD Roman shipwreck based on chloroplast DNA sequences. In: Fourth international symposium on biomolecular archaeology, Copenhagen. http://www.isba4.net/ISBA4_FINAL.pdf

Freeman RE, Simon PW (1983) Evidence for simple genetic control of sugar type in carrot (*Daucus carota* L.). J Am Soc Hort Sci 108:50–54

Gepts P (2004) Crop domestication as a long-term selection experiment. In: Janick J (ed) Plant breeding reviews. Wiley, New York, NY, USA, pp 1–44

Grzebelus D, Iorizzo M, Senalik D (2014) Diversity, genetic mapping, and signatures of domestication in the carrot (*Daucus carota* L.) genome, as revealed by diversity arrays technology (DArT) markers. Mol Breeding 33:625–637

Harlan JR (1971) Agricultural origins: centers and noncenters. Science 174:468–474

Harlan JR (1992) Crops and man. Crop Science Society of America, Madison, WI, USA

Heywood VH (1983) Relationships and evolution in the *Daucus carota* complex. Isr J Bot 32:51–65

Iorizzo M, Senalik D, Ellison S et al (2013) Genetic structure and domestication of carrot (*Daucus carota* subsp. *sativus*) (Apiaceae). Am J Bot 100:930–938

Iorizzo M, Ellison S, Senalik D et al (2016) A high-quality carrot genome assembly provides new insights into carotenoid accumulation and asterid genome evolution. Nat Genet 48:657–666

Jourdan M, Gagné S, Dubois-Laurent C et al (2015) Carotenoid content and root color of cultivated carrot: a candidate-gene association study using an original broad Unstructured population. Plos One 10:e016674

Just BJ, Santos C, Boiteux LS et al (2007) Carotenoid biosynthesis structural genes in carrot (*Daucus carota*): isolation, sequence-characterization, single nucleotide polymorphism (SNP) markers and genome mapping. Theor Appl Genet 114:693–704

Just BJ, Santos CF, Yandell BS et al (2009) Major QTL for carrot color are associated with carotenoid biosynthetic genes and interact epistatically in a domesticated x wild carrot cross. Theor Appl Genet 119:1155–1169

Keilwagen J, Lehnert H, Berner T et al (2017) The terpene synthase gene family of carrot (*Daucus carota* L.): identification of QTLs and candidate genes associated with terpenoid volatile compounds. Frontiers Plant Sci 8:1930

Kramer M, Bufler G, Ulrich D et al (2012) Effect of ethylene and 1-methylcyclopropene on bitter compounds in carrots (*Daucus carota* L.). Postharvest Biol Technol 73:28–36

Laferriere L, Gabelman WH (1968) Inheritance of color, total carotenoids, alpha carotene, and beta-carotene in carrots, *Daucus carota* L. Proc Am Soc Hort Sci 93:408–418

Laufer B (1919) Sino-Iranica. Anthropological series, vol 15. Field Museum of Natural History, Publication 201, Chicago, pp 451–454

Liu Y-J, Wang G-L, Ma J (2018) Transcript profiling of sucrose synthase genes involved in sucrose metabolism among four carrot (*Daucus carota* L.) cultivars reveals distinct patterns. BMC Plant Biol 18:8

Lu S, Van Eck J, Zhou X et al (2006) The cauliflower Or gene encodes a DnaJ cysteine-rich domain-containing protein that mediates high levels of -carotene accumulation. Plant Cell 18:3594–3605

Luby CH, Dawson JC, Goldman IL (2016) Assessment and accessibility of phenotypic and genotypic diversity of carrot (*Daucus carota* L. var. *sativus*) cultivars commercially available in the United States. Plos One 11:e0167865

Ma ZG, Kong XP, Liu LJ et al (2016) The unique origin of orange carrot cultivars in China. Euphytica 212:37–49

Ma J, Xu Z, Tan G et al (2017) Distinct transcription profile of genes involved in carotenoid biosynthesis among six different color carrot (*Daucus carota* L.) cultivars. Acta Biochim Biophys Sin 49:817–826

Maass D, Arango J, Wüst F et al (2009) Carotenoid crystal formation in arabidopsis and carrot roots caused by increased phytoene synthase protein levels. PLoS ONE 4:e6373

Machaj G, Bostan H, Macko-Podgórni A et al (2018) Comparative transcriptomics of root development in wild and cultivated carrots. Genes 9:431

Mackevic VI (1929) The carrot of Afghanistan. Bul Appl Bot Genet Plant Breed 20:517562

Macko-Podgórni A, Iorizzo M, Smółka K et al (2014) Conversion of a diversity arrays technology marker differentiating wild and cultivated carrots to a co-dominant cleaved amplified polymorphic site marker. Acta Biochim Pol 61:19–22

Macko-Podgórni A, Machaj G, Stelmach K et al (2017) Characterization of a genomic region under selection in cultivated carrot (*Daucus carota* subsp. *sativus*) reveals a candidate domestication gene. Frontiers Plant Sci 8:12

Madeira NR, Reifschneider FJB, Giordano LB (2008) Contribuição portuguesa à produção e ao consumo de hortaliças no Brasil: uma revisão histórica. Hort Bras 26:428

Maksylewicz A, Baranski R (2013) Intra-population genetic diversity of cultivated carrot (*Daucus carota* L.) assessed by analysis of microsatellite markers. Acta Biochim Pol 60:753–760

Meyer RS, Purugganan MD (2013) Evolution of crop species: genetics of domestication and diversification. Nat Rev Genet 14:840

Neuweiler E (1931) Die Pflanzenreste aus dem spätbronzezeitlichen Pfahlbau "sumpf" bei Zug. Vierteljahrschr Naturf Ges Zurich 76:116–132

Nielsen R (2005) Molecular signatures of natural selection. Annu Rev Genet 39:197

Nissan E (2014) Language, culture, computation computational linguistics and linguistics etymothesis, fallacy, and ontologies: an illustration from phytonymy. Springer, pp 207–364

Ou C-G, Mao J-H, Liu L-J et al (2016) Characterizing genes associated with flowering time in carrot (*Daucus carota* L.) using transcriptome analysis. Plant Biol 19:286–297

Rong J, Janson S, Umehara M et al (2010) Historical and contemporary gene dispersal in wild carrot (*Daucus carota* ssp. *carota*) populations. Ann Bot 106:285–296

Rong J, Lammers Y, Strasburg et al (2014) New insights into domestication of carrot from root transcriptome analyses. BMC Genom 15:895

Ross-Ibarra J, Morrell PL, Gaut BS (2007) Plant domestication, a unique opportunity to identify the genetic basis of adaptation. Proc Natl Acad Sci USA 104:8641–8648

Rubatzky VE, Quiros CF, Simon PW (1999) Carrots and related vegetable umbelliferae. CAB International, Wallingford, pp 2–9

Santos CA, Simon PW (2002) QTL analyses reveal clustered loci for accumulation of major provitamin A carotenes and lycopene in carrot roots. Mol Genet Genom 268:122–129

Santos CAF, Senalik D, Simon PW (2005) Path analysis suggests phytoene accumulation is the key step limiting the carotenoid pathway in white carrot roots. Genet Mol Biol 28:287–293

Shen Q, Ou C, Sun T et al (2018) QTL analysis of date of initial flowering main stalk length and seed weight per plant in carrot. Acta Hortic Sinica 3:571–578

Shi J, Lai J (2015) Patterns of genomic changes with crop domestication and breeding. Curr Opin Plant Biol 24:47–53

Shim S, Jorgensen R (2000) Genetic structure in cultivated and wild carrots (*Daucus carota* L.) revealed by AFLP analysis. Theor Appl Genet 101:227–233

Shinohara S (1984) Introduction and variety development in Japan. In: Vegetable seed production technology of Japan elucidated with respective variety development histories, particulars, vol 1. Shinohara's Authorized Agricultural Consulting Engineer Office 4-7-7, Tokyo, pp 273–282

Simon PW (2000) Domestication, historical development, and modern breeding of carrot. Plant Breed Rev 19:147–190

Simon PW, Freeman RE, Vieira JV et al (2008) Carrot. In: Prohens J, Nuez F (eds) Handbook of plant breeding, vol 2. Springer, New York, NY, pp 327–357

Small E (1978) Numerical taxonomic analysis of *Daucus carota* complex. Can J Bot 56:248–276

Smartt NW, Simmonds JW (1976) Evolution of Crop Plants, 2nd edn. Longman Sci. Technol, Harlow, pp 291–293

Soufflet-Freslon V, Jourdan M, Clotault J et al (2013) functional gene polymorphism to reveal species history: the case of the CRTISO gene in cultivated carrots. PLoS ONE 8(8):e70801

Stolarczyk J, Janick J (2011) Carrot: history and iconography Chron Hortic 51:13

Thellung A (1927) Die Abstammung der Gartenmöhre (*Daucus carota* subsp. *sativus*) und der Gartenrettichs (*Raphanus raphanistrum* subsp. *sativus*). Feddes Repertorium Specierum Novarum Regni Vegetabilis 46:1–7

Turner SD, Ellison SL, Senalik DA et al (2018) An automated, high-throughput image analysis pipeline enables genetic studies of shoot and root morphology in carrot (*Daucus carota* L.). https://doi.org/10.1101/384974

Vavilov NI (1926) Centres of origin of cultivated plants. Institut Botanique Appliqué et d'Amélioration des Plantes, Leningrad, USSR

Vavilov NI (1951) The origin, variation, immunity and breeding of cultivated plants. Chron Bot 13:1–366

Vergauwen D, Smet ID (2016) Down the rabbit hole—carrots, genetics, and art. Trends Plant Sci 21:895–898

Vilmorin M (1859) L'hérédité dans les végétaux. In: Vilmorin M (ed) Notice sur l'amelioration des plantes par la semis. Librairie Agricole, Paris, France, pp 5–29

Wang H, Ou CG, Zhuang FY et al (2014) The dual role of phytoene synthase genes in carotenogenesis in carrot roots and leaves. Mol Breed 34:2065

Xu Z-S, Feng K, Que F et al (2017) A MYB transcription factor, *DcMYB6*, is involved in regulating anthocyanin biosynthesis in purple carrot taproots. Sci Rep 7:45324

Yau Y, Simon PW (2003) A 2.5-kb insert eliminates acid soluble invertase isozyme II transcript in carrot

(*Daucus carota* L.) roots, causing high sucrose accumulation. Plant Mol Biol 53:151–162
Yau YY, Santos K, Simon P (2005) Molecular tagging and selection for sugar type in carrot roots using co-dominant, PCR-based markers. Mol Breed 16:1
Yildiz M, Willis DK, Cavagnaro PF et al (2013) Expression and mapping of anthocyanin biosynthesis genes in carrot. Theor Appl Genet 126:1689–1702
Zhang Z, Ersoz E, Lai C-Q et al (2010) Mixed linear model approach adapted for genome-wide association studies. Nature Genet 42:355–360
Zohary D, Hopf M (2000) Domestication of plants in the old world. Oxford University Press, Oxford, UK

Genetic Resources for Carrot Improvement

6

Charlotte Allender

Abstract

Plant genetic resources offer the essential raw material of genetic diversity for crop improvement. Globally, ex situ carrot germplasm collections are extensive with >13,400 listed by 62 different institutes. The majority of accessions conserved are of cultivated origin, and however, recent interest and recognition of the importance of crop wild relatives have led to an increase in the number of wild *Daucus* accessions conserved in genebanks. Carrot genetic resources can also be conserved in situ and on-farm methods which are particularly applicable to wild and landrace material. The scale of global *Daucus* collections means that the identification and use of core collections and subsets is helpful in order to reflect genepool variation with manageable numbers of samples.

C. Allender (✉)
School of Life Sciences, University of Warwick, Wellesbourne Campus, CV35 9EF Warwick, UK
e-mail: charlotte.allender@warwick.ac.uk

6.1 An Introduction to Genetic Resources for Crop Improvement

In common with other crops, carrot crop improvement and breeding programmes rely on the raw material of genetic diversity for the development and production of new varieties. Similarly, plant and crop scientists seeking to understand the biological underpinnings of phenotypic variation also rely on diverse genotypes to enable them to understand the underpinning mechanisms and genomic regions involved. Access to relevant genetic resources for crop improvement and research in carrot is therefore vital. This chapter briefly describes key collections of genetic resources for carrot as well as practical issues surrounding access and management for potential users.

The twentieth century heralded major developments in agricultural systems which impacted the type of cultivated material being grown on farms, and a threat to crop genepool diversity was recognised. Farmers sought to improve the agronomic and economic performance of their operations through adoption of new technologies and practices and began to operate at larger scales than was previously possible. Many switched from locally developed traditional varieties and landraces, in favour of more broadly adapted varieties which offered better uniformity, higher yields and were more amenable to mechanized cultivation. The widespread adoption of broadly

P. Simon et al. (eds.), *The Carrot Genome*, Compendium of Plant Genomes,
https://doi.org/10.1007/978-3-030-03389-7_6

adapted varieties and the loss of landraces triggered concern over the loss of crop genetic diversity, and efforts were made at a national and international level to safeguard crop genepool diversity through ex situ conservation of germplasm in genebanks. In 2010, there were estimated to be some 7.4 million accessions conserved in 1700 genebanks across the world (FAO 2010). This germplasm serves the purposes of conservation and as a tool for research and plant breeding. Genebanks operate at an international, regional, or national scale depending on their focus. There are no international genebanks besides that of the World Vegetable Center which has a specific focus on vegetable crops, and major collections of carrot germplasm can be found in the national genebanks of several different countries.

The need for a coherent mechanism to govern the exchange of carrot genetic resources is clear when it is bred and cultivated in many different countries. The international legislative framework underpinning the acquisition and use of plant genetic resources for crop improvement has shifted significantly over time. Prior to the Convention on Biological Diversity (CBD), crop genetic resources were regarded as the common heritage of mankind and as such could be freely accessed. The CBD created a legal framework of recognition of national sovereignty over genetic resources, and however, it was recognised that crop genetic resources exhibited a specific set of issues, given that crops have moved in tandem with people across the world and the international nature of plant breeding and agriculture. The issue of timely and legal access to the diverse resources required for plant breeding became potentially problematic. In response, the International Treaty on Plant Genetic Resources for Food and Agriculture (ITPGRFA) was adopted in 2001. This treaty facilitated the exchange, use and equitable sharing of benefits of genetic resources of 64 commonly cultivated crops, including *Daucus*. Access to germplasm conserved in countries which are party to the treaty is on the basis of a Standard Material Transfer Agreement (SMTA). This arrangement has replaced the many individual MTAs which existed prior to the ITPGRFA and has allowed clarification and standardization of the terms of access. The SMTA has facilitated the use of carrot genetic resources through the provision of a common set of terms and conditions governing us and benefit sharing; if a commercial benefit is derived from germplasm supplied under an SMTA, then users make a payment into a common fund to support agricultural development on conservation in the developing world.

6.2 Collections of Carrot Genetic Resources

6.2.1 Genebanks and Ex Situ Conservation

No single international genebank has responsibility for the conservation of global diversity in carrot. The UK Vegetable Genebank, founded in 1980, was designated at the global base collection for carrot germplasm by the International Board for Plant Genetic Resources (IBPGR, now Bioversity). A snapshot of carrot germplasm collections across the world can be obtained via the Genesys information system (https://www.genesys-pgr.org). This online database provides information on the holdings of participating institutions. A total of 58 institutions report carrot germplasm collections in Genesys. In addition to the UK Vegetable Genebank, significant collections are held by the USDA, the Plant Breeding and Acclimatization Institute in Poland and the German genebank at the Leibniz Institute of Plant Genetics and Crop Plant Research, Gatersleben (Table 6.1). A total of 6642 accessions are reported as being conserved via the accession passport data submitted to the Genesys database by participating genebanks and other organisations. However, Genesys does not contain information on every carrot germplasm collection at the current time. Important collections to note are the extensive collection held by the Vavilov Institute, Russia (3102 accessions (Khmelinskaya et al. 2013), carrot genetic resources conserved in the genebank of the Institute of Vegetables and Flowers, Chinese

Table 6.1 Breakdown of collections of *Daucus* germplasm conserved by country and institute where collection size is >20 accessions

Country	FAO institute code	Organisation name	No. accessions
Azerbaijan	AZE015	Genetic Resources Institute	25
Bulgaria	BGR001	Institute for Plant Genetic Resources 'K. Malkov'	105
Brazil	BRA020	Embrapa Clima Temperado	25
Switzerland	CHE001	Agroscope Changins	40
Czech Republic	CZE122	Gene bank, Crop Research Institute	387
Germany	DEU146	Genebank, Leibniz Institute of Plant Genetics and Crop Plant Research	493
Spain	ESP004	Centro Nacional de Recursos Fitogenéticos	101
	ESP027	Gobierno de Aragón. Centro de Investigación y Tecnología Agroalimentaria. Banco de Germoplasma de Hortícolas	78
United Kingdom	GBR004	Millennium Seed Bank Project, Seed Conservation Department, Royal Botanic Gardens, Kew, Wakehurst Place	169
	GBR006	Warwick Genetic Resources Unit—UK Vegetable Genebank	1457
Croatia	HRV041	Faculty of Agriculture, University of Zagreb	24
Hungary	HUN003	Institute for Agrobotany	208
Israel	ISR002	Israel Gene Bank for Agricultural Crops, Agricultural Research Organisation, Volcani Center	64
Poland	POL003	Plant Breeding and Acclimatization Institute	629
Portugal	PRT001	Portuguese Bank of Plant Germplasm	145
Romania	ROM007	Suceava Genebank	70
Sweden	SWE054	Nordic Genetic Resource Center	200
Taiwan	TWN001	World Vegetable Center	20
Ukraine	UKR008	Ustymivka Experimental Station of Plant Production	48
	UKR021	Institute of Vegetable and Melon Growing	330
USA	USA005	National Seed Storage Laboratory, USDA-ARS	100
	USA020	North Central Regional Plant Introduction Station, USDA-ARS, NCRPIS	1381
	USA974	Seed Savers Exchange	239
	USA995	National Center for Genetic Resources Preservation	100
		Total in other institutes with <20 accessions	204
		Total	6642

Data originate from the Genesys database (http://www.genesys-pgr.org; accessed 22 May 2018)

Academy of Agricultural Sciences (approximately 400 accessions, mostly open-pollinated varieties; Zhuang, personal comm.) and a collection of 112 accessions maintained in India by the National Bureau of Plant Genetic Resources (NBPGR 2018). In France, a different approach is used to conserve and maintain crop genetic resources; a national network 'Carrot and other *Daucus* genetic resources' manages 3131 accessions (including heritage and research lines) in a co-operative effort among public research organisations and commercial breeding companies. Global *Daucus* germplasm collections contain material which has been collected from

>75 countries, indicating a broad coverage of ecogeographic adaptation, and however, sampling depth is uneven with some areas (Europe, USA) extensively sampled and others lacking in depth (South America, Africa). Other carrot germplasm collections doubtless exist, and however, those detailed above most certainly represent the vast majority of available ex situ carrot genetic resources; the total number of accessions conserved across these collections is >13,400. This figure almost certainly does not represent the total number of unique and distinct accessions—there is likely to be a level of duplication of material between institutes.

Ex situ conservation of carrot genetic resources is generally through long-term storage of seed samples. Carrot has orthodox seed, which means that viability is maintained when seed are stored under conditions of low-moisture content and low temperature (Kew 2018). The FAO has developed a set of guidelines for the long-term conservation of germplasm (FAO 2014). These are a general set of standards which apply to species with orthodox seed storage behaviour and indicate that long-term storage should be carried out at temperatures of -18 ± 3 °C and a relative humidity of $15 \pm 3\%$. Humidity control is vital to prevent rehydration of seed during storage; this is normally achieved by the use of suitable packaging material or containers to prevent ingress of water vapour. Viability monitoring is essential and should be carried out every 10 years. Successful ex situ conservation of crop germplasm requires the regeneration of seed samples due to depletion of seed stocks through use or due to the inevitable gradual loss of viability during storage. Due to the highly outcrossing nature of carrot and its wild relatives, successful conservation of genetic diversity requires the production of seed from sufficiently large populations of plants to reduce the loss of allelic variation through stochastic genetic drift (FAO 2014; Le Clerc et al. 2003). This is relevant for open-pollinated varieties, but particularly for samples of landraces and wild populations which are likely to be even more heterogeneous and heterozygous. Currently, many new carrot varieties are produced as F1 hybrids, for reasons of uniformity, vigour and varietal protection. Given that the parental lines of these hybrids are not often donated to genebanks for commercial reasons, it is not possible to maintain the hybrid variety in a genebank collection using standard regeneration techniques, meaning that the combinations of allele represented by elite varieties cannot be maintained in the longer term, and new approaches will have to be considered.

6.2.2 Biological Status of Daucus Germplasm Held Ex Situ: Cultivated Versus Wild

Based on the Genesys data set (Table 6.1), 35% of global *Daucus* germplasm collections are made up of 'advanced cultivars'—material which is the result of some kind of formal crop improvement programme (Fig. 6.1). As noted earlier, the vast majority of this material is likely to be made up of open-pollinated varieties due to the difficulties of maintaining hybrid varieties. Samples identified as originating from wild populations make up 27% of the global collection (1311 accessions in total), with landrace samples comprising 17%. Samples of breeding lines and material of weedy origin make up a minority proportion of the global collection. There is a lack of information on some accessions due to incomplete associated passport data with 14% accessions lacking a clear biological status. The number of accessions in ex situ collections currently identified as wild has increased by 463 from 848 in 2011. The latter figure was taken from a survey of the USDA and European genebank holdings (Grzebelus et al. 2011). A total of 76% (1004 accessions) of germplasm classified as wild belongs to the species *Daucus carota* and its subtaxa. Other *Daucus* species are less well-represented (Table 6.2). Wild material is also conserved in collections not listed in Genesys, notably the French *Daucus* network and the Vavilov Institute. Heightened interest and recognition of crop wild relatives as

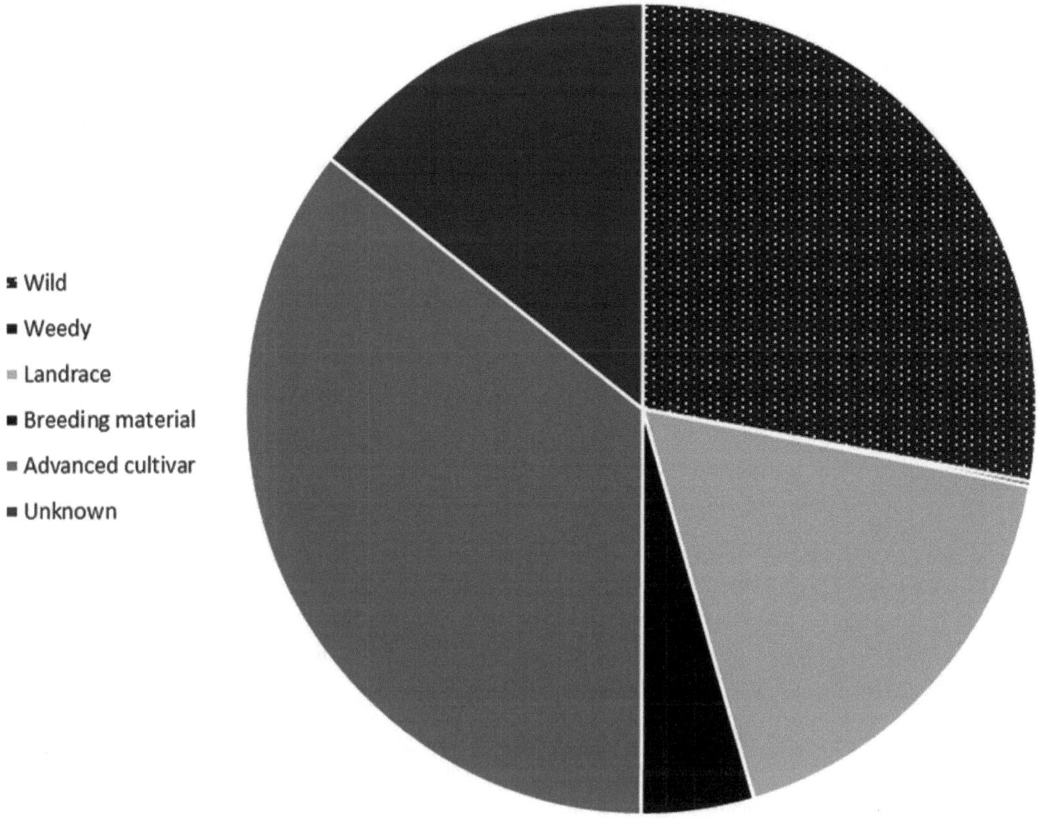

Fig. 6.1 Breakdown of *Daucus* germplasm by sample biological status. Data originate from the Genesys database (http://www.genesys-pgr.org; accessed 22 May 2018)

important resources for crop improvement have led to an increase in collection activity of wild *Daucus* species in recent years, and this has fed through into an increase in ex situ genebank holdings.

6.2.3 In Situ and on-Farm Conservation

Landraces are farmer-developed traditional varieties which are maintained on farm through saving and propagating seed of desired plants. They tend to have a higher degree of within-population variation than open-pollinated varieties created through formal crop improvement programmes and exhibit local adaptation to biotic and abiotic conditions. This makes them valuable resources for crop improvement as they may contain alleles which may be absent from existing formal breeding programmes due to intensive selection by plant breeders. It is, of course, possible to conserve them ex situ in genebanks, but on-farm maintenance of this type of material brings with it a continuation of the process of adaptation to local conditions. However, loss or genetic erosion of landraces can occur very easily as farmers may choose to cultivate modern varieties instead, and the product of many generations of selection and maintenance are lost. Although globally, the carrot is a significant vegetable, there are relatively few examples of carrot landraces documented in the scientific literature, unlike for major arable crops such as maize and rice. Perhaps the best studied is Polignano, a landrace from the south of Italy, which has been found to be at potential risk of genetic erosion (Renna et al. 2014). Likewise,

Table 6.2 **a** *Daucus* species conserved ex situ in genebanks. **b** Subtaxa of *D. carota* conserved in genebanks

a		b	
Daucus species	N	*D. carota* subtaxon	N
Daucus aureus	16	*azoricus*	1
Daucus bicolor	1	*carota*	308
Daucus broteri	38	*commutatus*	10
Daucus capillifolius	12	*drepanensis*	1
Daucus carota	1004	*fontanesii*	2
Daucus crinitus	41	*gadecaei*	3
Daucus durieua	1	*gummifer*	21
Daucus durieua Lange	1	*hispanicus*	4
Daucus glaber	7	*hispidus*	12
Daucus glochidiatus	2	*major*	10
Daucus guttatus	28	*maritimus*	59
Daucus halophilus	1	*maximus*	51
Daucus hispidifolius	1	*rupestris*	1
Daucus involucratus	4	*sativus*	20
Daucus littoralis	3	Unknown/blank	503
Daucus mauritii	1	Total	1004
Daucus montevidensis	1		
Daucus muricatus	40		
Daucus pusillus	41		
Daucus sahariensis	7		
Daucus syrticus	11		
Daucus tenuisectus	2		
Other/unknown/hybrid	48		
Total	1311		

Data originate from the Genesys database (http://www.genesys-pgr.org; accessed 22 May 2018). Taxonomic designation is that provided by the genebank to Genesys

carrots are among vegetable landraces disappearing from traditional farming systems in Morocco (Walters et al. 2018). In Turkey, Ipek et al. (2016) report that while 90% of orange carrots produced are F1 hybrid varieties, production of highly pigmented purple or black carrots with high levels of anthocyanins is based on traditional local varieties. Rather than being directly consumed, these purple or black carrots are used to produce food colourings and dyes. Traditional varieties and landraces have also been developed and maintained in locations far away from the accepted centre of domestication of the carrot in Central Asia. For example, in Chile, the Chiuchiu carrot is recognised by farmers for traits including adaptation to local conditions and postharvest storage (Pedreros et al. 2017).

While losses of crop landraces in general have been a global concern since the 1970s (FAO 2010), the advent of genebanks and the collection and conservation of landrace samples ex situ has addressed some of the concerns and allows ease of access for users. Adequate conservation on farms allows the continued development and adaptation of landrace type varieties, and however, this is often dependent on the landrace offering an economic return for the farmer compared to available modern varieties. In other words, conservation of landrace genetic resources on farms is dependent on continued utilisation. A parallel approach is undertaken by seed saving organisations such as the Seed Savers Exchange in the USA and Heritage Seed Library in the UK who rely on networks of gardeners and growers to maintain heritage and heirloom varieties.

In situ conservation of carrot crop wild relatives (CWR), similar to on-farm conservation of landraces offers an opportunity to conserve diversity in large populations, typically in national parks, nature reserves or other protected areas. Successful conservation requires an understanding of which species are present in a given area and the overlap of species distributions with protected areas (Castaneda-Alvarez et al. 2016). These strategies need to be developed at a national level as well as regional. In Europe, such strategies have been or are in the process of development in many countries (Labokas et al. 2018). *Daucus* is one of the focal crops of a major project co-ordinated by the Global Crop Diversity Trust which seeks to safeguard key CWR taxa through a programme of prioritisation, collection, conservation and pre-breeding. Material collected through this project is made available via project partners and by the Millennium Seed Bank, Royal Botanic Gardens Kew, UK.

6.3 Using Carrot Genetic Resources

While there are thousands of *Daucus* accessions conserved globally, a major barrier to use is the lack of available associated data on basic morphological or phenological traits such as annual/biennial growth habit, root colour and shape. There are two sets of standard descriptors in use; those produced by UPOV (2015) and IPGRI (1998). Collectively these kinds of data are termed characterisation data, and they can aid users of genetic resources by signposting in which sets of germplasm are most appropriate for their work. While characterisation data are straightforward to acquire, many genebanks lack the financial or other resources to undertake comprehensive characterisation programmes. Evaluation for more complex phenotypic traits such as disease resistance or drought tolerance is even more challenging in terms of the resources required and is often beyond the remit of genebanks funded only for conservation and management of their collections. Such work tends to be carried out in separate research projects.

As with many crops, given the scale of both global and national collections, characterisation and evaluation projects have been undertaken by consortia including both genebanks and academic research organisations and also including public and private sector research organisations. Trait screening of genetic resources allows germplasm to be identified, either to be used directly in breeding programmes or through the construction of research populations and lines, for example, biparental mapping populations, substitution lines and other research resources. Due to the numbers of accessions conserved, it is necessary to construct core collections and manageable numbers of accessions which nonetheless reflect genepool diversity. An example of this is the Carrot Diversity Set produced by the UK Vegetable Genebank—a set of 77 genebank accessions covering phenotypic diversity and geographic origin which is available on request.

Carrot genetic resources have been deployed for a range of purposes—pertinent examples include studies of domestication and phylogenetics (Grzebelus et al. 2014; Spooner et al. 2013), identification of sources of resistance to biotic stresses (Ellis et al. 1993; Nothnagel et al. 2017) and developmental studies (Rong et al. 2014). The advent of comparatively cheap and rapid sequencing technologies will doubtless see larger-scale studies of genetic variation and will aid both understanding of critical regions of genetic variation associated with key traits and the uptake of novel diversity into crop varieties in the future.

References

Castaneda-Alvarez NP, Khoury CK, Achicanoy HA, Bernau V, Dempewolf H, Eastwood RJ, Guarino L, Harker RH, Jarvis A, Maxted N, Muller JV, Ramirez-Villegas J, Sosa CC, Struik PC, Vincent H, Toll J (2016) Global conservation priorities for crop wild relatives. Nat Plants 2:16022

Ellis PR, Hardman JA, Crowther TC, Saw PL (1993) Exploitation of the resistance to carrot fly in the wild carrot species *Daucus capillifolius*. Ann Appl Biol 122:79–91

FAO (2010) The second report on the state of the world's plant genetic resources for food and agriculture. Rome

FAO (2014) Genebank standards for plant genetic resources for food and agriculture, Revised edn

Grzebelus D, Iorizzo M, Senalik D, Ellison S, Cavagnaro P, Macko-Podgorni A, Heller-Uszynska K, Kilian A, Nothnagel T, Allender C, Simon PW, Baranski R (2014) Diversity, genetic mapping, and signatures of domestication in the carrot (*Daucus carota* L.) genome, as revealed by Diversity Arrays Technology (DArT) markers. Mol Breed 33:625–637

Grzebelus D, Baranski R, Spalik K, Allender C, Simon PW (2011) *Daucus*. In: Kole C (ed) Wild crop relatives: genomic and breeding resources, vegetables. Springer, Berlin Heidelberg

Ipek A, Turkmen O, Fidan S, Ipek M, Karci H (2016) Genetic variation within the purple carrot population grown in Eregli district in Turkey. Turk J Agric Forest 40:570–576

IPGRI (1998) Descriptors for wild and cultivated Carrots (*Daucus carota* L.). Rome

Kew RBG (2018) Seed information database (SID). Version 7.1. http://data.kew.org/sid. Accessed 22 May 2018

Khmelinskaya T, Zvereva O, Artemyeva A (2013) Status of the Umbelliferae ssp. in Russia. In: Second meeting of the ECPGR working group on Umbellifer crops. St Petersburg

Labokas J, Maxted N, Kell S, Brehm JM, Maria Iriondo J (2018) Development of national crop wild relative conservation strategies in European countries. Genet Resour Crop Evol 65:1385–1403

Le Clerc V, Briard M, Granger J, Delettre J (2003) Genebank biodiversity assessments regarding optimal sample size and seed harvesting techniques for the regeneration of carrot accessions. Biodiv Conserv 12:2227–2236

NBPGR (2018) http://genebank.nbpgr.ernet.in. Accessed 22 May 2018

Nothnagel T, Budahn H, Kramer R (2017) Characterization of resistance to *Alternaria* spp. in wild relatives of carrot (*Daucus carota* L. ssp *carota*). In: Briard M (ed) International symposium on carrot and other Apiaceae, pp 251–257

Pedreros CAH, Segura VDP, Takasaki AKV, Aspe PP (2017) Attitudes and preferences towards conservation of traditional seeds: the carrot case in the indigenous community of San Francisco de Chiuchiu, northern Chile. Interciencia 42:839–845

Renna M, Serio F, Signore A, Santamaria P (2014) The yellow-purple Polignano carrot (*Daucus carota* L.): a multicoloured landrace from the Puglia region (Southern Italy) at risk of genetic erosion. Genet Resour Crop Evol 61:1611–1619

Rong J, Lammers Y, Strasburg JL, Schidlo NS, Ariyurek Y, de Jong TJ, Klinkhamer PGL, Smulders MJM, Vrieling K (2014) New insights into domestication of carrot from root transcriptome analyses. BMC Genom 15:895

Spooner D, Rojas P, Bonierbale M, Mueller LA, Srivastav M, Senalik D, Simon P (2013) Molecular phylogeny of *Daucus* (Apiaceae). Syst Bot 38:850–857

UPOV (2015) Carrot (*Daucus carota* L.). Guidelines for the conduct of tests for distinctness, uniformity and stability, TG/49/8 Corr. Geneva

Walters SA, Bouharroud R, Mimouni A, Wifaya A (2018) The deterioration of Morocco's vegetable crop genetic diversity: an analysis of the Souss-Massa region. Agric-Basel 8:49

7 Carrot Molecular Genetics and Mapping

Massimo Iorizzo, Shelby Ellison, Marti Pottorff and Pablo F. Cavagnaro

Abstract

Carrot (*Daucus carota* L.) is an important root vegetable crop that is consumed worldwide and is appreciated for its taste and nutritional content (e.g., provitamin A carotenoids, anthocyanins, vitamins, and other minerals). Carrot genetic research has improved vastly over the past few decades due to advancements in molecular genomic resources developed for carrot. The increasing availability of DNA sequences such as expressed sequence tags (ESTs), creation of a physical map, sequencing of the carrot genome, and the numerous advancements in DNA genotyping has enabled the study of phenotypic variation of crop traits through the development of genetic linkage maps, which enable the ability to identify QTLs and their underlying genetic basis. In addition, the creation of genetic and genomic tools for carrot has enabled the study of diversity within carrot populations and germplasm collections, enabled genome-wide association studies (GWASs), characterization of populations at the species level, and comparative genomics with other crops and model species. Combined, these tools will advance the breeding process for carrot by enabling a targeted approach to improving traits by utilizing marker-assisted selection (MAS) strategies.

M. Iorizzo (✉)
Department of Horticultural Sciences, Plants for Human Health Institute, North Carolina State University, 600 Laureate Way, Kannapolis, NC 28081, USA
e-mail: miorizz@ncsu.edu

S. Ellison
USDA-Agricultural Research Service, Vegetable Crops Research Unit, Department of Horticulture, University of Wisconsin-Madison, 1575 Linden Dr., Madison, WI 53706, USA

M. Pottorff
Plants for Human Health Institute, North Carolina State University, 600 Laureate Way, Kannapolis, NC 28081, USA

P. F. Cavagnaro
National Scientific and Technical Research Council (CONICET), National Institute of Agricultural Technology (INTA) E.E.A. La Consulta, San Carlos, Argentina

P. F. Cavagnaro
Faculty of Agricultural Sciences, National University of Cuyo, Mendoza, Argentina

7.1 Introduction

Classical genetic mapping in crop species, including carrot (*Daucus carota* subsp. *sativus*), was initially based on a relatively small number of qualitative traits most of which were related to pigmentation, morphological and physiological traits, response to disease, and other easily measured phenotypes (reviewed by Simon 1984; Tanksley 1983). Early application of molecular markers in genetic mapping studies used isozyme markers which was the most frequently used

P. Simon et al. (eds.), *The Carrot Genome*, Compendium of Plant Genomes,
https://doi.org/10.1007/978-3-030-03389-7_7

method of analysis to detect genetic variation between 1960 and 1980 (Tanksley 1983). The subsequent development of the polymerase chain reaction (PCR) technology provided a method to easily analyze DNA polymorphisms and led to the expansion of several DNA marker technologies (Mullis et al. 1986). These included PCR-based microsatellite or simple sequence repeat (SSR), random amplified polymorphic DNA (RAPD), and amplification fragment length polymorphism (AFLP) markers (Joshi et al. 1999). These markers have extensively been applied in carrot genetic studies (reviewed by Bradeen and Simon 2007). As more high throughput and less expensive, on a cost-by-marker basis, DNA sequencing and genotyping systems became available, and detection and screening of single nucleotide polymorphisms (SNPs) or sequence length polymorphisms (e.g., SSR) became more common. Rapid analysis of thousands of SNPs in a large number of individuals using systems such as the GoldenGate® assay from Illumina, Inc., or the KASPar assay from KBiosciences has resulted in the widespread use of SNP markers in genetic analyses and has facilitated large-scale QTL mapping studies and gene cloning in many crops (Rasheed et al. 2017), including carrot (reviewed by Iorizzo et al. 2017). With the completion of sequenced reference genomes for many species, the advent of high-throughput sequencing technologies now enables rapid and accurate resequencing of a large number of crop genomes to detect the genetic basis of phenotypic variation.

Comprehensive maps of genome variation and the development of new computational methods are rapidly facilitating the application of genome-wide association studies (GWASs) of economically important traits and are accelerating the identification and functional characterization of candidate genes in many crop species (Huang and Han 2014), including carrot. These advances will greatly accelerate crop improvement via genomics-assisted breeding in carrot and other species with such available resources.

7.2 Genetic Markers

7.2.1 Isozyme Markers

Isozymes are multiple forms of an enzyme that differ in amino acid sequence and control different chemical reactions based on different kinetic parameters or regulatory properties. Isozyme markers are generally codominant, allowing differentiation between heterozygous and homozygous individuals. Their application in plants has been limited by the paucity of loci that can be unambiguously scored. For carrot, fourteen isozyme markers were used to develop the first linkage map (Westphal and Wricke 1991). Eight of these isozyme markers were also used to assess the genetic variability within the *D. carota* complex (St. Pierre et al. 1990). The markers were able to discriminate wild and cultivated taxa and indicated that these two groups of taxa harbor the same level of genetic diversity in terms of the mean number of alleles per locus, the proportion of polymorphic loci, and the observed and expected heterozygosity.

7.2.2 Restriction Fragment Length Polymorphism (RFLP), Randomly Amplified Polymorphic DNA (RAPD), and Amplified Fragment Length Polymorphism (AFLP) Markers

As DNA-based markers proved to be very effective, in terms of cost, time, and outcome, molecular tools in plant genetic studies increased in the mid- and late 1990s and their application in carrot became common. RAPD, RFLP, and AFLP became the markers of choice in genetic studies from the mid-1990s to early 2000s. RAPD markers use short (10-nt) random primers to amplify DNA fragments, which, depending on the primer annealing sites, can reveal polymorphic PCR amplicons. RFLP includes digestion of

DNA samples using restriction enzymes, then the separation of restriction fragments by gel electrophoresis followed by hybridization with genomic DNA/cDNA probes. The presence of fragments of different lengths is due to the different position of the restriction sites within and among plants, and their analysis is used to determine DNA polymorphisms. AFLP markers effectively combine principles of both RAPD and RFLP in order to obtain reproducible results. Fragmented DNA generated as a result of restriction digestion is ligated with primer-recognition sequences, called 'adaptors,' and selective PCR amplification of these restriction fragments using labeled primers is performed, generating multiple amplicons varying in size, which are then separated on gel/capillary electrophoresis to detect fragment length polymorphisms. These three types of molecular markers were extensively used in carrot genetic studies, especially for linkage map construction and QTL mapping (reviewed by Bradeen and Simon 2007). Overall, across multiple studies in carrot, the polymorphic rate of RFLP and RAPD markers varied from 10 to 33%, respectively, yielding about 60 informative markers (Bradeen and Simon 2007). Further, out of 404 AFLP bands generated using seven primer combinations, 164 polymorphic fragments were identified for an observed polymorphic rate of 42% (Vivek and Simon 1999). Thus, AFLPs were more efficient than RFLPs and RAPD markers at generating markers for linkage map construction. Consistently, Nakajima et al. (1998) reported that while both AFLP and RAPD markers were useful for phylogenetic studies in *Daucus*, the AFLP system yielded more than four times as many useful markers per reaction. Conversely, an advantage of RFLPs over RAPDs and AFLPs is the fact that the former tends to yield codominant markers, which are particularly useful for robust linkage mapping, whereas RAPD and AFLP produce mainly dominant markers.

7.2.3 Diversity Arrays Technology (DArT) Markers

In the late 1990s and early 2000s, increasing amount of sequence information became available for model crops like barley (Wenzl et al. 2004), and new high-throughput genotyping technologies able to screen single nucleotide polymorphism (SNP) markers were developed and applied in plant genetic studies (LaFramboise 2009), making the early generation of DNA-based markers, including AFLPs, relatively low throughput. Yet discovering sequence polymorphisms in non-model species was difficult, which was particularly true for many crops with limited sequence resources such as carrot. In the early 2000s, Diversity Arrays Technology (DArT) represented a pioneering cost-effective high-throughput genotyping technology that did not require prior knowledge of the genome sequence (Kilian et al. 2012). Although the technology does not directly provide sequence information, it uses cloned fragments, which may be easily characterized by Sanger sequencing. Given these advantages, DArT markers have been widely applied in plant genetic studies and represented the first high-throughput genotyping platform for several non-model species, including carrot. A DArT array comprising 7680 DArT clones generated from 169 diverse genotypes including wild and cultivated germplasm was successfully used in carrot for population genetics and linkage map construction studies (Grzebelus et al. 2014). Across a diverse set of carrot germplasm, 866 markers were non-redundant, polymorphic, and present in over 95% of the samples. 79% of the markers were highly discriminating with a PIC value above 0.25. Approximately 50% of the DArT markers, 431, were polymorphic in a biparental population and were used to construct a genetic map and identify molecular markers associated with domestication (Grzebelus et al. 2014). Recently, a DArT marker associated with carrot root domestication, named 'cult,' was cloned and its sequence was annotated as *DcAHLc1*, a member of the AT-hook motif nuclear localized (AHL) family of plant regulatory genes which are involved in the regulation of organ development, including root tissue patterning (Macko-Podgórni et al. 2017).

7.2.4 Repetitive Sequence-Based Markers

Repetitive DNA sequences are present in all higher plants and can account for up to 90% of the genome size in some species. These repetitive DNA sequences account for major differences across genomes, both within and among species. Microsatellites represent a unique type of tandemly repeated genomic sequences, which are abundantly distributed across plant genomes and demonstrate high levels of allele polymorphism (Vieira et al. 2016). To date, both genomic and transcript sequences have been used to detect and design SSR markers in carrot. About 300 SSR markers were developed from genomic DNA sequences, including 144 SSRs detected on BAC-end sequences (BSSR) (Cavagnaro et al. 2009) and 156 SSRs were developed from an enriched repetitive sequence library (GSSR) (Cavagnaro et al. 2011). SSR markers have also been mined from expressed sequence tags (EST-ESSR) (Iorizzo et al. 2011). Frequency distributions of both repeat types and sequence motifs for each microsatellite origin, i.e., a library enrichment procedure (GSSRs), BAC-end derived (BSSRs), and EST-derived SSRs (ESSRs), varied markedly across these DNA fractions. Di- and tetranucleotide repeats are most common in GSSRs, while trinucleotides are most abundant in BSSRs and ESSRs. Within BSSRs, trinucleotide repeats occurred preferentially inside open reading frames (ORFs). The abundance of the trinucleotide repeats in ESTs and in ORFs has been attributed to a negative selection against frameshift mutations in the coding regions (caused by SSRs different from tri- or hexanucleotides) and to a positive selection for specific single amino acid stretches (Morgante et al. 2002). Besides the differences in structure and abundance, the polymorphic rate varies across these three types of SSR markers. Cavagnaro et al. (2011) reported that GSSRs were more polymorphic than BSSRs. Overall, nearly 77% of GSSRs and 52% of BSSRs were polymorphic in at least one F_2 family. ESSR polymorphism rate was 83% in a mapping population. Despite this observation, a direct comparison of the latter with the GSSR and BSSR polymorphism rate could not be made, since these ESSR markers were developed on the basis of a computational preselection for polymorphic SSR loci, and thus, a high polymorphic rate was expected. An additional set of 100 SSR markers has been described (Le Clerc et al. 2015). The abundance of SSR markers was also investigated at the genome level. After identifying SSR motifs based on the whole-genome DNA sequence of an orange inbred line, 'DC-27,' 57,519 SSR primer pairs were identified (Xu et al. 2014). Mononucleotide repeats were the most abundant, followed by di- and tetranucleotide repeats. A small number of additional SSR markers, including Inter Simple Sequence Repeat (ISSR) have been developed from various sequence resources (Bradeen et al. 2002; Niemann et al. 1997; Rong et al. 2010; Vivek and Simon 1999). In summary, over 66,000 SSRs have been detected in carrot and over 500 primer pairs have been used, empirically, for genetic studies.

Other repetitive sequence-based markers developed in carrot include those targeting transposable elements (TEs), also known as transposable display (TE display). Transposable elements account for 41% of the carrot genome, and DNA transposons (class II TE) are particularly abundant in carrot (Iorizzo et al. 2016). Miniature inverted-repeat transposable elements (MITEs) are a special type of class II non-autonomous elements with a maximum of a few hundred base pairs in size. Their dispersal, repetitiveness, and the fact that their mobilization is a source of polymorphism make them good candidates as molecular markers (Le and Bureau 2004). To date, molecular markers targeting two MITE superfamilies, *PIF/Harbinger-like* and *Stowaway-like*, have been developed in carrot and used to examine genetic diversity, develop linkage maps, and address cytogenetic questions. Grzebelus et al. (2006) used primers complementary to the terminal inverted repeats (TIRs) of *Master* transposable elements to characterize a new family of *PIF/Harbinger-like* TEs in carrot and demonstrated that the amplicon products were highly polymorphic. Following this study, a modified transposon display approach was used to characterize two *DcMaster*-like elements, *DcMaster-Krak* and *DcSto*, and observed that over 70% of the amplicons were polymorphic and could be used for genetic mapping (Budahn et al. 2014; Grzebelus et al. 2007; Grzebelus and Simon 2009), hybrid seed purity testing (Macko and Grzebelus 2008), and cytogenetic studies (Macko-Podgorni et al. 2013; Nowicka et al. 2016). The release of the carrot genome in 2016 (Iorizzo et al. 2016) opened the opportunity to develop a new generation of transposable element-based markers that specifically target genes and insertions within introns also known as intron length polymorphisms (ILPs) (Wang et al. 2005). ILPs can be detected by PCR with a pair of primers anchored in the exons flanking the intron of interest, which offers several advantages, including reliability, cost-efficiency, and the ability to detect codominance. Primers targeting 209 *Stowaway-like* (*DcSto*) MITE insertion sites within introns along the carrot genome have been designed and tested for genotyping in carrot (Stelmach et al. 2017). Over 47% of the DcS-ILP were polymorphic and successfully used to characterize carrot root-shape diversity and population structure. Due to the nature of the markers being codominant, locus specific, and highly reproducible, DcS-ILP markers could also be used for gene tagging and genetic map construction.

7.2.5 Conserved Orthologous Sequence (COS) Markers

COS markers are PCR-based markers developed from a set of single-copy conserved orthologous genes (Fulton et al. 2002). These markers have been utilized in *Daucus* species to resolve the taxonomy of some *Daucus* clades (Arbizu et al. 2014). Since the markers were designed to target genes that are single copy, they are largely used to elucidate phylogenies and to study comparative genomics across different species (Small et al. 2004). A set of carrot COS markers was developed by comparing carrot EST sequences against Arabidopsis, sunflower, and lettuce sequences (Arbizu et al. 2014). Out of 102 COS markers, a total of 94 (92%) were successfully used to assess the taxonomic relationships among carrot, lettuce, and sunflower. Therefore, these markers could also be useful for phylogenetic studies among Euasterid II species.

7.2.6 Single Nucleotide Polymorphism (SNP) Markers

Despite the advances made in carrot genetics using all of the markers mentioned above, the number of molecular assays required to identify informative DNA polymorphisms is still limiting in large-scale carrot genetic studies. Single nucleotide polymorphism (which includes single base changes) are present throughout the genome in genic and non-genic regions at a higher frequency than polymorphisms in repetitive sequences (e.g., detected by SSR, ILP, and TE display markers) or restriction enzyme sites (e.g., detected by DArT, RFLP, or AFLP markers) making this type of polymorphism a very powerful tool for large-scale studies. In the last decade, the rapid advancement of next-generation sequencing technologies (NGS) in conjunction with new bioinformatics tools and the development of high-throughput SNP genotyping platforms has provided essential genomic resources for accelerating the molecular understanding of biological properties. This rapid development has decreased the cost, improved the quality of large-scale genome surveys, and allowed speciality crops such as carrot to access these technologies (Egan et al. 2012). The first high-throughput transcriptome data for carrot, which was published in 2011 (Iorizzo et al. 2011), provided an opportunity to detect a massive number of SNPs and establish the first high-throughput SNP resource. Sequences from four different carrot genotypes were compared, and over 20,000 SNPs were detected in 7684 contigs with an average density of 1.4 SNPs/kb. In comparison, within the same sequence set, 8823 ESSRs located in 6995 contigs were detected, confirming the higher abundance of SNP markers in the genome. A subset of 4000 SNPs (K-SNPs) were used to design the first high-throughput SNP genotyping assay in carrot using the KASPar chemistry assay. KASPar utilizes a unique form of competitive allele-specific PCR combined with a homogeneous, fluorescence-based reporting system for the identification and measurement of genetic variation occurring at the nucleotide level to detect SNPs (http://www.lgcgroup.com/products/kasp-genotyping-chemistry/how-does-kasp-work/). In total, 3636 (91%) SNP markers were validated in carrot and were used to characterize the genetic diversity of carrot and patterns of domestication (Iorizzo et al. 2013). The large number of SNPs also enabled researchers, for the first time, to clearly resolve genetic differentiation among and within wild and cultivated carrot subpopulations (Iorizzo et al. 2013). Previous studies that used SSR, ISSR, and AFLP markers, found a low to moderate differentiation between subpopulations. The same set of SNPs were further used to develop the first SNP-based linkage map (Cavagnaro et al. 2014) and perform marker-trait association analyses (Cavagnaro et al. 2014; Iorizzo et al. 2016; Parsons et al. 2015). Despite the proliferation of several new ultra-high-throughput SNP genotyping platforms available today (Rasheed et al. 2017), the KASPar assay described above still represents the only large-scale SNP genotyping assay available for carrot.

Recently, SNPs were also mined from genotyping-by-sequencing (GBS) and resequencing data in carrot. Genotyping-by-sequencing (GBS) is a genome-wide reduced representation method that generates sequence variants, such as indels and SNPs, by utilizing next-generation sequencing technology, producing a powerful and cost-effective genotyping procedure (Elshire et al. 2011). The reduced representation of the genome is obtained by performing an enzymatic digestion with one or more restriction enzymes that are sensitive to methylation. DNA polymorphisms between the reduced genome representations of each genotype can be detected with or without using a reference genome (Scheben et al. 2017). To date, six studies have utilized GBS in carrot to identify SNP markers for phylogenetic, linkage map construction and marker-trait association analysis. For the initial digestion, five of these studies used *Ape*KI as the restriction enzyme (Arbizu et al. 2016; Ellison et al. 2017, 2018; Iorizzo et al. 2016; Macko-Podgórni et al. 2017) and one study used *Msl*I (Keilwagen et al. 2017). In all cases, the authors used the carrot reference genome (Iorizzo et al. 2016) to align the sequences and to detect the SNPs. The number of SNPs detected in these different studies varied from nearly 78,000 to 890,000. Arbizu et al. (2016) compared over 140 accessions of wild carrot (*D. carota* subsp. *carota*) and other related species and subspecies and detected 889,445 SNPs. After filtering using different criteria such as a minor allele frequency of 0.1–1, missing data <10%, he retained 10,814 SNPs for a phylogenetic analysis. In comparison, Keilwagen et al. (2017) detected 281,394 bi-allelic SNP markers among 85 cultivated accessions and after filtering for a minor allele frequency of <5% or >90% heterozygosity retained 168,663 SNPs for further analyses. Ellison et al. (2018) used GBS to detect SNPs across 676 carrot samples, and after filtering using <30% missing data, minor allele frequency <5%, >5× depth coverage, retained 39,710 SNPs. Overall, these studies demonstrated that the application of GBS in carrot is a powerful tool to investigate genetic studies. It also highlights that other than differences in the natural diversity existing in the germplasm, the number of SNPs detected and used in GBS studies is determined by other factors, including the type of restriction enzyme used for DNA digestion and/or the parameters used to detect or filter SNPs.

Resequencing represents the ultimate strategy to achieve the maximum number of SNPs. The new sequencing techniques not only increase sequencing throughput by several orders of magnitude but also enable simultaneous sequencing of a large number of samples using a multiplexed sequencing strategy (Craig et al. 2008; Cronn et al. 2008). These recent technical advances have paved the way for the development of a whole-genome sequencing-based high-throughput genotyping method that combines advantages of time and cost-effectiveness, dense marker coverage, high mapping accuracy and resolution, and an easier comparison of genomes and genetic maps among mapping populations and organisms. However, the cost and success of a resequencing study are determined by several factors, including those that are specific to the species, e.g., genome size and ploidy level, and others factors that are more technical, such as the availability of computing infrastructure and bioinformatic tools and expertise. Thus, establishing and validating resequencing experiments for each crop is a critical step to apply this approach for genetic studies. To date, the only resequencing study in carrot was performed by resequencing 35 carrot accessions which were representative of *D. carota* subspecies ($N = 31$) and outgroups ($N = 4$) (Iorizzo et al. 2016). Each accession was sequenced using the Illumina platform, at a median depth of 14×. Multiple bioinformatic tools including BWA (Li and Durbin 2009), SAMtools (Li et al. 2009), Picard MarkDuplicates (http://broadinstitute.github.io/picard/), and GATK (McKenna et al. 2010) were used to map the Illumina reads, detect SNPs, and apply multiple filters to remove low-quality SNP calls. The minimum depth for each allele was set to 5×. Using this strategy, 39,695,937 SNPs were detected and after filtering 1,393,431 SNPs were retained and used for further analyses. SNP calls were validated against a set of 3,202 previously characterized SNPs (Iorizzo et al. 2013) and indicated an over 98.8% accuracy rate. The SNPs were successfully used to establish phylogenetic relationships, perform cluster analysis, estimate nucleotide diversity, and identify signatures of selection. The same set of sequences was also used to perform a genome-wide comparative analysis of the repetitive sequences across all 35 accessions. This first study provided fundamental information to the carrot community to further use resequencing as a high-throughput genotyping method for genetic studies.

7.3 Carrot Genetic Maps

Construction of linkage maps represents a prerequisite to determining the genomic location of loci controlling agronomic traits targeted for quality improvement. They may also be used to establish the structure of chromosomes and to study the recombination frequency among homologous chromosomes. Sequence-based markers are the only category of markers directly transferrable from genetic linkage maps to genomic sequences, enabling analyses such as the study of candidate genes underlying important traits, and to establish comparative genomic studies. A summary of the genetic maps and traits mapped in carrot is reported in Table 7.1. To date, 19 mapping populations, mainly F_2 segregating populations, have been used for genetic mapping of traits in carrot. The first linkage map integrated biochemical isozymes and DNA-based markers, RFLPs and RAPDs, which generated a genetic map with 8 LGs, utilizing 55 markers with an average distance of 13.1 cM (Schulz et al. 1993). Two F_2 mapping populations, Brasilia × HCM and B493 × QAL, were used to develop the most dense carrot linkage maps, at the time, using AFLP markers combined with codominant SCAR markers (Santos and Simon 2004). The linkage maps included 277 and 242 dominant AFLP markers and 10 and 8 codominant markers assigned to the nine linkage groups, respectively. The linkage maps were further used to study the genetic inheritance of carotenoid accumulation in carrot root and detect 21 QTL associated with this trait (Santos and Simon 2002). Until 2002, this represented the most comprehensive QTL study in carrot. Just et al. (2007) integrated the first set of SNP markers into the B493 × QAL AFLP linkage map, previously developed by Santos and Simon (2002). The SNP markers anchored 22 genes related to the carotenoid pathway. Cavagnaro et al. (2011) further used the B493 × QAL map to integrate 49 SSR markers (BSSR, ESSR, and GSSR), making this the densest sequence-based genetic map for carrot utilizing 79 markers, enabling the integration of additional data. Iovene et al. (2011) anchored this linkage map to the corresponding pachytene chromosomes by FISH mapping of 17 map-anchored BACs, which established a landmark to further anchor genetic maps to carrot chromosomes. Other mapping efforts focused on identifying simply inherited loci for vernalization (*Vrn1* locus) and male fertility restoration (*Rf1* locus), using 355 AFLP, RAPD, SCAR, and SSR markers (BSSRs, ESSRs, and GSSRs), covering all 9 chromosomes with a total map length of 669 cM and an average marker distance of 1.88 cM (Alessandro et al. 2013). Yildiz et al. (2013) identified loci for anthocyanin and carotenoid pigmentation in population B1896 × B7262, using AFLP, SSR, and SNPs to construct a map with 279 marker data points. These included 2 phenotypic loci (*P1* and *Y2*), 237 AFLPs, 40 SSRs, 1 SCAR, 5 anthocyanin biosynthesis structural genes (*F3H*, *FLS1*, *LDOX2*, *PAL3*, and *UFGT*), and 3 anthocyanin transcription factors (*DcEFR1*, *DcMYB3*, and *DcMYB5*). Ali et al. (2013) generated a genetic map using a combination of RAPDs and SSRs to identify a new source of root-knot nematode resistance in PI 652188 (Ping Ding) × B7262. Budahn et al. (2014) generated a 781-cM genetic map using 285 RAPD, AFLP, SCAR, BSSR, ESSR, and GSSR markers and identified loci controlling fertility and flower development. This map was anchored to the Cavagnaro et al. (2011) map. Dunemann et al. (2014) used a linkage map that included 285 AFLP molecular markers located on nine linkage groups to locate *CENH3*, a centromeric histone. Le Clerc et al. (2015) generated two linkage maps for populations PC2 and PC3 using SSR markers, which segregated for resistance to *Alternaria dauci*. A consensus genetic map was generated and detected 11 QTLs for resistance to Alternaria.

By 2013, 19 carrot linkage maps were developed, though the number of codominant sequence-based markers was still very limited (<150). The advent of DArT and SNP molecular markers for carrot has brought the first high-throughput class of markers for genetic mapping of traits. Grzebelus et al. (2014) developed the first linkage map based on DArT markers which spanned 419.1 cM and included 431 non-redundant markers across nine LGs. The validation of the first set of SNP markers using the KASPar chemistry in 2013 opened the opportunity to advance carrot mapping studies (Iorizzo et al. 2013). Using this set of markers, Parsons et al. (2015) built three linkage maps including over 550 SNP markers and identified several QTL for resistance to *Meloidogyne incognita*. Cavagnaro et al. (2014) developed a dense genetic map using 894 SNP and SSR markers, and three major loci conditioning anthocyanin pigmentation in roots and petioles and 15 QTL for root anthocyanins were identified (Cavagnaro et al. 2014). To anchor the carrot genome, these SNP markers were used to develop the first high-density integrated linkage map (Iorizzo et al. 2016). The map was developed using SNP data from three mapping populations, 70349 (Cavagnaro et al. 2014), Br1091 × HM1 (Parsons et al. 2015), and 70796, and integrated 2,073 markers for the full dataset and 918 markers for the bin dataset covering 622 and 616 cM, respectively. In the bin map, each marker represents a true recombination event. This analysis revealed that in the three mapping populations, on average, one recombination event occurred every 388 kb.

A GBS approach was used in a genetic study on β-carotene accumulation in carrot roots; 37,361 novel SNPs were identified and used to create a genetic linkage map using 569 high-quality GBS-SNPs with an average of 1.3 cM distance (Ellison et al. 2014). Another GBS map was developed which integrated 394 markers and covered 450 cM was used to assess the quality of the carrot genome assembly (Iorizzo et al. 2016). GBS was used again in a separate mapping population to study β-carotene in which 33,712 high-quality SNPs were used to create a genetic map with nine linkage groups and an average of

Table 7.1 Summary of genetic mapping studies in carrot

Mapping population	Generation	Type of markers	No. of mapped markers		Mapped traits/genes	70,349	References
			Dominant	Codominant			
I0/1 lines	F_2	Isozyme, RFLP, RAPD	55				Schulz et al. (1993)
B9304 × YC7262	F_2	AFLP, SNP	6	1	Carotenoid accumulation	*Y2*	Bradeen and Simon (1998)
B9304 × YC7262	F_2	RFLP, RAPD, AFLP, SSR		10	Anthocyanin, carotenoid, sugar	*P1*, *Y2*, *Rs*	Vivek and Simon (1999)
Brasilia-1252 × B6274	F_2	RAPD	4	0	Resistance to *Meloidogyne javanica*	*Mj1*	Boiteu et al. (2000)
Brasilia × HCM	F_2	AFLP	164	0	Carotenoids, lycopene, and phytoene	16 QTLs	Santos and Simon (2002, 2004)
B493 × QAL	F_2	AFLP	141	0	Carotenoids, lycopene, and phytoene	16 QTLs	Santos and Simon (2002, 2004)
		Integration of carotenoid SNPs	22	22	Carotenoid biosynthesis genes, *Y2* mark	1 QTL	Just et al. (2007)
		Integration of DcMTD	51	0			Grzebelus et al. (2007)
		Integration of BSSR, ESSR, GSSR		55			Cavagnaro et al. (2011)
		FISH, 17 BAC clones					Iovene et al. (2011)
Samson (mutant) × Samson (WT)	F_2	RAPD, AFLP	236		Compressed lamina (COLA)	1 QTL	Nothnagel et al. (2005)
S269 × R268	F_2	AFLP, ISSR	139	0	Resistance to *Alternaria dauci*	3 QTLs	(Le Clerc et al. (2009)
Biennial × Criolla INTA	F_2	AFLP, BSSR, ESSR, GSSR, RAPD, SCAR	355	23	Vernalization, male fertility restoration	*Vrn*, *Rf1*	Alessandro et al. (2013)
B1896 × B7261	F_2	AFLP, SSR, SNP	279	38	Anthocyanin, carotenoid accumulation	*P1*, *Y2*	Yildiz et al. (2013)
70,349	F_2	DArT	866		Markers associated with domestication	2 markers	Grzebelus et al. (2013)

(continued)

Table 7.1 (continued)

Mapping population	Generation	Type of markers	No. of mapped markers		Mapped traits/genes	70,349	References
			Dominant	Codominant			
PI 652188 × B7262	F_3	RAPD, SSR		8	Resistance to *Meloidogyne javanica*	*Mj-2*	Ali et al. (2013)
DM19	F_2	AFLP, BSSR, ESSR, GSSR, RAPD, SCAR		285	Flower development and fertility	Cola locus	Budahn et al. (2014)
70,349	F_2, F_3, F_4	SNP, SSR		894	Anthocyanin glycosides, *RTPE*	15 QTLs	Cavagnaro et al. (2014)
Yellow × Cola	F_2	AFLP, RAPD, DcMTD, SSR, SNP		4	Centromeric histone *H3*	*CENH3*	Dunemann et al. (2014)
2569	F_4	GBS-SNP		811	β-carotene	1 QTL	Ellison et al. (2014)
PC2 and PC3	$F_{2:3}$	SSR		~100	Resistance to *Alternaria dauci*	11 QTLs	Le Clerc et al. (2015)
Br1091 × HM1	F_2	K-SNP		389	Resistance to *Meloidogyne incognita*	4 QTLs	Parsons et al. (2015)
SFF × HM2	F_2	AFLP, K-SNP, SSR	20	138	Resistance to *Meloidogyne incognita*	3 QTLs	Parsons et al. (2015)
HM3	F_5	SSR, K-SNP		70	Resistance to *Meloidogyne incognita*	3 QTLs	Parsons et al. (2015)
74,146	F_4	GBS-SNP		2999	β-carotene	1 QTL	Ellison et al. (2017)

DcMTD: *DcMaster* transposon display; BSSR: SSRs from BAC-end sequences; GSSR: library enrichment procedure, GBS-SNP: genotype-by-sequencing SNPs; K-SNP: KASPar assay SNP

one GBS marker every 11.3 kb (Ellison et al. 2017). In general, the density of the linkage maps and the number of codominant sequence-based markers in carrot are rapidly increasing, providing an opportunity to perform fine mapping QTL studies to identify candidate genes (Iorizzo et al. 2016) and study the recombination behavior of homologous chromosomes. For example, in carrot, segregation distortion, or the skewed frequency of genotypes from a typical Mendelian ratio in a segregating population, has been observed in multiple genetic mapping studies. Schulz et al. (1993) reported that 24% of the markers (RFLP, isozyme, and RAPD) used in their study deviated from the expected Mendelian ratios. A high segregation distortion of DArT markers was observed in an F_2 population, which resulted in very few markers mapping to chromosome 8 (Grzebelus et al. 2013). Using the same F_2 population, Cavagnaro et al. (2014) observed clusters of distorted SNP markers on CH1, CH8, and CH9; however, the majority of segregation distortion was observed on CH8. In a separate mapping study to identify nematode resistance in carrot, a significant segregation distortion was observed in the two mapping populations used. In the Br1091 × HM1 population, K-SNP markers on chromosomes 4 and 9 were skewed from the typical segregation of 1:2:1, and in the HM3 map, three chromosomes lacked segregating markers (Parsons et al. 2015). The segregation distortion observed in the two studies on CH8 could be considered a segregation distortion loci (SDL) in which distorted markers are clustered in the same chromosomal region (Xian-Liang et al. 2006). Lethal alleles (gamete genes) controlling skewed homologous recombination have been described in other crops like maize and rice (Cheng et al. 1996; Iwata et al. 1964; Yan et al. 2003). Screening of a large number of F_2 individuals from different mapping populations using SNPs surrounding the distorted region on carrot CH8 will facilitate the identification of candidate genes causing segregation distortion. Given the preliminary findings regarding segregation distortion, a comparative genetic and cytogenetic analysis could be undertaken to understand the effect of segregation distortion on the inheritance of the QTLs. This will enhance knowledge to breeders regarding which combination of crosses causes segregation distortion and the expected number of progenies that will inherit the loci where distortion exists.

7.4 Marker-Trait Association Mapping

As summarized above, utilizing linkage analysis to map genomic loci that have an effect on a trait of interest has been commonplace for the last 25 years. Since recombination rates are relatively low in mapping populations, tagging a region in linkage with a casual variant requires only a few genetic markers per chromosome. However, the downside to a small number of recombination blocks is that the mapping resolution can be very low. Other disadvantages of linkage mapping include the substantial amount of time and resources needed to generate mapping populations and that the identified QTL are limited to the diversity of the parents of the biparental population. Genome-wide association studies (GWASs) have emerged in the last decade as an alternative to linkage analysis to expose the genetic basis of quantitative traits. Such studies address the relationship between marker-based polymorphism and phenotypic variation in a diverse population, which in turn may increase the resolution of a study by using all ancestral recombination events (Myles et al. 2009). GWAS can take advantage of pre-existent germplasm populations, exploit multiple recombination events, and consider a greater diversity in alleles. Additionally, if the mapping resolution is high, associated SNPs can be used directly for marker-assisted selection.

GWAS is based on the principle of linkage disequilibrium (LD) or the non-random association between alleles at different loci. The genomic distance at which LD decays determines how many genetic markers are needed to tag a haplotype. The high effective recombination rate in outcrossing species, such as carrot, is expected to cause a fast decay of LD. The first report of genome-wide LD in carrot showed fast decay in wild samples (<1 kbp) and moderate rates (<10 kb) in cultivated samples (Ellison et al. 2018). LD decay rates appear even slower in domesticated samples around regions putatively under selection such as the *Y* region (Iorizzo et al. 2016), the '*cult*' region (Macko-Podgórni et al. 2017), the *Or* region (Ellison et al. 2018), and several carotenoid biosynthesis genes (Clotault et al. 2010; Soufflet-Freslon et al. 2013).

A potential pitfall of GWAS is the lack of power when performed in structured populations which can lead to an increase of false discovery rate. This occurs when phenotypic traits are correlated with underlying population structure at non-causal loci (Nordborg and Weigel 2008). *D. carota* L. genetic resources are known to be structured into at least six genetic groups (Ellison et al. 2018; Grzebelus et al. 2014; Iorizzo et al. 2013, 2016; Rong et al. 2014) according to their geographical origin and level of domestication. This can be a potential problem in carrot as many traits, such as carotenoid content, are associated with a particular genetic group and GWAS could lead to a false-positive detection. The effect of population structure can be estimated and added as a covariate in association models, which will limit false positives. Two commonly used ways to estimate population structure are the use of the Structure software (Pritchard et al. 2000) or conduct a principal component analysis (Falush et al. 2003; Price et al. 2006). Estimates of population structure as well as a kinship matrix are commonly used in a unified mixed model approach to account for relatedness between individuals (Yu et al. 2006).

To date only a few GWASs have been conducted in carrot. Prior to the availability of the carrot reference genome (Iorizzo et al. 2016), a candidate gene association study was conducted in 380 carrot genotypes, derived from the intercrossing of 67 cultivars for three generations, using 109 SNPs spread across 17 carotenoid biosynthesis genes (Jourdan et al. 2015). The strongest association with carotenoid content and color components was for the carotenoid genes *zeaxanthin epoxidase* (*ZEP*) and *carotenoid isomerase* (*CRTISO*). In 2017, a diverse set of 85 carrot cultivars and ~168,000 SNPs were used to identify 30 QTL for 15 terpenoid volatile organic compounds (Keilwagen et al. 2017). Genomic locations of known terpene synthase genes were positioned with respect to significant GWAS signals to suggest candidate terpene synthase genes for particular terpenoid compounds. More recently, ~40,000 SNPs were used in 674 wild and cultivated globally distributed carrots to analyze orange pigmentation (Ellison et al. 2018). A significant association for pigmentation was found on chromosome 3, in which the *Or* gene, which has been shown to be important for chromoplast development and the accumulation of carotenoids, was identified within the region on chromosome 3.

The statistical power to detect associations between DNA variants and a trait depends on the experimental sample size, the distribution of effect sizes and frequency of causal genetic variants segregating in the population, and the LD between genotyped DNA variants and causal variants (Visscher et al. 2017). Therefore, the potential of a GWAS to succeed depends on how many loci affecting the trait segregate in the population, the genetic architecture of the trait, the experimental sample size, and the variants that are used in the GWAS. Additionally, the accuracy at which a trait can be measured is imperative to the success of the GWAS. Since LD decays rapidly in carrot, GWAS has a great potential to identify linked or causal variants for traits of interest. Future GWAS projects in carrot will benefit from improved genotyping techniques, such as whole-genome sequencing, to increase SNP density across the genome.

7.5 Future Perspectives

A wide range of molecular markers have been developed and applied in carrot genetic and genomic studies (Fig. 7.1), which has accelerated knowledge in traits of agronomic interest and the domestication of carrot. Resequencing is a valuable approach for identifying SNPs in carrot, and given the relatively small size of its genome, generating resequencing data is cost-effective.

However, compared to some genotyping assays like array-based platforms (e.g., Affymetrix array), the resequencing approach requires additional resources such as computational infrastructures to store and analyze the raw sequences and bioinformatics expertise to process the data and identify SNPs. These additional costs associated with resequencing are usually not included in the overall genotyping price. Other potential issues include a high level of missing data and the absence of perfect bioinformatics tools for data imputation models. Whereby, high-throughput genotyping platforms such as Affymetrix or Illumina arrays provide multiple benefits including: (1) a range of multiplex levels providing rapid high-density genome scans; (2) robust allele calling with high call rates; (3) cost-effectiveness per data point when genotyping large numbers of SNPs and samples; (4) does not require extensive post-processing analysis and computational resources. A disadvantage to this approach is that it requires a relatively large up-front investment to build the array; however, the cost can be reduced by increasing the number of samples that will be genotyped. In other crops, such as potato and corn, to overcome these challenges, public and private breeding programs have established partnerships to develop the genotyping array. These partnerships increased the number of users and consequently samples, which reduced the costs to design the array and cost per sample. A genotyping platform is still not available for carrot. As part of the public–private partnership that supported the carrot sequencing and genomic efforts (see Chap. 11), establishing an array-based genotyping platform should be considered a high priority.

As new SNP markers associated with economically important traits in carrot are developed and validated, it will be critical to develop a low-density genotyping assay that specifically

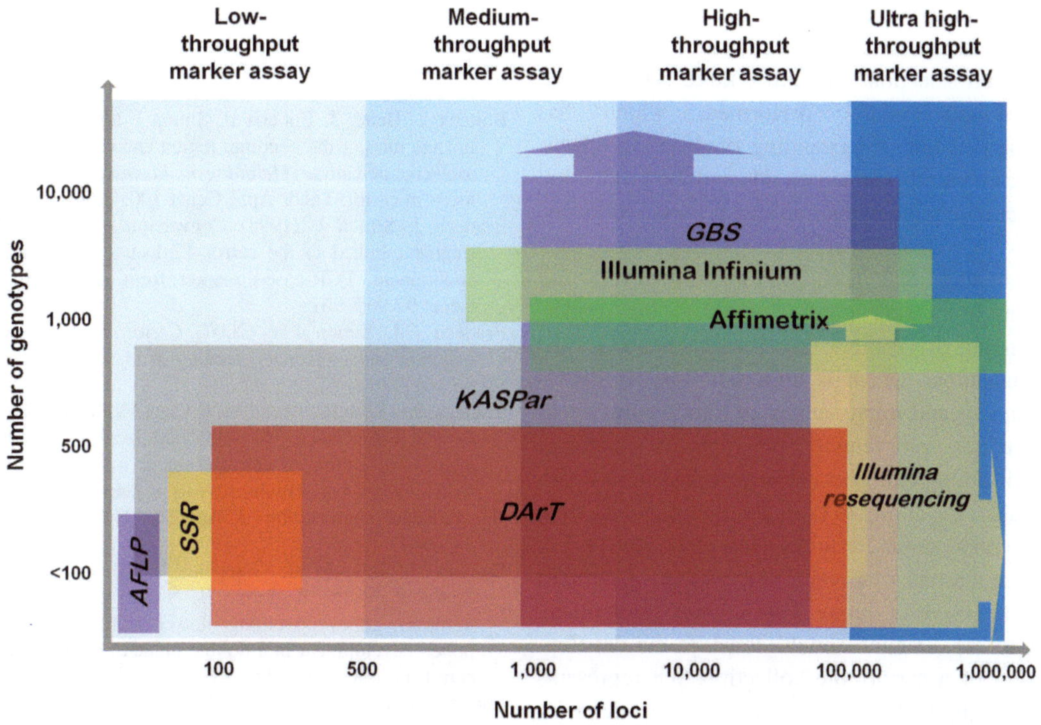

Fig. 7.1 Different platforms for genotyping, showing their relative high throughput in terms of number of samples and assays that can be used in a single run (developed using data from Rasheed et al. 2017). Assays labeled in italic are those platforms/assays that have been used in carrot genetic studies

targets those SNPs. Some single-marker methods have been developed in carrot and could be used for marker-assisted breeding. This includes an allele-specific PCR (AS-PCR) assay, cleaved amplified polymorphic sequences (CAPS), and sequence-tagged site (STS) markers (see Chap. 9). However, all of these methods have in common limitations of low throughput, high cost, and are labor intensive, which limits their application in carrot breeding programs. Multiple cost-effective low-density genotyping assays are currently available including KASPar, TaqMan, and semi-thermal asymmetric reverse PCR (STARP). These genotyping technologies could be used to develop a panel of allele-specific assays from functional genes or QTLs, which could be used in marker-assisted breeding in a high-throughput cost-effective fashion in carrot.

For future marker-trait associations in carrot, GWAS is a promising method to associate a genotype with a phenotype. Since LD decays very fast in carrot, the resolution is high and an association could be identified directly with the gene that controls a given phenotype; thereby, the marker could be directly used for marker-assisted selection. Currently, three GWAS have been successfully performed, which has enhanced our understanding of the biosynthesis of carotenoids and terpenoids and the production of orange pigmentation in carrot roots. All of the GWAS were conducted in a single location, each evaluating one specific trait. However, use of multi-location studies enables the understanding of genotype × environmental (G × E) interactions, which is important for the understanding and improvement of carrot cultivars in breeding programs. The use of accurate high-throughput phenotyping techniques, which enables the ability to evaluate multiple traits from several locations, requires a large investment. For these reasons, optimizing the number of samples to be used in future GWAS will be critical. A future direction should consider the development of a carrot core collection that represents the highest phenotypic and genotypic diversity for future GWAS and could be used to make more informative breeding decisions regarding the diversity of their breeding materials and the potential to exploit novel alleles. The use of a core collection in a GWAS would reduce the number of samples needed to be phenotyped and genotyped, without reducing the potential genetic gain. A core collection could then be used to evaluate the performance of economically important traits across multiple locations and with high accuracy.

References

Alessandro MS, Galmarini CR, Iorizzo M, Simon PW (2013) Molecular mapping of vernalization requirement and fertility restoration genes in carrot. Theor Appl Genet 126:415–423

Ali A, Matthews WC, Cavagnaro PF, Iorizzo M, Roberts PA, Simon PW (2013) Inheritance and mapping of *Mj-2*, a new source of root-knot nematode (*Meloidogyne javanica*) resistance in carrot. J Hered 105:288–291

Arbizu C, Ruess H, Senalik D, Simon PW, Spooner DM (2014) Phylogenomics of the carrot genus (*Daucus*, Apiaceae). Am J Bot 101:1666–1685

Arbizu CI, Ellison SL, Senalik D, Simon PW, Spooner DM (2016) Genotyping-by-sequencing provides the discriminating power to investigate the subspecies of *Daucus carota* (Apiaceae). BMC Evol Biol 16:234

Boiteux L, Belter J, Roberts P, Simon P (2000) RAPD linkage map of the genomic region encompassing the root-knot nematode (*Meloidogyne javanica*) resistance locus in carrot. Theor Appl Genet 100:439–446

Bradeen J, Simon P (1998) Conversion of an AFLP fragment linked to the carrot *Y2* locus to a simple, codominant, PCR-based marker form. Theor Appl Genet 97:960–967

Bradeen JM, Simon PW (2007) Carrot. In: Genome mapping and molecular breeding in plants. Springer, pp 161–184

Bradeen JM, Bach IC, Briard M, le Clerc V, Grzebelus D, Senalik DA, Simon PW (2002) Molecular diversity analysis of cultivated carrot (*Daucus carota* L.) and wild *Daucus* populations reveals a genetically nonstructured composition. J Am Soc Hort Sci 127:383–391

Budahn H, Barański R, Grzebelus D, Kiełkowska A, Straka P, Metge K, Linke B, Nothnagel T (2014) Mapping genes governing flower architecture and pollen development in a double mutant population of carrot. Frontiers Plant Sci 5:504

Cavagnaro PF, Chung S-M, Szklarczyk M, Grzebelus D, Senalik D, Atkins AE, Simon PW (2009) Characterization of a deep-coverage carrot (*Daucus carota* L.) BAC library and initial analysis of BAC-end sequences. Mol Genet Genomics 281:273–288

Cavagnaro PF, Chung S-M, Manin S, Yildiz M, Ali A, Alessandro MS, Iorizzo M, Senalik DA, Simon PW (2011) Microsatellite isolation and marker development in carrot-genomic distribution, linkage mapping, genetic diversity analysis and marker transferability across Apiaceae. BMC Genom 12:386

Cavagnaro PF, Iorizzo M, Yildiz M, Senalik D, Parsons J, Ellison S, Simon PW (2014) A gene-derived SNP-based high resolution linkage map of carrot including the location of QTL conditioning root and leaf anthocyanin pigmentation. BMC Genom 15:1118

Cheng R, Saito A, Takano Y, Ukai Y (1996) Estimation of the position and effect of a lethal factor locus on a molecular marker linkage map. Theor Appl Genet 93:494–502

Clotault J, Geoffriau E, Lionneton E, Briard M, Peltier D (2010) Carotenoid biosynthesis genes provide evidence of geographical subdivision and extensive linkage disequilibrium in the carrot. Theor Appl Genet 121:659–672

Craig DW, Pearson JV, Szelinger S, Sekar A, Redman M, Corneveaux JJ, Pawlowski TL, Laub T, Nunn G, Stephan DA (2008) Identification of genetic variants using bar-coded multiplexed sequencing. Nat Methods 5:887

Cronn R, Liston A, Parks M, Gernandt DS, Shen R, Mockler T (2008) Multiplex sequencing of plant chloroplast genomes using Solexa sequencing-by-synthesis technology. Nucleic Acids Res 36:e122

Dunemann F, Schrader O, Budahn H, Houben A (2014) Characterization of centromeric histone H3 (CENH3) variants in cultivated and wild carrots (*Daucus* sp.). Plos One 9:e98504

Egan AN, Schlueter J, Spooner DM (2012) Applications of next-generation sequencing in plant biology. Am J Bot 99:175–185

Ellison S, Iorizzo M, Senalik D, Simon P (2014) The next generation of carotenoid studies in carrot (*Daucus carota* L.). In: International symposium on carrot and other Apiaceae 1153, pp 93–100

Ellison S, Senalik D, Bostan H, Iorizzo M, Simon P (2017) Fine mapping, transcriptome analysis, and marker development for *Y2*, the gene that conditions beta-carotene accumulation in carrot (Daucus carota L.). G3: Genes, Genom, Genet. https://doi.org/10.1534/g3.117.043067

Ellison S, Luby C, Corak K, Coe K, Senalik D, Iorizzo M, Goldman I, Simon P, Dawson J (2018) Association analysis reveals the importance of the Or gene in carrot (*Daucus carota* L.) carotenoid presence and domestication. Genetics

Elshire RJ, Glaubitz JC, Sun Q, Poland JA, Kawamoto K, Buckler ES, Mitchell SE (2011) A robust, simple genotyping-by-sequencing (GBS) approach for high diversity species. Plos One 6:e19379

Falush D, Stephens M, Pritchard JK (2003) Inference of population structure using multilocus genotype data: linked loci and correlated allele frequencies. Genetics 164:1567–1587

Fulton TM, Van der Hoeven R, Eannetta NT, Tanksley SD (2002) Identification, analysis, and utilization of conserved ortholog set markers for comparative genomics in higher plants. Plant Cell 14:1457–1467

Grzebelus D, Simon PW (2009) Diversity of *DcMaster*-like elements of the *PIF/Harbinger* superfamily in the carrot genome. Genetica 135:347–353

Grzebelus D, Yau Y-Y, Simon PW (2006) *Master*: a novel family of *PIF/Harbinger*-like transposable elements identified in carrot (*Daucus carota* L.). Mol Genet Genom 275:450

Grzebelus D, Jagosz B, Simon PW (2007) The *DcMaster* transposon display maps polymorphic insertion sites in the carrot (*Daucus carota* L.) genome. Gene 390:67–74

Grzebelus D, Baranski R, Iorizzo M, Senalik D, Repinski S, Cavagnaro P, Macko-Podgorni A, Heller-Uszynska K, Kilian A, Nothnagel T (2013) Diversity arrays technology (DArT) platform for genotyping and mapping in carrot (*Daucus carota* L.). In: Plant and animal genome conference

Grzebelus D, Iorizzo M, Senalik D, Ellison S, Cavagnaro P, Macko-Podgorni A, Heller-Uszynska K, Kilian A, Nothnagel T, Allender C (2014) Diversity, genetic mapping, and signatures of domestication in the carrot (*Daucus carota* L.) genome, as revealed by diversity arrays technology (DArT) markers. Mol Breed 33:625–637

Huang X, Han B (2014) Natural variations and genome-wide association studies in crop plants. Annu Rev Plant Biol 65:531–551

Iorizzo M, Senalik DA, Grzebelus D, Bowman M, Cavagnaro PF, Matvienko M, Ashrafi H, Van Deynze A, Simon PW (2011) De novo assembly and characterization of the carrot transcriptome reveals novel genes, new markers, and genetic diversity. BMC Genom 12:389

Iorizzo M, Senalik DA, Ellison SL, Grzebelus D, Cavagnaro PF, Allender C, Brunet J, Spooner DM, Van Deynze A, Simon PW (2013) Genetic structure and domestication of carrot (*Daucus carota* subsp. *sativus*) (Apiaceae). Am J Bot 100:930–938

Iorizzo M, Ellison S, Senalik D, Zeng P, Satapoomin P, Huang J, Bowman M, Iovene M, Sanseverino W, Cavagnaro P et al (2016) A high-quality carrot genome assembly provides new insights into carotenoid accumulation and asterid genome evolution. Nat Genet 48:657

Iorizzo M, Ellison S, Senalik D, Stoffel K, Zeng P, Iovene M, Cavagnaro C, Yildiz M, Ashrafi H, Zheng Z, Cheng C, Spooner D, Van Deynze A, Simon P (2017) Recent advance in carrot genomics. In: ISHS acta horticulturae 1153: international symposium on carrot and other Apiaceae

Iovene M, Cavagnaro PF, Senalik D, Buell CR, Jiang J, Simon PW (2011) Comparative FISH mapping of *Daucus* species (Apiaceae family). Chromosome Res 19:493–506

Iwata N, Nagamatsu T, Omura T (1964) Abnormal segregation of waxy and apiculus coloration by a gametophyte gene belonging to the first linkage group in rice. Japan J Breed 14:33–39

Joshi SP, Ranjekar PK, Gupta VS (1999) Molecular markers in plant genome analysis. Curr Sci 77:230–240

Jourdan M, Gagné S, Dubois-Laurent C, Maghraoui M, Huet S, Suel A, Hamama L, Briard M, Peltier D, Geoffriau E (2015) Carotenoid content and root color of cultivated carrot: a candidate-gene association study using an original broad unstructured population. Plos One 10:e0116674

Just B, Santos C, Fonseca M, Boiteux L, Oloizia B, Simon P (2007) Carotenoid biosynthesis structural genes in carrot (*Daucus carota*): isolation, sequence-characterization, single nucleotide polymorphism (SNP) markers and genome mapping. Theor Appl Genet 114:693–704

Keilwagen J, Lehnert H, Berner T, Budahn H, Nothnagel T, Ulrich D, Dunemann F (2017) The terpene synthase gene family of carrot (*Daucus carota* L.): identification of QTLs and candidate genes associated with terpenoid volatile compounds. Frontiers Plant Sci 8:1930

Kilian A, Wenzl P, Huttner E, Carling J, Xia L, Blois H, Caig V, Heller-Uszynska K, Jaccoud D, Hopper C, Aschenbrenner-Kilian M, Evers M, Peng K, Cayla C, Hok P, Uszynski G (2012) Diversity arrays technology: a generic genome profiling technology on open platforms. In: Pompanon F, Bonin A (eds) Data production and analysis in population genomics: methods and protocols. Humana Press, Totowa, NJ, pp 67–89

LaFramboise T (2009) Single nucleotide polymorphism arrays: a decade of biological, computational and technological advances. Nucleic Acids Res 37:4181–4193

Le QH, Bureau T (2004) Prediction and quality assessment of transposon insertion display data. Biotechniques 36:222–228

Le Clerc V, Pawelec A, Birolleau-Touchard C, Suel A, Briard M (2009) Genetic architecture of factors underlying partial resistance to Alternaria leaf blight in carrot. Theor Appl Genet 118:1251–1259

Le Clerc V, Marques S, Suel A, Huet S, Hamama L, Voisine L, Auperpin E, Jourdan M, Barrot L, Prieur R (2015) QTL mapping of carrot resistance to leaf blight with connected populations: stability across years and consequences for breeding. Theor Appl Genet 128:2177–2187

Li H, Durbin R (2009) Fast and accurate short read alignment with Burrows–Wheeler transform. Bioinformatics 25:1754–1760

Li H, Handsaker B, Wysoker A, Fennell T, Ruan J, Homer N, Marth G, Abecasis G, Durbin R, Genome Project Data Processing S (2009) The sequence alignment/map format and SAMtools. Bioinformatics 25:2078–2079

Macko A, Grzebelus D (2008) DcMaster transposon display markers as a tool for diversity evaluation of carrot breeding materials and for hybrid seed purity testing. J Appl Genet 49:33–39

Macko-Podgorni A, Nowicka A, Grzebelus E, Simon PW, Grzebelus D (2013) *DcSto*: carrot *Stowaway*-like elements are abundant, diverse, and polymorphic. Genetica 141:255–267

Macko-Podgórni A, Machaj G, Stelmach K, Senalik D, Grzebelus E, Iorizzo M, Simon PW, Grzebelus D (2017) Characterization of a genomic region under selection in cultivated carrot (*Daucus carota* subsp. *sativus*) reveals a candidate domestication gene. Frontiers Plant Sci 8:12

McKenna A, Hanna M, Banks E, Sivachenko A, Cibulskis K, Kernytsky A, Garimella K, Altshuler D, Gabriel S, Daly M (2010) The genome analysis toolkit: a mapreduce framework for analyzing next-generation DNA sequencing data. *Genome Res* 20:1297–1303

Morgante M, Hanafey M, Powell W (2002) Microsatellites are preferentially associated with nonrepetitive DNA in plant genomes. Nat Genet 30:194

Mullis K, Faloona F, Scharf S, Saiki R, Horn G, Erlich H (1986) Specific enzymatic amplification of DNA in vitro: the polymerase chain reaction. Cold Spring Harb Symp Quant Biol 51:263–273

Myles S, Peiffer J, Brown PJ, Ersoz ES, Zhang Z, Costich DE, Buckler ES (2009) Association mapping: critical considerations shift from genotyping to experimental design. Plant Cell 21:2194–2202

Nakajima Y, Oeda K, Yamamoto T (1998) Characterization of genetic diversity of nuclear and mitochondrial genomes in Daucus varieties by RAPD and AFLP. Plant Cell Rep 17:848–853

Niemann M, Westphal L, Wricke G (1997) Analysis of microsatellite markers in carrot (*Daucus carota* L. *sativus*). J Appl Genet A 38:20–27

Nordborg M, Weigel D (2008) Next-generation genetics in plants. Nature 456:720

Nothnagel T, Ahne R, Straka P (2005) Morphology, inheritance and mapping of a compressed lamina mutant of carrot. Plant Breed 124:481–486

Nowicka A, Grzebelus E, Grzebelus D (2016) Precise karyotyping of carrot mitotic chromosomes using multicolour-FISH with repetitive DNA. Biol Plant 60:25–36

Parsons J, Matthews W, Iorizzo M, Roberts P, Simon P (2015) *Meloidogyne incognita* nematode resistance QTL in carrot. Mol Breed 35:114

Price AL, Patterson NJ, Plenge RM, Weinblatt ME, Shadick NA, Reich D (2006) Principal components analysis corrects for stratification in genome-wide association studies. Nat Genet 38:904

Pritchard JK, Stephens M, Donnelly P (2000) Inference of population structure using multilocus genotype data. Genetics 155:945–959

Rasheed A, Hao Y, Xia X, Khan A, Xu Y, Varshney RK, He Z (2017) Crop breeding chips and genotyping platforms: progress, challenges, and perspectives. Mol Plant 10:1047–1064

Rong J, Janson S, Umehara M, Ono M, Vrieling K (2010) Historical and contemporary gene dispersal in wild carrot (*Daucus carota* ssp. *carota*) populations. Ann Bot 106:285–296

Rong J, Lammers Y, Strasburg JL, Schidlo NS, Ariyurek Y, De Jong TJ, Klinkhamer PG, Smulders MJ, Vrieling K (2014) New insights into domestication of carrot from root transcriptome analyses. BMC Genom 15:895

Santos C, Simon P (2002) QTL analyses reveal clustered loci for accumulation of major provitamin A carotenes and lycopene in carrot roots. Mol Genet Genom 268:122–129

Santos CA, Simon PW (2004) Merging carrot linkage groups based on conserved dominant AFLP markers in F2 populations. J Am Soc Hort Sci 129:211–217

Scheben A, Batley J, Edwards D (2017) Genotyping-by-sequencing approaches to characterize crop genomes: choosing the right tool for the right application. Plant Biotechnol J 15:149–161

Schulz B, Westphal L, Wricke G (1993) Linkage groups of isozymes, RFLP and RAPD markers in carrot (*Daucus carota* L. sativus). Euphytica 74:67–76

Simon PW (1984) Carrot genetics. Plant Mol Biol Rep 2:54–63

Small RL, Cronn RC, Wendel JF (2004) Use of nuclear genes for phylogeny reconstruction in plants. Aust Syst Bot 17:145–170

Soufflet-Freslon V, Jourdan M, Clotault J, Huet S, Briard M, Peltier D, Geoffriau E (2013) Functional gene polymorphism to reveal species history: the case of the *CRTISO* gene in cultivated carrots. Plos One 8: e70801

St. Pierre MD, Bayer RJ, Weis IM (1990) An isozyme-based assessment of the genetic variability within the *Daucus carota* complex (Apiaceae: Caucalideae). Can J Bot 68:2449–2457

Stelmach K, Macko-Podgorni A, Machaj G, Grzebelus D (2017) Miniature inverted repeat transposable element insertions provide a source of intron length polymorphism markers in the carrot (*Daucus carota* L.). Frontiers Plant Sci 8:725

Tanksley SD (1983) Molecular markers in plant breeding. Plant Mol Biol Rep 1:3–8

Vieira MLC, Santini L, Diniz AL, Munhoz CDF (2016) Microsatellite markers: what they mean and why they are so useful. Genet Mol Biol 39:312–328

Visscher PM, Wray NR, Zhang Q, Sklar P, McCarthy MI, Brown MA, Yang J (2017) 10 years of GWAS discovery: biology, function, and translation. Am J Hum Genet 101:5–22

Vivek B, Simon P (1999) Linkage relationships among molecular markers and storage root traits of carrot (*Daucus carota* L. ssp. *sativus*). Theor Appl Genet 99:58–64

Wang X, Zhao X, Zhu J, Wu W (2005) Genome-wide investigation of iintron length polymorphisms and their potential as molecular markers in rice (*Oryza sativa* L.). DNA Res 12:417–427

Wenzl P, Carling J, Kudrna D, Jaccoud D, Huttner E, Kleinhofs A, Kilian A (2004) Diversity Arrays Technology (DArT) for whole-genome profiling of barley. Proc Natl Acad Sci USA 101:9915–9920

Westphal L, Wricke G (1991) Genetic and linkage analysis of isozyme loci in *Daucus carota* L. Euphytica 56:259–267

Xian-Liang S, Xue-Zhen S, Tian-Zhen Z (2006) Segregation distortion and its effect on genetic mapping in plants. Chin J Agric Biotechnol 3:163–169

Xu Z-S, Tan H-W, Wang F, Hou X-L, Xiong A-S (2014) CarrotDB: a genomic and transcriptomic database for carrot. Database 2014

Yan J, Tang H, Huang Y, Zheng Y, Li J (2003) Genetic analysis of segregation distortion of molecular markers in maize F2 population. Acta Genet Sinica 30

Yildiz M, Willis DK, Cavagnaro PF, Iorizzo M, Abak K, Simon PW (2013) Expression and mapping of anthocyanin biosynthesis genes in carrot. Theor Appl Genet 126:1689–1702

Yu J, Pressoir G, Briggs WH, Bi IV, Yamasaki M, Doebley JF, McMullen MD, Gaut BS, Nielsen DM, Holland JB (2006) A unified mixed-model method for association mapping that accounts for multiple levels of relatedness. Nat Genet 38:203

8 Carrot Molecular Cytogenetics

Marina Iovene and Ewa Grzebelus

Abstract

In this chapter, we review the contribution of cytogenetics to our understanding of the genome organization of the carrot (*Daucus carota* subsp. *sativus*) and its wild *Daucus* relatives. The genus *Daucus* includes about 40, mainly diploid, species with basic chromosome numbers ranging from $n = 8$ to $n = 11$. Early studies have suffered the difficulty to distinguish individual carrot chromosomes. Thanks to the development of carrot genomic resources, reliable chromosome identification and high-resolution karyotyping were obtained by using fluorescence in situ hybridization (FISH) and bacterial artificial chromosomes (BACs) as well as cocktails of repetitive sequences. These advances have contributed to study the organization and distribution of several repeat elements, such as miniature inverted–repeat transposable elements (MITEs) and retrotransposons, identify candidate centromeric and knob-associated repeats in carrot and other *Daucus* species, and begin uncovering syntenic chromosome regions between carrot and other *Daucus* species. Genome size analysis of about ten diploid species indicated a three-fold difference across *Daucus*. However, for many species, basic cytological data remain sketchy. Given the difficult taxonomy and the ongoing revision of the entire genus, we briefly argue that expanding such data as well as comparative cytogenetics studies in *Daucus* will contribute to clarify the phylogeny and perform a more effective exploitation and management of the *Daucus* germplasm.

8.1 Introduction

Daucus carota is a morphologically diverse species that comprises a complex of subspecies with weak crossing barriers and difficult taxonomical delineation (reviewed in Spooner et al. 2014). Cultivated carrot (*Daucus carota* subsp. *sativus*) is the only cultivated species of the genus. Traditionally, the genus *Daucus* has included about 20–25 species mainly centered in the Mediterranean region (Sáenz Laín 1981; Rubatzky et al. 1999). However, a series of molecular studies have reappraised the phylogenetic relationships among *Daucus* and extended the genus boundaries to other nine genera (Arbizu et al. 2014; Banasiak et al. 2016; Spalik and Downie 2007; Spalik et al. 2010; Spooner

M. Iovene (✉)
CNR, Institute of Biosciences and BioResources (CNR-IBBR), 80055 Portici, NA, Italy
e-mail: marina.iovene@ibbr.cnr.it

E. Grzebelus
Faculty of Biotechnology and Horticulture, Institute of Plant Biology and Biotechnology, University of Agriculture in Krakow, 31-425 Krakow, Poland

P. Simon et al. (eds.), *The Carrot Genome*, Compendium of Plant Genomes,
https://doi.org/10.1007/978-3-030-03389-7_8

et al. 2013, 2017). Following these revisions, which are described in Chap. 2, the genus *Daucus* includes now about 40 species and two main clades (Banasiak et al. 2016). These species are for the most part diploids, with basic chromosome numbers ranging from $n = 8$ to $n = 11$.

Carrot has a relatively small genome, estimated at 473 Mb per haploid genome, organized in nine pairs of chromosomes (Arumuganathan and Earle 1991). In the last decade, a growing number of studies have contributed to a tremendous development of the carrot genomic resources. These resources include genetic linkage maps with medium to high resolution (Cavagnaro et al. 2014; Grzebelus et al. 2014); at least two deep-coverage BAC libraries, one generated from a carrot inbred line (Cavagnaro et al. 2009) and another from a "double haploid" line whose genome was sequenced (Iorizzo et al. 2016); the transcriptome of carrot root and leaf tissues from four genetic backgrounds (Iorizzo et al. 2011); and a high-quality assembly of the carrot genome along with several resequenced genomes of cultivated and wild accessions with diverse origin (Iorizzo et al. 2016). These resources have been used to gain insight into the carrot genome organization and evolution, clarify the origin of domesticated carrots, and identify genomic regions, markers, and candidate genes associated with traits of interest (Ellison et al. 2017; Iorizzo et al. 2013, 2016; Macko-Podgórni et al. 2017; Rong et al. 2014).

Molecular cytogenetics is being applied to several genome-related projects of plants. Cytogenetic analyses are instrumental in resolving the order of contigs and tightly linked genetic markers, estimating gap sizes within sequenced genomic regions, integrating heterochromatic domains in genetic and physical maps (Cheng et al. 2001; Iovene et al. 2008b; Saski et al. 2017; Shearer et al. 2014; Szinay et al. 2008; Wang et al. 2006), and characterizing repetitive sequences as well as specific chromosomal structures, such as centromeres and knobs (Ávila Robledillo et al. 2018; Fransz et al. 2000; Gong et al. 2012; Tek et al. 2005). In addition, cytogenetics plays an important role in comparative genomics, by revealing the chromosome rearrangements underlying the karyotypic variation among related plant species of several families. Such studies have been reported for Brassicaceae (Lysak et al. 2005; Mandáková et al. 2017; for a review, see also Lysak et al. 2016), Solanaceae (Braz et al. 2018; Gaiero et al. 2017; Lou et al. 2010; Szinay et al. 2012), Cucurbitaceae (Han et al. 2015; Lou et al. 2014) and Poaceae (Betekhtin et al. 2014; Dong et al. 2018; Ma et al. 2010).

Thanks to the availability of genetic and genomic resources, molecular cytogenetics has been successfully applied to carrot, which has allowed a reliable identification of the carrot chromosomes, high-resolution karyotyping and the characterization of heterochromatic domains spanned by repetitive elements. However, for most *Daucus* non-*carota* species, the basic cytological data, such as chromosome number and genome size, remain sketchy. Cytogenetics could provide complementary tools useful to achieve a refined elucidation of the carrot genome organization and contribute to a better understanding of the relationships among *Daucus* species by uncovering chromosomal differences and the underlying mechanisms. In this chapter, we review the contribution of the past and recent cytogenetic researches to our understanding of the carrot genome organization. Prospective applications of cytogenetics to *Daucus* comparative genomics are briefly discussed.

8.2 Chromosome Numbers and Classical Cytogenetic Studies in *Daucus*

Classical cytogenetics has provided information on the chromosome number and ploidy status of carrots and several *Daucus* species. However, due to the difficulty to distinguish individual carrot chromosomes, conventional cytogenetics has given a limited contribution to carrot genome research. One of the first somatic chromosome count of carrot traces back to 1932 (Lindenbein 1932; reviewed by Sharma and Ghosh 1954). Subsequent cytotaxonomic studies have confirmed that both cultivated and wild forms of *D. carota* are diploid with nine pairs of chromosomes (Bell and Constance 1960; Sharma and Bhattacharyya 1959; Sharma and Ghosh 1954). Other species, namely *D. syrticus*, *D. sahariensis* (Aparicio Martínez 1989), and two members of the recently added *Tornabenea* genus (*D. annuus* and *D. insularis*, for which the chromosome number is known), have $2n = 18$ chromosomes (Grosso et al. 2008). Conversely, most *Daucus* species (including the majority of those recently added) have chromosome numbers of $2n = 20$ or 22 (Bell and Constance 1957, 1960, 1966; Constance et al. 1976; Rice et al. 2015). In addition, the inclusion of *Cryptotaenia elegans* and *Pseudorlaya* spp. under the genus expands the range of variation to $2n = 16$ (Suda et al. 2005; Vogt and Oberprieler 1994, 2009). *Daucus* are for the most part diploid species but at least five polyploids, that is, the tetraploid *D. glochidiatus* ($2n = 44$), the hexaploid *D. montanus* ($2n = 66$), and the tetraploid species *D. incognitus*, *D. melananthos,* and *D. pedunculatus* (all $2n = 44$) formerly under *Agrocharis* genus (Banasiak et al. 2016), exist (Constance et al. 1976; Constance and Chuang 1982). To our knowledge, the record of the base chromosome number for *Daucus* is incomplete (e.g., there is no report for *D. mauritii* and the recently added *D. dellacellae* and *D. mirabilis*). In addition, for a few species (e.g., *D. durieua*), different chromosome counts are reported (Luque and Lifante 1991 and references therein). This discrepancy could potentially arise from species misidentification due to the difficult taxonomy of *Daucus*, and it calls for the need of a reassessment of the chromosome numbers of wild *Daucus*, in the frame of the revised classification of the genus.

Several cytotaxonomic studies have described the karyotype of various accessions of *D. carota*. Bayliss (1975) reported that carrot mitotic chromosomes have average length of ~ 1.5 µm and are metacentric to submetacentric, except for the chromosome pair with a prominent satellite which has a submetacentric to subtelocentric centromere. However, an apparent intraspecific karyotype variation emerges among several cytological studies, with differences in the chromosome morphology and the number of the secondary constrictions (Hamal et al. 1986; Sharma and Bhattacharyya 1959; Sharma and Ghosh 1954; Subramanian 1986). These differences could be rather an artifact due to the difficulty to identify accurately the centromere position and specific chromosomes. Indeed, all studies agreed that carrot chromosomes were uniformly short and difficult to distinguish based on their shape. In addition, classical staining procedures such as C and Q banding have provided limited aid to the chromosome identification of carrot (Essad and Maunoury 1985; Kumar and Widholm 1984). A better discrimination with Giemsa C banding was achieved by using carrot prometaphase chromosomes (Schrader et al. 2003).

Since the dawn of plant in vitro culture, carrot has served as a model system to develop in vitro culture procedures and study the process of somatic embryogenesis and the behavior of cultured cells (for a review see Sussex 2008). Therefore, a number of studies have addressed the questions of what type of chromosomal changes occurred in in vitro cells, what conditions contributed to such instability, and whether aneuploidy and polyploidy arising in certain carrot cell lines were responsible for the decline of totipotency (Al-Safadi and Simon 1990; Bayliss 1973, 1975, 1977; Smith and Street 1974). Bayliss (1973, 1975) described carrot aneuploid cell lines with 17 chromosomes, supposedly the result of a translocation. In addition to aneuploid and polyploid lines, a haploid cell line (HA) developed from a haploid carrot seedling was described and karyotyped (Smith et al. 1981; also reviewed in Simon 1984). This HA suspension has provided a valuable tool in several studies related to embryo development in plants (Borkird and Sung 1987). Aneuploid and polyploid plants have been regenerated from both protoplast and cell cultures (Dudits et al. 1976; Grzebelus et al. 2012; Sung and Jacques 1980) as well as after protoplast fusion (Dudits et al. 1977; Lazar et al. 1981). Dudits et al. (1976) reported that the inflorescences of both tetraploids and hexaploids regenerated from carrot protoplasts had normal phenotype;

however, the meiotic stability of these materials was not analyzed. In fact, there are only a few reports on the analysis of the carrot meiotic chromosomes. This is partly because carrot flowers are minute and difficult to manipulate, which complicates the preparation of meiotic chromosomes. Zenkteler (1962) conducted a comparative analysis of the microsporegenesis of male-fertile versus male-sterile plants. Male-fertile plants had a regular meiosis, with nine bivalents up to metaphase I. Conversely, several irregularities occurred during the meiosis of male-sterile plants, including a cross-shape configuration at pachytene and multivalent pairing at diakinesis, both indicative of a heterozygous reciprocal translocation (Zenkteler 1962). Sinha and Sinha (1978) confirmed regular bivalent formation in the pollen mother cells of two fertile carrot varieties. Most paired chromosomes (mean values of 6.8 and 7.5, depending on the variety) formed rings, and the remaining chromosomes paired as rods. The average number of chiasma per chromosome arm was ~0.9 (Sinha and Sinha 1978).

8.3 Nuclear Genome Size

The amount of DNA in an unreplicated gametic nuclear genome (known as C-value) is commonly used to describe the nuclear genome size of a species (Bennett and Leitch 2011) and is expressed in pg or Mb (1 pg = 978 Mb; Doležel et al. 2003). The C-value is an important parameter in phylogenetic studies because it contributes to species identification and to uncover misclassifications in germplasm collections as well as polyploidization/aneuploidization events and large-scale structural rearrangements such as large deletions/duplications or insertions (Nowicka et al. 2016b; Sliwinska 2018). The circumscription of species and genera with difficult taxonomy, such as *Daucus*, could greatly be benefitted by the integration of molecular phylogenetic and morphometric studies with the nuclear genome size analysis. To obtain meaningful nuclear genome size data, it is necessary to evaluate a large number of accessions and individuals, especially when dealing with species complexes and genera, such as *D. carota* and the *Daucus* genus, that lack a comprehensive taxonomic treatment (Nowicka et al. 2016b).

Owens (1974) reported the first nuclear DNA content estimates of several *Daucus* species in his doctoral thesis, which were later reviewed by Bennett and Smith (1976) in their large compilation of plant DNA amounts. These first estimates were based on Feulgen microdensitometry using *Allium cepa* as a DNA standard, and covered five *D. carota* subspecies (including cultivated carrot) and seven non-*carota Daucus* species. The 2C-values ranged from 2 pg in *D. carota* subsp. *carota* to 11 pg in *D. montanus*, which are much higher than the flow cytometry (FCM) values published later. However, such high C-values for *Daucus* species were not reported in any following experiments, and because polyploidy and supernumerary chromosomes are not common in this genus, they likely represented overestimates reflecting technical shortcoming or species misidentification. Indeed, subsequent FCM-based studies reported consistent values of nuclear DNA content in 2C nuclei of (slightly less than) 1.0–1.1 pg (Arumuganathan and Earle 1991; Bennett and Leitch 1995; Bai et al. 2012; Pustahija et al. 2013). However, these studies relied on the analysis of a single *D. carota* population with the exception of the work of Pustahija et al. (2013), who evaluated three populations collected from western Balkan regions (Table 8.1). A comprehensive survey, based on a large number of *Daucus* species, accessions and plants per accessions, confirmed that the cultivated carrot has a mean 2C-value of 0.96 pg (Nowicka et al. 2016b). This study analyzed multiple individuals from 26 cultivated carrots, including cultivars and advanced breeding lines with orange or purple roots, and landraces producing orange, yellow, or purple roots (Table 8.1). In addition, in the same study, 14 wild *D. carota* subspecies (each including several accessions and individuals) had 2C-values ranging from ~0.9 to ~1.1 pg (Table 8.1; Nowicka et al. 2016b). In another study based on different accessions, Tavares et al. (2014) assessed the genome size of four subspecies of *D. carota* native to Portugal, each represented by two to six populations and up to six individuals per population. Similar to the results obtained by Nowicka et al. (2016b), the measurements were highly reproducible with low variation in 2C-values among individuals of the same population (Tavares et al. 2014). However, the 2C-values reported by Tavares et al. (2014) were slightly higher compared to other published estimates for the same subspecies, and varied from 1.21 to 1.26 pg/2C (Table 8.1). Genome size data are also available for another approximately ten wild *Daucus* species (Table 8.1; Nowicka et al. 2016b; Suda et al. 2005). Nowicka et al. (2016b) found three-fold difference for the genome size of nine diploid *Daucus* species, with 2C-values ranging from about 1 pg in *D. carota* species complex ($2n = 18$) to 3.02 pg in *D. littoralis* ($2n = 20$). Differences among accessions within the same taxon were usually small. The only exception was *D. guttatus*, which displayed large differences in the nuclear DNA content among accessions (1.49–2.83 pg; Table 8.1). This discrepancy likely reflects the complicated taxonomy of *D. guttatus*, which is indeed a species complex including four species (Arbizu et al. 2016). On the other hand, differences in the 2C-content among diploid *Daucus* species are not related to their different chromosome number (Nowicka et al. 2016b) and the origin of such diversity (e.g., differential activity of mobile elements) remains to be investigated.

8.4 Development of Chromosome-Specific Probes for Chromosome Identification and Integration of Genetic and Cytological Maps in Carrot

Efficient methods for reliable chromosome identification are the foundation for cytogenetic research in both animals and plants. In species with small to medium size genomes, fluorescence in situ hybridization (FISH), coupled with the use of large insert genomic libraries, represents a well-established tool for chromosome identification, karyotyping, and integration of the chromosomal features in the genetic linkage map of a species (Cao et al. 2016; Chao et al. 2018; Dong et al. 2000; Pedrosa-Harand et al. 2009; Zhang et al. 2010).

In carrot, a set of 15 bacterial artificial chromosomes (BACs) was selected by screening two carrot BAC libraries with various types of molecular markers, but mainly sequence-tagged site (STS) markers and SSR markers (Cavagnaro et al. 2009; Grzebelus et al. 2007; Just et al. 2007). Markers anchoring the BACs mapped to the nine carrot linkage groups (LGs), with one to three markers for each LG. The selected BACs were used as FISH probes for mitotic and meiotic chromosome identification and integration of the genetic linkage groups of carrot with the carrot pachytene chromosomes (Fig. 8.1; Iovene et al. 2011). This way, each carrot linkage group was assigned to a specific chromosome, and six of them were oriented according to the short (north)/long (south) arm of the corresponding chromosome. In the same work, these chromosome-specific BACs provided a framework for the localization of additional DNA sequences with unknown genetic position relative to the markers used in the initial library screening. These sequences included the rDNA gene clusters and other six BACs (adding up to 21 BACs), which were either not mapped or mapped in diverse, unrelated mapping populations. For example, the FISH signal of BAC 2B20, selected for a marker linked to the nematode resistance locus *Mj-1* from a different genetic map (Boiteux et al. 2000), was located on the long arm of chromosome 8 (LG9), distal to BAC 9K15 (containing the STS marker for *LCYE*, from LG9) and to the 5S rDNA (Fig. 8.1). Using two-color FISH, reliable identification of the carrot chromosomes was achieved by using a probe cocktail containing a subset of ten BACs (Iovene et al. 2011).

Several other BACs have been mapped on the carrot pachytene chromosomes in the frame of the carrot genome sequencing project (Iorizzo

Table 8.1 Summary of the nuclear genome size analysis in several *Daucus*

Species name (2*n*)[a]	2C DNA content (pg)		No. pop.[d]	No. ind.[e]	Reference
	FCM[b]	Range[c]			
D. carota (18)	1.03	0.98–1.10	4	nd	Arumuganathan and Earle (1991), Bennett and Leitch (1995), Bai et al. (2012), Pustahija et al. (2013)
D. carota ssp. *azoricus*	1.06	–	1	15	Nowicka et al. (2016b)
D. carota ssp. *carota*	1.06	0.95–1.24	5	47	Tavares et al. (2014), Nowicka et al. (2016b)
D. carota ssp. *commutatus*	0.98	–	1	13	Nowicka et al. (2016b)
D. carota ssp. *drepanensis*	0.99	–	1	15	Nowicka et al. (2016b)
D. carota ssp. *gadecaei*	0.98	–	1	12	Nowicka et al. (2016b)
D. carota ssp. *gummifer*	1.11	0.99–1.29	5	43	Tavares et al. (2014), Nowicka et al. (2016b)
D. carota ssp. *halophilus*	1.18	1.12–1.33	7	36	Tavares et al. (2014), Nowicka et al. (2016b)
D. carota ssp. *hispanicus*	0.95	–	1	10	Nowicka et al. (2016b)
D. carota ssp. *hispidifolius*	0.98	0.97–1.00	2	26	Nowicka et al. (2016b)
D. carota ssp. *hispidus*	1.09	–	1	10	Nowicka et al. (2016b)
D. carota ssp. *libanotifolia*	0.97	–	1	7	Nowicka et al. (2016b)
D. carota ssp. *major*	0.95	0.94–0.96	2	30	Nowicka et al. (2016b)
D. carota ssp. *maritimus*	0.96	–	1	16	Nowicka et al. (2016b)
D. carota ssp. *maximus*	1.12	0.99–1.26	6	63	Tavares et al. (2014), Nowicka et al. (2016b)
D. carota ssp. *sativus*	0.96	0.95–0.98	26	403	Nowicka et al. (2016b)
D. broteri (20)	2.07	1.91–2.22	2	19	Nowicka et al. (2016b)*
D. crinitus (22)	2.39	2.37–2.40	2	20	Nowicka et al. (2016b)
D. elegans (16)	0.94	–	1	3–6	Suda et al. (2005)
D. guttatus (20)	2.05	1.49–2.83	3	23	Nowicka et al. (2016b)*
D. involucratus (22)	1.80	1.79–1.81	3	35	Nowicka et al. (2016b)
D. littoralis (20)	3.02	–	1	9	Nowicka et al. (2016b)
D. montevidensis (22)	1.30	–	1	7	Nowicka et al. (2016b)
D. muricatus (20)	1.99	1.97–2.04	3	45	Nowicka et al. (2016b)
D. pusillus (22)	1.30	1.29–1.40	3	41	Nowicka et al. (2016b)

[a]Species names and somatic chromosome numbers (in brackets) reported as in the corresponding references. *D. elegans* was previously classified as *Cryptotaenia elegans*

[b]FCM, flow cytometric measurement using propidium iodide; mean 2C DNA content (pg) calculated for each taxon based on the data presented in the corresponding references

[c]Range of the average 2C-values as reported in the corresponding references

[d]No. pop., the total number of analyzed populations/accessions calculated based on the corresponding references

[e]No. ind., the total number of analyzed individuals per taxon calculated based on the corresponding references

*The data of Nowicka et al. (2016b) were revised to account for the fact that one accession of *D. broteri* (Ames 25879) is reclassified as *D. guttatus* in the U.S. National Plant Germplasm System

– No variation

nd No data

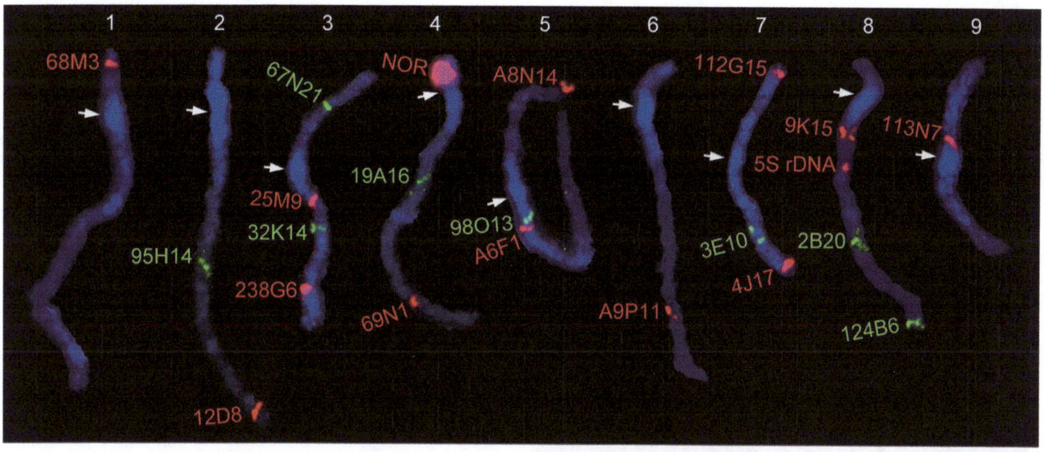

Fig. 8.1 Representative carrot pachytene chromosomes and their association with the nine genetic linkage groups (LGs) of the carrot based on FISH using LG-specific BAC clones (green and red signals). Chromosomes were counterstained with DAPI (blue). The numbering of the chromosomes is according to their decreasing length. BAC names are reported on the left of each chromosome. BAC clone information is provided in Table 1 of Iovene et al. (2011). Arrows indicate the centromeric regions. Reused and modified with kind permission of Springer Science+Business Media B.V. from Iovene et al. (2011)

et al. 2016). These included fourteen clones from the BAC library of carrot DH1 (the genotype whose genome was sequenced), which were identified to contain sequences that unambiguously aligned at the ends of the pseudomolecules of chromosome 1, 2, 4, 5, 6, 8, and 9. These BACs were FISH-mapped on carrot pachytene chromosomes along with a telomeric probe $(TTTAGGG)_n$, in order to evaluate the consistency and the coverage of the carrot genome assembly in the subtelomeric–telomeric regions (Iorizzo et al. 2016). Apart from BAC clones, the feasibility of using other sources or types of single/low copy sequences as carrot chromosome-specific markers has not been fully investigated. Recently, Macko-Podgórni et al. (2017) have successfully used a FISH probe made of bulked DNA fragments obtained through long-range PCR to map a region located at the distal region of the long arm of carrot chromosome 2. This FISH probe covered almost entirely the 37 kb long genomic region that the authors had identified to be under selection in cultivated carrot and to include a candidate gene for carrot domestication (Macko-Podgórni et al. 2017).

8.5 Carrot Pachytene-Based Karyotype and Candidate Centromeric- and Knob-Associated Tandem Repeats

A FISH-based karyotype of carrot was developed by measuring the length of each individual carrot pachytene chromosome in 24 best pollen mother cells of the inbred line B2566. The pachytene chromosomes were ordered from 1 to 9 according to their descending length. Each carrot pachytene was readily identified by using the chromosome-specific BACs described above. In addition, pachytene chromosomes could be distinguished, with relative ease, based on their length, arm ratio and DAPI staining pattern. The average length of the carrot pachytene complement measured $\sim 193 \pm 18$ µm, which represented about sevenfold the length of the somatic metaphase counterpart (Iovene et al. 2008a, 2011). Carrot chromosome 1 was about 27 µm long and covered approximately 14% of the total karyotype length, whereas the other chromosomes represented each 13 to 8% of the

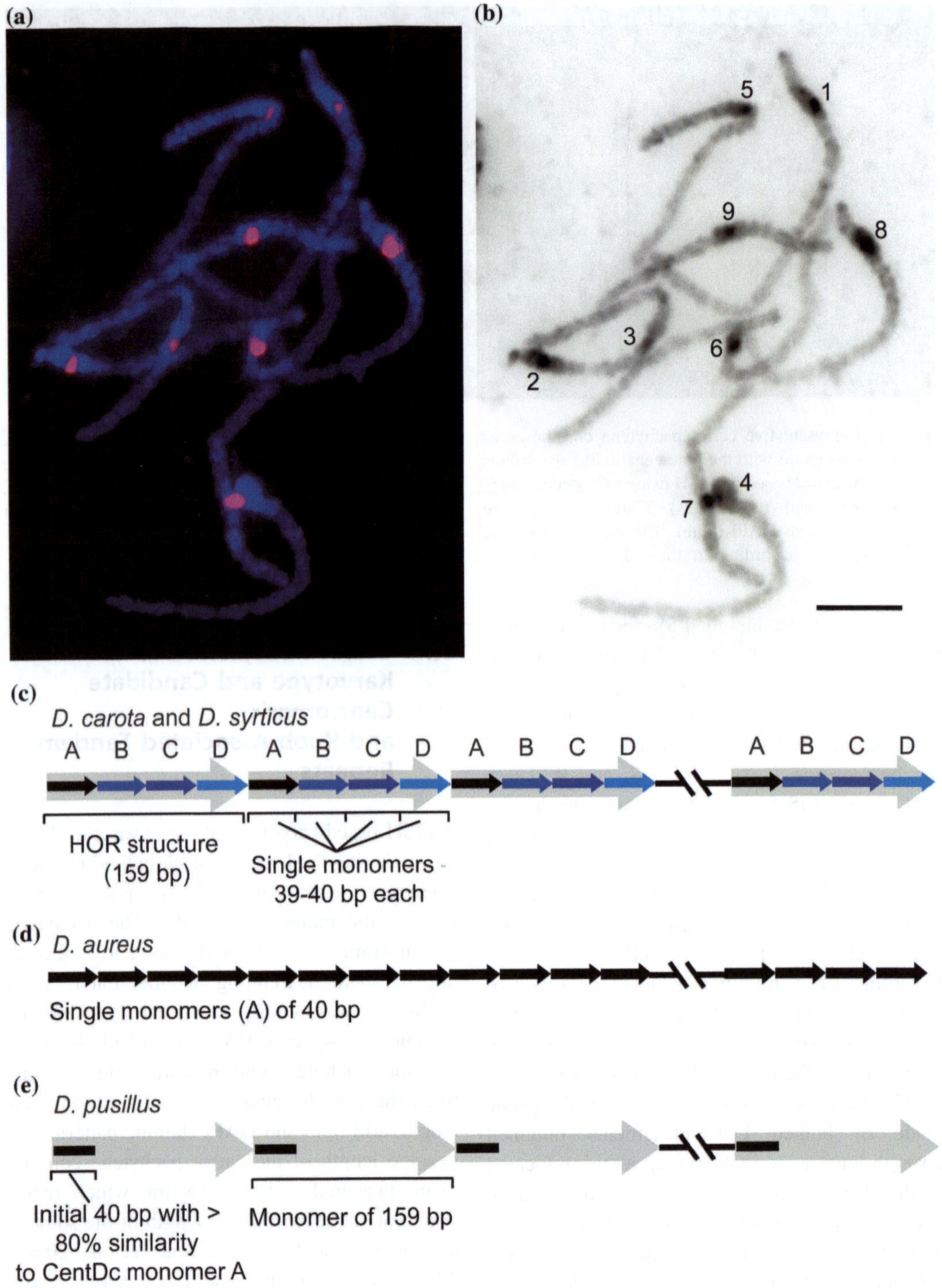
(a)
(b)
5
1
9
8
3
6
2
4
7
(c)
D. carota and D. syrticus
A B C D A B C D A B C D A B C D
HOR structure
(159 bp)
Single monomers -
39-40 bp each
(d)
D. aureus
Single monomers (A) of 40 bp
(e)
D. pusillus
Initial 40 bp with >
80% similarity
to CentDc monomer A
Monomer of 159 bp

◀ **Fig. 8.2** Localization and organization of CentDc-like repeats. **a** FISH mapping of CentDc repeat (red signals) on carrot pachytene chromosomes. Chromosomes are counterstained with DAPI (blue) and **b** presented as black and white image to enhance the visualization of the heterochromatic domains. Chromosomes are numbered according to their decreasing length. Bar = 5 μm. Reused and modified with kind permission of Springer Science +Business Media B.V. from Iovene et al. (2011). **c–e** Comparative organization of CentDc-like repeats among various *Daucus* spp. Note that the length of the CentDc cluster arrays is unknown. **c** Carrot and *D. syrticus* ($2n = 18$): CentDc repeat units of 159 bp represent higher-order repeat (HOR) structures, each made of four monomers (A, B, C, D) of 39–40 bp, which are arranged in the same order in adjacent HORs. The A, B, C, D monomers are represented by different colors to reflect the SNPs in their sequences. **d** *D. aureus* ($2n = 22$): CentDc-like repeat units of 40 bp (black arrows), most similar to CentDc monomer A. **e** *D. pusillus* ($2n = 22$): the initial portion (40 bp; thick black lines in the gray arrows) of the most abundant tandem repeat of *D. pusillus* (~159 bp long) shares >82% similarity with CentDc monomer A. The remaining portion of this 159 bp tandem repeat of *D. pusillus* differs from CentDc. Drawn based on the data published by Iorizzo et al. (2016)

karyotype length. Chromosome arm ratios (long/short) ranged from 1.2 to 10.6, but most chromosomes had a ratio within the range 1.2–4.9. Heterochromatic regions, which stain brightly with DAPI, represented a small fraction of all the chromosomes and were mainly located in the pericentromeric regions. However, centromeric regions of the carrot pachytene chromosomes did not have the obvious primary constrictions and the distinct differential staining that are instead observed in other species, such as tomato and maize. Related to this aspect, a candidate centromeric tandem repeat family (named CentDc) was identified in carrot, which allowed performing more accurate measurements (Fig. 8.2a, b). CentDc repeats were isolated from BAC 4H08, a clone that was initially selected for the *phytoene synthase 1* gene (PSY1; Cavagnaro et al. 2009) and it was expected to contain mainly low/single copy sequences. However, the FISH analysis revealed that BAC 4H08 hybridized to the centromeric regions of all carrot chromosomes. In addition, the FISH signals of this BAC overlapped with those generated by the carrot cot-1 DNA fraction, corroborating that it contained a dominant centromeric repeat of the carrot (Iovene et al. 2011). The partial sequencing of this BAC revealed the typical structure of a CentDc repeat unit with monomers of ~159 bp. Several lines of evidence suggested that each CentDc unit of ~159 bp represents, in turn, a higher-order repeat (HOR) structure, in that a typical 159 bp repeat motif is itself composed of four shorter monomers of 39–40 bp (Fig. 8.2c; Iorizzo et al. 2016; Iovene et al. 2011). Indeed, the 39–40 bp monomers (named A, B, C, and D) have accumulated several private polymorphisms each. These shorter monomers are repeated in the same order, that is, ABCD (Fig. 8.2c). The average pairwise similarity among these 39–40 bp monomers is lower than that among adjacent CentDc unit of ~159 bp (Iorizzo et al. 2016), which is a typical feature of HORs (Melters et al. 2013).

It is well established that the centromere function is determined epigenetically by the presence in the centromeric chromatin of a specialized histone H3 variant, known as CENPA in humans and CENH3 in plants (McKinley and Cheeseman 2016, for a review). Recently, Dunemann et al. (2014) developed an antibody against the carrot CENH3 based on the analysis of the CENH3 gene in carrot and three wild *Daucus* species. The immunofluorescence assays indicated that the anti-DcCENH3 antibody localizes to centromeres of carrot chromosomes as well as those of *D. glochidiatus*, indicative of the cross-reactivity of the *D. carota* antibody with CENH3 of other *Daucus* species (Dunemann et al. 2014). Unfortunately, to our knowledge, this anti-DcCENH3 antibody has not been used yet to confirm the association between CentDc repeats and the CENH3-containing nucleosomes. However, FISH signals derived from CentDc repeats hybridized at the most poleward position of each carrot chromosome at both meiotic metaphase I and mitotic anaphase, suggesting that CentDc repeats are indeed associated with the kinetochore complex (Iovene et al. 2011; Nowicka et al. 2016a).

In addition to the pericentromeric heterochromatin, small heterochromatic domains were consistently detected in other chromosomal regions. The short arm of chromosome 2 was almost entirely heterochromatic and ended with a terminal heterochromatic knob. The short arm of chromosome 4, also brightly stained by DAPI, was occupied by the 18S-25S rDNA sequences. Finally, the long arm of carrot chromosome 1 had a small heterochromatic knob located at ~39% from the end of the short arm. This knob is associated with another abundant satellite repeat family (named CL80) of carrot. CL80 repeat units are 169 bp long and their sequences are highly homogenized in the carrot genome (Iorizzo et al. 2016). An in silico search of the CL80 sequences throughout the carrot genome indicated that most CL80 repeats localize on chromosome 1 at the junction between superscaffold 7 and 8 of the corresponding assembled pseudomolecule (Iorizzo et al. 2016). Indeed, the FISH signal of CL80 overlapped with the knob on chromosome 1 and spanned the chromosomal region between carrot BACs 20G08 and 20P12, which were selected from the superscaffolds 7 and 8, respectively (Iorizzo et al. 2016).

8.6 Karyotyping Using Carrot Repetitive Sequences

In addition to rDNA sequences, other repetitive sequences identified in the carrot genome have been used as FISH probes to provide a reliable hybridization pattern for the identification of the carrot mitotic chromosomes as well as investigate their distribution along the carrot chromosomes. Nowicka et al. (2012) used as FISH probes random amplified polymorphic DNA (RAPD) amplicons obtained from a group of accessions representing carrot genetic diversity. These amplicons were abundant and non-polymorphic among carrot accessions, an indication of both their repetitive nature and sequence conservation in carrot (Nowicka et al. 2012). This way, the authors analyzed 13 RAPD products, ranging in size from 517 to 1758 bp. Four of such probes (B4A, C15A, n75, and T20B4) produced clear and reproducible hybridization patterns on most or all chromosomes. The majority of the signals were confined to the pericentromeric regions and had a prevalent dot-like hybridization pattern, suggesting that these sequences are organized in clusters comprising many copies (Nowicka et al. 2012). These probes had sequence similarity to coding portions of *gypsy* (C15A, n75) and *copia* (T20B4) retrotransposons of plant species distantly related to carrot, as well as to carrot BAC-end sequences (Cavagnaro et al. 2009). Simultaneous hybridization of either two RAPD-PCR probes in combination with CentDc repeats generated a specific FISH pattern that enabled individual chromosome identification.

Miniature inverted–repeat transposable elements (MITEs), which are particularly abundant and diversified in the carrot genome, have provided another source of FISH landmarks for the carrot chromosomes (Macko-Podgorni et al. 2013; Nowicka et al. 2016a). Both *Stowaway*-like (named *DcSto*) and *Tourist*-like (named *Krak*) elements, the two most abundant groups of MITEs in plants (Jiang et al. 2004), were identified in carrot. *Stowaway*-like *DcSto* elements are 300 bp long and are present in >4000 copies in the diploid carrot genome (Macko-Podgorni et al. 2013; Iorizzo et al. 2016). *Tourist*-like *Krak* elements are less than 400 bp long and have an estimated copy number in carrot of about 3600 (Grzebelus et al. 2007; Grzebelus and Simon 2009). However, Iorizzo et al. (2016) identified only about 400 *Krak* copies in the carrot assembled genome that carried intact terminal inverted repeats. FISH using *DcSto* and *Krak* resulted in a pattern of signals widely dispersed along all chromosome arms with intercalary and pericentromeric localization (Nowicka et al. 2016a; Fig. 8.3). Several *DcSto* signals were located in the euchromatic regions (Fig. 8.3). In addition, both MITE groups were not detected in the centromeric, telomeric, and nucleolar organizer regions. Iorizzo et al. (2016) found evidence that *DcSto* and *Krak* elements are randomly distributed in the carrot genome, and

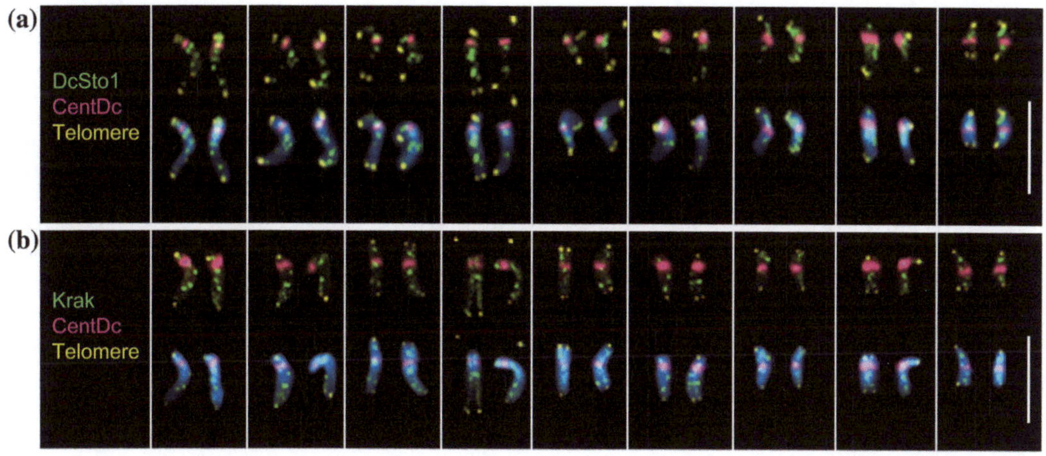

Fig. 8.3 Identification of the carrot mitotic chromosomes by FISH using a cocktail of repetitive sequences including the miniature inverted–repeat transposable element **a** DcSto1 probe or **b** Krak probe along with the centromeric (CentDc) and telomeric repeats. The top row in each panel shows the FISH hybridization signals; in the bottom row, the hybridization signals are superimposed on the DAPI-stained chromosomes. The chromosomes were paired based on their hybridization pattern and size and ordered according to their decreasing length. Bar = 5 µm. Reused and modified with kind permission of Springer Science+Business Media Dordrecht from Nowicka et al. (2016a)

not preferentially inserted into or near genes. The *DcSto* probe produced a pronounced dot-like banding pattern with stronger signals compared to *Krak*. Moreover, the intensity of the *DcSto* signals differed considerably among chromosomes, while the hybridization pattern of *Krak* was relatively uniform (Nowicka et al. 2016a; Fig. 8.3). Hybridization using *DcSto* or *Krak* probe, along with CentDc and the *Arabidopsis*-type telomeric probe, enabled the authors to distinguish the mitotic chromosome pairs (Fig. 8.3). In the same work, Nowicka et al. (2016a) investigated the distribution and the usefulness for karyotyping of repeat elements specific to *D. carota* (DCREs) previously identified by Cavagnaro et al. (2009). Out of eleven DCRE repeats screened as potential chromosome landmarks, three (DCRE9, DCRE16, and DCRE22) produced a specific FISH pattern on the carrot mitotic complement (Nowicka et al. 2016a; Fig. 8.4). These DCRE repeats had an estimated length of 388 bp (DCRE9), 677 bp (DCRE16), and 896 bp (DCRE22) and an estimated copy number in the carrot genome of 7340, 4621, and 2990, respectively (Cavagnaro et al. 2009). DCRE probes differed for their hybridization pattern and intensity of the signals (Nowicka et al. 2016a). In addition, for each probe, the intensity of the signals differed among the carrot chromosomes, an indication of different amounts of DCRE elements among chromosomes. Among the probes, DCRE9 produced the strongest dot-like pattern. Conversely, the DCRE22 probe generated the weakest signals, the majority of which were localized in pericentromeric regions (Fig. 8.4). The strongest DCRE22 signals were located on the NOR-bearing chromosome pair. Finally, DCRE16 hybridized to the pericentromeric region of all chromosomes with additional centromeric signals on two chromosome pairs (Fig. 8.4). The most pronounced DCRE16 signal was located on NOR-bearing chromosome pair. The differences in the strength of the FISH signals among the chromosomes, along with chromosome measurements and hybridization with CentDc, enabled to distinguish the carrot mitotic chromosomes (Nowicka et al. 2016a; Fig. 8.4).

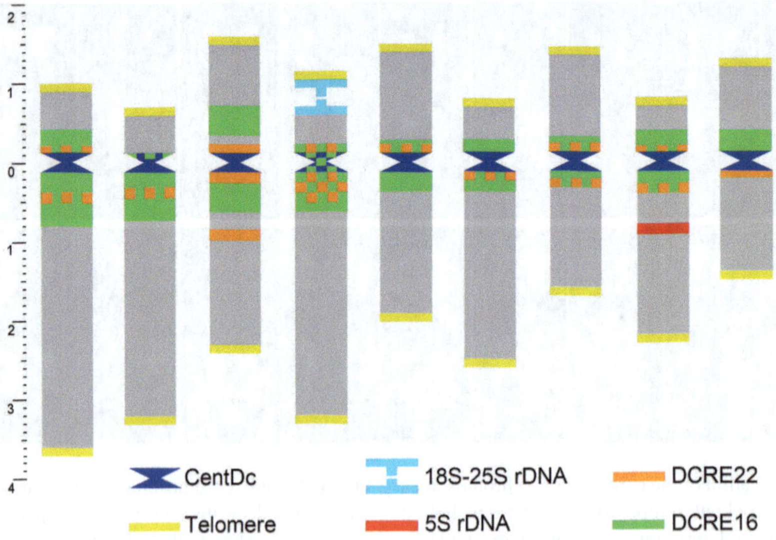

Fig. 8.4 Ideogram showing the FISH distribution of two *D. carota* repetitive elements (DCRE22 and DCRE16) on the carrot mitotic chromosomes in relation to the centromeric (CentDc), telomeric, and rDNA sequences. Chromosome length and arm ratio are based on the measurements reported in Table 1 of Nowicka et al. (2016a). Scale in micrometers. Reused and modified with kind permission of Springer Science+Business Media Dordrecht from Nowicka et al. (2016a)

8.7 Comparative Cytogenetics Among *Daucus*

Comparative cytogenetic tools have been instrumental to uncover large-scale chromosome changes and the mechanisms responsible for karyotype diversity among related species, especially within mammals (Ferguson-Smith and Trifonov 2007 for a review). In plants, comparative cytogenetics has mainly relied on the FISH mapping of chromosome-specific BACs from a given species on the chromosomes of its close relatives. In Brassicaceae, several favorable conditions have made it possible to develop painting probes covering long chromosome regions by pooling dozens of closely spaced BACs from *Arabidopsis thaliana* containing single/low copy sequences (Lysak et al. 2016 for a review). However, in most species from other plant families, a smaller number of BACs per chromosome has been usually used due to the difficulty to exclude repetitive DNA sequences from these cocktail probes (Fonsêca et al. 2016; Gaiero et al. 2017; Lou et al. 2010; Yang et al. 2014). Following the strategy of cross-species BAC-FISH, a subset of the carrot chromosome-specific BACs were applied to *D. crinitus* and *D. pusillus* (both $2n = 22$), belonging to *Daucus* clade I and II, respectively. This preliminary work began to uncover syntenic chromosome regions among these species (Iovene et al. 2011). For example, four clones located on carrot chromosome 3 (67N21 on 3S and 25M9, 32K14, and 238G6 on 3L) hybridized to two different chromosome pairs in both the wild species, with 67N21, 25M9, and 32K14 on a same chromosome and BAC 238G6 on a different chromosome. The analysis also indicated that the NOR-bearing chromosome of carrot (chromosome 4) is either not homologous to that of *D. crinitus*, or highly rearranged due to translocation(s). In addition, the 5S rDNA-bearing chromosome of the carrot (chromosome 8) is likely homologous to that of *D. crinitus* and *D. pusillus*. However, the relative order of the carrot chromosome 8-specific BACs and the 5S rDNA was different in any pairwise comparison among these three species, possibly suggesting the involvement of at least one inversion. In addition to single/low copy sequences, various repetitive sequences have been analyzed by FISH to gain insights into the origin of the genome size differences among related species and the evolutionary dynamics of specific repeat elements (Gong et al. 2012; Park et al. 2012). One of the first of such studies in *Daucus* analyzed the distribution of the rDNA gene clusters and the karyotypes in eight species with various phylogenetic distance to carrot (Iovene et al. 2008a). The cultivated carrot and its close relative *D. carota* subsp. *capillifolius* (both $2n = 18$) had similarly short chromosomes with uniform morphology (total mitotic karyotype length of 56 ± 6 and 59 ± 11 μm, respectively). *D. crinitus* ($2n = 22$) along with *D. littoralis* and *D. muricatus*, two species with $2n = 20$, had the longest karyotypes, about 1.7–2 times longer than that of *D. carota*. This figure likely reflects the two- to three-fold increase of their genome size compared to the carrot genome (Nowicka et al. 2016b; see also the genome size section of this chapter). Each species examined, including the tetraploid *D. glochidiatus*, had a single 5S rDNA and a single 18S-25S rDNA site (one chromosome pair each). The 18S-25S rDNA site was invariably terminally located. Conversely, the 5S rDNA locus was located interstitially on the long arm of a metacentric/submetacentric chromosome pair, except for *D. crinitus* in which it localized at the end of the short arm of a metacentric/submetacentric pair, suggesting the involvement of a chromosome rearrangement compared to the other species. Two additional repeats specific to carrot (that is, the carrot centromeric satellite repeat CentDc and the satellite repeat CL80 which hybridized to a knob on carrot chromosome 1) were analyzed in representative species with $2n = 18$, 20, and 22 of the two main *Daucus* clades (Iorizzo et al. 2016). This comparative analysis was carried out both in silico and cytologically. The analysis indicated that both CentDc-like and CL80-like repeats differed among species in terms of repeat sequence, structure, abundance, and distribution. However, there is an indication that origin of both repeats predated the divergence of the two *Daucus* clades (Iorizzo et al. 2016). CentDc-like repeats represented the most abundant tandem repeat in other species of the *Daucus* clade I. However, the structure of its monomers differed among these species (Iorizzo et al. 2016; Fig. 8.2c–e). In addition, the 40 bp monomers A of CentDc had a significant similarity with the initial 40 bp of the most abundant tandem repeat of *D. pusillus* ($2n = 22$, *Daucus* clade II; Fig. 8.2c–e). Similarly, CL80-like repeats were detected in species of both *Daucus* clades. The sequence of CL80 was conserved across *Daucus*, with a pairwise average similarity of >96% between any two species analyzed. However, the abundance and distribution of CL80 differed among the species. *D. guttatus* and *D. littoralis* (both $2n = 20$; clade II) were enriched of CL80 sequences but they were devoid of CentDc. FISH analysis detected CL80 signals at both subtelomeric and intercalary regions of each chromosome of *D. littoralis*, with intercalary signals likely spanning all centromeres. In *D. guttatus*, CL80 hybridized to the ends of most chromosomes and the pericentromeric regions of four chromosomes. In several ways, CL80 repeat resembles the *Oryza* satellite repeat CentO-C2/TrsC which localizes at several functional centromeres and subtelomeric regions in *O. rhizomatis* and exclusively at the subtelomeric regions in the related *O. officinalis* (Bao et al. 2006; Lee et al. 2005). Further analyses are necessary to characterize the DNA sequences associated with the centromeres of diverse *Daucus* species.

8.8 Conclusion and Perspectives

Carrot cytogenetics has advanced thanks to novel genomic resources and tools. This progress has contributed to the understanding of the carrot genome organization, by enabling reliable chromosome identification and high-resolution karyotyping, beginning to uncover the organization of several repeat elements including those spanning heterochromatic domains and identify candidate centromeric- and knob-associated repeats in carrot and related species. These studies indicate that *Daucus* is an appealing genus to study the evolutionary dynamics of satellite repeats as well as how and what type of mobile elements contributed to the genome size differences among *Daucus* species. Indeed, *Daucus* includes species for the most part diploid, with basic chromosome numbers ranging from $n = 8$ to $n = 11$ and up to a three-fold difference in genome size. However, for many species, basic cytological data such as chromosome number and genome size remain sketchy or to be confirmed. Such data are of great value especially given the ongoing taxonomic revision of the entire genus *Daucus* (Banasiak et al. 2016). In addition, new strategies in painting individual chromosomes or specific regions, based on probes made of pools of thousands of custom-synthesized oligonucleotides (Braz et al. 2018; Han et al. 2015), would greatly contribute to the elucidation of the chromosome rearrangements occurred during the evolution of the genus *Daucus*. This knowledge, in turn, would help clarify the phylogenetic relationships and perform a more effective exploitation and management of the *Daucus* germplasm.

Acknowledgements We are grateful to Prof. Domenico Carputo for his helpful suggestions on the manuscript.

References

Al-Safadi B, Simon PW (1990) The effects of gamma irradiation on the growth and cytology of carrot (*Daucus carota* L.) tissue culture. Environ Exp Bot 30:361–371

Aparicio Martínez A (1989) Números cromosomáticos de plantas occidentales, 487–507. An Jard Bot Madrid 45:483–494

Arbizu C, Ruess H, Senalik D et al (2014) Phylogenomics of the carrot genus (*Daucus*, Apiaceae). Am J Bot 101:1666–1685

Arbizu CI, Simon PW, Martínez-Flores F et al (2016) Integrated molecular and morphological studies of the Daucus guttatus complex (Apiaceae). Syst Bot 41:479–492

Arumuganathan K, Earle ED (1991) Nuclear DNA content of some important plant species. Plant Mol Biol Rep 9:208–218

Ávila Robledillo L, Koblížková A, Novák P et al (2018) Satellite DNA in *Vicia faba* is characterized by remarkable diversity in its sequence composition, association with centromeres, and replication timing. Sci Rep 8:5838

Bai C, Alverson WS, Follansbee A et al (2012) New reports of nuclear DNA content for 407 vascular plant taxa from the United States. Ann Bot 110:1623–1629

Banasiak LA, Wojewódzka J, Baczyński JP et al (2016) Phylogeny of Apiaceae subtribe Daucinae and the taxonomic delineation of its genera. Taxon 65: 563–585

Bao W, Zhang W, Yang Q et al (2006) Diversity of centromeric repeats in two closely related wild rice species, *Oryza officinalis* and *Oryza rhizomatis*. Mol Genet Genomics 275:421–430

Bayliss MW (1973) Origin of chromosome number variation in cultured plant cells. Nature 246:529–530

Bayliss MW (1975) The effects of growth *in vitro* on chromosome complement of *Daucus carota* L. suspension cultures. Chromosoma 51:404–411

Bayliss MW (1977) Factors affecting the frequency of tetraploid cells in a predominantly diploid suspension culture of *Daucus carota*. Protoplasma 92:109–115

Bell CR, Constance L (1957) Chromosome numbers in Umbelliferae. I. Am J Bot 44:565–572

Bell CR, Constance L (1960) Chromosome numbers in Umbelliferae. II. Am J Bot 47:24–32

Bell CR, Constance L (1966) Chromosome numbers in Umbelliferae. III. Am J Bot 53:512–520

Bennett MD, Leitch IJ (1995) Nuclear DNA amounts in angiosperms. Ann Bot 76:113–176

Bennett MD, Leitch IJ (2011) Nuclear DNA amounts in angiosperms: targets, trends and tomorrow. Ann Bot 107:467–590

Bennett MD, Smith JB (1976) Nuclear DNA amounts in angiosperms. Philos Trans R Soc B 274:227–274

Betekhtin A, Jenkins G, Hasterok R (2014) Reconstructing the evolution of *Brachypodium* genomes using comparative chromosome painting. PLoS One 9(12): e115108

Boiteux LS, Belter JG, Roberts PA, Simon PW (2000) RAPD linkage map of the genomic region encompassing the root-knot nematode (Meloidogyne javanica) resistance locus in carrot. Theor Appl Genet 100:439–446

Borkird C, Sung ZR (1987) Isolation and characterization of ABA-insensitive cell lines of carrot. Plant Physiol 84:1001–1006

Braz GT, He L, Zhao H et al (2018) Comparative oligo-FISH mapping: an efficient and powerful methodology to reveal karyotypic and chromosomal evolution. Genetics 208:513–523

Cao HX, Vu GT, Wang W et al (2016) The map-based genome sequence of Spirodela polyrhiza aligned with its chromosomes, a reference for karyotype evolution. New Phytol 209:354–363

Cavagnaro PF, Chung SM, Szklarczyk M et al (2009) Characterization of a deep-coverage carrot (*Daucus carota* L.) BAC library and initial analysis of BAC-end sequences. Mol Genet Genomics 281:273–288

Cavagnaro PF, Iorizzo M, Yildiz M et al (2014) A gene-derived SNP-based high resolution linkage map of carrot including the location of QTL conditioning root and leaf anthocyanin pigmentation. BMC Genom 15:1118

Chao YT, Chen WC, Chen CY et al (2018) Chromosome-level assembly, genetic and physical mapping of *Phalaenopsis aphrodite* genome provides new insights into species adaptation and resources for orchid breeding. Plant Biotechnol J (In press)

Cheng Z, Presting GG, Buell CR, Wing RA, Jiang J (2001) High-resolution pachytene chromosome mapping of bacterial artificial chromosomes anchored by genetic markers reveals the centromere location and the distribution of genetic recombination along chromosome 10 of rice. Genetics 157:1749–1757

Constance L, Chuang T-I (1982) Chromosome numbers of Umbelliferae (Apiaceae) from Africa south of the Sahara. Bot J Linn Soc 85:195–208

Constance L, Chuang T-I, Bell CR (1976) Chromosome numbers in Umbelliferae. V. Am J Bot 58:577–587

Doležel J, Bartoš J, Voglmayr H et al (2003) Nuclear DNA content and genome size of trout and human. Cytometry 51A:127–128

Dong F, Song J, Naess SK et al (2000) Development and applications of a set of chromosome specific cytogenetic DNA markers in potato. Theor Appl Genet 101:1001–1007

Dong G, Shen J, Zhang Q et al (2018) Development and applications of chromosome-specific cytogenetic BAC-FISH probes in *S. spontaneum*. Front Plant Sci 9:218

Dudits D, Kao KN, Constabel F, Gamborg OL (1976) Embryogenesis and formation of tetraploid and hexaploid plants from carrot protoplasts. Can J Bot 54:1063–1067

Dudits D, Gy Hadlaczky, Levi E et al (1977) Somatic hybridization of *Daucus carota* and *D. capillifolius* by protoplast fusion. Theor Appl Genet 51:127–132

Dunemann F, Schrader O, Budahn H, Houben A (2014) Characterization of centromeric histone H3 (CENH3) variants in cultivated and wild carrots (Daucus sp.). PLoS One 9(6):e98504

Ellison S, Senalik D, Bostan H et al (2017) Fine mapping, transcriptome analysis, and marker development for Y_2, the gene that conditions β-carotene accumulation in carrot (*Daucus carota* L.). G3: Genes Genomes Genet 7(8):2665–2675

Essad S, Maunoury C (1985) Banding C et biométrie appliqués à l'analyse du caryotype de carotte (*Daucus carota* L.). Agron EDP Sci 5:871–876

Ferguson-Smith MA, Trifonov V (2007) Mammalian karyotype evolution. Nat Rev Genet 8:950–962

Fonsêca A, Ferraz ME, Pedrosa-Harand A (2016) Speeding up chromosome evolution in *Phaseolus*: multiple rearrangements associated with a one-step descending dysploidy. Chromosoma 125:413–421

Fransz PF, Armstrong S, de Jong JH et al (2000) Integrated cytogenetic map of chromosome arm 4S of *A. thaliana*: structural organization of heterochromatic knob and centromere region. Cell 100:367–376

Gaiero P, van de Belt J, Vilaró F et al (2017) Collinearity between potato (*Solanum tuberosum* L.) and wild relatives assessed by comparative cytogenetic mapping. Genome 60:228–240

Gong Z, Wu Y, Koblížková A et al (2012) Repeatless and repeat-based centromeres in potato: implications for centromere evolution. Plant Cell 24:3559–3574

Grosso AC, Rodrigues L, Gomes I et al (2008) Preliminary data on microcharacters and chromosome number in *Tornabenea* species (Apiaceae) from Cape Verde Islands. Plant Biosyst 142:87–93

Grzebelus D, Simon PW (2009) Diversity of DcMaster-like elements of the PIF/Harbinger superfamily in the carrot genome. Genetica 135:347–353

Grzebelus D, Jagosz B, Simon PW (2007) The *DcMaster* transposon display maps polymorphic insertion sites in the carrot (*Daucus carota* L.) genome. Gene 390:67–74

Grzebelus E, Szklarczyk M, Baranski R (2012) An improved protocol for plant regeneration from leaf- and hypocotyl-derived protoplasts of carrot. Plant Cell Tiss Organ Cult 109:101–109

Grzebelus D, Iorizzo M, Senalik D et al (2014) Diversity, genetic mapping, and signatures of domestication in the carrot (*Daucus carota* L.) genome, as revealed by Diversity Arrays Technology (DArT) markers. Mol Breeding 33:625–637

Hamal IA, Langer A, Koul AK (1986) Nucleolar organizing region in the Apiaceae (Umbelliferae). Plant Syst Evol 154:11–30

Han Y, Zhang T, Thammapichai P et al (2015) Chromosome-specific painting in *Cucumis* species using bulked oligonucleotides. Genetics 200:771–779

Iorizzo M, Senalik DA, Grzebelus D et al (2011) *De novo* assembly and characterization of the carrot transcriptome reveals novel genes, new markers, and genetic diversity. BMC Genom 12:389

Iorizzo M, Senalik DA, Ellison SL et al (2013) Genetic structure and domestication of carrot (*Daucus carota* subsp. *sativus*) (Apiaceae). Am J Bot 100:930–938

Iorizzo M, Ellison S, Senalik D et al (2016) A high-quality carrot genome assembly reveals new insights into carotenoid accumulation and Asterid genome evolution. Nat Genet 48:657–666

Iovene M, Grzebelus E, Carputo D et al (2008a) Major cytogenetic landmarks and karyotype analysis in *Daucus carota* and other Apiaceae. Am J Bot 95:793–804

Iovene M, Wielgus SM, Simon PW et al (2008b) Chromatin structure and physical mapping of chromosome 6 of potato and comparative analyses with tomato. Genetics 180:1307–1317

Iovene M, Cavagnaro PF, Senalik D et al (2011) Comparative FISH mapping of *Daucus* species (Apiaceae family). Chromosome Res 19:493–506

Jiang N, Feschotte C, Zhang X et al (2004) Using rice to understand the origin and amplification of miniature inverted repeat transposable elements (MITEs). Curr Opin Plant Biol 7:115–119

Just BJ, Santos CAF, Fonseca MEN et al (2007) Carotenoid biosynthesis structural genes in carrot (*Daucus carota*):

isolation, sequence-characterization, single nucleotide polymorphism (SNP) markers and genome mapping. Theor Appl Genet 114:693–704

Kumar P, Widholm JM (1984) Techniques for chromosome analysis of carrot culture cells. Plant Mol Biol Rep 2:37–42

Lazar GB, Dudits D, Sung ZR (1981) Expression of cycloheximide resistance in carrot somatic hybrids and their segregants. Genetics 98:347–356

Lee HR, Zhang W, Langdon T et al (2005) Chromatin immunoprecipitation cloning reveals rapid evolutionary patterns of centromeric DNA in *Oryza* species. Proc Natl Acad Sci USA 102:11793–11798

Lindenbein W (1932) Karyologische studien an *Daucus carota*. Ber Deut Bot Ges 50:399–406 (cited in Sharma and Ghosh 1954)

Lou Q, Iovene M, Spooner DM et al (2010) Evolution of chromosome 6 of *Solanum* species revealed by comparative fluorescence in situ hybridization mapping. Chromosoma 119:435–442

Lou Q, Zhang Y, He Y et al (2014) Single-copy gene-based chromosome painting in cucumber and its application for chromosome rearrangement analysis in *Cucumis*. Plant J 78:169–179

Luque T, Lifante ZD (1991) Chromosome numbers of plants collected during Iter Mediterraneum I in the SE of Spain. Bocconea 1:303–364

Lysak MA, Koch MA, Pecinka A et al (2005) Chromosome triplication found across the tribe Brassicaceae. Genome Res 15:516–525

Lysak MA, Mandakova T, Schranz ME (2016) Comparative paleogenomics of crucifers: ancestral genomic blocks revisited. Curr Opin Plant Biol 30:108–115

Ma L, Vu GTH, Schubert V et al (2010) Synteny between *Brachypodium distachyon* and *Hordeum vulgare* as revealed by FISH. Chromosome Res 18:841–850

Macko-Podgorni A, Nowicka A, Grzebelus E et al (2013) *DcSto*: carrot *Stowaway*-like elements are abundant, diverse, and polymorphic. Genetica 141:255–267

Macko-Podgórni A, Machaj G, Stelmach K et al (2017) Characterization of a genomic region under selection in cultivated carrot (*Daucus carota* subsp. *sativus*) reveals a candidate domestication gene. Front Plant Sci 8:12

Mandáková T, Li Z, Barker MS et al (2017) Diverse genome organization following 13 independent mesopolyploid events in Brassicaceae contrasts with convergent patterns of gene retention. Plant J 91: 3–21

McKinley KL, Cheeseman IM (2016) The molecular basis for centromere identity and function. Nat Rev Mol Cell Biol 17:16–29

Melters DP, Bradnam KR, Young HA et al (2013) Comparative analysis of tandem repeats from hundreds of species reveals unique insights into centromere evolution. Genome Biol 14:R10

Nowicka A, Grzebelus E, Grzebelus D (2012) Fluorescent *in situ* hybridization with arbitrarily amplified DNA fragments differentiates carrot (*Daucus carota* L.) chromosomes. Genome 55:205–213

Nowicka A, Grzebelus E, Grzebelus D (2016a) Precise karyotyping of carrot mitotic chromosomes using multicolour-FISH with repetitive DNA. Biol Plant 60:25–36

Nowicka A, Sliwinska E, Grzebelus D et al (2016b) Nuclear DNA content variation within the genus *Daucus* (Apiaceae) determined by flow cytometry. Sci Hortic 209:132–138

Owens S (1974) An examination of the floral biology, breeding systems and cytology in species of the genus *Daucus* and related genera in the tribe Caucerlidea (Umbelliferae). PhD thesis, University of Reading (cited in Bennett and Smith 1976)

Park M, Park J, Kim S et al (2012) Evolution of the large genome in *Capsicum annuum* occurred through accumulation of single-type long terminal repeat retrotransposons and their derivatives. Plant J 69:1018–1029

Pedrosa-Harand A, Kami J, Gepts P et al (2009) Cytogenetic mapping of common bean chromosomes reveals a less compartmentalized small-genome plant species. Chromosome Res 17:405–417

Pustahija F, Brown SC, Bogunić F et al (2013) Small genomes dominate in plants growing on serpentine soils in West Balkans, an exhaustive study of 8 habitats covering 308 taxa. Plant Soil 373:427–453

Rice A, Glick L, Abadi S et al (2015) The Chromosome Counts Database (CCDB)—a community resource of plant chromosome numbers. New Phytol 206:19–26

Rong J, Lammers Y, Strasburg JL et al (2014) New insights into domestication of carrot from root transcriptome analyses. BMC Genom 15:895

Rubatzky VE, Quiros CF, Simon PW (1999) Carrots and related vegetable Umbelliferae. CABI Publishing, New York

Sáenz Laín C (1981) Research on Daucus L. (Umbelliferae). An Inst Bot AJ Cavanilles 37:481–533

Saski CA, Scheffler BE, Hulse-Kemp AM et al (2017) Sub genome anchored physical frameworks of the allotetraploid Upland cotton (*Gossypium hirsutum* L.) genome, and an approach toward reference-grade assemblies of polyploids. Sci Rep 7:15274

Schrader O, Ahne R, Fuchs J (2003) Karyoptype analysis of *Daucus carota* L. using Giemsa C-Banding and FISH of 5S and 18S-25S rRNA specific genes. Caryologia 56:149–154

Sharma AK, Bhattacharyya NK (1959) Further investigations on several genera of Umbelliferae and their interrelationships. Genetica 30:1–62

Sharma AK, Ghosh C (1954) Cytogenetics of some of the Indian umbellifers. Genetica 27:17–44

Shearer LA, Anderson LK, de Jong H et al (2014) Fluorescence *in situ* hybridization and optical mapping to correct scaffold arrangement in the tomato genome. G3: Genes Genomes Genet 4:1395–1405

Simon PW (1984) Carrot genetics. Plant Mol Biol Rep 2:54–63

Sinha BM, Sinha AK (1978) The chromosomes of certain species of *Umbelliferae*. Botanique 8:117–122

Sliwinska E (2018) Flow cytometry—a modern method for exploring genome size and nuclear DNA synthesis in horticultural and medicinal plant species. Folia Hortic 30:103–128

Smith SM, Street HE (1974) The decline of embryogenic potential as callus and suspension cultures of carrot (*Daucus carota* L.) are serially subcultured. Ann Bot 38:223–241

Smith J, Furner I, Sung RZ (1981) Nutritional and karyotypic characterization of a haploid cell culture of *Daucus carota* L. In vitro 17:315–321

Spalik K, Downie SR (2007) Intercontinental disjunctions in *Cryptotaenia* (Apiaceae, Oenantheae): an appraisal using molecular data. J Biogeogr 34:2039–2054

Spalik K, Piwczyński M, Danderson CA et al (2010) Amphitropic amphiantarctic disjunctions in Apiaceae subfamily Apioideae. J Biogeogr 37:1977–1994

Spooner D, Rojas P, Bonierbale M et al (2013) Molecular phylogeny of *Daucus* (Apiaceae). Syst Bot 38(3): 850–857

Spooner DM, Widrlechner MP, Reitsma KR et al (2014) Reassessment of practical subspecies identifications of the USDA *Daucus carota* L. germplasm collection: morphological data. Crop Sci 54:706–718

Spooner DM, Ruess H, Iorizzo M et al (2017) Entire plastid phylogeny of the carrot genus (*Daucus*, Apiaceae): concordance with nuclear data and mitochondrial and nuclear DNA insertions to the plastid. Am J Bot 104:296–312

Subramanian D (1986) Cytotaxonomical studies in South Indian Apiaceae. Cytologia 51:479–488

Suda J, Kyncl T, Jarolímová V (2005) Genome size variation in Macaronesian angiosperms: forty percent of the Canarian endemic flora completed. Plant Syst Evol 252:215–238

Sung ZR, Jacques S (1980) 5-Fluorouracil resistance in carrot cell cultures. Its use in studying the interaction of the pyrimidine and arginine pathways. Planta 148:389–396

Sussex IM (2008) The scientific roots of modern plant biotechnology. Plant Cell 20:1189–1198

Szinay D, Chang S, Khrustaleva L et al (2008) High-resolution chromosome mapping of BACs using multi-colour FISH and pooled-BAC FISH as a backbone for sequencing tomato chromosome 6. Plant J 56:627–637

Szinay D, Wijnker E, van den Berg R et al (2012) Chromosome evolution in *Solanum* traced by cross-species BAC-FISH. New Phytol 195:688–698

Tavares AC, Loureiro J, Castro S et al (2014) Assessment of *Daucus carota* L. (Apiaceae) subspecies by chemotaxonomic and DNA content analyses. Biochem Syst Ecol 55:222–230

Tek AL, Song J, Macas J et al (2005) Sobo, a recently amplified satellite repeat of potato, and its implications for the origin of tandemly repeated sequences. Genetics 170:1231–1238

Vogt R, Oberprieler C (1994) Chromosome numbers of North African phanerogams. IV. Candollea 49:549–570

Vogt R, Oberprieler C (2009) Chromosome numbers of North African phanerogams. IX. In: Marhold K (ed) IAPT/IOPB chromosome data 8. Taxon 58:1282–1283

Wang C-JR, Harper L, Cande ZW (2006) High-resolution single-copy gene fluorescence in situ hybridization and its use in the construction of a cytogenetic map of maize chromosome 9. Plant Cell 18:529–544

Yang L, Koo D, Li D et al (2014) Next-generation sequencing, FISH mapping and synteny-based modeling reveal mechanisms of decreasing dysploidy in *Cucumis*. Plant J 77:16–30

Zenkteler M (1962) Microsporogenesis and tapetal development in normal and male-sterile carrots (*Daucus carota*). Am J Bot 49:341–348

Zhang WL, Wai CM, Ming R et al (2010) Integration of genetic and cytological maps and development of a pachytene chromosome-based karyotype in papaya. Trop Plant Biol 3:166–170

Classical and Molecular Carrot Breeding

9

Philipp W. Simon

Abstract

Classical plant breeding approaches have succeeded in improving the productivity of the carrot crop for growers and the quality of the crop for consumers over the last century. A significant breeding focus has been on genetic control of male fertility to assure successful production of hybrid cultivars, with relatively little emphasis on formal studies of other reproductive traits such as seed yield and vernalization requirements, or on crop morphology. Another strong focus for carrot breeders has been selection for resistance to Alternaria leaf blight and root-knot nematodes. Future crop producers will likely face more challenging abiotic threats and additional biotic threats to the crop, and little effort has been directed to those traits. In an effort to improve carrot consumer quality, pigments and flavor compounds have received much attention by carrot breeders. With the expansion of carrot global markets, a broader range of consumer traits may require attention as carrot breeding programs move forward. The sequencing of the carrot genome provides an important foundation for a better understanding of the genetics of traits important for growers and consumers, for developing molecular tools to accelerate the breeding process, and for identifying genes of potential interest for gene editing. The breadth of genetic diversity in carrot germplasm is a valuable resource that will provide an important foundation for future carrot breeding. A better understanding of that diversity will be needed to take full advantage of it, and the carrot genome sequence will provide insights into that understanding.

9.1 Introduction

Cultivated carrot (*Daucus carota* ssp. *sativus* L.) is a diploid, outcrossing, insect-pollinated vegetable ($2n = 2x = 18$) that originated as a root crop in Central Asia around 1100 years ago. Carrot is the most widely grown member of the Apiaceae today. Storage root color and flavor were traits noted early in carrot domestication history, and root shape became an important trait to differentiate cultivars beginning around 500 years ago (see Chap. 5), but it was not until 85 years ago that the first genetic analysis of carrot was reported, describing white storage root color to be dominant over yellow, and under monogenic control (Borthwick and Emsweller 1933; Emsweller et al. 1935) (see Chap. 14).

P. W. Simon (✉)
USDA, Agricultural Research Service, Vegetable Crops Research Unit, Department of Horticulture, University of Wisconsin, 1575 Linden Dr., Madison, WI 53706, USA
e-mail: philipp.simon@ars.usda.gov

P. Simon et al. (eds.), *The Carrot Genome*, Compendium of Plant Genomes,
https://doi.org/10.1007/978-3-030-03389-7_9

Given the relatively slow growth of carrots in the field and maximum of one breeding cycle per year (Simon et al. 2008), molecular markers to facilitate selection are important to assure progress in carrot breeding programs. The development of dense molecular marker-based genetic maps described in Chap. 7 combined with precise, high-throughput phenotyping technologies provides plant breeders with detailed trait maps to routinely apply marker-assisted selection (MAS) in breeding programs. These detailed trait maps will contribute to not only more efficient MAS, but also more accurate identification of candidate genes that may become targets for editing approaches for carrot improvement.

9.2 Carrot Reproductive Biology and Seed Production

Carrot is categorized as a biennial crop since the crop of commerce is harvested in the vegetative phase of its life cycle. The transition from vegetative crop to flowering varies widely in diverse carrot germplasm, and the genetic and environmental bases underlying this transition are discussed in Chap. 3. Carrot cultivars are categorized as temperate and late flowering, or subtropical and early flowering, depending on their intended area of production. Vernalization of the vegetative crop, achieved by exposing it to cold temperatures to induce floral development, is required for floral initiation in temperate carrot cultivars. In contrast, carrots developed for subtropical or tropical climates typically require little or no exposure to cold temperatures for floral induction (Simon et al. 2008). Wild carrots from many global regions will flower with no apparent vernalization required. One gene influencing floral mutation, *Vrn1*, has been reported to date (Table 9.1) (Alessandro et al. 2013). *Vrn1* was mapped to chromosome 2 in a region spanning 0.36 cM. Several additional genes controlling carrot vernalization are expected to be discovered as a broader range of germplasm is evaluated, given the wide range of variation observed for this trait.

Phenotyping of carrot germplasm for vernalization requirements is not a trivial matter. To date, phenotyping has been done in open fields, so well-characterized, reliable environmental conditions are needed for effective selection of carrot breeding stocks. As the genetic control of floral induction becomes better understood, the development of well-defined phenotyping methods will be needed to ensure progress in carrot breeding programs.

The architecture of flowering plants and seed yield varies widely in diverse carrot germplasm. While adequate seed yield is crucial for producers of commercial seed, little has been published on the genetic control of variation in carrot seed productivity.

In contrast, since the initial discovery of cytoplasmic male sterility (CMS) in carrot in the late 1940s, the genetics and molecular basis of CMS have been extensively studied and is reviewed in Chap. 3. Most widely grown carrot cultivars in major global markets are hybrids, and reliable trait expression is mandatory for both male-sterile and male-fertile parents in the production of hybrid seed. Nuclear genes maintaining male sterility and restoring fertility in plants with male-sterile cytoplasm derived from wild and cultivated carrot have been characterized, nuclear genes controlling male sterility in plants with male-fertile cytoplasm have been reported, and their inheritance evaluated (Alessandro et al. 2013; Banga et al. 1964; Borner et al. 1995; Hansche and Gabelman 1963; Mehring-Lemper 1987; Thompson 1961) (Table 9.1). Alessandro et al. (2013) mapped *Rf1*, a nuclear restorer of cytoplasmic male sterility, to chromosome 9 within a 3.36 cM genomic region. It has been speculated that numerous additional restorers of CMS occur in carrot.

In addition to markers for nuclear restorer genes, molecular markers for the cytoplasm itself are important in breeding programs. Variation in the mitochondrial genome controls male fertility, and several studies have developed markers to differentiate male-sterile and male-fertile cytoplasms currently used in breeding programs (Bach et al. 2002; Nakajima et al. 1999)

Table 9.1 Genes of carrot: reproductive biology, morphology, and biotic stress resistance

Gene symbol (parentheses indicate suggested symbol)	Character description/trait	Mendelian inheritance reported	Placed on a linkage map
Reproductive biology			
Vrn1	Vernalization	Alessandro et al. (2013)	Alessandro et al. (2013)
Ms1-Ms-3 *Ms-4-Ms-5, ms,* *a, B, D, E,* *l, t,* *Rf1*	Nuclear restorers of CMS	Thompson (1961) Hansche and Gabelman (1963) Banga et al. (1964) Mehring-Lemper (1987) Alessandro et al. (2013)	Alessandro et al. (2013)
Gum1-2, Mar1-2, Gad1-2	Novel cytoplasms and sterility	Borner et al. (1995)	Borner et al. (1995)
STS1–STS6	Petaloid male-sterile and fertile cytoplasm		Nakajima et al. (1999)
14 primer pairs			Bach et al. (2002)
Morphology and growth			
(Cr)	Root cracking	Dickson (1966)	
Gls	Glabrous seed stalk	Morelock and Hosfield (1976)	
(sp1, sp2)	Spine formation	Nieuwhof and Garritsen (1984)	
Phen1	Small, dark green, annual	Schulz et al. (1994)	Schulz et al. (1994)
COLA	Compressed lamina	Budahn et al. (2014)	Budahn et al. (2014)
YEL	Yellow leaf	Nothnagel et al. (2005)	Budahn et al. (2014)
cult	Root thickening	Macko-Podgórni et al. (2017)	Macko-Podgórni et al. (2017)
5, 4, and 3 QTL 1, 5, and 3 QTL 6, 2, and 2 QTL	Shoot height, biomass, area Petiole number, width, and length Root length, biomass, and area	Turner et al. (2018)	Turner et al. (2018)
Disease and pest resistance			
(Ce)	Cercospora leaf spot	Angell and Gabelman (1968)	
Eh	Powdery mildew	Bonnet (1983)	
3 QTL	Alternaria leaf blight	Le Clerc et al. (2009)	Le Clerc et al. (2009)
11 QTL		Le Clerc et al. (2015)	Le Clerc et al. (2015)
(Mh-1, Mh-2)	*Meloidogyne hapla* root-knot nematodes	Wang and Goldman (1996)	
Mj-1	*M. javanica* root-knot nematodes	Simon et al. (2000)	Boiteux et al. (2000, 2004)

(continued)

Table 9.1 (continued)

Gene symbol (parentheses indicate suggested symbol)	Character description/trait	Mendelian inheritance reported	Placed on a linkage map
Mj-2	*M. javanica* root-knot nematodes	Ali et al. (2014)	Ali et al. (2014)
7 QTL	*M. incognita* root-knot nematodes	Parsons et al. (2015)	Parsons et al. (2015)
Storage root pigments, carbohydrates and terpenoids			
y	Yellow xylem and phloem	Laferriere and Gabelman (1968) Imam and Gabelman (1968) Buishand and Gabelman (1979) Simon (1996)	Just et al. (2007, 2009) Iorizzo et al. (2016)
y_1	Differential orange phloem/xylem		
y_2	Differential orange phloem/xylem		Bradeen and Simon (1998) Just et al. (2007, 2009) Yildiz et al. (2013) Ellison et al. (2017)
o, io	Orange xylem	Kust (1970)	
a, l	α-Carotene, lycopene	Umiel and Gabelman (1972)	
Rp	Reduced carotene content	Goldman and Breitbach (1996)	
16 QTL	Carotene content	Santos and Simon (2002)	Santos and Simon (2002)
Or	Carotene content	Ellison et al. (2018)	Ellison et al. (2018)
g	Petiole anthocyanins	Angell and Gabelman (1970)	
P_1	Root anthocyanins	Simon (1996) Cavagnaro et al. (2014)	Vivek and Simon (1999) Yildiz et al. (2013) Cavagnaro et al. (2014)
P_2	Node anthocyanins	Simon (1996) Cavagnaro et al. (2014)	Cavagnaro et al. (2014)
P_3	Root and petiole anthocyanins	Cavagnaro et al. (2014)	Cavagnaro et al. (2014)
Raa1	Acylated anthocyanins		
15 QTL	Anthocyanin content		
30 QTL	Volatile terpenoid content and composition	Keilwagen et al. (2017)	Keilwagen et al. (2017)
Rs	Reducing sugar	Freeman and Simon (1983) Vivek and Simon (1999) Yau and Simon (2003)	Vivek and Simon (1999) Yau and Simon (2003) Yau et al. (2005)

(Table 9.1). Additional cytoplasmic diversity has been noted in more diverse germplasm (Borner et al. 1995). Given the critical need for reliable fertility phenotypes, the extensive use of diverse carrot germplasm in carrot breeding programs that has not been characterized for CMS restorers or cytoplasms, and the fact that it can take up to one year to phenotype a plant, the development of additional molecular markers to characterize carrot male fertility restorer genes and cytoplasms is critical to progress in breeding programs.

9.3 Morphology and Growth

For carrot breeding programs that utilize both early-flowering and late-flowering germplasm in their breeding programs, a heavy emphasis is been placed upon selection for vernalization requirement to meet those production requirements in temperate or subtropical growing regions. Beyond variation in flowering tendency, carrots vary widely for growth rate and morphological characteristics, but little genetic analysis has been reported. Two carrot morphological traits—glabrous seed stalk (*gls*) (Morelock and Hosfield 1976) and tendency for root cracking (Dickson 1966)—were reported to be controlled by one gene, and spine formation (Neiuwhof and Garritsen 1984) followed a digenic pattern of inheritance (Table 9.1). None of these three traits have been placed on the carrot genetic map.

The *phen1* trait was observed in a mapping population derived from an open-pollinated cultivar (Schulz et al. 1994) (Table 9.1) where it was found to be under monogenic control, and it was the first trait mapped to the carrot genome. Two other leaf morphological traits, *COLA* (compressed leaf) first described by Nothnagel et al. (2005) on chromosome 4 and *YEL* (yellow leaf) on chromosome 1, are also monogenic traits, both discovered in crosses between cultivated and wild carrots (Budahn et al. 2014). Since MADS-box, alternative oxidase, and chalcone synthase genes had been associated with fertility and floral development in other plants, these genes were evaluated as candidates, but none colocated with *COLA* or *YEL*.

In another study that utilized a cross between cultivated and wild carrot, traits associated with domestication were evaluated. Macko-Podgórni et al. (2017) discovered a 37 kb genomic region on chromosome 2 that controlled root thickening or diameter which they evaluated with a marker referred to as *cult*. A candidate gene in that region, *DcAHLc1*, belongs to the AT-hook motif nuclear localized (AHL) family of plant regulatory genes. AHL genes are involved in the regulation of organ development, including root tissue patterning.

A recent study reported research on the genetic control of traits of interest to carrot breeders: canopy and root size and shape. Turner et al. (2018) evaluated shoot and root morphological characteristics and mapped five QTL for shoot height, seven for shoot biomass and area, seven for petiole width and length, seven for root length, and two for root biomass. Given the importance of these traits in carrot yield and cultivar classification, markers developed for these QTL may have some immediate application.

9.4 Disease and Pest Resistance

The inheritance of resistance has been reported for several carrot foliar diseases and root-knot nematodes (RKN) and discussed in Chap. 18. Monogenic resistance to two foliar diseases, Cercospora leaf spot (Angell and Gabelman 1968) and powdery mildew (Bonnet 1983), was reported, but resistance genes have not yet been mapped (Table 9.1).

Alternaria leaf blight (ALB) occurs worldwide and is regarded as the most important disease of carrots. Genetic control of resistance has been evaluated in several studies including two that mapped three and eleven QTL, respectively (Le Clerc et al. 2009, 2015) (Table 9.1), where relatively high heritability (75–78%) was reported.

Given the importance of ALB, markers to select for enhanced resistance will be of interest to breeders.

Galls formed on carrot roots caused by the attack of root-knot nematodes (RKN) disfigure them in production regions worldwide. Genetic resistance to three RKN species has been reported. Genetic resistance to *Meloidogyne hapla* has been studied and found to be controlled by two genes (Wang and Goldman 1996) which have not yet been mapped. A single dominant gene on chromosome 8, *Mj-1*, confers resistance to *M. javanica. Mj-1* has been mapped (Boiteux et al. 2000), and marker-assisted selection for resistance has been exercised (Boiteux et al. 2004). A second gene conferring additional *M. javanica* resistance, *Mj-2,* has also been mapped to chromosome 8 (Ali et al. 2014). The *Mj-1* gene, in addition to six additional QTL on chromosomes 1, 2, 4, and 9, confers resistance to *M. incognita* (Parsons et al. 2015), which is a common RKN species in the soils of most warmer carrot production regions of the world. Molecular markers to facilitate selection of *Mj-1* were reported (Boiteux et al. 2004).

9.5 Storage Root Quality Traits

Given their importance for human nutrition and consumer acceptance, genetic variation in the content and composition of carrot storage root pigments and flavor compounds has been studied quite extensively. Genetic control of root color due to carotenoids and anthocyanins, in particular, has been evaluated.

Relatively early studies on carotenoid-based colors of carrots discussed in Chap. 14 named the Y, Y_1, Y_2, O, IO, A, L, and Rp genes based on single gene inheritance patterns (Buishand and Gabelman 1979, 1980; Goldman and Breitbach 1996; Imam and Gabelman 1968; Laferriere and Gabelman 1968; Kust 1970; Simon 1996; Umiel and Gabelman 1972) (Table 9.1). More recently variation in the *Or* gene on chromosome 3 was discovered to be associated with orange storage root color in a diverse panel of carrots (Ellison et al. 2018). Candidate genes and/or closely linked markers developed for three of these genes (Y, Y_2, and Or) which dramatically alter the carotenoid composition and storage root color ranging from white to yellow to orange (Bradeen and Simon 1998; Ellison et al. 2017, 2018; Iorizzo et al. 2016) (Table 9.1). Candidate genes have been identified for the *Y* (Fig. 9.1) and *Or* genes, and molecular markers have been developed to facilitate breeding variation in storage root color for all three of these genes. These markers are also a valuable tool to provide insights into the domestication history of carrot (see Chap. 5). Given the important role of carrot carotenoids in human nutrition and their increasing use as a natural pigment, there has been some effort in breeding orange carrots for higher carotene content (Simon et al. 1989). In a cross between two orange carrots that had a fivefold difference in carotenoid content, 16 QTL were found to influence carotenoid content (Santos and Simon 2002).

Anthocyanin color varies widely among diverse carrot germplasm, and three genes controlling anthocyanin accumulation (P_1, P_3, and *Raa1*) have also been mapped (Cavagnaro et al. 2014; Yildiz et al. 2013) (Table 9.1) with candidate genes and/or closely linked markers identified for all three genes as discussed further in Chap. 15. Like carotenoids, anthocyanins also have a positive impact on human health, and they are also being extracted and extensively used as a natural food coloring. Studies evaluating the quantitative inheritance of anthocyanin content have reported 15 QTL contributing to that variation (Cavagnaro et al. 2014). As breeders incorporate genes conditioning carrot pigment content and composition, an understanding of the relative contribution of more of these genes, beyond the major genes listed above, will be important.

The two major attributes of fresh carrot flavor are sweetness and harsh flavor (Simon et al. 1980). Volatile terpenoids account for harsh flavor and Keilwagen et al. (2017) (Table 9.1)

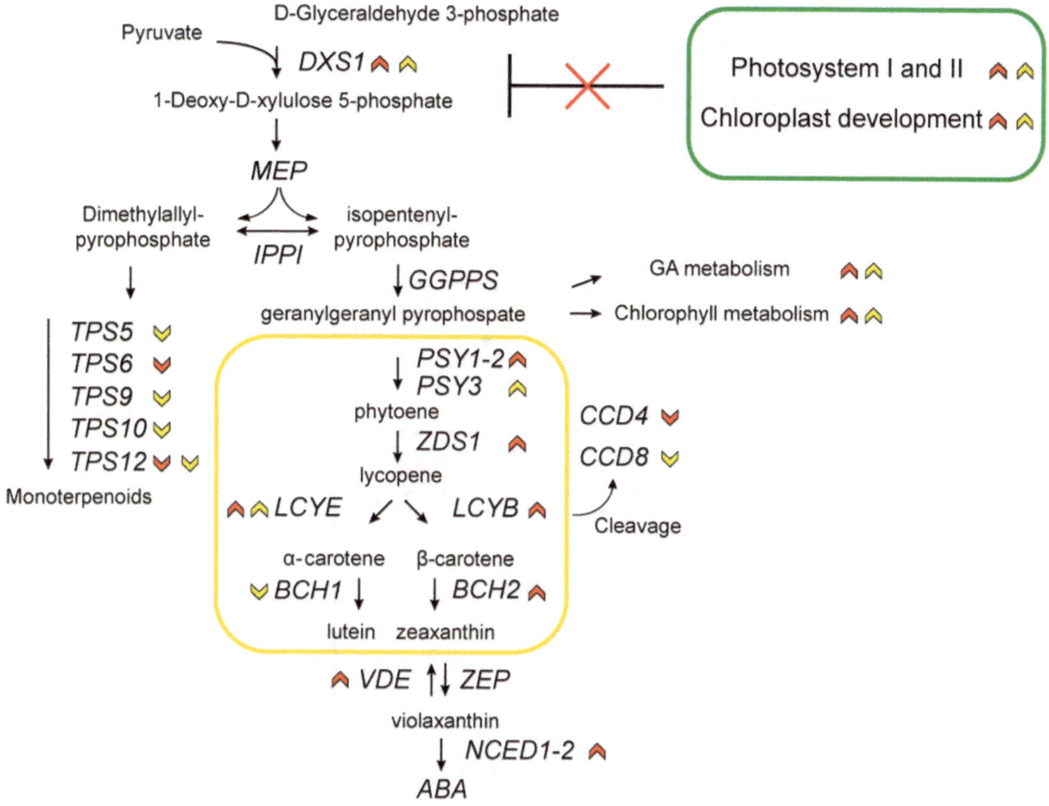

Fig. 9.1 *Y* gene of carrot that controls carotenoid accumulation in the storage root. Upward- and downward-pointing arrows indicate upregulated and downregulated genes, respectively, in the yellow versus white (yellow arrows) and dark orange versus pale orange (orange arrows) comparisons. The orange box delimits the isoprenoid biosynthetic branch that leads to the carotenoid pathway. As shown in the green box, the majority of the upregulated genes in yellow and dark orange roots are involved in the photosynthetic pathway; genes that are included are involved in the assembly and function of photosystems I and II and plastid development. We hypothesize that loss of the constitutive repression mechanisms conditioned by genes involved in de-etiolation and photomorphogenesis in non-photosynthetic tissue, such as carrot roots, induces overexpression of DXS1 and, consequently, activation of the metabolic cascade that leads to high levels of carotenoid accumulation in carrot roots. From Iorizzo et al. (2016)

recently identified 30 QTL conditioning 15 mono- and sesqui-terpenoids in carrot. Only 4 of the 30 QTL comprise terpene synthase candidate genes since these genes are clustered, but 65 candidate gene models were identified.

Carrots store little starch but up to 10% free sugars which contribute to sweet flavor (Simon 2000). A single gene, *Rs*, located on chromosome 2 conditions the ratio of reducing sugars (glucose and fructose) to sucrose in storage roots (Freeman and Simon 1983; Vivek and Simon 1999; Yau and Simon 2003) (Table 9.1). Invertase isozyme II has been identified as the candidate gene controlling this trait, where the mutation is caused by a 2.5-kb insertion into an intron of this gene (Yau and Simon 2003) (Fig. 9.2). Marker-assisted selection for sugar type has been demonstrated (Yau et al. 2005). Plants with the dominant wild-type allele store primarily glucose and fructose, while *rsrs* plants store primarily sucrose. Most carrot cultivars are *RsRs*, but the *rs* allele does occur in some fresh market genetic backgrounds where it can serve as a useful marker to monitor seed purity.

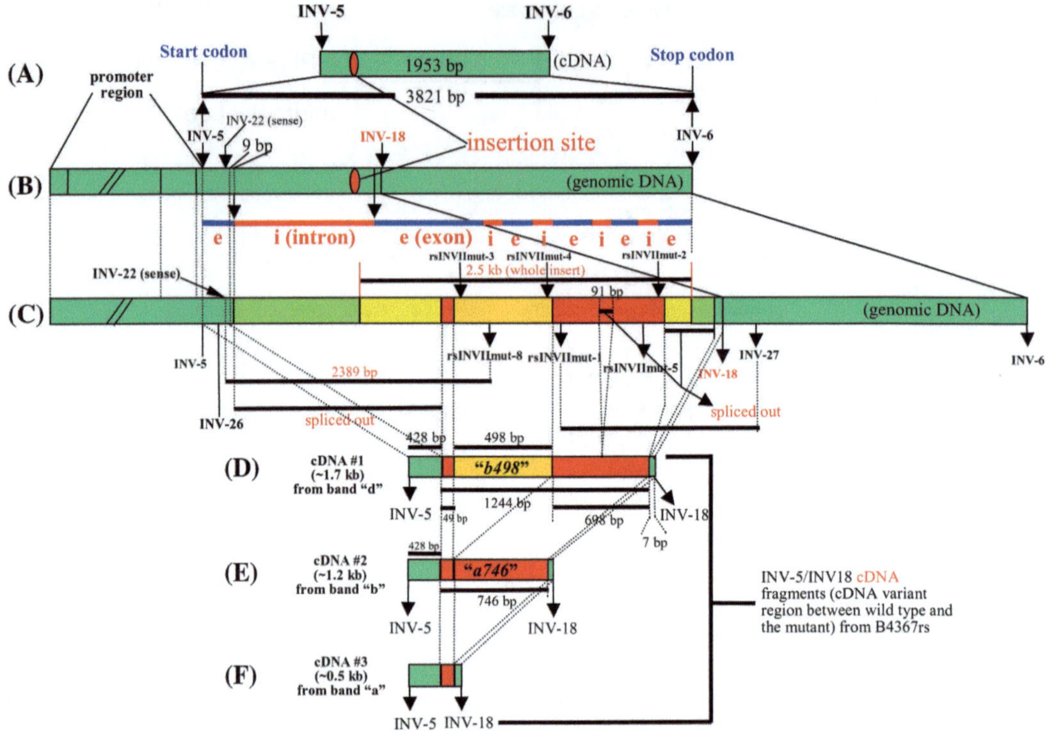

Fig. 9.2 *Rs* gene of carrot that controls sugar type in the storage root. Genomic DNA and cDNA of wild-type and mutant carrot acid soluble invertase isozyme II genes (diagrams are not drawn to scale). (A) Diagram of the cDNA of 1953 bps from the wild-type acid soluble invertase isozyme II gene from carrot line B4367Rs amplified using RT-PCR. Primers INV-5 and INV-6 contain the start and stop codons of the gene, respectively. The position of primers INV-5, INV-6, and INV-18 are indicated. (B) Diagram of genomic DNA of 3821 bps from wild-type acid soluble invertase isozyme II gene from carrot line B4367Rs. Symbols "i" and "e" stand for intron and exon, respectively. The first intron is in bright green, and the insertion site for the 2.5-kb insert is labeled with a red oval. Positions of primers INV-5, INV-6, INV-18, and INV-22 are noted. (C) Diagram of genomic DNA of the mutated acid soluble invertase isozyme II gene from carrot line B4367rs with the first intron highlighted. The green (including bright green) bars represent the wild-type carrot acid soluble invertase isozyme II gene. Bright green and yellow bars represent the sequences spliced out of the first intron during mRNA processing. Yellow, orange, and red bars comprise the 2.5-kb insert. Positions of primers INV-5, INV-6, INV-18, INV-22, INV-27, rsINVIImut-1 through rsINVIImut-5, and rsINVIImut-8 are indicated. (D–F) Diagram of cDNA fragments "a," "b," and "d" amplified using RT-PCR with primers INV-5 and INV-18 from line B4367rs. From Yau and Simon (2003)

9.6 Utilizing the Carrot Genome in Carrot Breeding

For most of the traits of importance to carrot breeders discussed above, biparental mapping populations were used to map genes controlling important traits, followed by the development of molecular markers linked to those genes to track them using MAS in breeding programs. For relatively simply inherited traits, this application of genomic tools will continue to be critically important, since breeding programs usually exercise selection in biparental populations. But those same approaches can be applied in genome-wide association studies (GWAS) to evaluate variation in broad-based germplasm collections, rather than biparental populations, especially when genetic control of the trait of interest in complex. Ellison et al. (2018) discovered the *Or* gene on chromosome 3 using GWAS, and the association between terpenoid biosynthetic genes and volatile terpenoid flavor components was accomplished using GWAS (Keilwagen et al. 2017). As discussed in Chap. 14, GWAS have also been used to evaluate the association between carotenoid pathway genes and carotenoid color (Jourdan et al. 2015). It is expected that the use of GWAS for molecular genetic mapping will be broadly applied by plant breeders in the future (Myles et al. 2009; Yu et al. 2006), and with the sequencing of the carrot genome, carrot will be no exception.

Regardless of whether candidate genes are identified with GWAS in diverse germplasm collections or in biparental populations, accurate phenotyping is essential to success in identifying candidate genes. To address that requirement, machine phenotyping has recently been developed to gather digital images for evaluating carrot top size and root shape (Turner et al. 2017, 2018). Both of these traits are difficult to phenotype accurately by manual analysis, but with the detail of digital imaging, extensive genetic data was collected and QTL mapped. This study sets the stage for fine-mapping of these traits and for identifying candidate genes. In these studies, the machine data was collected after harvest, which can be useful to evaluate storage root and canopy traits during storage for vernalization. But for many field traits such as biotic and abiotic stress, collection of machine data in the field, and with minimal human attendance, will be the intended approach for traits of interest for carrot breeders, as it is for other plant breeders.

The identification of candidate genes not only identifies the best genomic region in which to develop molecular markers to track a trait, but it also provides a breeding program able to utilize gene editing with the basic information to edit. Genome editing has been demonstrated to be successfully accomplished in carrot as described in Chap. 10, and as editing may well become a standard plant breeding technique, the carrot genome will play a critical role in the application of this technology.

References

Alessandro MS, Galmarini CR, Iorizzo M, Simon PW (2013) Molecular mapping of vernalization requirement and fertility restoration genes in carrot. Theor Appl Genet 126:415–423

Ali A, Matthews WC, Cavagnaro PF, Iorizzo M, Roberts PA, Simon PW (2014) Inheritance and mapping of Mj-2, a new source of root-knot nematode (*Meloidogyne javanica*) resistance in carrot. J Hered 105:288–291

Angell FF, Gabelman WH (1968) Inheritance of resistance in carrot, *Daucus carota* var. *sativa*, to the leafspot fungus, *Cercospora carotae*. J Am Soc Hort Sci 93:434–437

Angell FF, Gabelman WH (1970) Inheritance of purple petiole in carrot *Daucus carota* var. *sativa*. Hort Sci 5:175

Bach IC, Olesen A, Simon PW (2002) PCR-based markers to differentiate the mitochondrial genome of petaloid and male fertile carrot (*Daucus carota* L.). Euphytica 127:353–365

Banga O, Petiet J, Van Bennekom JL (1964) Genetical analysis of male-sterility in carrots, *Daucus carota* L. Euphytica 13:75–93

Boiteux LS, Belter JG, Roberts PA, Simon PW (2000) RAPD linkage map of the genomic region encompassing the root-knot nematode (*Meloidogyne javanica*) resistance locus in carrot. Theor Appl Genet 100:439–446

Boiteux LS, Hyman JR, Bach IC, Fonseca MEN et al (2004) Employment of flanking codominant STS markers to estimate allelic substitution effects of a nematode resistance locus in carrot. Euphytica 136:37–44

Borner T, Linke B, Nothnagel T, Scheike R et al (1995) Inheritance of nuclear and cytoplasmic factors affecting male sterility in *Daucus carota*. Adv Plant Breed 18:111–122

Borthwick HA, Emsweller SL (1933) Carrot breeding experiments. Proc Am Soc Hortic Sci 30:531–533

Bonnet A (1983) Source of resistance to powdery mildew for breeding cultivated carrots. Agronomie 3:33–37

Bradeen JM, Simon PW (1998) Conversion of an AFLP fragment linked to the carrot Y_2 locus to a simple, codominant, PCR-based marker form. Theor Appl Genet 97:960–967

Budahn H, Barański R, Grzebelus D, Kiełkowska et al (2014) Mapping genes governing flower architecture and pollen development in a double mutant population of carrot. Front Plant Sci 5:504

Buishand JG, Gabelman WH (1979) Investigations on the inheritance of color and carotenoid content in phloem and xylem of carrot roots (*Daucus carota* L.). Euphytica 28:611–632

Buishand JG, Gabelman WH (1980) Studies on the inheritance of root color and carotenoid content in red x yellow and red × white crosses of carrot, *Daucus carota* L. Euphytica 29:241–260

Cavagnaro PF, Iorizzo M, Yildiz M, Senalik D, Parsons J, Ellison S, Simon PW (2014) A gene-derived SNP-based high resolution linkage map of carrot including the location of QTL conditioning root and leaf anthocyanin pigmentation. BMC Genom 15:1118

Dickson MH (1966) The inheritance of longitudinal cracking in carrot. Euphytica 15:99–101

Ellison S, Senalik D, Bostan H, Iorizzo M, Simon PW (2017) Fine mapping, transcriptome analysis, and marker development for Y_2, the gene that conditions β-carotene accumulation in carrot (*Daucus carota* L.). G3: Genes, Genomes, Genet 7:2665–2675

Ellison S, Luby C, Corak K, Coe K et al (2018) Association analysis reveals the importance of the Or gene in carrot (*Daucus carota* L.) carotenoid presence and domestication. Genetics 210:1–12

Emsweller SL, Burrell PC, Borthwick HA (1935) Studies on the inheritance of color in carrots. Proc Am Soc Hortic Sci 33:508–511

Freeman RE, Simon PW (1983) Evidence for simple genetic control of sugar type in carrot (*Daucus carota* L.). J Am Soc Hortic Sci 108:50–54

Goldman IL, Breitbach DN (1996) Inheritance of a recessive character controlling reduced carotenoid pigmentation in carrot (*Daucus carota* L.). J Hered 87:380–382

Hansche PE, Gabelman WH (1963) Digenic control of male sterility in carrots, *Daucus carota* L. Crop Sci 3:383–386

Imam MK, Gabelman WH (1968) Inheritance of carotenoids in carrots, *Daucus carota*, L. Proc Am Soc Hortic Sci 93:419–428

Iorizzo M, Ellison S, Senalik D, Zeng P et al (2016) A high-quality carrot genome assembly provides new insights into carotenoid accumulation and asterid genome evolution. Nat Genet 48:657–666

Jourdan M, Gagne S, Dubois-Laurent C et al (2015) Carotenoid content and root color of cultivated carrot: a candidate-gene association study using an original broad unstructured population. PLoS ONE 10: e0116674

Just BJ, Santos CAF, Fonseca MEN, Boiteux LS et al (2007) Carotenoid biosynthesis structural genes in carrot (*Daucus carota*): isolation, sequence-characterization, single nucleotide polymorphism (SNP) markers and genome mapping. Theor Appl Genet 114:693–704

Just BJ, Santos CA, Yandell BS, Simon PW (2009) Major QTL for carrot color are positionally associated with carotenoid biosynthetic genes and interact epistatically in a domesticated × wild carrot cross. Theor Appl Genet 119:1155–1169

Keilwagen J, Lehnert H, Berner T, Budahn H, Nothnagel T, Ulrich D, Dunemann F (2017) The terpene synthase gene family of carrot (*Daucus carota* L.): identification of QTLs and candidate genes associated with terpenoid volatile compounds. Front Plant Sci 8:1930

Kust AF (1970) Inheritance and differential formation of color and associated pigments in xymlem and phloem of carrot, *Daucus carota*, L. PhD, University of Wisconsin

Laferriere L, Gabelman WH (1968) Inheritance of color, total carotenoids, alpha-carotene, and beta-carotene in carrots, *Daucus carota* L. Proc Am Soc Hortic Sci 93:408–418

Le Clerc V, Pawelec A, Birolleau-Touchard C, Suel A, Briard M (2009) Genetic architecture of factors underlying partial resistance to Alternaria leaf blight in carrot. Theor Appl Genet 118:1251–1259

Le Clerc V, Marques S, Suel A, Huet S, Hamama L, Voisine L, Auperpin E, Jourdan M, Barrot L, Prieur R (2015) QTL mapping of carrot resistance to leaf blight with connected populations: stability across years and consequences for breeding. Theor Appl Genet 128:2177–2187

Macko-Podgórni A, Machaj G, Stelmach K, Senalik D et al (2017) Characterization of a genomic region under selection in cultivated carrot (*Daucus carota* subsp. *sativus*) reveals a candidate domestication gene. Front Plant Sci 8:12

Mehring-Lemper M (1987) Genetisch-züchterische Untersuchungen zur Schaffung von Hybridsorten bei Möhren (*Daucus carota* L.), Dissertation, Universität Hannover

Morelock TE, Hosfield GL (1976) Glabrous seedstalk in carrot: inheritance and use as a genetic marker. Hort Sci 11:144

Myles S, Peiffer J, Brown PJ, Ersoz ES et al (2009) Association mapping: critical considerations shift from genotyping to experimental design. Plant Cell 21:2194–2202

Nakajima Y, Yamamoto T, Muranaka T, Oeda K (1999) Genetic variation of petaloid male-sterile cytoplasm of carrots revealed by sequence-tagged sites (STSs). Theor Appl Genet 99:837–843

Nieuwhof M, Garritsen F (1984) Inheritance of spine formation on seeds of carrot (*Daucus carota* L.). Euphytica 33:75–80

Nothnagel T, Ahne R, Straka P (2005) Morphology, inheritance and mapping of a compressed lamina mutant of carrot. Plant Breed 124:481–486

Parsons J, Matthews W, Iorizzo M et al (2015) *Meloidogyne incognita* nematode resistance QTL in carrot. Mol Breed 35:114

Santos CAF, Simon PW (2002) QTL analyses reveal clustered loci for accumulation of major provitamin A carotenes and lycopene in carrot roots. Mol Genet Genom 268:122–129

Schulz B, Westphal L, Wricke G (1994) Linkage groups of isozymes, RFLP and RAPD markers in carrot (*Daucus carota* L. *sativus*). Euphytica 74:67–76

Simon PW, Peterson CE, Lindsay RC (1980) Correlations between sensory and objective parameters of carrot flavor. J Agric Food Chem 28:549–552

Simon PW, Wolff XY, Peterson CE et al (1989) High Carotene Mass carrot population. HortScience 24:174

Simon PW (1996) Inheritance and expression of purple and yellow storage root color in carrot. J Hered 87:63–66

Simon PW (2000) Domestication, historical development, and modern breeding of carrot. Plant Breed Rev 19:157–190

Simon PW, Matthews WC, Roberts PA (2000) Evidence for simply inherited dominant resistance to *Meloidogyne javanica* in carrot. Theor Appl Genet 100:735–742

Simon PW, Freeman RE, Vieira JV, Boiteux LS, Briard M, Nothnagel T, Michalik B, Kwon Y-S (2008) Carrot: In: Prohens J, Carena MJ, Nuez F (eds) Handbook of crop breeding, Volume 1, Vegetable breeding. Springer, Heidelberg, pp 327–357

Thompson DJ (1961) Studies on the inheritance of male-sterility in the carrot, *Daucus carota* L. var. *sativa*. Proc Am Soc Hortic Sci 78:332–338

Turner SD, Maurizio PL, Valdar W, Yandell BS, Simon PW (2017) Dissecting the genetic architecture of shoot growth in carrot (*Daucus carota* L.) using a diallel mating design. G3: Genes, Genomes, Genet 8:411–426

Turner S, Ellison S, Senalik DA, Simon PW et al (2018) An automated, high-throughput image analysis pipeline enables genetic studies of shoot and root morphology in carrot (*Daucus carota* L.). Front Plant Sci 9:1703

Umiel N, Gabelman WH (1972) Inheritance of root color and carotenoid synthesis in carrot, *Daucus carota* L.: Orange vs. red. J Am Soc Hort Sci 97:453–460

Vivek BS, Simon PW (1999) Linkage relationships among molecular markers and storage root traits of carrot (*Daucus carota* L. ssp. *sativus*). Theor Appl Genet 99:58–64

Wang M, Goldman I (1996) Resistance to root knot nematode (*Meloidogyne hapla* Chitwood) in carrot is controlled by two recessive genes. J Hered 87:119–123

Yau Y, Simon PW (2003) A 2.5-kb insert eliminates acid soluble invertase isozyme II transcript in carrot (*Daucus carota* L.) roots, causing high sucrose accumulation. Plant Mol Biol 53:151–162

Yau YY, Santos K, Simon PW (2005) Molecular tagging and selection for sugar type in carrot roots with codominant, PCR-based markers. Mol Breed 16:1–10

Yildiz M, Willis DK, Cavagnaro PF, Iorizzo M, Abak K, Simon PW (2013) Expression and mapping of anthocyanin biosynthesis genes in carrot. Theor Appl Genet 126:1689–1702

Yu J, Pressoir G, Briggs WH, Bi IV, Yamasaki M, Doebley JF, McMullen MD, Gaut BS, Nielsen DM, Holland JB (2006) A unified mixed-model method for association mapping that accounts for multiple levels of relatedness. Nat Genet 38:203

Genetic Engineering of Carrot

10

Rafal Baranski and Aneta Lukasiewicz

Abstract

Carrot (*Daucus carota*) is one of the model species used in research for in vitro plant cell and tissue culture. The development of these techniques has enabled efficient cell and tissue proliferation and somatic embryogenesis under in vitro conditions, thus favoring the use of carrot for elucidating the mechanisms of horizontal gene transfer and gene function. Deployment of genetic engineering techniques has led to the development of carrots with improved traits, enhancing plant production for human health. The first product derived from genetically modified (GM) carrot cells cultured in a bioreactor has been approved for the treatment of human metabolic disease and for commercialization. This chapter describes methods of carrot genetic transformation using both vector and non-vector methods. Furthermore, we present reports of basic research in which carrot was used as a model to elucidate the function of heterologous genes and promoters, revealing selected mechanisms of plant metabolism, including the phenomenon of bacteria to plant gene transfer. Separate sections exemplify modified characteristics of GM carrot, including resistance to pathogens and the biosynthesis of recombinant proteins.

10.1 Introduction

Cultivated carrot (*Daucus carota* L. subsp. *sativis* Hoffm.) is the most commonly grown plant of high economic importance belonging to the Apiaceae family. Development of new carrot cultivars fulfills the demands of growers, industry, and consumers, and the genetic improvement of carrot using advanced biotechnology methods is an intriguing option with high potential. Numerous researches have been conducted on the optimization of carrot genetic engineering methods as well as creating carrots that exhibit new characteristics such as improved nutritional value, enhanced resistance to pathogens, and tolerance to abiotic stress (Fig. 10.1). Currently, there are no data available indicating that genetically modified (GM) carrots have ever been included in pre-registration field trials, which aim to introduce GM crops for commercialization. Advances in carrot genetic modification have largely resulted from the fact that carrot was a pioneer species utilized in research on the development of plant cell and tissue culture techniques in vitro. The first reports showing cell development in vitro and experimentally proving the hypothesis of plant cell totipotency was dated as early as

R. Baranski (✉) · A. Lukasiewicz
Faculty of Biotechnology and Horticulture,
Institute of Plant Biology and Biotechnology,
University of Agriculture in Krakow,
Al. 29 Listopada 54, 31-425 Krakow, Poland
e-mail: r.baranski@urk.edu.pl

P. Simon et al. (eds.), *The Carrot Genome*, Compendium of Plant Genomes,
https://doi.org/10.1007/978-3-030-03389-7_10

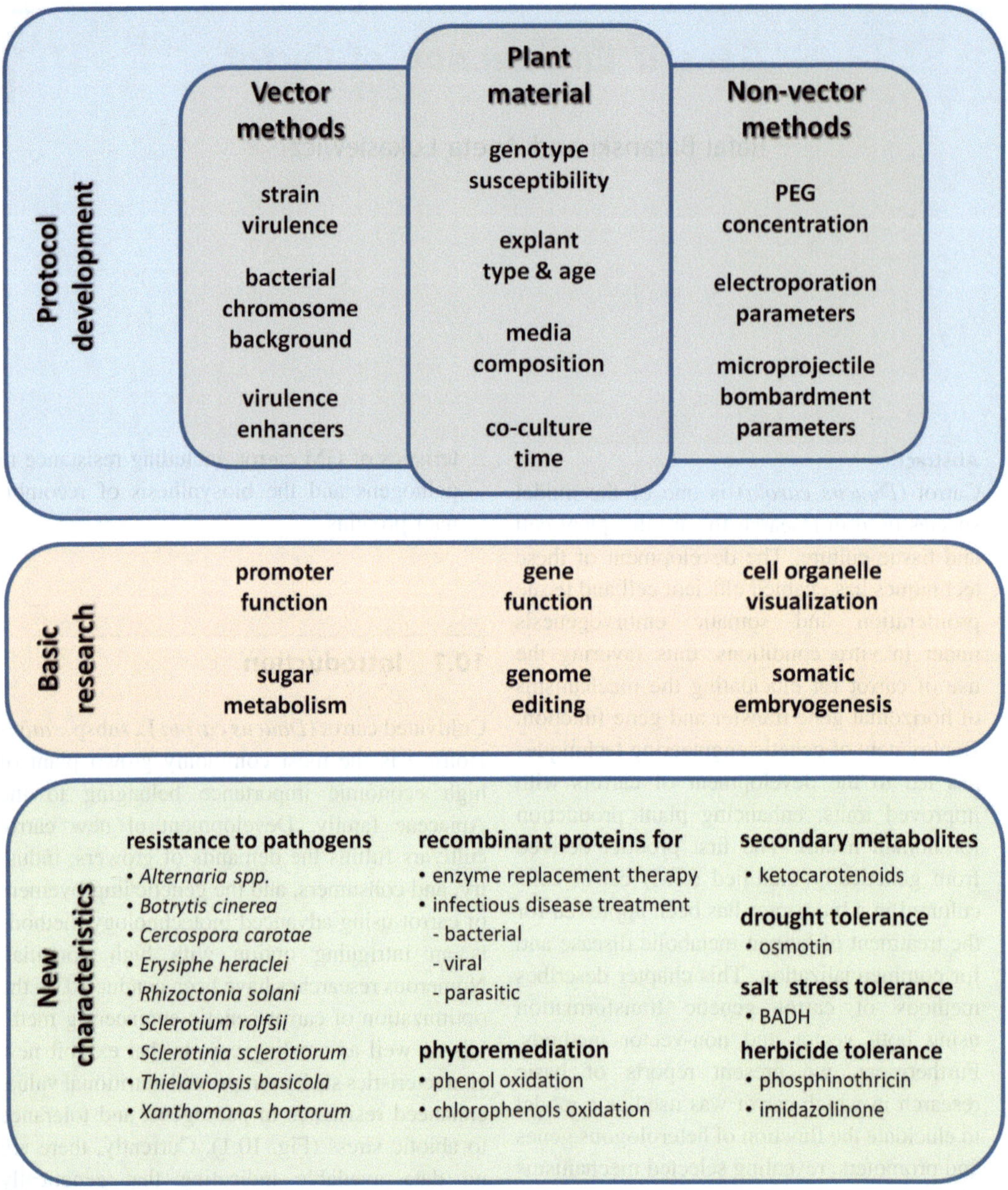

Fig. 10.1 Research aims in the optimization of carrot transformation protocols, utilization of transformation in basic research, and new characteristics obtained using GM carrot

1939–1958 (Gautheret 1939; Nobècourt 1939; Steward 1958; Steward et al. 1958). Since then, the development of cell and tissue culture techniques has improved exponentially and carrot was one of the main species utilized. Therefore, it is not surprising that this species was successfully used in the research on horizontal gene transfer that started in the 1980s. Consequently, carrot has served as one of the models for elucidating mechanisms of bacteria to plant gene transfer and promoter and gene function. The high potential of carrot cells to proliferate when cultured in vitro has also opened opportunities for the bioreactor production of recombinant proteins, in particular those of pharmaceutical significance. The first product of engineered carrot cells cultured in a bioreactor was approved for the treatment of Gaucher's disease in 2012 (FDA 2012) and is the only commercialized product of GM carrot.

10.2 Methods of Carrot Genetic Engineering

Both vector and non-vector methods have been developed and applied in carrot genetic engineering. Protocols utilizing *Agrobacterium tumefaciens* as a gene construct vector are the most common and have been widely used in both basic research and research aiming to develop carrots with new traits. The other vector, *Agrobacterium rhizogenes* (*Rhizobium rhizogenes*; Young et al. 2001), has been used mainly in the elucidation of the genetic mechanisms of the hairy root phenotype, which was utilized for the production of pharmaceutical metabolites. In contrast to vector methods, direct delivery of nucleic acids to carrot cells or protoplasts has been used less frequently.

The schematic presentation of the various genetic transformation protocols for carrot is shown in Fig. 10.2. In general, the main steps include: (1) the choice of initial plant material, (2) preparation of target explant, (3) choice of the gene construct delivery technique, (4) selection and production of GM tissue or plant.

10.2.1 Initial Plant Material

The seeds and the storage root of carrot are the two main plant organs used as the initial plant material for transformation. Seeds germinated in vitro may be used to produce sterile seedlings in which the juvenile plants can serve as explant donors. Production of sterile seedlings requires surface sterilization to eliminate microorganisms from the seed. In general, the seeds are normally washed in 70–95% ethanol for 30–120 s, then washed with a 1–10% sodium or calcium hypochlorite solution or a 20–50% commercial bleach for 15–45 min, then washed several times with sterile water (Luchakivskaya et al. 2011; Rosales-Mendoza et al. 2007; Simpson et al. 2016). A tea infuser can be used and convenient to completely immerse seeds in the different sterilizing solutions. Highly infected seeds may require a higher concentration or a longer exposure to the bleaching solution; however, it may negatively affect the germination of the seed. Additional steps may help such as the treatment with warm water (40 °C) or an overnight wash at room temperature to stimulate the germination or presence of microorganisms, after which the sterilization steps can be repeated. An additional bath using a fungicide solution before bleaching can be considered for additional protection when seeds are heavily infected (Aviv et al. 2002; Grzebelus et al. 2012).

The transformation of roots was primarily used, due to research on the development of hairy roots after inoculation with *A. rhizogenes*. In addition, root slices are also a convenient material for direct DNA delivery using microprojectile bombardment. By slicing the root, it is possible to produce many explants of hard tissue but also containing cambium, a meristematic tissue responsible for the secondary growth of the root. The ability of this meristematic region to proliferate is of particular importance for the multiplication and selection of rare GM events among a mass of untransformed root disc cells. Alternatively, root discs can be stimulated to massively produce callus when exposed to media enriched with auxins. Typically, 0.1–1.0 mg/l 2,4-dichlorophenoxyacetic acid (2,4-D) is sufficient to induce callogenesis and to maintain the continuous growth of the unorganized tissue, which can be used as a target explant for gene delivery. Alternatively, 2.0 mg/l of Dicamba can be used instead of 2,4-D (Luchakivskaya et al. 2011) or a combination of 2,4-D and a low concentration of cytokinin (e.g., 0.1 mg/l N6-[2-isopententl] adenine (2iP) (Noh et al. 2012) or 0.025–0.25 mg/l kinetin (Balestrazzi et al. 1991; Klimek-Chodacka et al. 2018).

The use of storage carrot roots for transformation requires that they are surface-sterilized.

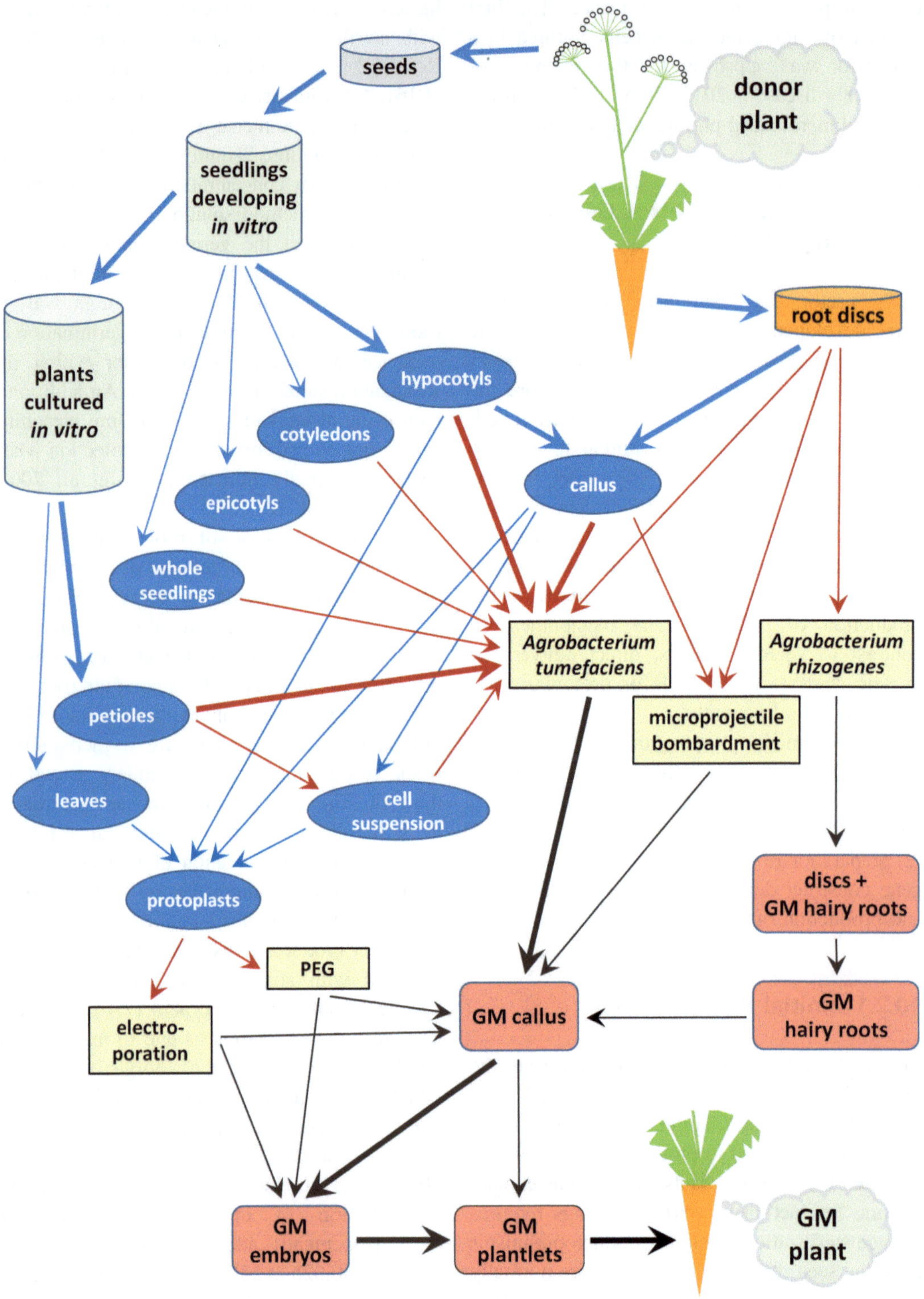

Fig. 10.2 Workflow showing carrot initial materials and explants suitable for genetic transformation using vector and non-vector methods reported in the literature. Bold arrows indicate preferred routes

Multi-stage protocols are commonly used. First, the storage root is washed and peeled which eliminates most soil residue and microorganisms attached to the skin. Then the root is sterilized in sodium or calcium hypochlorite solution or in a commercial bleach solution and then washed in sterile water (Baranski et al. 2006; Bercetche et al. 1987). However, this procedure is highly destructive and kills many layers of cells, and the symptoms are visible as tissue discoloration of the whole root surface. The destruction of the outer 5-mm tissue layer can be performed without any harm to the explant preparation if the cambium and proximal tissue remain unaffected. After slicing the root into 3- to 8-mm discs, the whole discs containing the core surrounded by the secondary cambium and the remaining layer of living tissue can be placed on a mineral medium with a growth regulator. The outer tissue destroyed by sterilization can be removed by excising after slicing or it can be left untouched since it does not prevent or limit callogenesis in the cambium layer. Disc explants can also be cut into several circular sectors or smaller fragments of about 1 cm^2 before incubation on mineral medium while ensuring that they contain secondary cambium capable of further callus development (Araujo et al. 2002). In addition, cutting the disc into cube fragments increases the surface developing callus as cambium cells are exposed not only at the upper disc surface but also on its sides; hence, callus production can be sped up. Callus is also a convenient source of cells for the establishment of a cell suspension culture or for protoplast release, both of which can be further transformed.

Currently, there are no successful reports of carrot transformation in planta. The floral-dip method, which is commonly used in the genetic transformation of Arabidopsis (Clough and Bent 1998) and has been used in other plant species (Niazian et al. 2017), has been unsuccessful in carrot, despite the fact that carrot has a well-developed inflorescence, making it a desired target for gene delivery. Carrot develops several branched stems, each containing more than 50 umbel inflorescences, which can contain about 50 exposed flowers. Thus, a single plant can produce ~30,000 seeds. The flowering period of a single umbel lasts for 7–10 days, and a plant develops flowers for approximately 30–50 days (Merfield et al. 2010; Rubatzky et al. 1999). The high seed yield from a single plant makes carrot a potential target for genetic transformation using in planta methods. However, attempts to transform carrot by immersing the umbels in *Agrobacterium* has remained unsuccessful, despite trying a broad range of factors including different bacteria strains, time of inoculation, vacuum infiltration, different temperatures applied to flowers at various developmental stages and the use of different cultivars (Gladysz and Baranski 2003). The floral-dip method has remained unsuccessful for carrot and other Apiaceae species; no GM events were found after screening 10,000 carrot seeds (Ghabouli et al. 2013). Currently, only sterile carrot explants grown in vitro have been reported as successfully transformed targets.

10.2.2 Target Explants

A wide range of carrot explants can be used for gene delivery depending on the initial plant material used (Table 10.1). The simplest procedure relies on the use of root discs that are directly exposed to a solid mineral medium or to a pure agar medium in a Petri dish. The whole seed-derived juvenile plants growing in vitro are rarely used as direct targets for genetic transformation (Luchakivskaya et al. 2011). They are usually grown for 4 weeks or longer to develop firm leaves, in which petiole fragments of about 1 cm in length can be excised and are easier to handle (Wally et al. 2006). Fragments of leaf blades can also be used for protoplast release (Dirks et al. 1996).

More frequently, seedlings are used as an explant source and virtually all seedling parts have been successfully transformed, i.e., hypocotyl, epicotyl, and cotyledons (Brodzik et al. 2009; Hardegger and Sturm 1998; Pawlicki et al. 1992; Tokuji and Fukuda 1999). The use of 3-week-old aseptic seedlings doubled the efficiency of hypocotyl transformation in

Table 10.1 Explant types and *Agrobacterium tumefaciens* strains used for carrot transformation

Explants	*A. t.* strain	Reference	Explants	*A. t.* strain	Reference
Hypocotyls	LBA4404	Thomas et al. (1989)	Petioles	LBA4404	Chen and Punja (2002)
		Chen and Punja (2002)			Jayaraj and Punja (2008)
		Marquet-Blouin et al. (2003)			Wally et al. (2008, 2009a, b)
		Brodzik et al. (2009)		GV2260	Pawlicki et al. (1992)
		Kim et al. (2009)	Roots	LBA4404	Hardegger and Sturm (1998)
		Noh et al. (2012)		GV3101	Hardegger and Sturm (1998)
		Hardegger and Sturm (1998)		GV2260	Pawlicki et al. (1992)
	C58C1	Porceddu et al. (1999)	Callus	LBA4404	Yau et al. (2008)
	GV3101	Hardegger and Sturm (1998)			Annon et al. (2014)
	A281	Balestrazzi et al. (1991)			Klimek-Chodacka et al. (2018)
	EHA101	Tokuji and Fukuda (1999)			Hardegger and Sturm (1998)
	GV2260	Pawlicki et al. (1992)		GV3101	Hardegger and Sturm (1998)
	AM109	Guan et al. (2009)		A281	Wurtele and Bulka (1989)
Cotyledons	GV2260	Pawlicki et al. (1992)		A348	Wurtele and Bulka (1989)
Epicotyls	GV3101	Monreal-Escalante et al. (2016)		EHA101	Wurtele and Bulka (1989)
	EHA105	Gilbert et al. (1996)		GV3850	Wurtele and Bulka 1989
	MOG101			n.a.	Permyakova et al. (2015)
Hypocotyls and cotyledons	GV3101	Maass et al. (2009)	Cell suspension	LBA4404	Mikschofsky et al. (2009)
		Arango et al. (2010, 2014)			Hardegger and Sturm (1998)
Hypocotyls and epicotyls	LBA4404	Rosales-Mendoza et al. (2007, 2008)		GV3101	Hardegger and Sturm (1998)
	GV3101	Simpson et al. (2016)		GV3850	Scott and Draper (1987)
Whole plants	GV3101	Luchakivskaya et al. (2011)		n.a.	Imani et al. (2002)

n.a.—not available

comparison with 2-week-old seedlings; however, a lower efficiency was observed for hypocotyls excised from seedlings that were older than 3 weeks (Pawlicki et al. 1992). However, 4-week-old plants were preferred over 1-week-old plants whose hypocotyls remained untransformed (Rosales-Mendoza et al. 2007).

Hypocotyls can also serve as a source of protoplasts. As carrot seedlings are fragile, their etiolation in the dark is often provoked. Such elongated hypocotyls enable the production of a higher mass of tissue necessary for protoplast isolation, particularly to ensure sufficient protoplast density in the final suspension (Grzebelus et al. 2012). Hypocotyls can also be exposed to mineral media supplemented in 2,4-D for the induction and growth of callus tissue (Pawlicki et al. 1992), which is similar to the case of root discs. Callus develops at the ends of injured hypocotyl segments and can also generate along the segment surface, however, with a lower efficiency. The production of callus from hypocotyls or root discs usually takes one to a few months, requiring several subcultures of the tissue to a fresh medium. Callus is a highly convenient material that can be easily propagated, divided into smaller fragments, and can be exposed to various conditions, making it a useful source of explant for transformation using vector or non-vector methods. Callus soft structure is highly advantageous for the establishment of a cell suspension when incubated in a liquid medium on gyratory shaker and is analogous to callus derived from root discs. Hence, callus tissue and cell suspension derived from hypocotyls or other explants can be used for protoplast isolation. Somatic embryos can also be induced from callus. In addition, incubation at a lower temperature restricts the callus growth rate, which slows down the growth rate between subcultures, thus enabling longer periods without human intervention, reducing labor and costs. All these features make the use of callus a valuable material for direct or indirect gene delivery (Wurtele and Bulka 1989).

Both cell suspensions and protoplasts can be targets for gene delivery through the use of vector and non-vector methods. The maintenance of a cell suspension culture is the most laborious, requiring frequent medium replacement and constant oxygenation by shaking; thus, additional equipment is necessary. However, cell suspensions are a valuable target for *Agrobacterium*-mediated transformation. Bacteria have an unlimited access to individual cells or small cell aggregates in the suspension and can easily attach to the plant cell wall. To enhance the suitability of the cell suspension for genetic transformation, Imani et al. (2002) proposed cell cycle synchronization using the fluorodesoxyuridine (FDU)/thymidine system, which resulted in a higher efficiency of transgenic events. Cells were incubated in the presence of 0.1 μM FDU for 24 h that arrested the cell cycle at the G1 phase, and then, 10 μM of thymidine was added to initiate the transition from the G1 to S phase. Consequently, the fraction of cells at the same stage of cell cycle was increased. Furthermore, cell proliferation activity in cell suspension resulted in a continuous increase of free cells that can be easily reprogrammed to somatic embryogenesis by replacing the medium with a fresh media without auxins. However, the efficiency of somatic embryogenesis may be unsatisfactory as the process is sensitive to a low cell density, which can be expected after selection is applied to eliminate non-GM cells (Higashi et al. 1998).

The gene transfer process is often performed immediately after explant excision and their exposure to mineral media. This procedure saves time and does not require repetitive subcultures. Alternatively, some authors pre-incubate explants for a few days to promote growth and limit stress applied by excision and change of the environment. Such pre-incubation was applied to hypocotyls (Hardegger and Sturm 1998; Pawlicki et al. 1992). In general, pre-incubation of seedling or juvenile plant-derived explants increases the transformation efficiency when a vector method is used. During the initial incubation, such explants swell and the expanding tissue breaks the cuticle, which is a natural barrier normally preventing *Agrobacterium* penetration. As the fragmented cuticle expands, bacteria enter the intercellular spaces, which favor bacteria attachment to the cell wall (Tokuji and Fukuda 1999). At the same time, explant cells initiate stress response mechanisms due to stress induced by wounding and the changing environment. Consequently, the amount of phenolic compounds elevate and are secreted, enabling a more effective binding of phenolics to *Agrobacterium* virA receptors, which promotes agroinfection (Balestrazzi et al. 1991).

Scott and Draper (1987) developed a method which enhanced cell proliferation using a feeder, or nurse layer that was later patented (Hauptmann et al. 1997). The feeder layer is prepared by diluting a 2-week-old carrot cell suspension of 1.3×10^6 cells/ml in a 0.8% agar using 70 ml suspension per one liter medium and poured into a Petri dish where it solidifies. The cell suspension is then covered by guard and transfer discs, i.e., two discs of filter or blotting paper covering the solidified medium and incubated in white light for 3 days. Fresh cell suspension (0.1–0.2 ml) is then applied to the upper (transfer) disc and incubated for 5–7 days prior to transformation. The transfer disc with attached and dividing cells can be easily removed and placed to a fresh medium when needed. A positive effect on transformation efficiency was also observed when the medium was enriched in a peptide plant hormone, phytosulfokine. This hormone stimulates cell proliferation, thus favoring callus development that was observed during hypocotyl transformation (Matsubayashi et al. 2002, 2004).

Independent to the explant type, explants are exposed to a mineral medium in vitro and are stimulated to cell division. The choice of medium composition is usually limited to either Murashige and Skoog (MS) (Murashige and Skoog 1962) or Gamborg B5 (Gamborg et al. 1968). Both media can be used with a complete amount of macro- and micronutrients or the concentration can be reduced by half. Gamborg B5 medium stimulates callus development with two- and threefold higher rates than MS medium when hypocotyl and root explants were exposed (Hardegger and Sturm 1998). Root discs can also be placed into a medium with highly diluted salts or even on water agar plates without any additional nutrients or growth regulators. Such minimalist medium can be used in experiments in which root discs are inoculated with *A. rhizogenes*. The developing hairy roots uptake nutrients stored in the root disc tissue and then are excised within 3–4 weeks; thus, the use of mineral medium to feed the disc explant is not necessary (Araujo et al. 2002).

Despite the sterilization of initial material, endogenous microorganisms may still exist inside tissues and their presence can be observed during explant culture. This problem may occur in any explant type but is most frequently observed in root disc and seedling explants. The appearance of bacterial colonies around the explant is the primary indicator of endogenous infestation. However, this problem is not a concern, and during the later stages of transformation, explants are exposed to various antibiotics to kill *Agrobacterium* or to select transgenic events. The presence of antibiotics in the medium often kills endogenous bacteria or at least prevents their development.

10.2.3 Techniques of Gene Construct Delivery

10.2.3.1 *Agrobacterium tumefaciens*-Mediated Transformation

Virtually all transformation methods have been used to engineer the carrot genome; however, *A. tumefaciens*-mediated transformation has been the most frequently reported. The first successful carrot transformation using this vector was reported over 30 years ago when a suspension culture was co-incubated with a culture of *A. tumefaciens* (Scott and Draper 1987). In this pioneering work, a high number of putative transgenic events were obtained as 60% of cell colonies developed on the selection medium containing 100 mg/l kanamycin. Gene transfer to the carrot genome was confirmed in callus and plants developed through use of somatic embryogenesis when the neomycine phosphotransferase II (*npt*II) gene presence was detected by Southern blotting. The acquired resistance to kanamycin and additionally the synthesis of nopaline by these materials confirmed the integration of functional bacterial genes. The authors also showed that GM plants with a normal phenotype can develop from *A. tumefaciens*-mediated cell suspension stimulated to somatic embryogenesis.

Early studies on carrot transformation were devoted to test the effects of various factors that might influence gene transfer to carrot cells. These included testing different bacterial strains and inoculum preparations as well as donor and target plant materials. A comparison of the different protocols has remained inconclusive, since different plant genotypes, explant types, and bacterial strains were used by the research groups. Hardegger and Sturm (1998) concluded that by using available protocols it was not possible to ensure the effective development of independent transgenic events. Until now, none of the protocols can be considered as versatile; however, some protocols predominate despite slight modifications implemented.

Hypocotyl explants remain the main target for *Agrobacterium*-mediated transformation. However, Thomas et al. (1989) reported that hypocotyls did not respond when they were exposed directly to *A. tumefaciens*. Therefore, pre-culture of hypocotyls may be implemented to overcome the problem. The most common procedure of *A. tumefaciens*-mediated carrot transformation involves excision of hypocotyls from 2- to 4-week-old seedlings (Brodzik et al. 2009; Monreal-Escalante et al. 2016); however, 1-week-old seedlings were also used (Arango et al. 2010; Hardegger and Sturm 1998). The hypocotyls are cut into ca. 1-cm-long fragments and either pre-incubated for 2–3 days on a mineral medium with growth regulators (Kim et al. 2009; Noh et al. 2012) or used directly for inoculation. *A. tumefaciens* inoculum is prepared from an overnight culture resuspended in the same medium as used for explant incubation. Additionally, phenolic compounds like acetosyringone can be included, although a stimulating effect of phenolic compounds on carrot transformation is questionable (Hardegger and Sturm 1998; Pawlicki et al. 1992; Wurtele and Bulka 1989). Co-cultivation is initiated by submerging hypocotyls in inoculum for 5–20 min, and then hypocotyls are incubated on a solid medium for 2 days. Co-cultivation is rarely prolonged past 5 days (Marquet-Blouin et al. 2003). Next, the hypocotyls are transferred to a fresh medium with antibiotics for 2–4 weeks for callus development. At this culture stage, two different strategies can be implemented. The first strategy uses antibiotics to kill the bacteria; claforan (200 mg/l), cefotaxim (200 mg/l), timentin (300 mg/l), or vancomycin (200 mg/l) are usually applied, but their concentrations can also be doubled. After 2 weeks, explants are transferred to a fresh medium supplemented with a selection agent, either an antibiotic or herbicide, depending on the introduced resistance gene (Arango et al. 2010; Brodzik et al. 2009; Hardegger and Sturm 1998). The alternative strategy and simpler procedure relies on the immediate placing of explants after co-cultivation on a medium containing agents eliminating bacteria and enabling selection of the transformants simultaneously (Kim et al. 2009; Noh et al. 2012). Several subsequent subcultures aim in the development and selection of transgenic callus and the formation of somatic embryos which later develop into plants.

The amount of tissue collected by hypocotyl excision can be limited; therefore, some researchers do not differentiate between seedling parts during explant preparation and use hypocotyls together with cotyledons (Arango et al. 2010, 2014; Maass et al. 2009) or epicotyls (Monreal-Escalante et al. 2016; Rosales-Mendoza et al. 2007; Simpson et al. 2016). Furthermore, Luchakivskaya et al. (2011) used the whole 14-day-old plants that were vacuum-infiltrated with inoculum and co-cultivated for 2 days. Then they were cut into fragments and stimulated for callus development using a selection medium with cefotaxim and kanamycin and 2 mg/l Dicamba instead of 2,4-D. Other experiments showed that petiole segments were more prone to develop transgenic callus than hypocotyls, 3.3% versus 1.4%, respectively (Chen and Punja 2002). Consequently, 5- to 10-mm-long petiole segments of 4- to 6-week-old plants were recommended as target explants (Jayaraj and Punja 2008; Wally et al. 2006, 2008; Wally and

Punja 2010). Similar to the case of hypocotyls, petioles' response to *A. tumefaciens* highly depends on the plant genotype, which ranged from 0 to 47% of explants developing callus (Pawlicki et al. 1992).

Callus tissue, either of hypocotyl or of storage root origin, can be conveniently used for *A. tumefaciens*-mediated transformation. Two procedures can be recommended. Fragments of callus tissue are collected in a Petri dish and submerged in the inoculum (Klimek-Chodacka et al. 2018), or alternatively, small aliquots are applied to callus clumps (Annon et al. 2014; Yau et al. 2008). The use of acetosyringone in the inoculum was also reported (Annon et al. 2014; Klimek-Chodacka et al. 2018). Independent of the method, the co-cultivation for 2–3 days is terminated by spreading callus on a fresh medium with antibiotics. Selection agents can be used simultaneously or in the next subculture. Theoretically, only transgenic cells should survive on the selection medium and subsequently small new GM callus clumps should develop on the surface of decaying initial material. However, if too much tissue is exposed to the medium after co-cultivation, the growth of GM callus may be arrested, and also the availability of the selection agent to upper layers of cells is limited. Therefore, it is essential to evenly distribute a thin layer of cells on the selection medium, and then transfer it to fresh selection media until single well-visible callus clumps of high growth potential are observed. The cell transfer to a fresh media can be simplified by using a filter or blotting paper disc placed on the medium surface and spreading the cells over the disc. Then the whole paper disc is transferred (Yau et al. 2008).

Cell suspension is infrequently used for carrot transformation. This may be due to the lengthy time required for its establishment and frequent interventions to keep it continuously growing by replacing the liquid medium. Nevertheless, large numbers of cells directly exposed to *A. tumefaciens* make cell suspensions a valuable target for transformation. The protocol relies on mixing a well-growing cell suspension with the inoculum and further co-cultivation on a gyrating shaker. Then cells are poured directly onto the solid medium or on a paper disc laying on the solid medium. Either the solid medium can be supplemented with a selection agent or the selection is done in the next round of the paper disc transfer (Hardegger and Sturm 1998; Imani et al. 2002; Scott and Draper 1987). Cell cycle synchronization can be additionally provoked using the FDU/thymidine system (Imani et al. 2002; Mikschofsky et al. 2009).

The comparison of data provided by various authors is difficult as they usually use different bacterial strains and plant cultivars. Predominantly, the octopine LBA4404 and nopaline GV3101 *A. tumefaciens* strains were used. These strains differ in their chromosomal background and virulence helper plasmids. Their direct comparison showed that the use of GV3101 strain leads to a higher transformation efficiency when applied to various explants (Hardegger and Sturm 1998). The most remarkable differences were observed for seedling root explants, 5% for GV3101 versus 95% for LBA4404, and storage root slices, 2% for GV3101 versus 41% for LBA4404, while efficiencies using hypocotyl explants were almost independent of the bacteria strain. Other works showed that the pGV3850 helper plasmid favors gene transfer in comparison with the pTiA6 plasmid when they were inserted into the same C58 strain (Wurtele and Bulka 1989). Despite the fact that the binary plasmid effect on the transformation efficiency of petioles of juvenile plants was reported, this effect was found meaningless for explants from older plants (Gilbert et al. 1996; Pawlicki et al. 1992). The effect of the type of bacteria strain is additionally modified, depending on the plant genotype being targeted for transformation. Such bacteria strain × plant genotype interaction was reported, which showed that transformation of 'Nantes Scarlet' was two times more efficient when using the LBA4404 strain than the C58C1 strain (Takaichi and Oeda 2000). In contrast, the efficiency of the transformation of 'Kuradogo-sun' was six times higher when using C58C1 rather than the LBA4404 strain. These observations were analogous to earlier results by Gilbert et al. (1996), who reported that 'Nanco' was about six times more susceptible to EHA105 than to MOG101, while 'Danvers Half Long' was 10 times more susceptible to MOG101 than to EHA105. The main effect of plant genotype was observed when the frequency of transgenic callus derived from hypocotyls ranged from 0.9 to 5.8% among four carrot cultivars (Thomas et al. 1989). The difference between the other two cultivars in the number of transgenic plants developed was twofold (Takaichi and Oeda 2000) and between the three other cultivars was over a threefold change (Wally et al. 2006). Less pronounced differences in cultivar response to *A. tumefaciens* were reported in a study using vacuum-infiltrated plants (Luchakivskaya et al. 2011).

10.2.3.2 *Agrobacterium Rhizogenes*-Mediated Transformation

Plants or plant explants infected by *A. rhizogenes* develop hairy roots from cells, which acquire *rol*, and *aux* genes located at the T-DNA or, in case of agropine strains, at the T_L-DNA and T_R-DNA of bacteria Ri plasmid. The newly developed roots usually have a characteristic hairy phenotype and the ability to grow on the medium without growth regulators after their excision from the host tissue. They are also highly branched, do not exhibit geotropism, and thus are easy to identify (Chilton et al. 1982; Willmitzer et al. 1982). In carrot, hairy roots are free of hairs, so morphologically they resemble a branched seedling root system. For these reasons, they are often described as 'adventitious roots' or 'transformed roots,' particularly when wild *A. rhizogenes* strains not possessing any binary plasmids are used for carrot transformation.

Experiments targeting the induction of hairy root development were initiated in the late 1980s. One protocol was established, although some researchers adjust it to fit their needs. The target for inoculation is the cambium, a meristematic tissue of storage roots that has a high potential for neoplasy. To inoculate cambium, discs of a surface-sterilized storage root are prepared and then placed on Petri dishes. They can be exposed to a mineral medium (Cardarelli et al. 1987b), water agar (Fründt et al. 1998; Cardarelli et al. 1985), or filter paper moistened with water (Epstein et al. 1991). Exogenous auxins are usually not required as hairy roots are naturally induced, due to the expression of bacterial *aux* genes introduced to explant cells. However, hairy root development is much more pronounced when the inoculation is performed at the root disc apical surface, i.e., the surface being closer to the root tip before the root was sliced. Therefore, discs are preferentially orientated to face the apical side up. A reversed orientation significantly reduces the transformation efficiency if no exogenous auxins are applied. This rule is less important when the basal positive strains (Bas^+), e.g., A4, 1855, 15834, and TR105, are used for inoculating carrot. The activity of the *aux* genes of Bas^+ strains is higher than those of Bas^- strains, and this mitigates the effect of unidirectional auxin transport in a root disc. The application of Bas^- strains to basal side of the root disc imposes the necessity of using exogenous auxin, which is added to the inoculum (Bercetche et al. 1987; Ryder et al. 1985). Nevertheless, auxins are often added to the inoculum which may increase the number of hairy roots, in particular if less susceptible carrot genotype to *A. rhizogenes* is used. Also, a stimulating effect was observed when discs were pre-incubated in the medium enriched in auxin (Guivarc'h et al. 1993). Further enhancement was observed by applying acetrosyringone, when present either in the inoculum or in the culture medium. This phenolic compound more effectively stimulated hairy root induction on explants of less susceptible carrot genotypes and in combination with NAA (Baranski et al. 2006; Guivarc'h et al. 1993).

The inoculum is spread on the disc surface, ensuring it covers cambium cells, which are preferentially transformed (Bercetche et al. 1987; Boulanger et al. 1986). Discs with bacteria are co-incubated usually in the dark for several weeks without any need for bacteria elimination unlike in protocols with *A. tumefaciens* being exposed to antibiotics after 2–3 days. The appearance of hairy roots along cambium ring is observed in 10–14 days, and during the next 1–3 weeks they elongate to a few centimeters in length. Too long incubation time causes the root tips to attach to the Petri dish lid, restricting their further undisturbed growth. Hairy roots of 1–3 cm in length can be easily excised from the disc and grown separately without need for exogenous growth regulators. Depending on the aim, hairy roots are then subcultured using a solid or liquid mineral media. When cultured on the surface of a solid medium, they rapidly spread in all directions due to a highly branching phenotype and form a mass of intertwined roots. The use of liquid medium requires constant agitation ensuring oxygenation, which accelerates hairy root growth. The mean yield of biomass after 30 days of culture increased 45 times (Araujo et al. 2006).

The response to *A. rhizogenes* inoculation is highly dependent on the carrot genotype. Highly susceptible cultivars develop a vast number of hairy roots on virtually all inoculated root discs. Less susceptible genotypes produce fewer hairy roots, and the frequency of responding root discs can be several times lower. There are usually not much differences in the response between root discs originating from the same storage root; however, such variation is observed when different storage roots are used, even of the same cultivar. The efficiency depends also on the bacterial strain used. Strains can differ in the Ri plasmid harboring *aux* and *rol* genes but also may have different chromosomes. Both genetic elements significantly interfere with the transformation process; e.g., LBA9402 and A4 strains tend to be less virulent than their counterparts LBA1334 and A4T, the latter being derivatives with the same Ri plasmids but possessing a chromosome from the C58 *A. tumefaciens* strain (Baranski et al. 2006).

10.2.3.3 Microprojectile Bombardment

The microprojectile bombardment method, also known as particle bombardment, uses biological ballistics or biolistic transformation, which was developed in the late 1980s by Sanford et al. (1987). In this method, microcarriers made from gold or more cost-effective tungsten are coated with DNA and fired at high velocity into cells or tissues (Klein et al. 1987). This method of transformation is applicable especially in monocot plants resistant to *Agrobacterium*. Although the most commonly used method of carrot transformation is made with vectors, successful transformation with direct DNA delivery has also been achieved.

The main starting material for microprojectile bombardment in carrot is callus, derived from cell suspension or pre-cultured on filter paper (Deroles et al. 2002) or stems and petioles cut into 0.5- to 5-mm fragments and placed on a mineral medium with hormones for callus induction (Kumar et al. 2004; Rojas-Anaya et al. 2009). For callus derived from cell suspension, a short pre-culture period of up to 6 days improved the transformation efficiency (Deroles et al. 2002). Root discs of about 3 mm thick placed on the moistened filter paper prior to bombardment were also used (Hibberd et al. 1998). Regardless of the target plant material, the downstream protocol steps and factors influencing microprojectile bombardment are similar. The tungsten or gold particles of 0.4–1.6 μm in diameter are coated with DNA in the presence of $CaCl_2$ and spermidine. The type and size of particles are important factors affecting transformation efficiency. In carrot, 1-μm gold particles ensured better transformation efficiency than 1.6-μm gold or tungsten particles (Deroles et al. 2002) that is consistent with studies on other plant species, where higher efficiency of stable transformation was obtained when 0.7- to 1.0-μm gold particles were used as carriers (Kikkert et al. 2005). Also, the amount of DNA and volume of particles per shot were estimated for the best transformation efficiency. A significant difference was observed between 2 and 5 μl of particles, the latter being more effective; however, DNA quantity had no impact on transformation (Deroles et al. 2002). Other important parameters affecting successful transformation are helium pressure, which varied depending on the reports from about 500 to 8963 kPa, and a shooting distance, usually ranging from 6 to 14 cm (Deroles et al. 2002; Hibberd et al. 1998; Kumar et al. 2004; Rojas-Anaya et al. 2009). The results for optimization of the shooting distance are rather consistent, and the highest transformation efficiency was obtained at the 12 cm distance;

however, the results for helium pressure are inconclusive. Deroles et al. (2002) observed the highest efficiency at 600 kPa, while Kumar et al. (2004) at 7584 kPa, albeit the pressure considered the optimum by Deroles et al. (2002) was the highest, which they used. Another explanation for these differences could be the fact that the two research groups used different carrot genotypes. Deroles et al. (2002) observed that the efficiency of microprojectile bombardment could be influenced by the use of different genetic backgrounds of plant materials. After transformation, the bombarded material is incubated for a few days and then transferred to a selection medium with either antibiotics or an herbicide. Medium used for the selection is usually solid, but also semi-solid or liquid medium has also been used successfully (Rojas-Anaya et al. 2009). The presence of the *uidA* reporter gene in the gene construct allows for fast assessment of transformation effectiveness, and commonly, it is evaluated 24 h after the bombardment (Deroles et al. 2002).

10.2.3.4 DNA Uptake by Protoplasts

An alternative non-vector method of plant transformation is DNA uptake by protoplasts induced by either electroporation or polyethylene glycol (PEG) treatment. Protoplasts can be isolated from leaves, petioles, callus, or suspension cell culture; however, for carrot electroporation the most common starting material is cell suspension (Bates et al. 1988, 1990; Boston et al. 1987; Langridge et al. 1985). For the PEG method, both cell suspension (Dröge et al. 1992, Rasmussen and Rasmussen 1993, Gallie 1993) and petioles have been used (Aviv et al. 2002, Dirks et al. 1996). These two methods of DNA direct delivery to protoplasts have several common points and sometimes are used simultaneously in order to increase the transformation frequency. Besides the plant genotype and gene construct, other factors that should be taken into consideration prior to transformation are the density of protoplasts, amount of DNA, addition of carrier DNA, and ion presence. The most common concentration of protoplasts used for electroporation and PEG treatment is about 10^5–10^6 cells/ml, although Rasmussen and Rasmussen (1993) found 10^6 protoplasts/ml as the optimal density and either the increase or decrease of the density had an adverse impact on transformation frequency. Among the main factors affecting the efficiency of electroporation are voltage and time of the pulses. Generally, there are two approaches that can be used. In the first, longer pulses (1–50 ms) but with a low voltage (200–800 V/cm) are applied; in the second one, pulses are shorter (5–200 μs), but the voltage is higher (2–10 kV/cm) (Bates et al. 1988). The time of the pulses and voltage can be lower when the pulses are applied in series with short intervals (Langridge et al. 1985). Bates et al. (1988) showed that increasing the voltage from 250 to 750 V resulted in higher expression of chloramphenicol acetyltransferase (CAT) with the maximum obtained when the pulses lasted for 8 ms; however, the increased voltage reduced protoplast viability. The impact of voltage and pulses period on transformation is strongly associated with the medium composition. When salt-free medium is used, the voltage should be increased to at least 2 kV/cm and pulses should be shorter than 1 ms. Transformation can be ineffective in a low-salt medium when the voltage is low and the pulses are longer, and also no transgene expression was observed and protoplasts lost their viability when HEPES-buffered saline (HBS) medium lacked Ca^{2+} (Bates et al. 1988). Some reports indicate that the electroporation carrying on ice allows cell membrane pores to stay open for a longer period, hence making electroporation more effective (Langridge et al. 1985; Neumann et al. 1982). The additional application of heat shock prior to transformation was also successful (Shillito et al. 1985). Nevertheless, the conclusions concerning the effect of temperature on carrot protoplast transformation are inconsistent. In experiments conducted by Langridge et al. (1985), the cold treatment was crucial for electroporation while Bates et al. (1988) did not observe any improvement; the best results they obtained were at room temperature. Another factor which could influence the protoplast transformation process is addition of the salmon sperm, which could even double the transgene expression (Bates et al. 1988) or calf thymus DNA, although in some experiments addition of the latter decreased the expression (Boston 1987). The concentration of DNA used has varied from 10 to 40 μg (Bates et al. 1988, 1990; Langridge et al. 1985). Bates et al. (1988) showed that doubling the DNA concentration from 20 to 40 μg resulted in a twofold change in CAT expression. The process of electroporation could be combined with PEG treatment, which also increases the transformation efficiency (Boston et al. 1987).

Similar to electroporation, PEG mediates reversible changes to the cell membrane causing pores. The unquestionable advantage of PEG-mediated transformation is that this method does not require any additional equipment. The optimal PEG concentration seems to be 22–25% (Gallie 1993; Rasmussen and Rasmussen 1993). However, that optimal range could still be too high for stable transformation and lead to the degradation of transformed cells within a few days (Baranski et al. 2007b). PEG should be added immediately to the prepared protoplast-DNA solution as the last component. By delaying the addition of PEG by 20 min, decreased GUS activity almost three times, whereby adding PEG to protoplast suspension before DNA reduced GUS activity from 293 to 0.46 (expressed as pmol 4-methyl-umbelliferone/μg protein/h)(Rasmussen and Rasmussen 1993). Studies conducted on different plant species have indicated that the use of high molecular mass PEGs increases the number and area of cell membrane pores (Chakrabarty et al. 2008). Other authors have suggested that even the source of PEG may be important for transformation efficiency (Yoo et al. 2007). In carrot, a molar mass of used PEGs differed between 4000 and 8000 but the transformation efficiency remained at a similar level (Ballas et al. 1987). Divalent cations, Mg^{2+} and Ca^{2+}, are important components added to the protoplast solution which affects the membrane integrity and permeability and, thus, the transformation efficiency. The significance of Ca^{2+} for a successful carrot transformation is more pronounced. Removal of Ca^{2+} from the solution resulted in a decreased transgene expression by almost 90% while removal of Mg^{2+} by approximately 20% (Gallie 1993). The impact of DNA concentration on the efficiency of PEG-mediated transformation has not been assessed, and the DNA amounts used by authors varied from 10 to 100 μg (Rasmussen and Rasmussen 1993; Dirks et al. 1996; Aviv et al. 2002). It can be expected that similar to an electroporation-mediated transformation, higher amounts of DNA favor an increase in the transformation efficiency. In a transformation study using mRNA, the concentration of mRNA used was lower than those of DNA, ranging from 0.2 to 10 μg; even the lowest amount was sufficient for a successful transformation. However, by increasing mRNA amount up to 5 μg, the luciferase activity also increased, and the relationship was linear. The presence of carrier DNA improves transformation efficiency; e.g., the addition of 100 μg of salmon sperm DNA increased luciferase activity fivefold (Gallie 1993).

Additionally, the medium composition may affect the transformation process as PEG-mediated transformation is genotype dependent and requires an adjustment of culture conditions to the genotype used. Dirks et al. (1996) observed fast cell division and obtained high plating efficiency in the medium supplemented with 0.1 and 0.2 mg/l zeatine. The alginate used as an embedding matrix promoted cell division and influenced the response to a hormonal treatment, which resulted in a high number of embryos.

Although a wide range of factors were tested in studies on non-vector carrot transformation and there are similarities in the protocols used, they cannot be considered to be universal. As presented above, results obtained by one treatment may not be repeatable if a different genotype is used; hence prior to the transformation process, conditions should be determined experimentally and adjusted to the specific plant material used. The review of the literature indicates that vector methods of gene delivery are favored over non-vector methods for carrot transformation.

10.2.4 Selection and Development of Transgenic Plants

Selection of putative transgenic events is done using selection agents, using either antibiotics or herbicides. The concentration of the selection agents in the medium should be substantial for an effective selection of transformants, but there is no clear evidence showing optimal concentrations. Kanamycin is used in the range from 50 (Monreal-Escalante et al. 2016) to 300 mg/l (Wurtele and Bulka 1989), and 100 mg/l is the most common as higher kanamycin amounts may prevent transgenic cell development, considerably decreasing the transformation efficiency. In contrast, 50 mg/l of kanamycin does not guarantee the elimination of non-GM cells. Such cells may also grow at 100 mg/l, but the tissue development is much slower allowing for the selection of vigorously growing GM tissue. Non-GM plantlets show reduced growth and albinism at this concentration but may survive several subcultures until they decay. Apparently, carrot susceptibility to kanamycin is genotype dependent and should be verified. Concentrations above 50 mg/l may also reduce the efficiency of somatic embryogenesis (Hardegger and Sturm 1998). This can be overcome by using 25 mg/l kanamycin during somatic embryogenesis and then transferring embryos or plantlets to the selection medium with a much higher antibiotic concentration (Gilbert et al. 1996). An unambiguous differentiation between GM and non-GM tissues is achieved using 10 mg/l (Arango et al. 2010) to 100 mg/l hygromycin (Guan et al. 2009), or 1–10 mg/l phosphinothricin, although carrot genotype susceptibility should also be verified. The successful strategy of increasing the concentration of phosphinothricin was also reported (Wally et al. 2006; Jayaraj and Punja 2008).

Regenerated GM plantlets may be weak and require an acclimation to ex vitro conditions. Their survival rate may be lower than non-GM seed-derived or somatic embryo-derived plantlets, but the acclimation using substrates with reduced amounts of macro- and microelements, organic matter, and pH close to neutral favored a successful transfer rate (Mikschofsky et al. 2009).

Plants with deformed organs or atypical morphology are usually eliminated during subculturing in vitro, unless the research aim is the elucidation of a specific gene or promoter function. The morphology of GM plants is typical for in vitro-derived plants; they develop a deformed storage root due to the altered growth in the mineral medium in vitro and root injury during transfer to soil. Thus, storage root characteristics are not able to be assessed in T_0 plants, requiring seed production by self-fertilization. Since carrot is a biennial species, this process is long and requires vernalization to induce flowering. Plants generated using *A. rhizogenes*-mediated transformation may result in an abnormal morphology, which includes wrinkled leaves, curved petioles, dwarfism, and the development of fewer meristems. Distortions affecting reproduction were reported such as annuality and reduced pollen viability, the later significantly lowering seed production (Baranski et al. 2006; Limami et al. 1998; Tepfer 1984).

10.3 Carrot as a Model to Elucidate Promoter and Gene Functions

Carrot has been one of the prime model species suitable for elucidating gene function using a transgenic approach. This was largely due to the in vitro systems developed for carrot, ensuring effective cell division and tissue growth, in the 1980s when the era of plant genetic engineering had begun. In particular, the ease of root disc preparation and in vitro culture as well as a high root cambium potential for proliferation favored carrot to be included as one of the main targets suitable for the delivery of heterologous genes and their regulatory elements.

10.3.1 Mechanism of Hairy Root Development

Carrot became a model significantly contributing to the elucidation of the mechanisms of agroinfection and the role of bacterial genes in the process of hairy root disease. Carrot root discs conducive to the development of a large number of hairy roots after *A. rhizogenes* inoculation made this species a desirable object for T-DNA delivery. Due to the ease of infection by *A. tumefaciens*, carrot was used in early studies revealing the principal role of the T-DNA right border in gene transfer (Jen and Chilton 1986). The Ri plasmid of *A. rhizogenes* may contain two T-DNAs with a series of four *rol* genes in the T_L-DNA (Altamura 2004). These genes were identified by observing the response of carrot root discs transformed with plasmids having modified open reading frames. By this approach, the role and significance of *rol* genes for the occurrence of hairy root phenotype were described. The *rol*B gene was found to be the most important as it was able to ensure the hairy root phenotype even if other *rol* genes were not introduced (Boulanger et al. 1986; Capone et al. 1989, 1994; Cardarelli et al. 1987a; Serino et al. 1994). The second fragment, T_R-DNA of the Ri plasmid contains a series of genes for the synthesis of auxins and opines. The presence of auxins is critical for the initiation of hairy root development and that the activity of *rol*B promoter was auxin-dependent. Using carrot somatic embryos, it was also shown that its activity changed during specific developmental stages (Di Cola et al. 1997) and that nuclear proteins are involved in the activation of *rol* genes (Fujii 1997). In consequence, the enrichment of *A. rhizogenes* inoculum in exogenous auxins is recommended to enhance the activation of *rol* genes and thus hairy root development (Bercetche et al. 1987; Cardarelli et al. 1987b). The application of auxins occurred important when using basal attenuated *A. rhizogenes* strains, i.e., the strains of polar virulence inducing hairy roots at the apical side of the root disc and not inducing them at the basal side due to the unidirectional auxin flux toward the apical part (Cardarelli et al. 1985; Ryder et al. 1985). Further experiments led to the identification of functional genes responsible for opine synthesis (Hansen et al. 1991). The hairy roots of carrots were also used as a model organ. Souza et al. (2007) evaluated hairy roots under nitrogen stress and observed changes in enzymatic activities; elevated amounts of ammonium stimulated senescence and the activity of glutamate dehydrogenase increased while the activity of glutamine synthase decreased.

10.3.2 Promoter Activity

Further studies focused on evaluating heterologous promoters that could ensure high expression of the introduced genes of interest. Among the first experiments, various promoters of either bacterial or plant origin were fused to reporter genes, which were used to assess their suitability for driving gene expression in carrot. These included promoters of the *Cauliflower mosaic virus* (CaMV) 35S, *A. tumefaciens* nopaline synthase (*nos*) and mannopine synthase (*mas*), and maize endosperm *zein* genes. All promoters successfully activated the expression of fused chloramphenicol acetyltransferase (*cat*) gene or β-glucuronidase (GUS) gene (*uidA*). However, GUS expression was four times higher when the *nos* promoter was used instead of the CaMV 35S promoter (Boston et al. 1987; Rathus et al. 1993). An enhanced expression was observed by using the *Agrobacterium* octopine synthase enhancer sequence fused to the *CaMV 35S* promoter (Rathus et al. 1993). Further modification of the promoter sequence by inserting the G-box element with GC-rich flanking sequences considerably improved the *CaMV 35S* promoter activity, raising the observed GUS level about 13 times (Ishige et al. 1999). A doubled CaMV 35S (*d35S*) ensured 2–3 times higher expression of GUS than the single *35S* promoter in callus, and in leaves and roots of in vitro plants as well as in plants acclimated to growth ex vitro. A similar expression level was observed using the *Arabidopsis* ubiquitin promoter (*UBQ3*) versus the *35S* promoter; however, the maize ubiquitin *ubi-1* was less active (Punja et al. 2007; Wally et al. 2008).

Additional attention was given to research in tissue-specific and organ-specific promoters, in particular root-specific promoters. The *A. rhizogenes* root-specific *rol*D promoter was 15–18 times less active in carrot roots than a constitutive *d35S* promoter, which was almost inactive in leaves of greenhouse-grown plants, although the expression was observed in leaves of plantlets cultured in vitro. GUS activities in leaves and roots of the in vitro-cultured plants were at the same expression level although, in the greenhouse-grown plants, the activity was almost four times higher in roots, confirming the root-specific activity of the *rolD* promoter. The *A. rhizogenes* agropine synthase promoter was almost non-active in leaves and roots (Punja et al. 2007; Wally et al. 2008). Two other plant promoters were evaluated in carrot for their function as root-specific regulators: the *pDJ3S* promoter of yam (*Dioscorea japonica*) storage protein discorin 3 subunit and the *pMe1* promoter of cassava (*Manihot esculenta*) gene of unknown function. The *pMe1* promoter showed similar activity to *35S*, and the *pDJ3S* promoter was almost two times more active when assessed in carrot root. The latter exhibited also high specificity to root organ as it was responsible for about seven times higher GUS expression in roots than in leaves and stem. The *pMe1* promoter was also root-specific, but the expression in this organ was only about 50% higher than in other organs. Histochemical staining revealed also that *pDJ3S* was mainly active in the secondary xylem while *pMe1* in all root tissues except of the secondary xylem (Arango et al. 2010). Another promoter of sweet potato (*Ipomoea batatas*) *SRD1* gene also conferred the root-specific activity when introduced to carrot and was not active in leaves. The GUS activity was over 10 times higher when the gene was controlled by *SRD1* promoter in comparison with *35S*. The *SRD1* promoter ensured a high gene expression in all carrot root tissues, in contrast to the *pMe1* and *pDJ3S* promoters, and its activity was developmentally related as older and thicker storage roots showed more intense GUS staining (Noh et al. 2012).

10.3.3 Metabolic Pathways and Physiological Processes

Carrot is a common vegetable possessing high amounts of pro-vitamin A carotenoids accumulating in its storage root. Wild carrot develops white roots devoid of carotenoids, and thus, carrot became an intriguing model to study metabolic pathway of carotenoid biosynthesis. The enhanced carotenoid biosynthesis and accumulation were obtained by inserting the *Erwinia herbicola crt*B gene coding for phytoene synthase (PSY) and fused to the plastid transit peptide aiming in enzyme targeting chromoplasts. Its expression increased phytoene biosynthesis in orange carrot root and in consequence increased β-carotene accumulation which amount was doubled (Hauptmann et al. 1997). When the *crtB* gene under the control of yam root-specific promoter was inserted into wild carrot, the root color changed from white to intense yellow. This color was due to elevated amounts of carotene intermediates, i.e., phytoene, phytofluene, and ζ-carotene, and lycopene accompanying β-carotene; the later was in much lower amounts than in typical orange root, and α-carotene was not present. Thus, overexpression of PSY enabled increased carotenoid levels in carrot and their sequestration in chromoplasts in the crystalline form, analogously as it happens in orange carrots, although carotenoids composition was different (Maass et al. 2009). The *Arabidopsis CYP97A3* gene coding for the carotene hydroxylase was also expressed in orange carrot. The *AtCYP97A3* is fully functional in contrast to carrot *DcCYP97A3* of orange cultivars. This carrot mutant gene contains a premature stop codon resulting in a non-functional hydroxylase protein. This causes a restricted α-carotene to lutein conversion and high α/β carotene ratio. The overexpression of *AtCYP97A3* changed carotenoid composition in carrot by reducing the amount of α-carotene. Simultaneously, total carotenoids were also reduced that resulted from a lower PSY protein level despite unaltered *PSY* gene expression, and thus, it was concluded that carotene hyrdoxylase overexpression negatively affected PSY protein translation (Arango et al. 2014). The role of precursors on carotenoid biosynthesis was also evaluated by inserting the *Arabidopsis* deoxyxylulose 5-phosphate synthase (*DXS*) and reductoisomerase (*DXR*) genes of the methylerythritol 4-phosphate (MEP) pathway. The *DXS* overexpression enhanced PSY transcript levels and thus, on average, doubled carotenoid amounts while the *DXR* overexpression had no significant effect. Similar effects were observed for changes in chlorophyll contents in leaves (Simpson et al. 2016).

The role of the algal *Haemotococcus pluvialis* β-carotene ketolase (*bkt*) was assessed in carrot. This enzyme converts β-carotene to keto-carotenoids not present in carrot, canthaxanthin and astaxanthin, which have strong antioxidant activity, and are valuable nutraceuticals in human diet as well as feed supplements used in cultures of pink-colored fish, salmon and trout. The *bkt* gene was fused to the ribulose bisphosphate carboxylase-oxygenase (RuBisCO) signal peptide to ensure the enzyme activity in plastids. The expression of carrot β-carotene hydroxylases was up-regulated in leaves and roots. Ketocarotenoids, mainly astaxanthin, adonirubin, and canthaxanthin were accumulated up to 2400 μg/g root dry weight with simultaneous reduction of carotenes (Jayaraj et al. 2008). Plants containing high amounts of keto-carotenoids grew better when exposed to high UV-B irradiation and leaves showed less injury when H_2O_2 or methyl viologen stress was applied. It was concluded that high antioxidant and free-radical scavenging activity of keto-carotenoids prevented cells from oxidative stress (Jayaraj and Punja 2008).

Carrot was also used to elucidate the role of heterologous genes in physiological processes like a transmembrane transport and programmed cell death. The expression of ATPase introduced to carrot cells in the antisense orientation blocked the vacuolar ATPase A subunit specific to the tonoplast. In consequence, proton exchanged was disturbed that affected water uptake into the vacuole. The plants showed also leaf morphological aberrations (Gogarten et al. 1992). Active transmembrane transport of Ca^{2+} ions was assessed by introducing the *Arabidopsis* cation exchanger 1 (CAX1) transporter. The CAX1 expressing plants had significantly enhanced selective transport of Ca^{2+} ions in the roots while the transport of other divalent cations remained unaffected (Park et al. 2004). The transgenic cells expressing the antisense *tip1b* gene sequence of topoisomerase I had a lower activity of this enzyme. The activity of ascorbate peroxidase and ascorbate content was also reduced. Cells proliferation slowed down, and apoptosis was observed (Locato et al. 2006).

10.3.4 Genetic Rearrangements

The maize transposable elements, the activator (*Ac*) element and a defective dissociation (*Ds*) element, were introduced into carrot hairy roots. The selected hairy roots possessed both elements, but in one-fourth of these roots the *Ac* was excised indicating the *Ac*/*Dc* system activity in carrot (Van Sluys et al. 1987). Later, analogous system with the acetolactate synthase (*ALS*) gene conferring chlorsulfuron resistance and fused to the *Dc* element separating the *ALS* sequence and the promoter was introduced into carrot callus. The *Ac* element was delivered using a separate vector by co-transformation. Herbicide-resistant calli selected after transformation indicated that the *Dc* element was excised enabling the expression of *ALS*. The *Dc* elements were found in new loci of which almost 30% were located within gene sequences (Ipek et al. 2006b). The expected transposition or excision was not found in F_1 plants obtained by hybridization of parents possessing either the *Ac* transposase gene or the *Ds* element. However, in callus obtained from these F1 plants the *Ds* element was transposed indicating that the correct expression and splicing of the transposase was tissue-specific (Ipek et al. 2006a).

10.3.5 Cell and Organelle Tagging

The assessment of GUS activity by histochemical staining was commonly utilized in evaluation of transgenesis in carrot, but this approach requires the delivery of exogenous substrate for enzymatic reaction by incubating a piece of tissue for several hours at high temperature that is destructive to plant tissue. A non-destructive reporter system verified in carrot utilized green fluorescent protein (GFP). Fluorescence of GFP expressed in transgenic cells can be visually detected when tissue is exposed to UV light. Early detection of carrot GM cells and protoplasts can be performed by observing fluorescence under the fluorescence microscope equipped with filters ensuring correct emission and excitation wavelengths fitting characteristics of particular fluorescent protein variant. Hibberd et al. (1998) mentioned that after microprojectile bombardment GFP expression could be found in chromoplasts although no results were presented. The *smGFP* gene variant controlled by the *35S* promoter was delivered to carrot explants together with another gene conferring resistance to kanamycin. The putatively transgenic calli that were resistant to kanamycin exhibited also green fluorescence due to GFP expression, which was confirmed by Western blotting. Hence, GFP was proposed to be a suitable reporter for non-invasive detection of transgenic carrot callus (Yau et al. 2008). This selection approach confirmed earlier observations that GFP fluorescence can be non-destructively observed in carrot hairy roots after transformation with the 35S::*mGFP5-er* gene construct and enables selection of these roots that stably expressed the introduced transgene by using a hand-held UV lamp (Baranski et al. 2006). The mGFP5-er is a hybrid protein with the signal peptide targeting membranes. It tags outer cell membrane, endoplasmic reticulum, and nucleus which emit then intense green fluorescence. Transient green fluorescence was observed in protoplasts within a few hours after their transformation that enabled fast assessment of the transformation efficiency. Fluorescence was observed during first cell divisions as well as later during the culture when cell aggregates developed with a stable gene insertion and expression that was manifested also at later stages of development. In transgenic carrot plants, GFP fluorescence was observed in petioles, leaves, stem, and flowers. In contrast, petals did not emitted green fluorescence and variation in fluorescence intensity was also observed among different tissues indicating various activities of the 35S promoter depending on carrot plant tissue. Particularly intense fluorescence was seen in leaf marginal meristems, vascular bundles along petioles and stem, as well as flowers in the style. Thus, GFP can be used also for monitoring transgenic status of plants at any developmental stage, including generative organs (Baranski et al. 2007b).

In our current research, other fluorescent proteins than GFP have been also successfully delivered into carrot callus cells. Cyan fluorescence protein (CFP), yellow fluorescence protein (YFP), and red fluorescence protein (mCherry) genes with fused signal sequences targeting mitochondria are stably expressed in callus and protoplasts derived from its cells. These fluorescent markers serve to distinguish components used for protoplast fusion and the detection of two different fluorescent signals in one cell confirms its hybrid status. This approach has been found useful for optimization of fusion parameters, in particular when the electrofusion is performed using a microchamber suitable for live visualization of the fusion process (unpublished).

10.4 Role of Carrot Genes Elucidated by Using Carrot Transformation

10.4.1 Promoter Activity

Carrot is a model species used to demonstrate the process of somatic embryogenesis (Steward 1958). Both cell suspension and callus tissue can be stimulated for somatic embryogenesis by 2,4-D treatment. The removal of 2,4-D and further cell culture in a medium free of plant growth regulators initiate development of somatic embryos that are capable to convert to plants. Some genes involved in this process were identified in carrot and characterized. Among them, the expression of early somatic embryogenesis 1 (*C-ESE1*) gene was suppressed using the transgenic approach by inserting an additional copy in a sense orientation. This gene is active in primordial cells at early developmental stages of the embryo and codes for a cell wall glycoprotein. Its suppressed expression modified cell attachment so cell arrangement was altered causing delayed embryo formation (Takahata et al. 2004). At a later developmental stage, the late embryogenesis abundant proteins (*LEA*) genes activate and they confer tissue tolerance to desiccation. Expression of the *C-ABI3* transcription factor showed that it was involved in regulation of *LEA* genes expression (Shiota and Kamada 2000). Also the carrot *CAREB1* transcription factor was found to be involved in regulation of somatic embryogenesis. Carrot cells were transformed with the reporter construct containing the *GUS* gene under the LEA protein *Dc3* promoter and with the effector constructs containing one of the two *CAREB1 or CAREB2* transcription factor genes under the control of the *35S* promoter. Both transcription factors were expressed, bound to the *Dc3* promoter, and transactivated the GUS expression. The use of mutant promoters revealed that both transcription factors interacted with the ABRE motif indicating on ABA-dependent regulation. The *CAREB1* overexpressing embryos had typical morphology when cultured in a low-sucrose (1%) medium but grew slowly in the presence of 3% sucrose, and the *CAREB1* expression was elevated in torpedo- and cotyledon-shaped somatic embryos that could not elongate. The *CAREB2* expression was at the same level at all developmental stages but was induced by abscisic acid treatment and in drought stress, unlike *CAREB1*. Hence, despite similar DNA-binding activity a different role of these transcription factors was proposed with *CAREB1* being involved in ABA-related somatic embryo development regulation pathway (Guan et al. 2009).

The thaumatin-like protein (DcTLP) is a pathogenesis-related protein of a high homology to the tobacco osmotin protein, which is involved in the regulation of water uptake in drought stress. Carrot callus was transformed with a gene construct containing the *dcTLP* promoter driving the expression of *GUS* reporter gene. Under stimulated drought conditions GUS activity was induced and a prolonged stress increased GUS expression indicating that the *dcTLP* promoter is indeed induced by drought but not by signal molecules like abscisic acid, salicylic acid, and jasmonic acid (Jung et al. 2005).

10.4.2 Gene Function

Sucrose metabolism, and in particular the role of sucrose synthase isoforms, was investigated in carrot plants transformed with the construct containing the *Susy*Dc1* sucrose synthase gene fragment in the antisense orientation and controlled by the *35S* promoter. The selected plants had decreased sucrose synthase activity in storage roots, but its activity in leaves remained unaltered. Although the carbohydrate composition was affected, the pronounced changes were observed in plant morphology as plants developed much smaller roots and leaves. Plants with the lowest sucrose synthase activity had the lowest biomass. Thus, this gene variant was more related to the regulation of plant growth rather than to partitioning sucrose to different organs (Tang and Sturm 1999). Analogous approach with antisense sequences was applied to reveal function of carrot acid invertases which hydrolyze sucrose to glucose and fructose. It has been postulated that both the vacuolar invertase and the cell wall invertase are involved in sucrose breakdown and partitioning to plant organs but the activity of the latter affects additionally plant development. The suppressed expression of both invertases caused abnormal development of the first leaves; thus, morphological distortions occurred at early stage of plant development that could be reversed if monosaccharides were available to the plant. Also a modified sugar content and composition was observed in roots and leaves of plants with suppressed invertases. Plants with the inactive cell wall invertase had enhanced leaf growth, and their leaves contained more bi- and polysaccharides while the root growth was slower and roots contained less saccharides. The dry weight of leaves was 17 times higher than that of the roots in contrast to control plants with three times higher biomass of roots than leaves. Plants with suppressed vacuolar invertase developed morphologically normal roots while leaf biomass was also elevated exceeding the root biomass by 50%. Such plants had also higher carbohydrate accumulation in the leaves and lower in roots (Tang et al. 1999).

Anthocyanin pigments are synthesized in some carrot cultivars, which are manifested as a purple-black pigment in the root and or petiole. Enhanced expression of flavonoid pathway genes was observed in purple carrots (Xu et al. 2014; Yildiz et al. 2013). Carrot callus containing anthocyanins was transformed using a gene construct with the elements of CRISPR/Cas9 system to target the flavanone 3-hydroxylase (*F3H*) gene involved in naringenin hydroxylation, which results in intermediates necessary for subsequent anthocyanidin and anthocyanin biosynthesis. In particular, the construct contained the bacterial Cas9 protein under the control of 35S promoter and guide RNA (gRNA) sequences homologous to the second exon of the *F3H* gene and under the control of *AtU3* promoters. The expressed Cas9 protein of DNA cleavage activity forms a complex with gRNA and binds to plant DNA at the gRNA hybridization site. Various indel mutations were generated by the Cas9/gRNA complex in the *F3H* sequence that was confirmed by sequencing. Transgenic cells with the *F3H* knockouts were not able to synthesize anthocyanins, and in consequence, the developing calli remained discolored. The use of this precise editing system for gene knockout provided evidence on the critical functional role of F3H in anthocyanin biosynthesis (Klimek-Chodacka et al. 2018).

10.5 New Carrot Characteristics Obtained via Genetic Engineering

10.5.1 Resistance to Pathogens

Several research projects have been devoted to enhancing carrot resistance to diseases. However, developing resistance to both leaf and root pathogens was challenging. The strategies relied on the introduction of heterologous genes from other species that were controlled by constitutive promoters, ensuring their expression in all organs throughout the plant life. Predominantly, a single gene was introduced into the carrot genome and rarely a combination of two genes. In the case of the latter, genes were inserted by explant co-cultivation with two *A. tumefaciens* strains, each containing a single gene of interest (Jayaraj and Punja 2007; Wally et al. 2009b). To determine whether the introduced genes enhanced resistance, the micropropagated GM plants were acclimatized to ex vitro conditions, then were grown in either a greenhouse or climate-controlled chamber where they were inoculated, and evaluated for resistance to selected pathogens (Jayaraj and Punja 2007; Wally et al. 2009a). Until now, only Takaichi and Oeda (2000) evaluated T_0 GM plants as well as their T_1 progeny after self-pollination and PCR selection of transgenic seedlings. However, most resistance assays were performed in laboratory conditions in which plant organs, and not the growing plants, were inoculated. Such assays enable the evaluation of resistance against two or more pathogens simultaneously. However, the scoring results from detached organs kept in fully controlled conditions may substantially differ from those obtained using the whole plants during the vegetative stage. Field evaluation of carrot GM plants was reported only by Melchers and Stuiver (2000), but they did not present results in detail. Laboratory-based resistance assays for root pathogens were carried out using the whole, harvested, and cleaned storage roots (Wally et al. 2009a; Wally and Punja 2010) or root system of young plantlets (Imani et al. 2006). For leaf pathogens, either the whole detached leaves (Chen and Punja 2002; Punja 2005; Wally et al. 2009a, 2009b; Wally and Punja 2010), leaf segments or leaflets (Baranski et al. 2007a, 2008), or petioles (Baranski et al. 2007a, 2008; Punja and Raharjo 1996; Takaichi and Oeda 2000) were inoculated. In such assays, the development of disease symptoms can be evaluated at various time points and the whole assay is completed in 2–4 weeks. Most attention was given to *Alternaria* spp. and *Botrytis cinerea* infesting leaves and roots and to *Sclerotinia sclerotiorum* invading roots. GM carrots were also challenged in various research projects to four other fungal pathogens, *Cercospora carotae*, *Erysiphae heraclei*, *Sclerotium rolfsii*, and *Thielaviopsis basicola*, bacteria *Xanthomonas hortorum* pv. *carotae*, and *Carrot virus Y* (Table 10.2).

Chitinases and glucanases are enzymes commonly recognized as pathogenesis-related (PR) proteins of a broad-spectrum activity against various fungal phytopathogens (Punja et al. 2007). Thus, genes of plant and microbial origins coding for chitinases were introduced to carrot to enhance resistance to various fungi. Detached organs of plants expressing either petunia *Chl* or wheat *Ch383* acidic chitinases had disease symptoms at similar severity levels as the non-GM control when inoculated by *A. radicina*, *B. cinerea*, *R. solani*, *S. rolfsi*, or *S. sclerotiorum* (Punja and Raharjo 1996; Wally et al. 2009b). In contrast, significant resistance enhancement was observed when basic type chitinase was expressed and which accumulate intercellularly thus exhibit stronger antifungal activity than extracellularly accumulated acidic chitinases. The expression of tobacco, barley, and *Trichoderma harzianum* basic chitinase genes enhanced carrot resistance to most fungal pathogens evaluated. The disease symptoms developed much slower and covered a smaller leaf area inoculated with *A. dauci, A. radicina, B. cinerea, R. solani,* and *S. rolfsi*, despite various intensities (Table 10.2). The enhanced resistance was more pronounced in assays with the leaf rather than root pathogens (Punja and Raharjo 1996).

The expression of a wheat glucanase gene did not affect carrot resistance (Wally et al. 2009b) and the effect of tobacco glucanase was not clear as evaluated plants were co-transformed with chitinase gene, in which introduction to carrot highly enhanced resistance to *B. cinerea*, and moderately to *R. solani* and *S. rolfsi*, although not to *A. radicina* nor *T. basicola* (Melchers and Stuiver 2000; Punja and Raharjo 1996). Co-expression of two genes, barley chitinase (*Chi-2*) and wheat lipid-transfer protein (*ltp*), highly reduced disease symptoms of *A. radicina* and *B. cinerea*. The severity of these symptoms for most resistant plants was estimated at about 10% of symptoms for the non-GM control. Actually, the combined expression of both genes ensured more effective resistance than each of these genes when expressed alone. This response results from the involvement of chitinase and lipid-transfer protein in complementary mechanisms, fungal cell wall degradation, and lipid movement, respectively (Jayaraj and Punja 2007).

A very high level of resistance was achieved by introducing the *Arabidopsis NPR1* gene, which is a regulator of the salicylic acid-mediated systemic acquired resistance (SAR). Transgenic plants expressing *NPR1* acquired enhanced sensitivity to elicitors, i.e., salicylic acid, jasmonic acid, and 2,6-dichloroisonicotinic acid as well as to a suspension of *S. sclerotinium* cell walls. These elicitors activated expression of endogenous pathogenesis-related genes in the *NPR1* expressing carrot plants, but not in the control, indicating activation of the systemic acquired resistance mechanism. Carrots highly responding to elicitors showed also a very high resistance level when inoculated with *B. cinerea*, *E. heraclei*, and *X. hortorum*, and moderate resistance level to *A. radicina* and *S. sclerotiorum* (Wally et al. 2009a). A bacterial protein, the *Pseudomonas fluorescence* Microbial Factor 3 (MF3) homologous to FK506-binding protein was also suggested to induce systemic acquired resistance. The *mf3* expressing carrots were indeed less susceptible to *A. dauci*, *A. radicina*, and *B. cinerea* as revealed in detached leaflet and petiole assays; the disease symptoms were reduced up to 40% in comparison with the control (Baranski et al. 2007a).

Table 10.2 Carrot engineering aimed in enhanced resistance to phytopathogens

Pathogen	Introduced gene[a]	Resistance assay	Resistance level[b]	Ref. [c]
Alternaria (section Porri)				
A. dauci	*chit36*	leaflets	M	2
	Chl + Glu-1	plants	H	6
	hly	petioles	H	9
	tlp	leaves	H	3, 7
	mf3	leaflets	M	1
Alternaria (section Radicina)				
A. carotiincultae	*tlp*	leaves	N	3, 7
A. petroselini	*tlp*	leaves	H	3, 7
A. radicina	*Chl (a)*	petioles	N	8
	Chl	petioles	N	8
	Chi-2	plants	H	5
	chit36	petioles	H	2
	Chl + Glu-1	plants	H	6
	Chi-2 + ltp	plants	VH	5
	ltp	plants	H	5
	tlp	leaves	H	3, 7
	mf3	petioles	M	1
	NPR1	plants, storage roots	M	10
	Prx114	storage roots	VH	12
Botrytis cinerea	*chit36*	leaflets	H	2
	Chl (a)	petioles	N	8
	Chl	petioles	H	8
	Chi-2	leaves	VH	5
	Ch383 (a)	leaves	N	11
	Chi-2 + ltp	leaves	VH	5
	Prx114 + Ch383 (a)	leaves	M	11
	Glu-638 + Ch383 (a)	leaves	N	11
	ltp	leaves	H	5
	tlp	leaves	M	3, 7
	HvBI-1	leaves	H	4
	mf3	leaflets	M	1
	Glu-638	leaves	N	11
	Prx114	leaves	H	11
	NPR1	leaves	H	10
Cercospora carotae	*Chl + Glu-1*	plants	H	6
Erysiphe heraclei	*Chl + Glu-1*	plants	H	6
	hly	plants, T_1 progeny	H	9
	NPR1	leaves	VH	10
	Prx114	leaves	N	12

(continued)

Table 10.2 (continued)

Pathogen	Introduced gene[a]	Resistance assay	Resistance level[b]	Ref. [c]
Rhizoctonia solani	*Ch1 (a)*	petioles	N	8
	Ch1	petioles	M	8
	tlp	leaves	M	3, 7
Sclerotinia sclerotiorum	*tlp*	leaves	S	3, 7
	Ch383 (a)	leaves	N	11
	Glu-638	leaves	N	11
	Prx114	leaves	M	11
	NPR1	leaves	M	10
	Prx114 + Ch383 (a)	leaves	H	11
	Glu-638 + Ch383 (a)	leaves	N	11
Sclerotium rolfsii	*Ch1 (a)*	petioles	N	8
	Ch1	petioles	M	8
	tlp	leaves	H	3, 7
Thielaviopsis basicola	*Ch1 (a)*	petioles	N	8
(syn. *Chalara elegans*)	*Ch1*	petioles	N	8
	HvBI-1	root system	M	4
Xanthomonas hortorum pv. *carotae*	*NPR1*	plants	VH	10

[a]Genes: *Ch1* tobacco chitinase, *Ch1 (a)* petunia acidic chitinase, *Ch383 (a)* wheat acidic chitinase, *Chi-2* barley chitinase, *Chit36 Trichoderma harzianum* chitinase, *Glu-1* tobacco β-1,3-glucanase, *Glu-638* wheat β-1,3-glucanase, *hly* human lysozyme, *HvBI-1* barley BAX inhibitor-1, *ltp* wheat lipid-transfer protein, *mf3 Pseudomonas fluorescence* microbial factor 3, *NPR1* Arabidopsis regulatory protein, *Prx114* rice cationic peroxidase, *tlp* rice thaumatin-like protein
[b]Resistance levels: VH—very high (estimated severity of disease symptoms <20% of the non-GM control), H—high (<50%), M—moderate (<80%), S—slight (>80%), N—none (at the level of the control); estimated by the authors of this chapter
[c]References: 1—Baranski et al. (2007a), 2—Baranski et al. (2008), 3—Chen and Punja (2002), Punja (2005), 4—Imani et al. (2006), 5—Jayaraj and Punja (2007), 6—Melchers and Stuiver (2000), 7—Punja (2005), 8—Punja and Raharjo (1996), 9—Takaichi and Oeda (2000), 10—Wally et al. (2009a), 11—Wally et al. (2009b), 12—Wally and Punja (2010)

The rice thaumatin-like protein (TLP) belongs to pathogenesis-related proteins which affects membrane permeability, signal transduction and has a hydrolyzing activity of β-1,3-glucans. It thus shows an antifungal activity by hydrolyzing pathogen cell wall polymers. The *ltp*-expressing carrots, depending on the transgenic line, had enhanced resistance to *A. dauci*, *A. petroselini*, *A. radicina*, *B. cinerea*, *R. solani*, *S. rolfsi*, and *S. sclerotiorum*, although some differences were observed in response depending on the carrot line. The resistance level obtained in the *tlp* expressing carrots was estimated to be higher in comparison with the tobacco chitinase or human lysozyme (*hlp*)-expressing plants (Chen and Punja 2002; Punja 2005). However, the later were evaluated for disease symptoms caused by *A. dauci* in a petiole assay and *E. heraclei* after the infestation of the whole plants, and thus, the results cannot be directly compared (Takaichi and Oeda 2000). Carrots expressing *hlp,* and which had a chitinolytic activity, exhibited partial resistance to foliar diseases caused by these pathogens, and the resistance level correlated with the HLP protein level in the tissue in both T_0 and T_1 plants.

The barley BAX inhibitor-1 protein is a broad-spectrum cell death inhibitor that protects against abiotic and biotic stress. *B. cinerea* induces programmed cell death, and thus, blocking this mechanism may reduce disease severity. Disease symptoms on leaves inoculated with *B. cinerea* were indeed reduced in comparison with the control, and the mycelium development was almost completely inhibited in the selected lines. Also, roots of transgenic young plants exposed to *T. basicola* were less diseased although the resistance level to this pathogen was, on average, lower than in case of *B. cinerea*. However, individual plants showed no symptoms that indicated on a high efficacy of the inhibitor in the resistance response (Imani et al. 2006).

10.5.2 Tolerance to Abiotic Stress

10.5.2.1 Tolerance to Drought and Salinity

Drought, salinity, or UV exposure are factors limiting crop yield and its quality, and plant tolerance to these abiotic stresses exaggerating with climate change become a crucial issue for ensuring food security. Hence, there is a need for developing plants, which could withstand unfavorable environmental conditions. Enhanced tolerance to abiotic stress was achieved in transgenic carrot plants expressing heterologous genes coding for various compounds of physiological significance.

Osmotin is a protein acting as the osmoprotectant; thus, it is often involved in enhancing tolerance to drought, but it also plays important roles in defense mechanisms to other abiotic and biotic stresses. Carrot was transformed with a shortened tobacco osmotin gene under the control of 35S promoter and the transgenic plants were assessed for physiological changes. After drought treatment the transgenic plants recovered faster than control plants, and they had higher relative water content, less ion leakage, and lower levels of lipid peroxidation. Tolerance to water deficiency was also demonstrated by Shiota and Kamada (2000) who obtained ABA-dependent desiccation-tolerant non-embryogenic carrot cells after transformation with the *C-ABI3* gene.

Soil salinity is an accelerating problem in agricultural production. Plants exposed to salt stress may accumulate osmoprotectants, e.g., glycine betaine and β-alanine betaine which help to maintain cell homeostasis. Betaine aldehyde dehydrogenase (BADH) is involved in betaine synthesis (Rathinasabapathi et al. 2001); therefore, enhanced expression of BADH may contribute to salt tolerance in plants. Carrot was transformed with the chloroplast transformation vector pDD-*Dc-aadA/badh* which integrates the *aad*A and *badh* genes into the plastid 16S-23S spacer region. Transformed cells accumulated up to 54-fold more betaine than the non-transformed control, and transgenic plants showed increased salt tolerance being able for growing in saline soil with up to 400 mM NaCl (Kumar et al. 2004).

Tolerance to stress can be mediated by the biosynthesis of compounds exhibiting free-radical scavenging activity. Ketocarotenoids are strong antioxidants, and hence, plants possessing their higher level were supposed to be more tolerant to oxidative stress. Carrots synthesize various carotenoids but not ketocarotenoids. Jayaray and Punja (2008) transformed carrot with the *Haematococcus pluvialis* β-carotene ketolase (*bkt*) gene, where expression leads to astaxanthin biosynthesis via echinenone and canthaxanthin intermediates, and thus, they modified the metabolite profile in carrot roots, which accumulated ketocarotenoids, on average, 23% in total carotenoids. The transgenic plants exposed to UV-B light or oxidative extracts exhibited better antioxidant and free-radical scavenging activities, the leaves did not accumulate H_2O_2, and plant growth was not reduced by stress conditions.

10.5.2.2 Tolerance to Herbicides

Broomrape (*Orobanche* sp.) is a root parasite limiting plant productivity. Its control is difficult and requires the use of a highly toxic methyl bromide which is expensive and not inert to environment and thus used mainly for the protection of economically important crops. Aviv et al. (2002) proposed an alternative broomrape biocontrol in carrot production by introducing resistance to imidazolinone herbicides. To prove this concept, they transformed carrot with a mutant *A. thaliana* acetolactate synthase (*ALS*) gene controlled by its own promoter. The T_0 transgenic plants were heterozygous for *ALS* and after backcrossing the first generation segregated for resistance to imazapyr. Seeds exposed to 0.15 μM imazapyr germinated and survived in the expected 1:1 ratio as the non-transformed seedlings were unable to develop. Transgenic carrot plants tolerated imazapyr at concentrations up to 500 μM while 100 μM was high enough to control broomrape.

Phosphinothricin (PPT), the active ingredient of a broad-spectrum herbicides Basta® and Liberty®, is an inhibitor of glutamine synthase which is the main enzyme of nitrogen metabolism in plants. After phosphinothricin application the level of glutamine is reduced that results in increased ammonia accumulation and finally cell death. The phosphinothricin-N-acetyltransferase (*pat*) gene controlled by the 35S promoter was introduced into the carrot genome. The *pat* sequence was isolated from *Streptomyces viridochromogenes* and modified by additional DNA fragments for expression in plants. Transformants were selected on the medium enriched with 20 mg/l PPT. Besides development of PPT-resistant plants, Dröge et al. (1992) also tracked the path of PPT metabolisms in transformed and untransformed plants. They showed that in the control plants PPT undergoes conversion to three metabolites while in the transgenic plants PPT is rapidly acetylated; hence, its further metabolism is blocked. Resistance to PPT was also received by introducing the *Streptomyces hygroscopicus bar* gene. Transformed callus cultures were selected on medium with 10 mg/l of PPT. The obtained plants were first sprayed with 0.2% Liberty®, and subsequently, the withstanding plants were sprayed with 0.4% Liberty® that finally enabled selection of herbicide-resistant plants. It was also observed that resistance levels varied between the obtained carrot lines, and the differences in resistance could be a result of transgene copy number (Jayaraj and Punja 2007).

10.5.2.3 Phytoremediation

Hairy roots obtained after transformation with *A. rhizogenes* are useful for production of recombinant proteins and secondary metabolites but can be also considered as a system for the assessment of substance toxicity and remediation. Phenol and its chloro-derivatives are environmental contaminants released from industrial waste, and they may be also residues of chemicals used commonly in agricultural production. Biological systems with high ability for phytoremediation without release of harmful by-products are considered more sustainable and less expensive alternative to conventional methods for environment decontamination. Carrot hairy root culture obtained after transformation with a wild *A. rhizogenes* strain tolerated phenol up to 1000 μM, but chlorophenols were more toxic and only concentrations above 50 μM reduced hairy root growth index. After 120-h culture, phenol and trichlorophenol were almost completely removed from the medium confirming high potential of carrot hairy roots for phytoremediation (Araujo et al. 2002). Further studies revealed that medium with the concentration of phenols below 1000 μM has stimulating effect on growth of hairy roots and that the removal of dichlorophenol was faster than that of phenol, 83.0 and 72.7% within 72 h, respectively (Araujo et al. 2006). Oxidation of phenol and its derivatives depends on substrate chemical structure, and an additional chlorine or hydroxyl group can lower the affinity of peroxidase, which affects resistance to these compounds. Nevertheless, peroxidases extracted from carrot hairy root culture were able to oxidize phenol, 2-chlorophenol, guaiacol, and catechol with a similar efficiency (Araujo et al. 2004).

10.5.3 Recombinant Proteins

Transgenic plants expressing recombinant proteins are considered as an alternative source of inexpensive and accessible pharmaceuticals. Carrot is an attractive candidate for the production of biopharmaceuticals since it is present worldwide, can be stored for a long period, and can be eaten raw; hence, biological activity of metabolites or immunogenicity of antigens present in carrot can be retained if root processing is omitted. Also the naturally occurring high level of soluble proteins in carrot could led to a higher level of accumulation of desired protein (Luchakivskaya et al. 2011). Carrot is widely used in research as a potential source of compounds for disease treatments although it rarely has been applied (Table 10.3). So far the most promising research results were obtained for carrot-derived enzymes suitable for enzyme replacement therapy (ERT) in metabolic diseases and one product is already available on the market. The common method for delivering genes coding for heterologous proteins to carrot is *A. tumefaciens*-mediated transformation of hypocotyl explants. In most cases, the gene construct consisted of the gene of interest under the 35S promoter control and the kanamycin resistance gene. In some constructs, additional sequences were included, for example, endoplasmic reticulum retention signal (SEKDEL) which prevents degradation of foreign proteins in cytoplasm (Lindh et al. 2009). Calli obtained after hypocotyl transformation was further stimulated for somatic embryogenesis and ultimately developed plants were used as a source of desired proteins.

10.5.3.1 Recombinant Enzymes

Gaucher's disease is an inherited genetic disorder causing storage of glucocerebroside in cells due to the lack of enzyme called glucocerebrosidase and affecting liver, spleen, and bone cells. The first recombinant enzyme, used to treat Gaucher's disease, was produced in Chinese hamster ovary cells (Cerezyme®), but it required further in vitro glycan modification (Grabowski et al. 1995). Shaaltiel et al. (2007) developed protein expression system (ProCellEx™) in carrot cells cultured in a bioreactor characterized by a greater glycan efficacy, longer half-life of glucocerebrosidase, and more cost-effective production (Tekoah et al. 2015). Since glucocerebrosidase does not require in vitro modification in ProCellEx™, this system is advantageous over protein production in mammalian cell lines. To produce the glucocerebrosidase known as taliglucerase alfa in carrot cells, a modified human gene was used. Its fragment coding for a native signal peptide was replaced by the *Arabidopsis thaliana* endochitinase gene and an additional signal from tobacco chitinase A was inserted at its C-terminal end. After successful clinical trials taliglucerase alfa (Elelyso® Protalix Biotherapeutics) was the first plant-based human recombinant protein approved for ERT in Gaucher's disease (FDA 2012) and is still available on the market (Rosales-Mendoza and Tello-Olea 2015).

Other three carrot-derived recombinant proteins, a human alpha galactosidase A (PRX-102), an anti-tumor necrosis factor (PRX-106) and a chemically modified human deoxyribonuclease I (DNase I) resistant to inhibition by actin (PRX-110), have been listed in a review article by Rosales-Mendoza and Tello-Olea (2015) as products under development by Protalix Biotherapeutics. However, the company emended that these recombinant proteins are produced in tobacco and not carrot cell systems (Tekoah et al. 2015). In earlier works, the human glutamic acid decarboxylase (*GAD65*) gene was expressed in carrot. GAD65 is the major autoantigen in insulin-dependent diabetes mellitus (type 1 diabetes). The protein obtained in planta folded properly but the level of GAD65 in carrot was low (0.012%), which could be due to targeting GAD65 to organelle membranes (Mason et al. 1996). A tenfold higher expression was obtained when the cytosolic isoform of GAD was used (Ma et al. 1997; Porceddu et al. 1999).

10.5.3.2 Vaccines and Interferons

Interferons are proteins with antibacterial and antiviral properties inhibiting cell proliferation and stimulating the immune system. Hence, interferons have been used in a wide range of treatments, e.g., influenza, hepatitis, and several types of cancer. Carrot was transformed with a gene construct containing sequence of the human interferon alpha 2b fused with the *Nicotiana plumbagenifolia* calreticulin apoplast targeting signal under the control of *35S* promoter or root-specific *Mll* promoter. The calreticulin

Table 10.3 Recombinant proteins produced in GM carrot for the treatment of human diseases

Deficiency/pathogen	Disease	Recombinant protein name	Content	Comment	References
Deficiency	Diabetes	Glutamic acid decarboxylase	0.012%	Concentration of GAD65 was quantified by radioimmunoassay	Porceddu et al. (1999)
	Gaucher's disease	Glucocerebrosidase	n.a.	Available on the market as Elelyso®	Shaaltiel et al. (2007)
Bacteria	Cholera	B subunit of cholera toxin (CTB)	up to 0.48% FW	Presence of CTB confirmed by ELISA	Kim et al. (2009)
	Diphtheria, Tetanus, Pertussis	Diphtheria toxin, Tetanus toxin fragment C, S1 subunit of pertussis toxin	n.a.	Specific immune responses in mice sera	Brodzik et al. (2009)
	Tuberculosis	MPT64	n.a.	Southern blot analysis confirmed presence of *MPT64* in carrot genome	Wang et al. (2001)
		ESAT6 and CFP10	0.056% TSP Esat6 0.024% TSP CFP10	Cell-mediated and humoral responses in mice	Uvarova et al. (2013)
		ESAT6-CFP10 fusion protein	0.035% TSP	Cell-mediated and humoral responses in mice	Permyakova et al. (2015)
	E. coli	B subunit of heat-labile toxin (LTB)	3 μg/g root FW	Mucosal and systemic antigen response in mice	Rosales-Mendoza et al. (2007)
	H. pylori	Urease B (UreB)	up to 25 μg/g root FW	Mucosal and systemic humoral response in mice	Zhang et al. (2009)
	Chlamydia infection	Major outer membrane protein (MOMP)	3% TSP	Induced antibodies production and T cell response	Kalbina et al. (2011)
Virus	Hepatitis B	Hepatitis B antigen (HBsAg)	25 ng/g cell culture FW	Presence of HBsAg confirmed by ELISA	Imani et al. (2002)
	HIV infection	P24 capsid protein	41 μg/g TSP	Induced antigenicity in the sera of HIV-positive patient	Lindh et al. (2009)
	Influenza, hepatitis, cancer	Human interferon alpha 2b	16.5×10^3 IU/g root FW	Antiviral activity in in vitro studies	Luchakivskaya et al. (2011)
	Measles	Hemagglutinin protein (H)	2 μg/g plant FW	Humoral and cellular immune responses in mice	Marquet-Blouin et al. (2003)

(continued)

Table 10.3 (continued)

Deficiency/pathogen	Disease	Recombinant protein name	Content	Comment	References
		[L4T4]2 epitope	n.a.	Humoral responses in mice	Bouche et al. (2003)
	Plague	F1-V fusion protein	0.3% TSP	IgG antibodies predominant humoral response	Rosales-Mendoza et al. (2011)
	Rabies	G protein of rabies virus	up to 1.4% TSP	Induced antibodies accumulation in mice	Rojas-Anaya et al. (2009)
Parasite	Porcine cysticercosis	Glutathione s-transferase HP6 and TSOL18 fusion protein	14 μg/g callus DM	Induced IgG and IgA anti-HP6/Tsol18 in mice	Monreal-Escalante et al. (2016)

DM—dry matter, FW—fresh weight, TSP—total soluble proteins, n.a.—not available

apoplast targeting signal is considered as a factor influencing proper processing and biological activity of pharmaceuticals (Peng et al. 2005). The activity of the interferon was higher in leaves when the *35S* promoter was used. The root-specific *Mll* promoter ensured higher interferon expression in roots, but the expression in leaves was still observed. It was estimated that a daily dose of 8–10 g of transgenic carrot root would be enough for influenza treatment and 60–120 g for hepatitis B treatment (Luchakivskaya et al. 2011).

Although none of the carrot-derived vaccines have been tested in humans, the results obtained using animal models have been promising and indicated that carrot might be used as a potential source for vaccine production for diseases caused by parasites, bacteria, or viruses. The expression of glutathione s-transferase-HP6 and TSOL18 fusion protein in carrot cells could be a source of oral vaccine for porcine cysticercosis, the disease caused by parasite *Taenia solium*. The recombinant *hp6/tsol18* gene was designed based on *T. solium* genome sequence and optimized for in planta expression. The amount of protein obtained in carrot was up to 14 μg per gram of callus dry weight. Tests on mice showed that subcutaneous injection and oral vaccination with freeze-dried callus or total soluble protein extracted from it induced both anti-HP6/TSOL18 IgG and IgA antibodies and induced specific humoral response at the level similar to that obtained using the *E. coli* produced antigen (Monreal-Escalante et al. 2016).

Tuberculosis is one of the diseases with the highest mortality rate. Currently, there is the BCG vaccine available with no other alternatives for disease prophylaxis. In the first attempts for the development of oral vaccine using transgenesis, the *MPT64* gene was introduced to carrot cells. The gene presence was confirmed by Southern blot analysis, although no further research nor assessment of immunological response in animal model was conducted (Wang et al. 2001). Another approach included the usage of ESAT6 and CFP10 proteins, which are virulence factors of *Mycobacterium tuberculosis*. First, ESAT6 and CFP10 were successfully produced in carrot as separate proteins and oral administration of vaccine to mice induced humoral and cell-mediated response; however, the ESAT6 protein was also toxic to peripheral blood mononuclear cells (Uvarova et al. 2013). In further studies, Permyakova et al. (2015) created a fusion protein consisting of ESAT6, CFP10, and human deltaferon dIFN, the latter acting as an adjuvant. The fusion protein induced cell-mediated and humoral response in mice when administered orally or by injection, although the oral administration resulted in lower level of antibodies. It was assumed that either carrot cell wall is not completely digested in the gastrointestinal tract or the enzymes present in it caused protein degradation. Western blot analysis of the fusion protein showed a band of lower molecular weight than expected which could indicate that the fusion protein was degraded; nevertheless, the cytotoxic effect of the single ESAT6 protein on peripheral blood mononuclear cells was not observed.

The heat-labile toxin (LT) consisting of two subunits, A and B, is a major pathogenic factor of enterotoxigenic *E. coli*. The subunit B is used for vaccine production due to its immunoprotective properties and lack of toxicity. Rosales-Mendoza et al. (2007) adapted the LT subunit B (LTB) for the expression in plant and added the SEKDEL signal to the LTB sequence at its 3' end. Transgenic carrot root contained 3 μg LTB/g fresh weight. Three doses administered intragastrically to mice induced mucosal and systemic antigen response although the IgG- and IgA-specific antibody levels were higher when pure LTB was used rather than the carrot-derived one. Another LT-like protein with the amino acid sequence identical in 80% to LT is the cholera toxin (CT) causing intestinal infectious disease. Kim et al. (2009) used a similar approach as described above; the gene encoding B subunit of CT (CTB) was optimized for expression in plants, and the SEKDEL signal was added. The amount of recombinant CTB produced in carrot ranged from 0.28% to 0.48% of total soluble proteins and CTB formed functional pentameric structure as was shown by ganglioside binding analysis.

Helicobacter pylori is a human pathogen, classified as class I human carcinogen by World Health Organization, and it is estimated that half of the world's population is a *H. pylori* carrier (Gerrits et al. 2006). The urease B (UreB) is considered as candidate protein for the development of vaccines against *H. pylori*, and its gene was successfully introduced into the carrot genome. The amount of protein obtained in transgenic carrot root was up to 25 μg/g fresh weight, which was sufficient for inducing anti-UreB immune response in mice (Zhang et al. 2009).

Yersinia pestis is a gram-negative bacterium and etiological factor of the plaque, which could be easily spread among humans. Hence, it is necessary to create easily accessible and rapid immunization method. Available vaccines consist of F1 and V proteins responsible for phagocytosis resistance or for changes in host defense mechanism. Expression of the F1-V fusion protein was obtained in carrot at the 0.3% level of the total soluble proteins. A carrot-derived vaccine successfully induced serum production of anti-F/V IgG antibodies in mice although response was predominantly humoral (Rosales-Mendoza et al. 2011).

The major outer membrane protein (MOMP) is considered potential protein for vaccine development against *Chlamydia trachomatis*. Since the full-length protein was not expressed in *E. coli*, a chimeric MOMP consisting of large regions of the VS2 and VS4 domains was developed. This altered protein was successfully produced in *E. coli* and plant systems such as Arabidopsis and carrot. The yield obtained in carrot cells was up to 3% of total soluble proteins. MOMP induced the production of neutralizing antibodies and T cell responses (Kalbina et al. 2011; Rosales-Mendoza and Tello-Olea 2015).

Carrot was also used as a tool for producing diphtheria-tetanus-pertussis (DTP) vaccine. The expression cassettes contained gene fragments: the coding region of diphtheria toxin (*DT*), tetanus toxin fragment C (*TetC*) and S1 subunit of pertussis toxin (*PTX S1*); also the His6 tag, FLAG tag and SEKDEL were present. The expression of the designed cassette led to a high level of protein accumulation and induced specific immune responses in mice sera at the level required for in vivo protection (Brodzik et al. 2009).

Vaccines for hepatitis B are produced in transgenic yeast containing the small hepatitis B virus surface protein. The sequence encoding the same protein was introduced into carrot cells under the control of mannopine synthase (*MAS*) promoter. The expression of hepatitis B antigen (HBsAg) was observed in cell culture and in plants obtained via somatic embryogenesis. *MAS* promoter is auxin sensitive, and thus, additional application of IAA or NAA enhanced the HBsAg expression, especially in cell suspension culture since cells were immersed in liquid medium with readily available hormones. The level of HBsAg obtained in carrot cells was about 25 ng per g of fresh weight (Imani et al. 2002).

Cases of death caused by measles were reduced by implementation of vaccination program, although the developing countries where the access to vaccines is limited are still affected; measles caused death of almost 90,000 people in 2016 (WHO 2018). Common vaccines against measles contain antibodies to hemagglutinin protein (H) (Bouche et al. 2002). Carrot was transformed with a sequence corresponding to the measles hemagglutinin protein. The immune capability was tested in mice injected with 2 or 3 doses of the extract obtained from transgenic plants. The antibody level after immunization with leaf extract was, on average, four times higher in comparison with root extract. The results showed that the levels of antibodies in mice vaccinated with carrot extract were similar to those obtained after vaccination with the H protein produced in mammalian cells, although the carrot-derived protein induced IgG1 and IgG2A antibody subclasses, while the mammalian cells-derived H protein generated only the IgG1 subclass (Marquet-Blouin et al. 2003). Bouche et al. (2003) proposed a different approach, and they transformed carrot with the $[L4T4]_2$ chimeric polyepitope from B cell epitope of the H protein combined with the T cell epitope of the tetanus toxoid, under the control of double 35S promoter. Carrot-derived vaccines showed high neutralizing potential against virus of Asian and African origin as well as against wild-type mutated virus (Bouche et al. 2003, 2005).

The G protein of rabies virus was expressed in carrot plants. Immunodetection analysis of transformants revealed the presence of protein with higher molecular weight than expected which was probably due to in planta glycosylation. The yield of G protein obtained was up to 1.4% of total soluble proteins, and oral administration resulted in the presence of rabies antibodies in mice (Rojas-Anaya et al. 2009).

Currently, there is no available effective vaccine for human immunodeficiency virus (HIV) transmitting mainly through mucosal tissues. Potentially, the application of HIV-1 capsid protein p24 which enhances immune responses in these tissues could enable prophylaxis. Carrot was transformed with constructs containing the *p24* and SEKDEL sequence. The protein content obtained was 41 μg/g total soluble proteins, and the extracts obtained from carrot showed antigenicity in the sera from HIV-positive patient (Lindh et al. 2009; Rosales-Mendoza and Tello-Olea 2015).

10.6 Conclusions

In this chapter, we summarize the available published reports in which carrot genetic engineering was implemented. Two main research areas involving carrot modification have been summarized, which includes the development and optimization of efficient protocols suitable for gene delivery to carrot cells and the development of GM carrot plants. In general, these works were historically the earliest and preceded basic research on elucidating gene functions, mainly heterologous genes and promoters of other species but also endogenous carrot genes. Those researches were also carried out to help understand selected metabolic and physiological processes. Finally, carrot was modified to express new characteristics. Modified carrot with introduced PR genes or genes conferring enhanced tolerance to abiotic stress like salinity and herbicide potentially may have significance for plant production. Also carrots with new composition of metabolites like carotenoid compounds seem attractive option of potential value for human health. However, until now, none of GM carrot was introduced into agricultural practice or even a subject of extensive field trials. In contrast to field production, research using contained culture of GM carrot cells in bioreactors has been well advanced. Such cell systems proved to be efficient in the biosynthesis of recombinant proteins of pharmaceutical significance. For 6 years now, a commercial product is available in treatment of Gaucher's disease.

New prospects emerge with the development of new genome-editing techniques enabling precise targeting of DNA sites that become alternative to conventional genetic transformation methods, which lead to genetic changes randomly located in the genome. Additionally, techniques implementing RNA/protein complex delivery to plant cells may direct gene modification leaving no signs of exogenous genetic material. Such plants differ from the wild type by only a small nucleotide change or may have a new DNA insert at the target site. They become indistinguishable from spontaneous mutants using standard procedures used by approval and control agencies; thus, the identification of the nature of their development may remain not detectable (Glass et al. 2018; Svitashev et al. 2016). Such genome-edited plants are considered as non-regulated in the USA (USDA Press 2018). Other countries such as Canada, Australia, India, and Argentina have expressed policies toward accepting genome-edited plants as non-GMO (Schuttelaar 2015) while the European Union faces an opposite direction. It seems that the old GMO definition is not adequate to current technologies and knowledge (Sprink et al. 2016) will not be changed in near future preventing any liberation of a stringent regulatory process of GMO authorization, hence commercialization in the EU. The recent decision of the EU Court of Justice clarifying the interpretation of the GMO definition adopted in EU legislation in favor of GMO opponents and indicating that genome-edited plants by using new technologies should be considered as GMO (ECJ 2018). These decisions and policies most probably will stimulate research on precise editing of carrot genome, in which outcomes may then lead to introduction of such mutated carrot to agriculture outside EU. Research on targeted mutagenesis in carrot using genome-editing tools like CRISPR/Cas9/Cpf1 has been initiated. The first report showing successful modification of the target gene in the carrot cell model system was published in January 2018 (Klimek-Chodacka et al. 2018), and further research aimed at the development of gene-edited plants is ongoing in various research groups.

Acknowledgements The support of the Ministry of Science and Higher Education of the Republic of Poland is gratefully acknowledged.

References

Altamura MM (2004) *Agrobacterium rhizogenes rol*B and *rol*D genes: Regulation and involvement in plant development. Plant Cell Tiss Organ Cult 77:89–101

Annon A, Rathore K, Crosby K (2014) Overexpression of a tobacco osmotin gene in carrot (*Daucus carota* L.) enhances drought tolerance. In Vitro Cell Dev Biol Plant 50(3):299–306

Arango J, Salazar B, Welsch R, Sarmiento F, Beyer P, Al-Babili S (2010) Putative storage root specific promoters from cassava and yam: cloning and evaluation in transgenic carrots as a model system. Plant Cell Rep 29(6):651–659

Arango J, Jourdan M, Geoffriau E, Beyer P, Welsch R (2014) Carotene hydroxylase activity determines the levels of both alpha-carotene and total carotenoids in orange carrots. Plant Cell 26(5):2223–2233

Araujo BS, Charlwood BV, Pletsch M (2002) Tolerance and metabolism of phenol and chloroderivatives by hairy root cultures of *Daucus carota* L. Environ Pollut 117:329–335

Araujo BS, Dec J, Bollag JM, Pletsch M (2006) Uptake and transformation of phenol and chlorphenols by hairy root cultures of *Daucus carota*, *Ipomoea batatas* and *Solanum aviculare*. Chemosphere 63:642–651

Araujo BS, Oliveira JO, Machado SS, Pletsch M (2004) Comparative studies of the peroxidases from hairy roots of *Daucus carota*, *Ipomoea batatas* and *Solanum aviculare*. Plant Sci 167:1151–1157

Aviv D, Amsellem Z, Gressel J (2002) Transformation of carrots with mutant acetolactate synthase for *Orobanche* (broomrape) control. Pest Manag Sci 58:1187–1193

Balestrazzi A, Carbonera D, Cella R (1991) Transformation of *Daucus carota* hypocotyls mediated by *Agrobacterium tumefaciens*. J Genet Breed 45:135–140

Ballas N, Zakai N, Loyter A (1987) Transient expression of the plasmid pCaMVCAT in plant protoplasts following transformation with polyethylene glycol. Exp Cell Res 170:228–234

Baranski R, Klocke E, Nothnagel T (2007a) Enhancing resistance of transgenic carrot to fungal pathogens by the expression of *Pseudomonas fluorescence* Microbial Factor 3 (MF3) gene. Physiol Mol Plant Pathol 71:88–95

Baranski R, Klocke E, Nothnagel T (2008) Chitinase CHIT36 from *Trichoderma harzianum* enhances resistance of transgenic carrot to fungal pathogens. J Phytopathol 156:513–521

Baranski R, Klocke E, Ryschka U (2007b) Monitoring the expression of green fluorescent protein in carrot. Acta Physiol Plant 27:239–246

Baranski R, Klocke E, Schumann G (2006) Green fluorescent protein as an efficient selection marker for *Agrobacterium rhizogenes* mediated carrot transformation. Plant Cell Rep 25:190–197

Bates GW, Carle SA, Piastuch WC (1990) Linear DNA introduction into carrot protoplasts by electroporation undergoes ligation and recircularization. Plant Mol Biol 14:899–908

Bates GW, Piastuch W, Riggs CD, Rabussay D (1988) Electroporation for DNA delivery to plant protoplast. Plant Cell Tiss Organ Cult 12:213–218

Bercetche J, Chriqui D, Adam S, David C (1987) Morphogenetic and cellular reorientations induced by *Agrobacterium rhizogenes* (strain 1855, 2659 and 8196) on carrot, pea and tobacco. Plant Sci 52:195–210

Boston RS, Becwar MR, Ryan RD, Goldsbrough PB, Larkins BA, Hodges TK (1987) Expression from heterologous promoters in electroporated carrot protoplasts. Plant Physiol 83:742–746

Bouche FB, Erlt OT, Muller CP (2002) Neutralizing B cell response in measles. Viral Immunol 15(3):451–471

Bouche FB, Marquet-Blouin E, Yanagi Y, Steinmetz A, Muller CP (2003) Neutralising immunogenicity of a polyepitope antigen expressed in a transgenic food plant: a novel antigen to protect against measles. Vaccine 21:2065–2072

Bouche FB, Steinmetz A, Yanagi Y, Muller CP (2005) Induction of broadly neutralizing antibodies against measles virus mutants using a polyepitope vaccine strategy. Vaccine 23:2074–2077

Boulanger F, Berkaloff A, Richaud F (1986) Identification of hairy root loci in the T-regions of *Agrobacterium rhizogenes* Ri plasmids. Plant Mol Biol 6:271–279

Brodzik R, Spistin S, Pogrebnyak N, Bandurska K, Portocarrero C, Andryszak K, Koprowski H, Golovkin M (2009) Generation of plant-derived recombinant DTP subunit vaccine. Vaccine 27(28): 3730–3734

Capone I, Frugis G, Costantino P, Cardarelli M (1994) Expression in different populations of cells of the root meristem is controlled by different domains of the *rol*B promoter. Plant Mol Biol 25:681–691

Capone I, Spanò L, Cardarelli M, Bellincampi D, Petit A, Costantino P (1989) Induction and growth properties of carrot roots with different components of *Agrobacterium rhizogenes* T-DNA. Plant Mol Biol 13:43–52

Cardarelli M, Mariotti D, Pomponi M, Spanò L, Capone I, Costantino P (1987a) *Agrobacterium rhizogenes* T-DNA genes capable of inducing hairy root phenotype. Mol Gen Genet 207:475–480

Cardarelli M, Spanò L, de Paolis A, Mauro ML, Vitali G, Costantino P (1985) Identification of the genetic locus responsible for non–polar root induction by *Agrobacterium rhizogenes* 1855. Plant Mol Biol 5:385–391

Cardarelli M, Spanò L, Mariotti D, Mauro ML, van Sluys MA, Costantino P (1987b) The role of auxin in hairy root induction. Mol Gen Genet 208:457–463

Chakrabarty B, Ghosal AK, Purkait MK (2008) Effect of molecular weight of PEG on membrane morphology and transport properties. J Membrane Sci 309:209–221

Chen WP, Punja ZK (2002) Transgenic herbicide- and disease-tolerant carrot (*Daucus carota* L.) plants obtained through *Agrobacterium*-mediated transformation. Plant Cell Rep 20:929–935

Chilton MD, Tepfer D, Petit A, David C, Casse-Delbart F, Tempé J (1982) *Agrobacterium rhizogenes* inserts T-DNA into plant roots. Nature 295:432–434

Clough SJ, Bent AF (1998) Floral dip: a simplified method for *Agrobacterium*-mediated transformation of *Arabidopsis thaliana*. Plant J 16:735–743

Deroles S, Smith MAL, Lee C (2002) Factors affecting transformation of cell cultures from three dicotyledonous pigment-producing species using microprojectile bombardment. Plant Cell Tiss Organ Cult 70:69–76

Di Cola A, Poma A, Spanò L (1997) *rol*B expression pattern in the early stages of carrot somatic embryogenesis. Cell Biol Int 21:595–600

Dirks R, Sidorov V, Tulmans C (1996) A new protoplast culture system in *Daucus carota* L. and its application for mutant selection and transformation. Theor Appl Genet 93:809–815

Dröge W, Broer I, Puhler A (1992) Transgenic plants containing the phosphinothricin-N-acetyltransferase gene metabolize the herbicide L-phosphinothricin (glufosinate) differently from untransformed plants. Planta 187:142–151

ECJ (2018) Organisms obtained by mutagenesis are GMOs and are, in principle, subject to the obligations laid down by the GMO Directive. Court of Justice of the EU, case C-528/16. Press release No. 111/18 of 25 July 2018

Epstein E, Nissen SJ, Sutter EG (1991) Indole-3-acetic acid and indole-3-butyric acid in tissues of carrot inoculated with *Agrobacterium rhizogenes*. J Plant Growth Regul 10:97–100

FDA (2012) Elelyso for injection, Approval letter. US Food and Drug Administration. https://www.accessdata.fda.gov/drugsatfda_docs/appletter/2012/022458s000ltr.pdf. Accessed 6 Sept 2018

Fründt C, Meyer AD, Ichikawa T, Meins F (1998) A tobacco homologue of the Ri-plasmid orf13 gene causes cell proliferation in carrot root discs. Mol Gen Genet 259:559–568

Fujii N (1997) Pattern of DNA binding of nuclear proteins to the proximal *Agrobacterium rhizogenes rol*C promoter is altered during somatic embryogenesis of carrot. Gene 201:55–62

Gallie DR (1993) Introduction of mRNA to protoplasts using polyethylene glycol. Plant Cell Rep 13:119–122

Gamborg OL, Miller RA, Ojima K (1968) Nutrient requirements of suspension culture of soybean root cells. Exp Cell Res 50:151–158

Gautheret R (1939) Sur la possibilité de réaliser la culture indéfinie des tissues de tubercules de carotte. C R Soc Biol Paris 208:118–120

Gerrits MM, van Vliet AH, Kulpers EJ, Kusters JG (2006) *Helicobacter pylori* and antimicrobial resistance: molecular mechanisms and clinical implication. Lancet Infect Dis 6:699–709

Ghabouli M, Bahrami AR, Shahriari FA, Zolala J, Mohammadi A (2013) Studying the efficiency of floral dip method for genetic transformation of apiaceaeplants. J Hort Sci (Agricultural Sciences and Technology) 27(2):139–147

Gilbert MO, Zhang YY, Punja ZK (1996) Introduction and expression of chitinase encoding genes in carrot following *Agrobacterium*-mediated transformation. Vitro Cell Dev Biol Plant 32:171–178

Gladysz K, Baranski R (2003) Attempts to in planta genetic transformation of carrot. Polish J Nat Sci Suppl 1:236–237

Glass Z, Lee M, Li Y, Xu Q (2018) Engineering the delivery systems for CRISPR-based genome editing. Trends Biotechnol 36(2):173–185

Gogarten JP, Fichmann J, Braun Y, Morgan L, Styles P, Taiz SI, Delapp K, Taiz LB (1992) The use of antisense messenger-RNA to inhibit the tonoplast H^+ ATPase in carrot. Plant Cell 4:851–864

Grabowski GA, Barton NW, Pastores G, Dambrosia JM, Banerjee TK, McKee MA, Parker C, Schiffmann R, Hill SC, Brady RO (1995) Enzyme therapy in type 1 Gaucher disease: comparative efficacy of mannose-terminated glucocerebrosidase from natural and recombinant sources. Ann Intern Med 122:33–39

Grzebelus E, Szklarczyk M, Baranski R (2012) An improved protocol for plant regeneration from leaf- and hypocotyl-derived protoplasts of carrot. Plant Cell Tiss Organ Cult 109(1):101–109

Guan Y, Ren H, Xie H, Ma Z, Chen F (2009) Identification and characterization of bZIP-type transcription factors involved in carrot (*Daucus carota* L.) somatic embryogenesis. Plant J 60(2):207–217

Guivarc'h A, Caissard JC, Brown S, Marie D, Dewitte W, Vanonckelen H, Chriqui D (1993) Localization of target-cells and improvement of *Agrobacterium*-mediated transformation efficiency by direct acetosyringone pretreatment of carrot root disks. Protoplasma 174:10–18

Hansen G, Larribe M, Vaubert D, Tempe J, Biermann BJ, Montoya AL, Chilton MD, Brevet J (1991) Agrobacterium rhizogenes pRi8196 T-DNA-mapping and DNA-sequence of functions involved in mannopine synthesis and hairy root differentiation. Proc Natl Acad Sci USA 88:7763–7767

Hardegger M, Sturm A (1998) Transformation and regeneration of carrot (*Daucus carota* L.). Mol Breed 4:119–127

Hauptmann R, Eschenfeldt WH, English J, Brinkhaus FL (1997) Enhanced carotenoid accumulation in storage organs of genetically engineered plants. The United States Patent No. 5618988

Hibberd JM, Linley PJ, Khan MS, Gray JC (1998) Transient expression of green fluorescent protein in various plastid types following microprojectile bombardment. Plant J 16:627–632

Higashi K, Daita M, Kobayashi T, Sasaki K, Harada H, Kamada H (1998) Inhibitory conditioning for carrot somatic embryogenesis in high-cell-density cultures. Plant Cell Rep 18:2–6

Imani J, Baltruschat H, Stein E, Jia G, Vogelsberg J, Kogel KH, Hückelhoven R (2006) Expression of barley BAX Inhibitor-1 in carrots confers resistance to *Botrytis cinerea*. Mol Plant Pathol 7:279–284

Imani J, Berting A, Nitsche S, Schaefer S, Gerlich WH, Neumann KH (2002) The integration of a major hepatitis B virus gene into cell-cycle synchronized carrot cell suspension cultures and its expression in regenerated carrot plants. Plant Cell Tiss Organ Cult 71:157–164

Ipek A, Ipek M, Simon PW (2006a) Association of reversible inactivation of the maize transposable element *Ds* with tissue-specific processing of the *35S:TPase* transcript in carrot (*Daucus carota* L.). J Hortic Sci Biotech 81:819–826

Ipek A, Masson P, Simon PW (2006b) Genetic transformation of an *Ac/Ds*-based transposon tagging system in carrot (*Daucus carota* L.). Eur J Hortic Sci 71:245–251

Ishige F, Takaichi M, Foster R, Chua NH, Oeda K (1999) A G–box motif (GCCACGTGCC) tetramer confers high-level constitutive expression in dicot and monocot plants. Plant J 18:443–448

Jayaraj J, Devlin R, Punja Z (2008) Metabolic engineering of novel ketocarotenoid production in carrot plants. Transgenic Res 17(4):489–501

Jayaraj J, Punja Z (2008) Transgenic carrot plants accumulating ketocarotenoids show tolerance to UV and oxidative stresses. Plant Physiol Biochem 46 (10):875–883

Jayaraj J, Punja ZK (2007) Combined expression of chitinase and lipid transfer protein genes in transgenic carrot plants enhances resistance to foliar fungal pathogens. Plant Cell Rep 26:1539–1546

Jen GC, Chilton MD (1986) The right border region of pTiT37 T-DNA is intrinsically more active than the left border region in promoting T-DNA transformation. Proc Natl Acad Sci USA 83:3895–3899

Jung YC, Lee HJ, Yum SS, Soh WY, Cho DY, Auh CK, Lee TK, Soh HC, Kim YS, Lee SC (2005) Drought-inducible but ABA-independent thaumatin-like protein from carrot (*Daucus carota* L.). Plant Cell Rep 24:366–373

Kalbina I, Wallin A, Lindh I, Engstrom P, Andersson S, Strid K (2011) A novel chimeric MOMP antigen expressed in *Escherichia coli*, Arabidopsis thaliana, and *Daucus carota* as a potential *Chlamydia trachomatis* vaccine candidate. Protein Expres Purif 80(2):194–202

Kikkert JR, Vidal JR, Reisch BI (2005) Stable transformation of plant cells by particle bombardment/biolistics. Methods Mol Biol 286:61–78

Kim Y-S, Mi-Y Kim, Kim T-G, Yang M-S (2009) Expression and assembly of cholera toxin B subunit (CTB) in transgenic carrot (*Daucus carota* L.). Mol Biotechnol 41:8–14

Klimek-Chodacka M, Oleszkiewicz T, Lowder LG, Qi Y, Baranski R (2018) Efficient CRISPR/Cas9-based genome editing in carrot cells. Plant Cell Rep 37:575–586

Klein TM, Wolf ED, Wu R, Sanford JC (1987) High velocity microprojectiles for delivering nucleic acids into living cells. Nature 327:70–73

Kumar S, Dhingra A, Daniell H (2004) Plastid-expressed betaine aldehyde dehydrogenase gene in carrot cultured cells, roots, and leaves confers enhanced salt tolerance. Plant Physiol 136:2843–2854

Langridge WHR, Li BJ, Szalay AA (1985) Electric field mediated stable transformation of carrot protoplasts with naked DNA. Plant Cell Rep 4:355–359

Limami AM, Sun LY, Douat C, Helgeson J, Tepfer D (1998) Natural genetic transformation by *Agrobacteriumm rhizogenes*. Plant Physiol 118:543–550

Lindh I, Wallin A, Kalbina I, Sävenstrand H, Engström P, Andersson S, Strid A (2009) Production of the p24 capsid protein from HIV-1 subtype C in Arabidopsis thaliana and *Daucus carota* using an endoplasmic reticulum-directing SEKDEL sequence in protein expression construct. Protein Expr Purif 66:45–51

Locato V, Balestrazzi A, De Gara L, Carbonera D (2006) Reduced expression of *top1β* gene induces programmed cell death and alters ascorbate metabolism in *Daucus carota* cultured cells. J Exp Bot 57:1667–1676

Luchakivskaya Yu, Kischenko O, Gerasymenko I, Olevinskaya Z, Simonenko Yu, Spivak M, Kuchuk M (2011) High-level expression of human interferon alpha-2b in transgenic carrot (*Daucus carota* L.) plants. Plant Cell Rep 30:407–415

Ma SW, Zhao DL, Yin ZQ, Mukerjee R, Singh B, Qin HY (1997) Transgenic plants expressing autoantigens fed to mice to induce oral immune tolerance. Nature Med 3:793–796

Maass D, Arango J, Wust F, Beyer P, Welsch R (2009) Carotenoid crystal formation in Arabidopsis and carrot roots caused by increased phytoene synthase protein levels. PLoS ONE 4(7):e6373

Marquet-Blouin E, Bouche FB, Steinmetz A, Muller CP (2003) Neutralizing immunogenicity of transgenic carrot (*Daucus carota* L.)-derived measles virus hemagglutinin. Plant Mol Biol 51:459–469

Mason HS, Ball JM, Shi J-J, Jiang X, Estes MK, Arntzen CJ (1996) Expression of Norwalk virus capsid protein in transgenic tobacco and potato and its oral immunogenicity in mice. Proc Natl Acad Sci USA 93:5335–5340

Matsubayashi Y, Goto T, Sakagami Y (2004) Chemical nursing: phytosulfokine improves genetic transformation efficiency by promoting the proliferation of surviving cells on selective media. Plant Cell Rep 23:155–158

Matsubayashi Y, Ogawa M, Morita A, Sakagami Y (2002) An LRR receptor kinase involved in perception of a peptide plant hormone, phytosulfokine. Science 119:877–878

Melchers LS, Stuiver MH (2000) Novel genes for disease-resistance breeding. Curr Opin Plant Biol 3:147–152

Merfield CN, Hampton JG, Wratten SD, Prapanoppasin P, Yeeransiri P (2010) The effect of plant density on seed yield and quality of carrot (*Daucus carota* L.). In: McGill CR, Rowarth JS (eds) Seed symposium: seeds for futures. Agronomy Society of New Zealand Special Publication 13(14):75–83

Mikschofsky H, Mann G, Broer I (2009) Soil adaptation of transgenic in vitro carrot plantlets. J Agr Sci 147:43–49

Monreal-Escalante E, Govea-Alonso DO, Hernández M, Cervantes J, Salazar-González JA, Romero-Maldonado A, Rosas G, Garate T, Fragoso G, Sciutto E, Rosales-Mendoza S (2016) Towards the development of an oral vaccine against porcine cysticercosis: expression of the protective HP6/TSOL18 antigen in transgenic carrots cells. Planta 243(3):675–685

Murashige T, Skoog F (1962) A revised medium for rapid growth and bioassays with tobacco tissue cultures. Physiol Plant 15:473–497

Neumann E, Schaefer-Riddder M, Wang Y, Hofschneider PH (1982) Gene transfer into mouse lyoma cells by electroporation in high electric fields. EMBO J 1:841–845

Niazian M, Sadat Noori SA, Galuszka P, Mortazavian SMM (2017) Tissue culture-based *Agrobacterium*-mediated and *in planta* transformation methods. Czech J Genet Plant Breed 53(4):133–143

Nobécourt P (1939) Sur la pérennité et l'augmentation de volume des cultures de tissues végétaux. Compt. Rendus Soc. Biol. Lyon 130:1270–1271

Noh SA, Lee HS, Huh GH, Oh MJ, Paek KH, Shin JS, Bae JM (2012) A sweetpotato SRD1 promoter confers strong root-, taproot-, and tuber-specific expression in Arabidopsis, carrot, and potato. Transgenic Res 21 (2):265–278

Park S, Kim CK, Pike LM, Smith RH, Hirschi KD (2004) Increased calcium in carrots by expression of an *Arabidopsis* H^+/Ca^{2+} transporter. Mol Breed 14:275–282

Pawlicki N, Sangwan RS, Sangwan-Norreel BS (1992) Factors influencing the *Agrobacterium tumefaciens*-mediated transformation of carrot (*Daucus carota* L.). Plant Cell Tissue Org 31:129–139

Peng R-H, Yao Q-H, Xiong A-S, Cheng Z-M, Li Y (2005) Codon-modifications and an endoplasmic reticulum-targeting sequence additively enhance expression of an Aspergillus phytase gene in transgenic canola. Plant Cell Rep 25:124–132

Permyakova NV, Zagorskaya AA, Belavin PA, Uvarova EA, Nosareva OV, Nesterov AE, Novikovskaya AA, Zav'yalov EL, Moshkin MP, Deineko EV (2015) Transgenic carrot expressing fusion protein comprising *M. tuberculosis* antigens induces immune response in mice. Biomed Res Int 2015(2): 1–11

Porceddu A, Falorni A, Ferradini N, Cosentino A, Calcinaro F, Faleri C, Cresti M, Lorenzetti F, Brunetti P, Pezzotti M (1999) Transgenic plants expressing human glutamic acid decarboxylase (GAD65), a major autoantigen in insulin-dependent diabetes mellitus. Mol Breed 5:553–560

Punja ZK (2005) Transgenic carrots expressing a thaumatin-like protein display enhanced resistance to several fungal pathogens. Can J Plant Pathol 27:291–296

Punja ZK, Jayaraj J, Wally O (2007) Carrot. In: Pua EC, Davey MR (eds) Biotechnology in agriculture and forestry: transgenic crops IV, vol 59. Springer, Berlin, pp 277–294

Punja ZK, Raharjo SHT (1996) Response of transgenic cucumber and carrot plants expressing different chitinase enzymes to inoculation with fungal pathogens. Plant Dis 80:999–1005

Rasmussen JO, Rasmussen OS (1993) PEG mediated DNA uptake and transient GUS expression in carrot, rapeseed and soybean protoplasts. Plant Sci 89:199–207

Rathinasabapathi B, Fouad WM, Sigua CA (2001) *β*-Alanine betaine synthesis in the Plumbaginaceae. Purification and characterization of a trifunctional, *S*-adenosyl-l-methionine-dependent *N*-methyltransferase from *Limonium latifolium* leaves. Plant Physiol 126:1241–1249

Rathus C, Bower R, Birch RG (1993) Effects of promoter, intron and enhancer elements on transient gene-expression in sugarcane and carrot protoplasts. Plant Mol Biol 23:613–618

Rojas-Anaya E, Loza-Rubio E, Olivera-Flores MT, Gomez-Lim M (2009) Expression of rabies G protein in carrots (*Daucus carrota*). Transgenic Res 18:911–919

Rosales-Mendoza S, Soria-Guerra RE, López-Revilla R, Moreno-Fierros L, Alpuche-Solis AG (2008) Ingestion of transgenic carrots expressing the *Escherichia coli* heat-labile enterotoxin B subunit protects mice against cholera toxin challenge. Plant Cell Rep 27:79–84

Rosales-Mendoza S, Soria-Guerra RE, Moreno-Fierros L, Han Y, Alpuche-Solis AG, Korban SS (2011) Transgenic carrot tap roots expressing an immunogenic F1-V fusion protein from *Yersina pestis* are immunogenic in mice. J Plant Physiol 168:174–180

Rosales-Mendoza S, Soria-Guerra RE, Olivera-Flores MTD, López-Revilla R, Argullo-Astorga GR, Jimenez-Bremont JF, Garcia-de la Cruz RF, Loyola-Rodriguez JP, Alpuche-Solís AG (2007) Expression of *Escherichia coli* heat-abile enterotoxin B subunit (LTB) in carrot (*Daucus carota* L.). Plant Cell Rep 26:969–976

Rosales-Mendoza S, Tello-Olea MA (2015) Carrot cells: a pioneering platform for biopharmaceuticals production. Mol Biotechnol 57:219–232

Rubatzky VE, Quiros CF, Simon PW (1999) Carrots and related vegetable Umbelliferae. CABI Publ, New York

Ryder MH, Tate ME, Kerr A (1985) Virulence properties of strains of *Agrobacterium* on the apical and basal surfaces of carrot root discs. Plant Physiol 77:215–221

Sanford JC, Klein TM, Wolf ED, Allen N (1987) Delivery of substances into cells and tissues using a particle bombardment process. J Part Sci Tech 5:27–37

Schuttelaar (2015) The regulatory status of New Breeding Techniques in countries outside the European Union. Schuttelaar & Partners. Version: June 2015. http://www.nbtplatform.org/background-documents/rep-regulatory-status-of-nbts-oustide-the-eu-june-2015.pdf. Accessed 6 Sept 2018

Scott RJ, Draper J (1987) Transformation of carrot tissues derived from proembryogenic suspension cells: a useful model system for gene expression studies in plants. Plant Mol Biol 8:265–274

Serino G, Clerot D, Brevet J, Costantino P, Cardarelli M (1994) *rol* genes of *Agrobacterium rhizogenes* cucumopine strain; sequence, effects and pattern of expression. Plant Mol Biol 26:415–422

Shaaltiel Y, Bartfeld D, Hashmueli S, Baum G, Brill-Almon E, Galili G, Dym O, Boldin-Adamsky SA, Silman I, Sussman JL, Futerman AH, Aviezer D (2007) Production of glucocerebrosidase with terminal mannose glycans for enzyme replacement therapy of Gaucher's disease using a plant cell system. Plant Biotechnol J 5:579–590

Shillito RD, Saul MW, Paskowski SJ, Muller M, Potrykus L (1985) High efficiency direct gene transfer to plants. Bio/Technology 3:1099–1103

Shiota H, Kamada H (2000) Acquisition of desiccation tolerance by cultured carrot cells upon ectopic expression of C-ABI3, a carrot homolog of ABI3. J Plant Physiol 156:510–515

Simpson K, Quiroz L, Rodriguez-Concepcion M, Stange C (2016) Differential contribution of the first two enzymes of the MEP pathway to the supply of metabolic precursors for carotenoid and chlorophyll biosynthesis in carrot (*Daucus carota*). Front Plant Sci 7. https://doi.org/10.3389/fpls.2016.01344

Souza SR, Souza ES, Berbara RLL, Fernandes MS, Stark EMLM (2007) Enzymes of nitrogen metabolism and proteases activity in hairy roots of clover and carrots, with and without arbuscular mycorrhizal fungi. J Plant Nutr 30:1185–1204

Sprink T, Eriksson D, Schiemann J, Hartung F (2016) Regulatory hurdles for genome editing: process- vs. product-based approaches in different regulatory contexts. Plant Cell Rep 35:1493–1506

Steward FC (1958) Growth and organized development at cultured cells. II. Interpretation of the growth from free cell to carrot plant. Am J Bot 45:709–713

Steward FC, Mapes MO, Smith J (1958) Growth and organized development at cultured cells. I. Growth and division of freely suspended cells. Am J Bot 45:693–703

Svitashev S, Schwartz Ch, Lenderts B, Young JK, Cigan AM (2016) Genome editing in maize directed by CRISPR-Cas9 ribonucleoprotein complexes. Nat Commun 7:13274. https://doi.org/10.1038/ncomms13274

Takahata K, Takeuchi M, Fujita M, Azuma J, Kamada H, Sato F (2004) Isolation of putative glycoprotein gene from early somatic embryo of carrot and its possible involvement in somatic embryo development. Plant Cell Physiol 45:1658–1668

Takaichi M, Oeda K (2000) Transgenic carrots with enhanced resistance against two major pathogens, *Erysiphe heraclei* and *Alternaria dauci*. Plant Sci 153:135–144

Tang GQ, Lüscher M, Sturm A (1999) Antisense repression of vacuolar and cell wall invertase in transgenic carrot alters early plant development and sucrose partitioning. Plant Cell 11:177–189

Tang GQ, Sturm A (1999) Antisense repression of sucrose synthase in carrot (*Daucus carota* L.) affects growth rather than sucrose partitioning. Plant Mol Biol 41:465–479

Tepfer D (1984) Transformation of several species of higher plants by *Agrobacterium rhizogenes*: Sexual transmission of the transformed genotype and phenotype. Cell 37:959–967

Thomas JC, Guiltinan MJ, Bustos S, Thomas T, Nessler C (1989) Carrot (*Daucus carota*) hypocotyls transformation using *Agrobacterium tumefaciens*. Plant Cell Rep 8:354–357

Tokuji Y, Fukuda H (1999) A rapid method for transformation of carrot (*Daucus carota* L.) by using direct somatic embryogenesis. Biosci Biotechol Biochem 63:519–523

Tekoah Y, Shulman A, Kizhner T, Ruderfer I, Fux L, Nataf Y, Bartfeld D, Ariel T, Gingis-Velitski S, Hanania U, Shaaltiel Y (2015) Large-scale production of pharmaceutical proteins in plant cell culture – the protalix experience. Plant Biotechnol J 13:1199–1208

USDA Press (2018) Secretary Perdue issues USDA statement on plant breeding innovation. USDA Office of Communications. Press release No. 0070.18 of 28 Mar 2018

Uvarova EA, Belavin PA, Permyakova NV, Zagorskaya AA, Nosareva OV, Kakimzhanova AA, Deineko EV (2013) Oral Immunogenicity of plant-mad *Mycobacterium tuberculosis* ESAT6 and CFP10. BioMed Res Int. https://doi.org/10.1155/2013/316304

Van Sluys MA, Tempe J, Fedoroff N (1987) Studies on the introduction and mobility of the maize *Activator* element in *Arabidopsis thaliana* and *Daucus carota*. EMBO J 6:3881–3889

Wally O, Punja ZK (2010) Enhanced disease resistance in transgenic carrot (*Daucus carota* L.) plants over-expressing a rice cationic peroxidase. Planta 232(5):1229–1239

Wally O, Jayaraj J, Punja ZK (2006) Carrot (*Daucus carota* L.). In: Wang K (ed) Methods in molecular biology, vol. 344: *Agrobacterium* protocols, 2/e vol.2. Humana Press Inc., Totowa, NJ

Wally O, Jayaraj J, Punja ZK (2008) Comparative expression of beta-glucuronidase with five different promoters in transgenic carrot (*Daucus carota* L.) root and leaf tissues. Plant Cell Rep 27(2):279–287

Wally O, Jayaraj J, Punja ZK (2009a) Broad-spectrum disease resistance to necrotrophic and biotrophic pathogens in transgenic carrots (*Daucus carota* L.) expressing an Arabidopsis NPR1 gene. Planta 231 (1):131–141

Wally O, Jayaraj J, Punja ZK (2009b) Comparative resistance to foliar fungal pathogens in transgenic carrot plants expressing genes encoding for chitinase, β-1,3-glucanase and peroxidise. Eur J Plant Pathol 123:331–342

Wang LJ, Ni DA, Chen YN, Lee ZM (2001) The expression of *Mycobacterium tuberculosis* MPT64 protein in transgenic carrots. Acta Bot Sin 43:132–137

WHO (2018) Immunization, vaccines and biological. http://www.who.int/immunization/diseases/measles. Accessed: 5 Sept 2018

Willmitzer L, Sanchez-Serrano J, Bushfield E, Schell J (1982) DNA from *Agrobacterium rhizogenes* is transferred to and expressed in axenic hairy root plant tissue. Mol Gen Genet 186:16–22

Wurtele E, Bulka K (1989) A simple, efficient method for the *Agrobacterium*-mediated transformation of carrot callus cells. Plant Sci 61:253–262

Xu ZS, Huang Y, Wang F, Song X, Wang GL, Xiong AS (2014) Transcript profiling of structural genes involved in cyanidin-based anthocyanin biosynthesis between purple and non-purple carrot (*Daucus carota* L.) cultivars reveals distinct patterns. BMC Plant Biol 14:262–271

Yau Y-Y, Davis SJ, Ipek A, Simon PW (2008) Early identification of stable transformation events by combined use of antibiotic selection and vital detection of green fluorescent protein (GFP) in carrot (*Daucus carota* L.) Callus. Agric Sci China 7(6):664–671

Yildiz M, Willis DK, Cavagnaro PF, Iorizzo M, Abak K, Simon PW (2013) Expression and mapping of anthocyanin biosynthesis genes in carrot. Theor Appl Genet 126:1689–1702

Yoo SD, Cho YH, Sheen J (2007) Arabidopsis mesophyll protoplasts: a versatile cell system for transient gene expression analysis. Nat Protoc 2(7):1565–1572

Young JM, Kuykendall LD, Martínez-Romero E, Kerr A, Sawada HA (2001) A revision of Rhizobium Frank 1889, with an emended description of the genus, and the inclusion of all species of Agrobacterium Conn 1942 and Allorhizobium undicola de Lajudie,1998 as new combinations: *Rhizobium radiobacter*, *R. rhizogenes*, *R. rubi*, *R. undicola*, and *R. vitis*. Int J Syst Evol Microbiol 51:89–103

Zhang H, Liu M, Li Y, Zhao y, He H, Yang G, Zheng C (2009) Oral immunogenicity and protective efficacy in mice of a carrot-derived vaccine candidate expressing UreB subunit against *Helicobacter pylori*. Protein Expres Purif 69:127–131

The Carrot Nuclear Genome and Comparative Analysis

11

Massimo Iorizzo, Alicja Macko-Podgórni, Douglas Senalik, Allen Van Deynze and Philipp W. Simon

Abstract

The first draft of the carrot genome of an orange inbred line, "DC-27," was published in 2014. However, the genome assembly was fragmented and not assembled to the chromosome level, which limited its application for comprehensive genetic and genomic analyses. In 2016, a high-quality chromosome level, genome assembly of a doubled-haploid orange carrot DH1 was published, which rapidly advanced carrot genetic and genomic studies. The sequenced genome enabled the ability to identify candidate genes underlying important agronomic and nutrition-related traits such as root development, the accumulation of terpenoids, β-carotenes, and anthocyanins. Genome-level contributions include the clarification of phylogenetic relationships within carrot germplasm and the elucidation of the evolutionary history within the Euasterid II and Euasterid I clades. In this chapter, a description of the history of efforts made to characterize the carrot genome in the pre- and post-genomic era and the partners involved in the development of the high-quality carrot genome assembly are also described.

11.1 Introduction

The Carrot Genome in the Pre-genomic Era

Prior to whole-genome sequencing projects, studies to explore the structure and the content of the carrot genome were made. Based on flow cytometry, the size of the carrot genome was estimated ~473 Mb (Arumuganathan and Earle 1991). Early cytogenetic analysis demonstrated that carrot is a diploid species ($2n = 2x = 18$) with nine pairs of chromosomes (Simon 1984). A chromosome karyotype was established and indicated that four pairs of chromosomes are metacentric and five pairs are submetacentric (Iovene et al. 2011, see Chap. 8). Iovene et al. (2011) integrated the cytogenetic and genetic maps and each linkage group was assigned to a specific chromosome, and six of them were

M. Iorizzo (✉)
Department of Horticultural Sciences, Plants for Human Health Institute, North Carolina State University, 600 Laureate Way, Kannapolis, NC 28081, USA
e-mail: miorizz@ncsu.edu

A. Macko-Podgórni
Faculty of Biotechnology and Horticulture, Institute of Plant Biology and Biotechnology, University of Agriculture in Krakow, Krakow, Poland

D. Senalik · P. W. Simon
USDA ARS Vegetable Crops Research Unit, University of Wisconsin-Madison, 1575 Linden Dr., Madison, WI 53706, USA

A. Van Deynze
University of California, Davis, Seed Biotechnology Center and Plant Breeding Center, Davis, CA, USA

P. Simon et al. (eds.), *The Carrot Genome*, Compendium of Plant Genomes,
https://doi.org/10.1007/978-3-030-03389-7_11

oriented according to the short (north)/long (south) arm of the corresponding chromosome.

Based on a DNA association curve and thermal denaturation, it was estimated that the carrot genome consists of approximately 46% repetitive sequences, and the GC content is 37–38% (Simon 1984). The first global characterization of carrot repetitive DNA was based on BAC-end sequences (BES) and indicated that transposable elements (TE) represented the largest fraction of repetitive DNA (11.9%) (Cavagnaro et al. 2009). Class I retrotransposons, represented by two of the most numerous LTR superfamilies, *Ty1/copia* and *Ty3/gypsy*, have been identified as the most abundant TEs in both genomic BES (Cavagnaro et al. 2009) and transcriptome sequences (Iorizzo et al. 2011). Besides Class I retrotransposons, multiple families of Class II DNA transposons have been identified and characterized. The first thoroughly analyzed TE of carrot, was named *Tdc1*, a non-autonomous member of the *CMC* (*CACTA*) superfamily identified as an insertion in the 5′ flanking region of phenylalanine ammonia-lyase (*gDcPAL1*) (Ozeki et al. 1997). Characterization of related elements based on amplification and comparative analysis of conserved domains showed that three subfamilies of *Tdc* elements (*Tdc A*, *Tdc B*, and *Tdc C*) were present in the carrot genome (Itoh et al. 2003). Similarly, *DcMaster*, a member of the *PIF/Harbinger* superfamily, causing a knock-out mutation in the *acid-soluble invertase isozyme II* gene (Grzebelus et al. 2006; Yau and Simon 2003) was used to identify multiple-related TE including *KrakL1*, *KrakL2*, *Midi*, and the most abundant *Tourist*-like *DcMaster*-related MITEs, *Krak* (Grzebelus and Simon 2009). More recently, nine families of carrot *Stowaway* MITEs, named *DcSto1-DcSto9* were described (Macko-Podgorni et al. 2013).

Attempts to estimate gene content in the carrot genome were also made. A comparison of the translated BES with protein databases indicated that approximately 10% of the carrot genome represents coding sequences (Cavagnaro et al. 2009) and the number of genes was estimated to be ~25,000, with a frequency of one gene per 19 kb of genomic DNA.

These analyses provided an overall picture of the carrot genome structure and repetitive DNA fraction, and represented a foundation to initiate whole-genome sequencing projects.

The First Draft of the Carrot Genome In the last 10 years, the rapid advancement of next-generation sequencing (NGS) technologies has provided essential genomic resources for accelerating the molecular understanding of biological properties and has supported breeding efforts in several crops, including carrot. A first draft assembly of the carrot genome was released by Xu et al. (2014), who sequenced an orange inbred line named "DC-27" using Illumina and 454 Roche sequencing technologies. The genome assembly included >185,000 contigs and scaffolds covering about 372 Mb, with a scaffold N50 (sequence length of the shortest contig at 50% of the total genome length) >4.8 kb (Table 11.1). The authors predicted >75,000 genes with an average length of 833 nucleotide bases long. Although this resource has been useful to identify genes involved in the flavonoid pathway, transcription factors and candidate SSR markers (Wang et al. 2018), the genome assembly was highly fragmented and not assembled at the chromosome level. This limited its application for comprehensive comparative genomic and marker trait association studies and the identification of candidate genes controlling economically important traits.

To develop a high-quality chromosome-scale assembly of the carrot genome, a collaborative project between USDA/ARS-Madison, Wisconsin, the University of California, Davis, and an international consortium of private companies began in 2012. In this chapter, the background history of this collaboration, the strategy used to assemble the genome at the chromosome level will be reviewed, and its content and examples of the application for comparative genomic analysis will be summarized.

Table 11.1 Statistics of the "DC-27" and DH1 carrot genomes and gene predictions

Assembly feature	"DC-27"		DH1	
	Number	Size	Number	Size
Assembled sequences	185,376	371.6 Mb (78%)[a]	4826	421.5 Mb (88%)[a]
Scaffolds/superscaffolds	168,741	371.3 Mb	3468	419.5 Mb
N50 scaffolds/superscaffolds		0.04 Mb		36.6 Mb
Longest scaffold/superscaffold		0.19 Mb		30.2 Mb
Contigs	185,376	348.8 Mb	30,938	386.8 Mb
N50 Contigs		4.2 kb		31.2 kb
Anchored sequences	NA*	NA	60	361.1 Mb
Anchored and oriented sequences	NA	NA	50	353 Mb
GC content		34.4		34.8
Fraction Ns		22.7 Mb (6.1%)		34.7 Mb (8.2%)
Genome annotation				
Total repetitive sequences	NA	NA		193.7 Mb
Gene models	31,891	–	32,113	108.2 Mb
Genes in pseudomolecules	NA	NA	30,824 (96.0%)	
Noncoding RNAs	1559	151.5 kb	1386	188.9 kb

*NA data not available

[a] The percentage is calcualted considering the estimated carrot genome size 473 Mb (Arumuganathan and Earle 1991).

11.2 Background History of the Public and Private Partners Involved

The public–private partnership that supported the carrot genome project is just one example of how common goals can drive outcomes from what seem to be competing interests. Public scientists strive to develop research, presenting and publishing their findings which will advance science in their field. Similarly, private companies have product-oriented goals that will be marketed with the protection of intellectual property through trade secrets, i.e., hybrid lines, plant variety protection and patents. In plant breeding, these products include know-how, capacity and efficiency to develop, release, and market crop varieties. Similarly, public plant breeders often collect and characterize novel and sometimes more exotic germplasm to introgress genetic diversity through pre-breeding. In both public and private arenas, plant varieties are protected and licensed with royalty streams or payments supporting further research. Both institutions have a common goal to make their programs as efficient as possible to understand the biology, genetics, molecular and cellular processes, and environmental influences that shape crop traits. Understanding these systems is a daunting task that often can be best addressed by collaborating with what may seem to be competitors. Pre-competitive research towards the development of comprehensive genomic tools such as a well-assembled, genetically anchored and annotated reference and pan-genomes are excellent programs that benefit from public–private partnerships, even when published. The benefit is not only monetary, where costs are shared among federal, state, local, and private funding from multiple companies, but better programs are developed with access to in-kind resources and expertise from complementary industries, e.g., genomics, sequencing technologies, germplasm, bioinformatics, stakeholder-driven goals, etc. The broader impact of working in public–private partnerships is that it benefits the training of students and employees in the respective fields, who in turn, have a much better understanding of stakeholder-driven research while interacting with possible future employers. For partnerships to succeed, there must be open communication and a clear understanding of the nature of the collaboration, such as the objectives, timelines, deliverables, use and access of project results, reporting, publication, and the understanding of potential intellectual properties prior to the project initiation. The above ingredients are indeed what led to the establishment of a consortium between public researchers from USDA/ARS-Madison, Wisconsin, the University of California, Davis, and scientists from leading carrot seed companies including Bejo BV, Carosem, Monsanto Co., Nunhems BV, Rijk Zwaan BV, Sumika Co., Takii & Company LTD, and Vilmorin Co. The consortium began in 2011, initially as a modest project on developing genetic markers and defining genetic diversity in carrot, which was followed by the funding of the Carrot Genome. The outcome of this partnership has increased funding in carrot research. For example, there has been a continuation of privately funded research for carrot, successful funding of a national USDA Specialty Crops Research Initiative grant, which included stakeholders from industry, academia, and government. Additionally, the consortium has accelerated carrot breeding programs in both private and public institutions. The carrot genome consortium has served as a model for public–private partnerships and will continue to do so.

11.3 From Short Reads to Chromosome-Scale Sequences

11.3.1 Genetic and Genomic Resources

Plant Material and Whole-Genome Sequencing Since carrot is an outcrossing species (also see Chaps. 3 and 9), the genome of a carrot plant is highly heterozygous, which negatively affects the quality of the genome assembly, especially when using short-read Illumina sequencing technologies. New long-read sequencing technologies such as PacBio or Nanopore are improving the quality of heterozygous genome assemblies (Jiao and Schneeberger 2017). However, in 2013, when the carrot project was initiated, short-read Illumina sequencing technology was the most cost-effective sequencing strategy used for de novo genome assembly projects. To facilitate the genome assembly, a doubled-haploid carrot with a completely homozygous genome obtained from an orange Nantes type carrot, DH1, was used for genome sequencing. Seeds from a self-pollinated DH1 plant were kindly provided by Rijk Zwaan Seeds, Inc. The homozygous nature of the DH1 plant was validated using 3636 SNPs (Iorizzo et al. 2013). Over 99% of the SNPs were monomorphic; the few polymorphic SNPs were found to be false positives after additional analysis.

The DH1 whole genomic sequence (WGS) was generated exclusively using the Illumina platform HiSeq 2000, at the Beijing Genome Institute (BGI), Shenzhen, China. Seven paired-end libraries with insert sizes of 170, 280, and 800 nt and 2, 5, 10, 20, and 40 kb were prepared for sequencing. In total, 147.2 Gb sequence data were generated and after filtering, approximately 88 Gb (186× coverage) of high-quality sequence data were used for the de novo assembly (Iorizzo et al. 2016).

WGS data were used to estimate the carrot genome size using a k-mer analysis. Based on this method, the DH1 genome size was estimated to be 473.3 Mb. This was consistent with an earlier flow-cytometry analysis which estimated the genome size as 473 Mb (Arumuganathan and Earle 1991).

BAC-End Sequences and Linkage Maps In addition to DH1 WGS sequences, paired-end BAC sequences (PE-BACs) from 29,875 clones (insert size: 148 ± 70 kb, 0.04 Gb) from a DH1-BAC library were used in the carrot assembly. These sequences were provided by Rijk Zwaan seeds, Inc. An integrated consensus genetic map was developed for guiding the construction of superscaffolds and pseudomolecules and to identify and correct chimeric sequences. The consensus genetic map was developed by integrating genetic maps from three populations, 70349, Br1091 × HM1, and 70796 (also see Chap. 7). Each linkage map consisted of a "full" dataset including all the segregating markers mapped in each population and a "bin" dataset, consisting of markers representing unique recombination events. The integrated maps included 2073 markers for the full dataset and 918 markers for the bin dataset, covering 622 and 616 cM, respectively.

11.3.2 Genome Assembly

The carrot genome assembly strategy included three phases or steps as shown in Fig. 11.1. In general, the bioinformatics analysis used for the carrot genome assembly was previously used in other genome sequencing projects. However, for carrot, several manually curated analyses were performed during the assembly process to ensure the quality of the final assembly.

Assembly Phase I Filtered Illumina reads were assembled using SOAPdenovo (Luo et al. 2012). This generated the assembly v1.0 that included contigs and scaffolds. A contig refers to a contiguous genomic sequence that does not contain unknown bases ("N"s). A scaffold is a portion of the genome sequence reconstructed from end-sequenced whole-genome shotgun fragments (Illumina PE) and composed of contigs with associated gaps. The carrot assembly v1.0 resulted in 8096 contigs and scaffolds covering >424 Mb with an N50 of 787 kb.

Assembly Phase II An integrated approach was used to build superscaffolds, identify and, correct chimeric scaffolds or contigs. First, three sources of sequences were aligned against the carrot assembly v1.0. This included: (1) 8057 PE-BACs

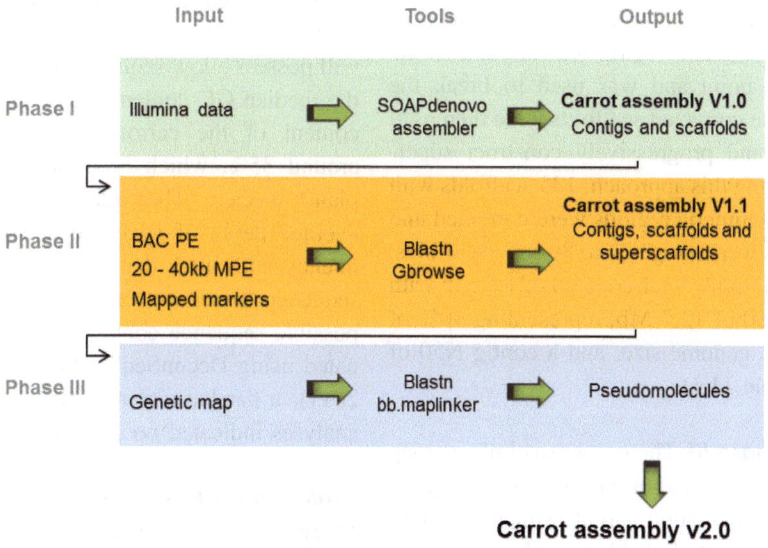

Fig. 11.1 Scheme of the carrot genome assembly pipeline. In Phase I, quality filtered Illumina data from eight insert libraries were assembled using SOAPdenovo producing the carrot assembly v1.0, which included contigs and scaffolds. In Phase II, unambiguously aligned sequences from mapped molecular markers, BAC-end sequences and 20 and 40 kb Illumina MPE were visualized in Gbrowse to manually inspect and correct chimeric regions and construct superscaffolds. This process produced the carrot assembly v1.1, which included contigs, scaffolds, and superscaffolds. In Phase III, the integrated linkage map was used to anchor superscaffolds and construct the nine carrot pseudomolecules (chromosomes). The final assembly named carrot assembly v2.0 includes pseudomolecules and the remaining unanchored contigs, scaffolds, and superscaffolds. Figure from Iorizzo et al. (2016)

unambiguously (both ends, length > 300 nt) aligned (>95% coverage and >99% similarity) to the carrot assembly v1.0; (2) 20 and 40 kb MPE Illumina data that unambiguously (100% coverage) aligned to the carrot assembly v1.0; (3) 1980 markers from the newly developed integrated map that unambiguously aligned (>90% coverage and >95% similarity) to the carrot assembly v1.0. Next, superscaffolding was initiated with scaffolds containing sequences of mapped markers. Scaffold-to-scaffold connections supported by at least two PE-BACs were used to build superscaffolds. During this process, the quality of each scaffold was manually verified by visually inspecting the coverage of large insert libraries (20 and 40 kb) and the consistency of the marker order along the linkage map was verified. Possible chimeric scaffolds were identified as: (1) scaffolds containing sequences of markers mapped to different LGs or to distal locations of the same LG; (2) scaffolds with regions not covered by MPE sequences. Those regions were then manually inspected. The midpoint between the closest unambiguously aligned PE sequences flanking the chimeric region was defined as the misassembly point and was used to break the assembly. The corrected scaffolds were then used to continue and progressively construct superscaffolds. Using this approach, 135 scaffolds with one or more chimeric regions were corrected and 881 scaffolds were merged into 89 superscaffolds. The carrot assembly v1.1 covered 421.5 Mb with an overall N50 of 12.7 Mb, representing 90% of the estimated genome size, and a contig N50 of 31.2 kb (Table 11.1).

Assembly Phase III The consensus linkage map was used to anchor and orient LGs, following the chromosome orientation and classification established by Iovene et al. 2011 (see Chap. 8). Scaffolds and superscaffolds were assigned to a chromosome if they contained at least one SNP marker from the "full" set linkage map. At least three markers mapped in the "bin" set linkage map representing unique recombination events were required to orient each sequence. However, some scaffolds had just one mapped marker and therefore were not oriented. In total, 60 sequences were anchored to the nine linkage groups, and 52 sequences were oriented (Table 11.1). The total length of anchored sequences was 361.2 Mb, which accounts for 84.8% of the carrot genome assembly (425.6 Mb). The average ratio between genetic-to-physical distance was 576.9 kb/cM, with one recombination event every 388 kb. As expected, genes and transposable elements were more abundant at the telomeric and pericentromeric regions of the chromosomes, respectively (Fig. 11.2).

11.3.3 Assembly Quality Verification

The reliability of reference sequence data is crucial for the interpretation of downstream structural and functional genomic analysis. Thus, multiple analyses were carried out to evaluate the quality of the final carrot genome assembly.

GC Content and Sequence Contamination Sequence contamination can influence the median GC content across the genome. Usually, the genomic regions with high or low GC content will possess a low sequencing depth compared to the median GC content region. The average GC content of the carrot genome was estimated around 35%, which is similar to that of other plant species. The relationship between the average depth of coverage and the % GC frequency indicated that there were no obvious sequence biases or contaminations. Presence of possible sequence contamination was also evaluated using DeconSeq (Schmieder and Edwards 2011), a database of non-plant genomes. These analyses indicated no sequence contamination.

Evaluation of Sequence Assembly Consistency Two sets of paired-end data were used to evaluate the correctness of the assembled sequences. This included: (1) 454 PE data (0.23 Gb, 0.5 M reads) from DH1 with an estimated insert size of 8.3 ± 2.3 kb; (2) 4717 PE-BACs that were not used to join scaffolds into superscaffolds during phase II of the assembly process. The coordinates of the mapping locations of the 454 PE and PE BAC reads

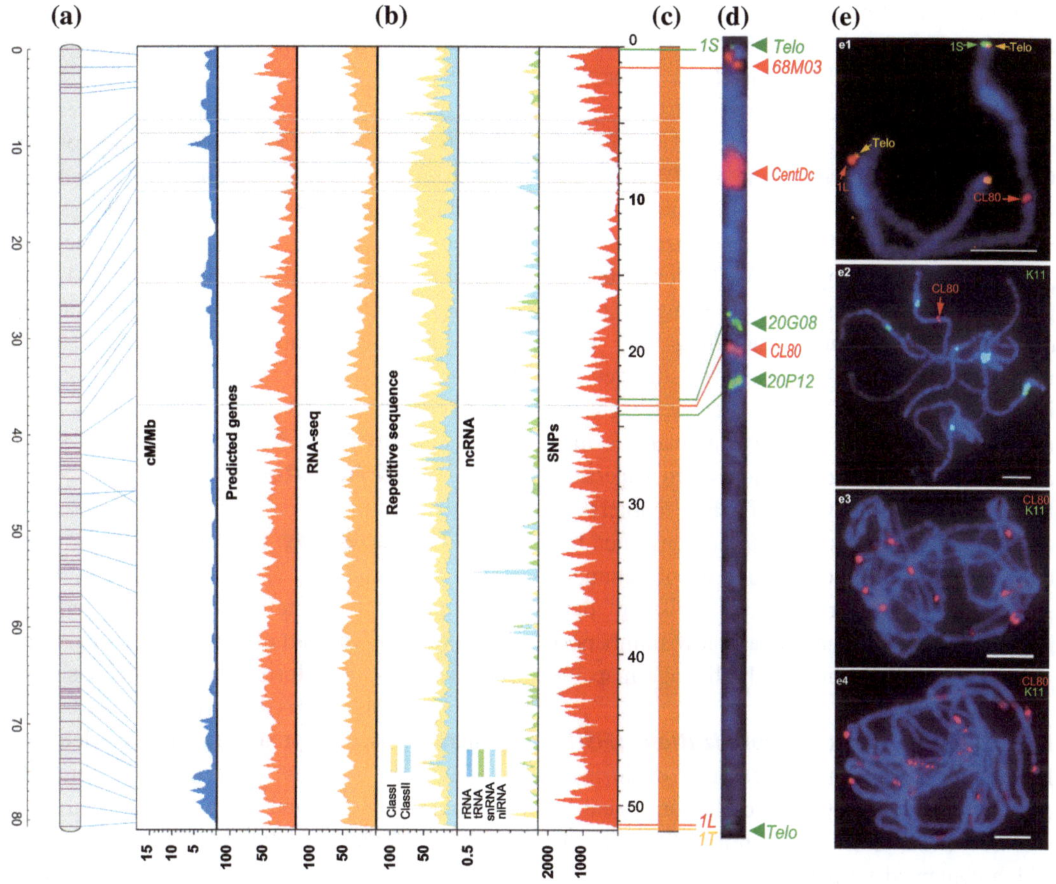

Fig. 11.2 Multi-dimensional topography of carrot chromosome 1. **a** Carrot integrated linkage map. Vertical bar to the left indicates genetic distance in cM. Lines connect a subset of markers to the assembled pseudomolecule. **b** From left to right: linkage map distance (cM/Mb), predicted genes (% nt/200 kb), transcriptomes from 20 different DH1 tissue types (% nt/200 kb), class I and class II repetitive sequences (% nt/200 kb), non-coding RNA (% nt/200 kb), and SNPs detected comparing resequencing data from 35 diverse carrot genotypes (number of SNPs per 100 kb). Genes and transposable elements were more abundant at the telomeric and pericentromeric regions of the chromosomes, respectively. **c** DNA pseudomolecules. Gaps between superscaffolds are indicated by gray horizontal lines. Locations of BAC probes hybridized to pachytene chromosome 1 (see panel e1) are identified by horizontal green and red lines and labeled on the right. Horizontal orange lines indicate the location of the telomeric repeats (Telo, T). **d** A digitally straightened carrot chromosome 1 (Chr 1) probed with a telomeric (TTTAGGG) 4 oligo-probe (T, green signals), CL80 and Cent-Dc repeats (red signals), BAC 68M03 (red signal), which is specific to carrot Chr 1, and BACs 20G08 and 20P12 (green signals) flanking the CL80 satellite array. **e** FISH mapping of a telomeric oligo-probe (Telo, yellow signal), the CL80 repeat (red signal) and BAC clone specific to the termini of the short (1S, green signal), and the long (1L, red signals) arm of carrot Chr 1. (e1). FISH mapping of CL80 (red signal) and Cent-Dc (green signal) repeats on the pachytene complements of DH1. (e2). FISH experiments indicated that the assembly extends into telomeric and subtelomeric regions, and confirmed that the probes flanking CL80 repeat and their location in the genome assembly are highly co-linear, further supporting the quality and high physical coverage of the carrot genome assembly. Modified figure from Iorizzo et al. (2016)

were used to estimate the distance between the PE pairs. The fraction of PE data that aligned within the expected library insert size reflects the fraction of assembled sequences that was consistently contiguous and correctly assembled. Overall, 99.4 and 95.6% of the 454 PE reads and PE-BACs, respectively, unambiguously aligned within the estimated library insert size.

The order of the superscaffolds along the nine pseudomolecules was verified using a newly developed linkage map including 394 GBS markers. The results confirmed that the genome is highly co-linear with the new genetic map. Overall, 82.2% of adjacent markers matched the orientation of scaffolds and superscaffolds. Further analysis of the markers in discordant alignments (17.8%) demonstrated that they were not a consequence of chimeric sequence but rather reflected missing data, insufficient marker density, and low numbers of genotypes used to generate the linkage map.

Fluorescence In Situ Hybridization (FISH) FISH experiments were carried out to evaluate the consistency and the coverage of the carrot genome assembly into telomeric regions. BAC clones which contained sequences that aligned near the ends of pseudomolecules corresponding to Chr 1, 2, 4, 5, 6, 8, and 9 were used as probes for the FISH experiments. As expected, BAC probes located near the end of the pseudomolecules localize near telomere probes (Fig. 11.2).

Gene Space Coverage Three sets of sequences were aligned against the carrot genome assembly v2.0 to evaluate the completeness of genic space. This included: (1) 58,751 consensus carrot expressed sequence tags (ESTs) from Iorizzo et al. (2011); (2) >500 M Illumina reads from 20 libraries, representing multiple tissues (petiole, hypocotyl, phloem, xylem, callus, flower, leaves at several developmental stages, and germinating seeds) and treatments (water stress, etiolation); (3) 258 ultra-conserved genes from the core eukaryotic genes dataset using CEGMA v2.4 (Parra et al. 2007). Using this approach, approximately 94% of the ESTs, 98% of RNA-Seq data, and 99.9% of core eukaryotic genes aligned to the carrot genome assembly, providing evidence that the assembly covers the majority of gene space.

Together, the assembly statistics and verification provided evidence that the DH1 carrot genome assembly was of high quality. Compared to the "DC-27" genome assembly, the DH1 genome assembly was the first chromosome-scale assembly, representing a 900 and 7.4 fold increase in terms of scaffold and contig N50, and a 10% increase in genome coverage (Table 11.1).

11.4 Genome Characterization and Annotation

11.4.1 Repetitive Sequences

A comprehensive identification and annotation of carrot repeats was part of the DH1 carrot genome sequencing project (Iorizzo et al. 2016). Using de novo, homology-based, structure-based, and clustering-based approaches, mobile elements, and tandem repeats were mined from the assembled genome and row reads. In total, 46% of the reference genome was annotated as repetitive DNA and the number correlated well with the previous reassociation-based evaluation of the repeat content.

TEs accounted for 97.9% of total repeats and were predominated by class I retrotransposons (66.7%), of which *Ty1/Copia* (37.7%) and *Ty3/Gypsy* (21.5%) were the most abundant groups, while non-LTR elements, SINEs and LINEs, accounted for 6.6 and 0.5% of total repeats, respectively. Ca. 8% of repeats were assigned as unclassified retrotransposons. Further analysis of LTR elements revealed a high frequency of incompletely sequenced LTR retrotransposons, suggesting that their fraction, especially related to high-copy families, might be underestimated (Macko-Podgórni et al. unpublished).

Class II DNA transposons occupied 13.6% of the assembled genome (29.7% of total repeats) and were represented by all superfamilies known in plant genomes. *CMC* (*CACTA*) and *hAT* superfamilies were the most abundant, reaching

9.7 and 9.0% of carrot repetitive DNA, respectively. *Mutators* and *Tc1/Mariners* accounted for 3.7% of repetitive DNA, while *Helitron*s and *PIF/Harbinger*s occupied about 1% of repeats. Detailed analysis of the previously described MITEs, *Krak*s, and *DcSto*s (Grzebelus and Simon 2009; Grzebelus et al. 2006; Macko-Podgorni et al. 2013) showed that they were present in 404 and 4028 intact copies in the DH1 genome, respectively. Based on the phylogenetic analysis, copies that met a commonly accepted 80-80-80 criterion were classified into nine *Krak* (*Tourist*-like) and 14 *DcSto* (*Stowaway*-like) families, respectively. More than 50% of MITE copies were inserted within a range of 2 kb from the nearest gene. However, statistical analysis did not rule out their random distribution, indicating that they were not preferentially targeted toward genic regions. All *DcSto* families experienced bursts of amplification at different time points during the carrot genome evolution. Phylogenetic relationships among individual copies revealed families characterized by uniform divergence time for most copies indicating rapid amplification from a few master copies, and families that have experienced multiple, independent bursts of mobilization. These observations suggest that various sources of transposase were engaged in the mobilization of different *DcSto* families.

Comparative analysis of *DcSto* and *Stowaway*-like elements identified in other *Asterid* species showed that MITEs were relatively more abundant and diverse in the carrot genome. While *Solanaceae* elements were interrelated at both intra- and inter-specific levels, suggesting that they colonized *Solanaceae* species genomes before the divergence of potato, tomato, and pepper ca. 36 Mya; the carrot *Stowaway*s were not related to *Solanaceae* MITEs. Moreover, most carrot *DcSto* families were unique, with no relation to other carrot *DcSto* elements. This suggested a lineage-specific evolution of carrot *Stowaways* and parallel expansion of multiple families followed by their diversification.

Identification of tandem repeats (TRs) in the assembled genome indicated that they accounted for 3.6% of a carrot genome. However, graph-based analysis of raw reads enabled detailed characterization of TRs and revealed significant underestimation of this DNA fraction. Four major TR families accounted for at least 7% of the carrot genome. Among them, the previously identified CentDc was present along with three new groups named CL8, CL80, and CL81. Further characterization of CentDc and CL80 sequences in DH1 and other *Daucus* species revealed a clear pattern of divergent evolution within the *Daucus* genus (also see Chap. 8).

11.4.2 Gene Prediction and Annotation

Prior to gene prediction, all repetitive sequences were masked using RepeatMasker (http://www.repeatmasker.org). Gene prediction was based on the integration of de novo gene prediction and evidence-based predictions. Training for de novo gene prediction was carried out with a set of genes from *Arabidopsis thaliana* and *Solanum lycopersicum*. Evidence-based prediction included protein-based homology searches from five closely related or model species and carrot transcriptome data-aided prediction. Transcriptome sequences included 58,751 carrot ESTs (Iorizzo et al. 2011) and >500 M Illumina reads from 20 libraries, representing multiple tissue and treatments. All gene models produced by de novo prediction, protein-homology searches, and prediction and transcript-based evidence were integrated using GLEAN (Elsik et al. 2007). This approach yielded 32,113 non-redundant gene models. The majority of the gene predictions (98.7%) had cDNA-EST expression evidence, which demonstrated the accuracy of gene prediction. The mean coding sequence size was 1183 nt, which was higher than carrot "DC-27," but similar to other annotated genomes with an average of 4.99 exons per gene.

Functional annotation of the predicted genes was based on their homology with multiple databases. Putative gene functions were assigned according to the best match of the alignments to SwissProt and TrEMBL databases. The motifs and domains of genes were determined by InterProScan (Zdobnov and Apweiler 2001). To annotate genes involved in biosynthetic pathways, gene models were aligned against KEGG proteins. Using this approach, about 89% of the genes have either known homologs or can be functionally classified. Gene models with no match in these databases were labeled "hypothetical proteins."

Non-protein-coding sequences were annotated based on de novo prediction using INFERNAL (Nawrocki and Eddy 2013) for candidate microRNAs, snRNAs, and tRNAscan-SE (Lowe and Eddy 1997) to tRNA. Ribosomal RNA (rRNA) sequences were identified by homologous searches using complete rRNA sequences from three closely related species, *Panax ginseng*, *P. quinquefolius*, and *Thapsia garganica*. This approach identified 31 nuclear rRNAs, 248 miRNAs, 564 tRNAs, and 532 snRNAs.

11.4.3 Curated Annotations

Regulatory and Resistance Genes Given the importance of transcription factors and resistance genes on the control of economically important traits, TF and R genes were annotated using PlantTFcat (http://plantgrn.noble.org/PlantTFcat/) (Dai et al. 2013) and MATRIX-R pipeline (Sanseverino et al. 2013), two bioinformatic tools specifically developed to annotate those genes. These tools combined InterProScan and a comprehensive prediction logic, based on relationships between gene families and conserved domains enabling the classification of plant TF and R genes in different subfamilies with high coverage and sensitivity. Based on this approach, 634 R genes were classified into 10 subgroups and 3267 TFs were classified into 90 subgroups, all of which were annotated.

Flavonoid, Isoprenoid Pathways and Terpene Synthase Secondary metabolites such as flavonoids (anthocyanin and flavones) and isoprenoids (carotenoid and terpenoid) played an important role throughout the history of carrot domestication and production (see also Chaps. 1 and 14–16). To establish a solid genomic framework and further study genes controlling the accumulation of these metabolites, a curated annotation of candidate genes involved in the flavonoid and isoprenoid pathway was part of the genome characterization. Prior to the release of the carrot genome, only 7 genes in the flavonoid pathway (Iorizzo et al. 2011) and 24 genes were in the isoprenoid pathway (Iorizzo et al. 2011; Just et al. 2007), were annotated. A homology-based analysis using BLAST against the KEGG database was integrated with an orthologous and phylogenetic analysis. Using this approach, 97 and 68 genes involved in the flavonoid/anthocyanin, and isoprenoid biosynthetic pathways were annotated, respectively. A branch of the isoprenoid pathway leads directly into the biosynthesis of terpenoids, which influences the taste and flavor of carrots; some of which may also play a role as bioactive molecules with an impact on human physiology and health (see Chap. 16). Terpene synthases were annotated based on the presence of TPS conserved domains DDXXD and DXDD (Martin et al. 2010). In total, 30 TPS genes were annotated and classified into five clades based on a phylogenetic analysis; the results were further used to refine the TPS gene model structures and to study their expression in multiple tissues (Keilwagen et al. 2017).

11.5 Comparative Analysis and Genome Evolution

The available sequences of a growing number of plant genomes provide the means to extract biological knowledge through the detection of similarities and differences within and between genomes of closely or more distantly related species. Indeed, the best method to reconstruct the evolutionary past of any species is by using a comparison with its living relatives. Using such comparative approaches: (1) knowledge can be transferred from model to non-model organisms; (2) insights can be gained into the evolution of specific genes or entire metabolic and signaling pathways; (3) genes of importance for niche-specific plant adaptations can be identified;

(4) large-scale genomic events such as whole-genome duplications (WGDs) can be unveiled. In this context, the density of the "genome community" will exponentially improve our ability to characterize a genome and associate genes with functions. Carrot belongs to the Euasterid II clade, a member of the Asterid clade, which encompasses about 32,000 species. This clade includes other important crops such as lettuce, sunflower, and more closely related members of the Apiaceae family such as celery, parsley, and cilantro (Bremer 2009). Prior to the release of the carrot genome, there were no sequenced genomes for species in the Euasterid II lineage; whereas the Euasterid I lineage had several sequenced genomes, including tomato (Sato et al. 2012), potato (Xu et al. 2011), pepper (Qin et al. 2014), coffee (Denoeud et al. 2014), and sesame (Wang et al. 2014). Since 2016, the genomes of lettuce (Reyes-Chin-Wo et al. 2017) and *Panax ginseng* (Kim et al. 2018) were released, which will provide new insights into evolution and speciation of the Euasterid II crops.

11.5.1 Euasterid II Divergence Time

A phylogenomic analysis among 13 plant genomes estimated that carrot diverged from grape ~113 million years ago (Mya), diverged from kiwifruit ~101 Mya, and diverged from potato and tomato ~90.5 Mya, confirming the previous estimates which dated the Asterid crown group to the Early Cretaceous and its radiation in the Late–Early Cretaceous (Bremer 2009) (Fig. 11.3). Further divergence between carrot and lettuce, both members of the Euasterid II clade, likely occurred ~72 Mya. Recently, a comparative analysis between carrot and *Panax ginseng* indicated that ginseng diverged from carrot ~50 Mya (Kim et al. 2018).

11.5.2 Whole-Genome Duplication

On the basis of transversions at fourfold degenerate sites (4DTv) obtained from the 8239 paralogous gene pairs, two new whole-genome duplications (WGDs) specific to the carrot lineage, named Dc-α and Dc-β, were identified and were superimposed on the earlier γ paleohexaploidy event shared by all eudicots (Iorizzo et al. 2016). Time estimates of the two WGDs indicated that Dc-α and Dc-β occurred ~43 and ~70 Mya, respectively (Iorizzo et al. 2016). The estimated timing of the Dc-β WGD around the Cretaceous–Paleogene (K–Pg) boundary supported the hypothesis that a WGD burst occurred around that time, perhaps reflecting a selective polyploid advantage in comparison to diploid progenitors (Vanneste et al. 2014). These findings also suggested a possible co-occurrence of the Dc-β WGD with the carrot–lettuce divergence. Characterization of the lettuce genome identified a whole-genome triplication that was placed 40–45 Mya distinct from the carrot Dc-β WGD (Reyes-Chin-Wo et al. 2017) (Fig. 11.3). Considering the time divergence estimates between carrot and *Panax*, 50 Mya and the estimated occurrence of Dc-β WGD, 70 Mya, a co-occurrence of WGD in these two lineages would be expected. Two WGD were detected in the *Panax* lineage, Pg-α and Pg-β, which were estimated to have occurred 28 and 2–3 Mya, respectively, independently from carrot Dc-α and Dc-β. Further comparative analysis among multiple members of the Euasterid II clade is needed to clarify the co-occurrence of WGDs and the evolutionary history and speciation of members of this clade.

11.5.3 Paleopolyploid History of the Carrot Genome

Reconstruction of carrot chromosomal blocks descending from the seven ancestral core eudicot chromosomes indicated that the carrot genome has gone through at least 60 chromosome fusions or translocations, possibly due to multiple WGD events. The two lineage-specific WGDs were evident from the distribution of the 4DTV of carrot paleohexaploid paralogous genes, whereas genes from the shared eudicot γ WGT were largely lost, likely owing to extensive genome fractionation. Comparative analysis with grape, tomato, and coffee combined with depth analysis of duplicated blocks harboring paralogous genes

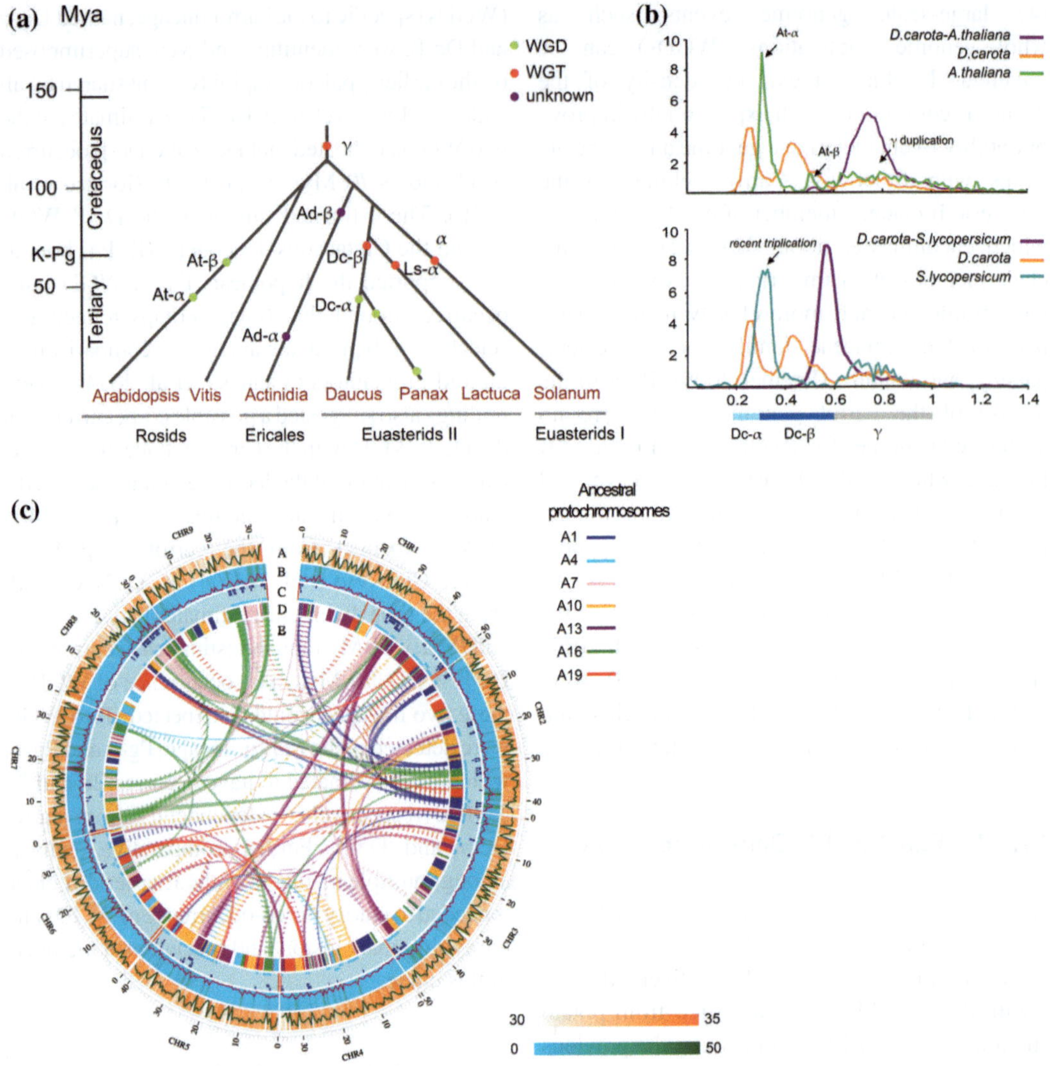

Fig. 11.3 Genome evolution of Euasterid II species. **a** Evolutionary relationships of the eudicot lineage reconstructed from the phylogenetic analysis (Iorizzo et al. 2016; Kim et al. 2018). Dots indicate whole-genome duplication (WGD) ages. For *A. thaliana*, kiwi, lettuce, panax, and Solanaceae species, WGD age estimates were obtained from the literature (Kim et al. 2018; Reyes-Chin-Wo et al. 2017; Vanneste et al. 2014). Green dots indicate confirmed WGDs. Red dots indicate confirmed WGTs. The polyploidization level of kiwi (purple dots) awaits confirmation. **b** Age distribution of 4DTv for genes from *D. carota*, *A. thaliana* and *S. lycopersicum* genomes. *X*-axis indicates 4DTv values; *Y*-axis indicates percentage of gene pairs in syntenic/co-linear blocks. The γ peak represents the ancestral hexaploidization event shared by core eudicots, Dc-α and Dc-β represent carrot-specific WGD events. (c) Representation of carrot-specific genome duplications. Circle A: GC (%) content; B: tandem duplication density (# per 0.5 Mb); C: Retained genes in carrot-specific WGDs (Dc-α in cyan, Dc-β in blue); D: Chromosomal blocks descending from the seven ancestral core eudicot protochromosomes (colored as in C); E: Duplicated segments derived from Dc-α events (dashed links, pairwise) and Dc-β events (solid links, triplicates). Modified figure from Iorizzo et al. (2016)

under the Dc-α 4DTV and Dc-β peaks indicated that a WGT (Dc-β) followed by a WGD (Dc-α) contributed to diversification of the 9 carrot chromosomes from the 21 chromosome intermediate ancestor (Iorizzo et al. 2016).

Whole-genome duplications significantly contribute to the diversity of genes and functions inherited in each genome including genes controlling economically important traits. Characterization of Dc-α and Dc-β duplicated blocks demonstrated that extensive gene fractionation has occurred during the evolutionary history of the carrot genome. Dc-α ohnologs are significantly enriched ($P \leq 0.01$) in protein domains involved in selective molecule interactions such as protein binding and dimerization functions, supporting the gene dosage hypothesis (Lang et al. 2010), which predicts that categories of genes encoding interacting products will likely be over-retained.

To further study the impact of genome duplication on carrot gene diversity, a comparative analysis of the regulatory genes (RG) was performed. De novo prediction and comparative analysis of regulatory genes across 11 genomes indicated that genomes that experienced WGDs after the γ paleohexaploidization event harbored more regulatory genes. In carrot, large-scale duplications represented the most common mode of regulatory gene expansion. Approximately, 33% of these genes were retained after the two carrot WGDs, which demonstrated the evolutionary impact of large-scale duplications on plant regulatory network diversity (Lang et al. 2010). Considering that only regulatory gene families with at least a total of 100 predicted genes across all plant species, 27 RG families in the carrot genome were over-represented relative to all species, six RG families were over-represented relative to species encompassing the Euasterid clade, and 23 RG families were over-represented relative to species encompassing the Asterid clade. Interestingly, transcription factor (TF) MADS type 1 and MADS-MIKC subfamilies were among the most under-represented RG families in carrot, relative to all the other species analyzed in this study. Considering the importance of this TF family's involvement in key developmental processes in plants, particularly the development of reproductive organs (Smaczniak et al. 2012), carrot represents a unique genetic model to further investigate how its genome compensated for a low diversity of MADS genes during its evolution, and the implications on the development of the reproductive system.

Six regulatory gene families have expanded in carrot as a consequence of lineage-specific duplications. The expanded families include a zinc-finger (ZF-GFR) regulatory gene family, the JmjC, TCP, and GeBP families, the B3 superfamily, and response regulators. The over-represented regulatory gene subgroups shared orthologous relationships with functionally characterized genes involved in cytokinin signaling, which can influence the circadian clock as well as plant morphology and architecture. For example, the expanded JmjC, response regulator, and B3-domain subgroups share ancestry with the *A. thaliana* REF6; PRR5, PRR6, and PRR7; and VRN1 genes, respectively, which regulate flowering time (Levy et al. 2002; Nakamichi et al. 2007; Noh et al. 2004) and are of major importance in plant adaption and survival.

11.6 Carrot Genomics Perspectives

A major milestone for carrot research and crop improvement was achieved with the public availability of the high-quality carrot genome sequence (Iorizzo et al. 2016). Since its publication in Nature Genetics in 2016, the DH1 genome assembly has been used in multiple experiments enabling genome-wide studies to identify candidate genes controlling terpenoid (Keilwagen et al. 2017), beta-carotene (Ellison et al. 2017, 2018) anthocyanin accumulation (Iorizzo et al. 2018), and root development (Macko-Podgórni et al. 2017), to study the transcriptome profile associated with flowering, carotenoid accumulation, and root development (Machaj et al. 2018; Oleszkiewicz et al. 2018; Ou et al. 2017; Perrin et al. 2017; Que et al. 2018), to clarify phylogenetic relationships within the *Daucus* germplasm (Arbizu et al. 2016;

Mezghani et al. 2018) and elucidate the evolutionary history within the Euasterid II (Kim et al. 2018; Reyes-Chin-Wo et al. 2017) and Euasterid I clades (Dong et al. 2018), to cite few examples.

While this is a major advance; however, one should not ignore the fact that this genome inherited the challenges and problems associated with short-reads-based de novo assembly, such as low contiguity at contig level, partial coverage of the gene prediction, and their structure. For example, the contig N50 is only 32 kb long and within very long scaffolds, each contig is separated by gaps ("Ns"); therefore, any unknown sequences that account for 40 Mb of the assembly could contain missing relevant genes. Also, gene predictions based on short-reads trascriptome sequences suffer multiple problems such as incomplete model structure, or a lack of or incorrect isoform predictions. To overcome these challenges, efforts are underway to develop an improved DH1 reference carrot genome assembly and gene prediction using PacBio long reads (Bostan et al. 2018). Thus, shortly after publishing the first reference sequence of the carrot DH1 genome, access to an improved genome information to advance large-scale genomic studies and functional genetics in carrot is already ongoing.

While access to a single reference genome was a critical step for the carrot community, it reveals only part of the total genomic composition of a given species. To extend our knowledge about the carrot genome, including the association between genes and phenotypes, concepts should be considered at a higher level than that of the single genome, such as a community genome scale. To date, the concept of "pan-genome" analyses has firmly taken hold also in plant sciences, where an increasing number of studies address the problem of characterizing the "core" and "dispensable" components of the genomes of crop species (Hirsch et al. 2014; Montenegro et al. 2017; Zhou et al. 2018). The core genome refers to the gene-rich parts of the genome found in every haplotype, whereas the dispensable genome refers to genomic components found in fewer haplotypes of a given species.

The de novo assembly of key and divergent genetic resources to analyze the pan-genome and resequencing of diversity panels at high coverage has the potential to distinguish key differences between individuals and to uncover more SNPs, genes, and structural variants than those detected on the basis of a single reference genome. For carrot, one goal would be to develop a pan-genome by selecting representative genotypes from multiple divergent carrot accessions. Six genetically distinct subpopulations have been identified within the global domesticated and wild carrot collection (Ellison et al. 2018). To ensure the representation of superior alleles in the pan-genome, one approach could include the identification of genotypes harboring favorable phenotypes within each subpopulation for de novo assembly, and then deep resequencing of additional related genotypes. Currently, efforts are underway to characterize genomic diversity of the global domesticated carrot population by resequencing of more than 700 carrot accessions. The resources generated from these sequencing efforts will help in the identification of regions linked to important agronomic traits, design effective marker-assisted selection (MAS) strategies and facilitate functional genetic studies. These sequencing efforts are expected to enhance carrot yield, nutritional value, and resistance to biotic and abiotic stresses.

References

Arbizu CI, Ellison SL, Senalik D, Simon PW, Spooner DM (2016) Genotyping-by-sequencing provides the discriminating power to investigate the subspecies of *Daucus carota* (Apiaceae). BMC Evol Biol 16:234

Arumuganathan K, Earle ED (1991) Nuclear DNA content of some important plant species. Plant Mol Biol Rep 9:208–218

Bostan H, Ellison S, Senalik D, Simon P, Iorizzo M (2018) Improving the Carrot Genome Assembly and Gene Prediction: Strategies to Overcome Challenges from Short Read Genome Assemblies. XXVI Plant & Animal Genome, San Diego, California, USA

Bremer B (2009) In: Hedges SB, Kumar S (eds) The timetree of life. Oxford University Press, New York, pp 177–187

Cavagnaro PF, Chung S-M, Szklarczyk M, Grzebelus D, Senalik D, Atkins AE, Simon PW (2009)

Characterization of a deep-coverage carrot (*Daucus carota* L.) BAC library and initial analysis of BAC-end sequences. Mol Genet Genomics 281:273–288

Dai X, Sinharoy S, Udvardi M, Zhao PX (2013) PlantTFcat: an online plant transcription factor and transcriptional regulator categorization and analysis tool. BMC Bioinformatics 14:321

Denoeud F, Carretero-Paulet L, Dereeper A, Droc G, Guyot R, Pietrella M, Zheng C, Alberti A, Anthony F, Aprea G, Aury J-M, Bento P, Bernard M, Bocs S, Campa C, Cenci A, Combes M-C, Crouzillat D, Da Silva C, Daddiego L, De Bellis F, Dussert S, Garsmeur O, Gayraud T, Guignon V, Jahn K, Jamilloux V, Joët T, Labadie K, Lan T, Leclercq J, Lepelley M, Leroy T, Li L-T, Librado P, Lopez L, Muñoz A, Noel B, Pallavicini A, Perrotta G, Poncet V, Pot D, Priyono, Rigoreau M, Rouard M, Rozas J, Tranchant-Dubreuil C, VanBuren R, Zhang Q, Andrade AC, Argout X, Bertrand B, de Kochko A, Graziosi G, Henry RJ, Jayarama, Ming R, Nagai C, Rounsley S, Sankoff D, Giuliano G, Albert VA, Wincker P, Lashermes P (2014) The coffee genome provides insight into the convergent evolution of caffeine biosynthesis. Science 345:1181–1184

Dong AX, Xin HB, Li ZJ, Liu H, Sun YQ, Nie S, Zhao ZN, Cui RF, Zhang RG, Yun QZ, Wang XN, Maghuly F, Porth I, Cong RC, Mao JF (2018) High-quality assembly of the reference genome for scarlet sage, Salvia splendens, an economically important ornamental plant. GigaScience 7:giy068

Ellison S, Luby C, Corak K, Coe K, Senalik D, Iorizzo M, Goldman I, Simon P, Dawson J (2018) Association analysis reveals the importance of the Or gene in carrot (*Daucus carota* L.) carotenoid presence and domestication Genetics:(accepted). Dawson J

Ellison S, Senalik D, Bostan H, Iorizzo M, Simon P (2017) Fine mapping, transcriptome analysis, and marker development for Y2, the gene that conditions beta-carotene accumulation in carrot (*Daucus carota.* L.). G3: Genes, Genomes, Genet:117.043067

Elsik CG, Mackey AJ, Reese JT, Milshina NV, Roos DS, Weinstock GM (2007) Creating a honey bee consensus gene set. Genome Biol 8:R13

Grzebelus D, Simon PW (2009) Diversity of *DcMaster*-like elements of the *PIF/Harbinger* superfamily in the carrot genome. Genetica 135:347–353

Grzebelus D, Yau Y-Y, Simon PW (2006) *Master*: a novel family of *PIF/Harbinger*-like transposable elements identified in carrot (*Daucus carota* L.). Mol Genet Genomics 275:450

Hirsch CN, Foerster JM, Johnson JM, Sekhon RS, Muttoni G, Vaillancourt B, Peñagaricano F, Lindquist E, Pedraza MA, Barry K (2014) Insights into the maize pan-genome and pan-transcriptome. Plant Cell 121–135

Iorizzo M, Cavagnaro P, Bostan H, Zhao Y, Zhang J, Simon PW (2018) A cluster of MYB transcription factors regulates anthocyanin biosynthesis in carrot (*Daucus carota* L.) root and petiole. Frontiers. Plant Sci 9

Iorizzo M, Ellison S, Senalik D, Zeng P, Satapoomin P, Huang J, Bowman M, Iovene M, Sanseverino W, Cavagnaro P (2016) A high-quality carrot genome assembly provides new insights into carotenoid accumulation and asterid genome evolution. Nature Genet 48:657

Iorizzo M, Senalik DA, Ellison SL, Grzebelus D, Cavagnaro PF, Allender C, Brunet J, Spooner DM, Van Deynze A, Simon PW (2013) Genetic structure and domestication of carrot (*Daucus carota* subsp. sativus) (Apiaceae). Amer J Bot 100:930–938

Iorizzo M, Senalik DA, Grzebelus D, Bowman M, Cavagnaro PF, Matvienko M, Ashrafi H, Van Deynze A, Simon PW (2011) De novo assembly and characterization of the carrot transcriptome reveals novel genes, new markers, and genetic diversity. BMC Genomics 12:389–389

Iovene M, Cavagnaro PF, Senalik D, Buell CR, Jiang J, Simon PW (2011) Comparative FISH mapping of Daucus species (Apiaceae family). Chromosome Res 19:493–506

Itoh Y, Hasebe M, Davies E, Takeda J, Ozeki Y (2003) Survival of Tdc transposable elements of the En/Spm superfamily in the carrot genome. Mol Genet Genomics 269:49–59

Jiao W-B, Schneeberger K (2017) The impact of third generation genomic technologies on plant genome assembly. Curr Opin Plant Biol 36:64–70

Just BJ, Santos CAF, Fonseca MEN, Boiteux LS, Oloizia BB, Simon PW (2007) Carotenoid biosynthesis structural genes in carrot (*Daucus carota*): isolation, sequence-characterization, single nucleotide polymorphism (SNP) markers and genome mapping. Theor Appl Genet 114:693–704

Keilwagen J, Lehnert H, Berner T, Budahn H, Nothnagel T, Ulrich D, Dunemann F (2017) The terpene synthase gene family of carrot (*Daucus carota* L.): identification of QTLs and candidate genes associated with terpenoid volatile compounds. Frontiers Plant Sci 8

Kim NH, Jayakodi M, Lee SC, Choi BS, Jang W, Lee J, Kim HH, Waminal NE, Lakshmanan M, Van Nguyen B (2018) Genome and evolution of the shade-requiring medicinal herb Panax ginseng. Plant Biotechnol J 16:1904–1917

Lang D, Weiche B, Timmerhaus G, Richardt S, Riaño--Pachón DM, Corrêa LGG, Reski R, Mueller-Roeber B, Rensing SA (2010) Genome-wide phylogenetic comparative analysis of plant transcriptional regulation: a timeline of loss, gain, expansion, and correlation with complexity. Genome Biol Evol 2:488–503

Levy YY, Mesnage S, Mylne JS, Gendall AR, Dean C (2002) Multiple roles of Arabidopsis VRN1 in vernalization and flowering time control. Science 297:243–246

Lowe TM, Eddy SR (1997) tRNAscan-SE: a program for improved detection of transfer RNA genes in genomic sequence. Nucleic Acids Res 25:955–964

Luo R, Liu B, Xie Y, Li Z, Huang W, Yuan J, He G, Chen Y, Pan Q, Liu Y, Tang J, Wu G, Zhang H,

Shi Y, Liu Y, Yu C, Wang B, Lu Y, Han C, Cheung DW, Yiu S-M, Peng S, Xiaoqian Z, Liu G, Liao X, Li Y, Yang H, Wang J, Lam T-W, Wang J (2012) SOAPdenovo2: an empirically improved memory-efficient short-read de novo assembler. GigaScience 1:18–18

Machaj G, Bostan H, Macko-Podgórni A, Iorizzo M, Grzebelus D (2018) Comparative Transcriptomics of Root Development in Wild and Cultivated Carrots. Genes 9:431

Macko-Podgórni A, Machaj G, Stelmach K et al (2017) Characterization of a genomic region under selection in cultivated carrot (*Daucus carota subsp. sativus*) reveals a candidate domestication gene. Front. Plant Sci 8:12

Macko-Podgorni A, Nowicka A, Grzebelus E, Simon PW, Grzebelus D (2013) DcSto: carrot Stowaway-like elements are abundant, diverse, and polymorphic. Genetica 141:255–267

Martin DM, Aubourg S, Schouwey MB, Daviet L, Schalk M, Toub O, Lund ST, Bohlmann J (2010) Functional annotation, genome organization and phylogeny of the grapevine (*Vitis vinifera*) terpene synthase gene family based on genome assembly, FLcDNA cloning, and enzyme assays. BMC Plant Biol 10:226–226

Mezghani N, Ruess H, Tarchoun N, Ben Amor J, Simon PW, Spooner DM (2018) Genotyping-by-sequencing reveals the origin of the Tunisian relatives of cultivated carrot (*Daucus carota*). Genet Res Crop Evol 65:1359–1368

Montenegro JD, Golicz AA, Bayer PE, Hurgobin B, Lee H, Chan CKK, Visendi P, Lai K, Doležel J, Batley J (2017) The pangenome of hexaploid bread wheat. Plant J 90:1007–1013

Nakamichi N, Kita M, Niinuma K, Ito S, Yamashino T, Mizoguchi T, Mizuno T (2007) Arabidopsis clock-associated pseudo-response regulators PRR9, PRR7 and PRR5 coordinately and positively regulate flowering time through the canonical CONSTANS-dependent photoperiodic pathway. Plant Cell Physiol 48:822–832

Nawrocki EP, Eddy SR (2013) Infernal 1.1: 100-fold faster RNA homology searches. Bioinformatics 29:2933–2935

Noh B, Lee S-H, Kim H-J, Yi G, Shin E-A, Lee M, Jung K-J, Doyle MR, Amasino RM, Noh Y-S (2004) Divergent roles of a pair of homologous jumonji/zinc-finger–class transcription factor proteins in the regulation of Arabidopsis flowering time. Plant Cell 16:2601–2613

Oleszkiewicz T, Klimek-Chodacka M, Milewska-Hendel A, Zubko M, Stróż D, Kurczyńska E, Boba A, Szopa J, Baranski R (2018) Unique chromoplast organisation and carotenoid gene expression in carotenoid-rich carrot callus. Planta 248:1455–1471

Ou CG, Mao JH, Liu LJ, Li CJ, Ren HF, Zhao ZW, Zhuang FY (2017) Characterising genes associated with flowering time in carrot (*Daucus carota* L.) using transcriptome analysis. Plant Biol 19:286–297

Ozeki Y, Davies E, Takeda J (1997) Somatic variation during long term subculturing of plant cells caused by insertion of a transposable element in a phenylalanine ammonia-lyase (PAL) gene. Mol Gen Genet 254:407–416

Parra G, Bradnam K, Korf I (2007) CEGMA: a pipeline to accurately annotate core genes in eukaryotic genomes. Bioinformatics 23:1061–1067

Perrin F, Hartmann L, Dubois-Laurent C, Welsch R, Huet S, Hamama L, Briard M, Peltier D, Gagné S, Geoffriau E (2017) Carotenoid gene expression explains the difference of carotenoid accumulation in carrot root tissues. Planta 245:737–747

Qin C, Yu C, Shen Y, Fang X, Chen L, Min J, Cheng J, Zhao S, Xu M, Luo Y, Yang Y, Wu Z, Mao L, Wu H, Ling-Hu C, Zhou H, Lin H, González-Morales S, Trejo-Saavedra DL, Tian H, Tang X, Zhao M, Huang Z, Zhou A, Yao X, Cui J, Li W, Chen Z, Feng Y, Niu Y, Bi S, Yang X, Li W, Cai H, Luo X, Montes-Hernández S, Leyva-González MA, Xiong Z, He X, Bai L, Tan S, Tang X, Liu D, Liu J, Zhang S, Chen M, Zhang L, Zhang L, Zhang Y, Liao W, Zhang Y, Wang M, Lv X, Wen B, Liu H, Luan H, Zhang Y, Yang S, Wang X, Xu J, Li X, Li S, Wang J, Palloix A, Bosland PW, Li Y, Krogh A, Rivera-Bustamante RF, Herrera-Estrella L, Yin Y, Yu J, Hu K, Zhang Z (2014) Whole-genome sequencing of cultivated and wild peppers provides insights into Capsicum domestication and specialization. Proc Natl Acad Sci USA 111:5135–5140

Que F, Wang G-L, Li T, Wang Y-H, Xu Z-S, Xiong A-S (2018) Genome-wide identification, expansion, and evolution analysis of homeobox genes and their expression profiles during root development in carrot. Functional & Integrative Genomics 18:685–700

Reyes-Chin-Wo S, Wang Z, Yang X, Kozik A, Arikit S, Song C, Xia L, Froenicke L, Lavelle DO, Truco M-J, Xia R, Zhu S, Xu C, Xu H, Xu X, Cox K, Korf I, Meyers BC, Michelmore RW (2017) Genome assembly with in vitro proximity ligation data and whole-genome triplication in lettuce. Nature Comm 8:14953

Sanseverino W, Hermoso A, D'Alessandro R, Vlasova A, Andolfo G, Frusciante L, Lowy E, Roma G, Ercolano MR (2013) PRGdb 2.0: towards a community-based database model for the analysis of R-genes in plants. Nucleic Acids Res 41:D1167–D1171

Sato S, Tabata S, Hirakawa H, Asamizu E, Shirasawa K, Isobe S, Kaneko T, Nakamura Y, Shibata D, Aoki K, Egholm M, Knight J, Bogden R, Li C, Shuang Y, Xu X, Pan S, Cheng S, Liu X, Ren Y, Wang J, Albiero A, Dal Pero F, Todesco S, Van Eck J, Buels RM, Bombarely A, Gosselin JR, Huang M, Leto JA, Menda N, Strickler S, Mao L, Gao S, Tecle IY, York T, Zheng Y, Vrebalov JT, Lee J, Zhong S, Mueller LA, Stiekema WJ, Ribeca P, Alioto T, Yang W, Huang S, Du Y, Zhang Z, Gao J, Guo Y, Wang X, Li Y, He J, Li C, Cheng Z, Zuo J, Ren J, Zhao J, Yan L, Jiang H, Wang B, Li H, Li Z, Fu F, Chen B, Han B, Feng Q, Fan D, Wang Y,

Ling H, Xue Y, Ware D, Richard McCombie W, Lippman ZB, Chia J-M, Jiang K, Pasternak S, Gelley L, Kramer M, Anderson LK, Chang S-B, Royer SM, Shearer LA, Stack SM, Rose JKC, Xu Y, Eannetta N, Matas AJ, McQuinn R, Tanksley SD, Camara F, Guigó R, Rombauts S, Fawcett J, Van de Peer Y, Zamir D, Liang C, Spannagl M, Gundlach H, Bruggmann R, Mayer K, Jia Z, Zhang J, Ye Z, Bishop GJ, Butcher S, Lopez-Cobollo R, Buchan D, Filippis I, Abbott J, Dixit R, Singh M, Singh A, Kumar Pal J, Pandit A, Kumar Singh P, Kumar Mahato A, Dogra V, Gaikwad K, Raj Sharma T, Mohapatra T, Kumar Singh N, Causse M, Rothan C, Schiex T, Noirot C, Bellec A, Klopp C, Delalande C, Berges H, Mariette J, Frasse P, Vautrin S, Zouine M, Latché A, Rousseau C, Regad F, Pech J-C, Philippot M, Bouzayen M, Pericard P, Osorio S, Fernandez del Carmen A, Monforte A, Granell A, Fernandez-Muñoz R, Conte M, Lichtenstein G, Carrari F, De Bellis G, Fuligni F, Peano C, Grandillo S, Termolino P, Pietrella M, Fantini E, Falcone G, Fiore A, Giuliano G, Lopez L, Facella P, Perrotta G, Daddiego L, Bryan G, Orozco M, Pastor X, Torrents D, van Schriek MGM, Feron RMC, van Oeveren J, de Heer P, daPonte L, Jacobs-Oomen S, Cariaso M, Prins M, van Eijk MJT, Janssen A, van Haaren MJJ, Jo S-H, Kim J, Kwon S-Y, Kim S, Koo D-H, Lee S, Hur C-G, Clouser C, Rico A, Hallab A, Gebhardt C, Klee K, Jöcker A, Warfsmann J, Göbel U, Kawamura S, Yano K, Sherman JD, Fukuoka H, Negoro S, Bhutty S, Chowdhury P, Chattopadhyay D, Datema E, Smit S, Schijlen EGWM, van de Belt J, van Haarst JC, Peters SA, van Staveren MJ, Henkens MHC, Mooyman PJW, Hesselink T, van Ham RCHJ, Jiang G, Droege M, Choi D, Kang B-C, Dong Kim B, Park M, Kim S, Yeom S-I, Lee Y-H, Choi Y-D, Li G, Gao J, Liu Y, Huang S, Fernandez-Pedrosa V, Collado C, Zuñiga S, Wang G, Cade R, Dietrich RA, Rogers J, Knapp S, Fei Z, White RA, Thannhauser TW, Giovannoni JJ, Angel Botella M, Gilbert L, Gonzalez R, Luis Goicoechea J, Yu Y, Kudrna D, Collura K, Wissotski M, Wing R, Schoof H, Meyers BC, Bala Gurazada A, Green PJ, Mathur S, Vyas S, Solanke AU, Kumar R, Gupta V, Sharma AK, Khurana P, Khurana JP, Tyagi AK, Dalmay T, Mohorianu I, Walts B, Chamala S, Brad Barbazuk W, Li J, Guo H, Lee T-H, Wang Y, Zhang D, Paterson AH, Wang X, Tang H, Barone A, Luisa Chiusano M, Raffaella Ercolano M, D'Agostino N, Di Filippo M, Traini A, Sanseverino W, Frusciante L, Seymour GB, Elharam M, Fu Y, Hua A, Kenton S, Lewis J, Lin S, Najar F, Lai H, Qin B, Qu C, Shi R, White D, White J, Xing Y, Yang K, Yi J, Yao Z, Zhou L, Roe BA, Vezzi A, D'Angelo M, Zimbello R, Schiavon R, Caniato E, Rigobello C, Campagna D, Vitulo N, Valle G, Nelson DR, De Paoli E, Szinay D, de Jong HH, Bai Y, Visser RGF, Klein Lankhorst RM, Beasley H, McLaren K, Nicholson C, Riddle C, Gianese G, Sato S, Tabata S, Mueller LA, Huang S, Du Y, Li C, Cheng Z, Zuo J, Han B, Wang Y, Ling H, Xue Y, Ware D, Richard McCombie W, Lippman ZB, Stack SM, Tanksley SD, Van de Peer Y, Mayer K, Bishop GJ, Butcher S, Kumar Singh N, Schiex T, Bouzayen M, Granell A, Carrari F, De Bellis G, Giuliano G, Bryan G, van Eijk MJT, Fukuoka H, Chattopadhyay D, van Ham RCHJ, Choi D, Rogers J, Fei Z, Giovannoni JJ, Wing R, Schoof H, Meyers BC, Khurana JP, Tyagi AK, Dalmay T, Paterson AH, Wang X, Frusciante L, Seymour GB, Roe BA, Valle G, de Jong HH, Klein Lankhorst RM (2012) The tomato genome sequence provides insights into fleshy fruit evolution. Nature 485:635

Schmieder R, Edwards R (2011) Fast identification and removal of sequence contamination from genomic and metagenomic datasets. PLoS ONE 6:e17288–e17288

Simon P (1984) Carrot genetics. Plant Mol Biol Rep 2:54–63

Smaczniak C, Immink RGH, Muiño JM, Blanvillain R, Busscher M, Busscher-Lange J, Dinh QDP, Liu S, Westphal AH, Boeren S, Parcy F, Xu L, Carles CC, Angenent GC, Kaufmann K (2012) Characterization of MADS-domain transcription factor complexes in Arabidopsis flower development. Proc Natl Acad Sci USA 109:1560–1565

Vanneste K, Baele G, Maere S, Van de Peer Y (2014) Analysis of 41 plant genomes supports a wave of successful genome duplications in association with the Cretaceous-Paleogene boundary. Genome Res 24:1334–1347

Wang F, Wang G-L, Hou X-L, Li M-Y, Xu Z-S, Xiong A-S (2018) The genome sequence of 'Kurodagosun', a major carrot variety in Japan and China, reveals insights into biological research and carrot breeding. Mol Genet Genomics 293:861–871

Wang L, Yu S, Tong C, Zhao Y, Liu Y, Song C, Zhang Y, Zhang X, Wang Y, Hua W, Li D, Li D, Li F, Yu J, Xu C, Han X, Huang S, Tai S, Wang J, Xu X, Li Y, Liu S, Varshney RK, Wang J, Zhang X (2014) Genome sequencing of the high oil crop sesame provides insight into oil biosynthesis. Genome Biol 15:R39–R39

Xu X, Pan S, Cheng S, Zhang B, Mu D, Ni P, Zhang G, Yang S, Li R, Wang J, Orjeda G, Guzman F, Torres M, Lozano R, Ponce O, Martinez D, De la Cruz G, Chakrabarti SK, Patil VU, Skryabin KG, Kuznetsov BB, Ravin NV, Kolganova TV, Beletsky AV, Mardanov AV, Di Genova A, Bolser DM, Martin DMA, Li G, Yang Y, Kuang H, Hu Q, Xiong X, Bishop GJ, Sagredo B, Mejía N, Zagorski W, Gromadka R, Gawor J, Szczesny P, Huang S, Zhang Z, Liang C, He J, Li Y, He Y, Xu J, Zhang Y, Xie B, Du Y, Qu D, Bonierbale M, Ghislain M, del Rosario Herrera M, Giuliano G, Pietrella M, Perrotta G, Facella P, O'Brien K, Feingold SE, Barreiro LE, Massa GA, Diambra L, Whitty BR, Vaillancourt B, Lin H, Massa AN, Geoffroy M, Lundback S, DellaPenna D, Robin Buell C, Sharma SK, Marshall DF, Waugh R, Bryan GJ, Destefanis M, Nagy I, Milbourne D,

Thomson SJ, Fiers M, Jacobs JME, Nielsen KL, Sønderkær M, Iovene M, Torres GA, Jiang J, Veilleux RE, Bachem CWB, de Boer J, Borm T, Kloosterman B, van Eck H, Datema E, te Lintel Hekkert B, Goverse A, van Ham RCHJ, Visser RGF (2011) Genome sequence and analysis of the tuber crop potato. Nature 475:189

Xu Z-S, Tan H-W, Wang F, Hou X-L, Xiong A-S (2014) CarrotDB: a genomic and transcriptomic database for carrot. Database: The Journal of Biological Databases and Curation:bau096.

Yau Y-Y, Simon PW (2003) A 2.5-kb insert eliminates acid soluble invertase isozyme II transcript in carrot (*Daucus carota* L.) roots, causing high sucrose accumulation. Plant Mol Biol 53:151–162

Zdobnov EM, Apweiler R (2001) InterProScan—an integration platform for the signature-recognition methods in InterPro. Bioinformatics 17:847–848

Zhou Z, Lundstrøm I, Tran-Dien A, Duchêne S, Alikhan N-F, Sergeant MJ, Langridge G, Fotakis AK, Nair S, Stenøien HK (2018) Pan-genome analysis of ancient and modern Salmonella enterica demonstrates genomic stability of the invasive para C lineage for millennia. Current Biol 28:(2420–2428) e2410

Carrot Organelle Genomes: Organization, Diversity, and Inheritance

12

David M. Spooner, Philipp W. Simon, Douglas Senalik and Massimo Iorizzo

Abstract

Cultivated carrot (*Daucus carota* subsp. *sativus*) is one of about 25–40 related wild species in the genus *Daucus* depending on the classification. It is part of a widely distributed and taxonomically complex family Apiaceae (Umbelliferae) containing 466 genera and 3820 species that is one of the largest families of seed plants. Members of the Apiaceae, particularly the genus *Daucus*, have been the subject of intensive recent molecular studies on the structure and genetics of plastids and mitochondria. This chapter summarizes organellar (plastids and mitochondria) structure, function, mutational rates, and inter-organelle DNA transfer in the Apiaceae and inheritance in the genus *Daucus*, with a wider focus on the Apiaceae and the sister family Araliaceae, and places these data in the context of other studies in the angiosperms.

D. M. Spooner (✉) · P. W. Simon · D. Senalik
USDA-Agricultural Research Service, Vegetable Crops Research Unit, Department of Horticulture, University of Wisconsin-Madison, 1575 Linden Dr., Madison, WI 53706, USA
e-mail: David.Spooner@ars.usda.gov

M. Iorizzo
Department of Horticultural Sciences, North Carolina State University, 600 Laureate Way, Kannapolis, NC 28081, USA

12.1 Plastid Structure, Mutational Rates, and Inheritance in Angiosperms

Palmer (1985) provided an early review of plastid structure and gene content, documenting, in angiosperms, (1) its relatively small size (generally 120–160 knt); (2) high copy number (as many as 1000 per cell); (3) quadripartite circular structure comprising two inverted repeats (IR), flanking a large single-copy (LSC) region and a small single-copy (SSC) region; (4) labile structure of the IR region variously shrinking and expanding in different lineages with the junction between the inverted repeat and the large single-copy region located in a generally fixed position within the 276-nt *rps 19* gene; (5) repertoire of a complete set of rRNA, tRNA, and protein-encoding genes (Fig. 12.1); (6) only rare modifications of this basic structure in parasitic plants with reduced gene content, deletion of the IR region in the Fabaceae, or extensive gene rearrangements in the Geraniaceae. In summary, most of the over 200 angiosperm chloroplast genomes examined at that time were overwhelmingly similar in size, conformation, repeat structure, gene content, and gene order and arrangement, with the predominant mode of structural evolution consisting of small deletions and insertions occurring in intergenic spacers, 5′ and 3′ untranslated regions, and in the few introns found in their genes.

P. Simon et al. (eds.), *The Carrot Genome*, Compendium of Plant Genomes,
https://doi.org/10.1007/978-3-030-03389-7_12

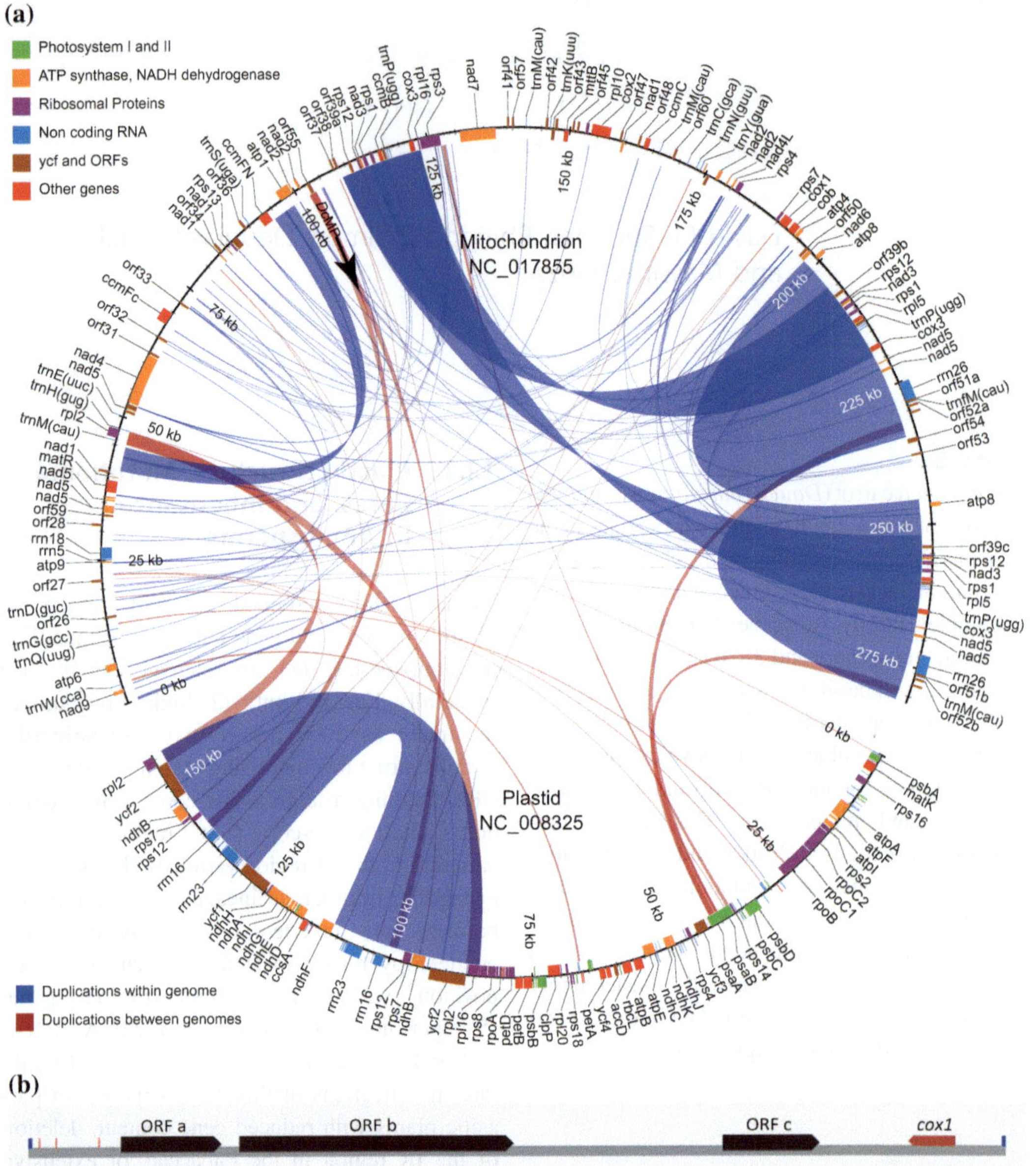

Fig. 12.1 Structure of the carrot mitochondrial and plastid genomes and inter-organelle DNA transfer; genome coordinates every 25 kb are listed inside the figure. **a** Mitochondrial (top) and plastid (bottom) genomes (visualized using Circos version 0.69-6; Krzywinski et al. 2009) and gene annotations of *Daucus carota*; these circularized genomes are drawn open to show gene transfers between them. For the plastid, only genes over 300 nt are annotated for space limitations, but these are collinear with those fully annotated in Ruhlman et al. (2006). Duplications within (blue) and between (red) genomes are shown by connected lines or ribbons. The direction of all duplications between genomes is presumed to be from plastid to mitochondrion except DcMP from mitochondrion to plastid (Iorizzo et al. 2012a, b) as labeled by the arrow. Organellar sequences and gene annotations were obtained from NCBI accessions NC_017855 (mitochondrion) and NC_008325 (plastid). Duplicated regions were detected using BLAST+ version 2.6.0 megablast program (Camacho et al. 2009) with minimum alignment length of 50, minimum percentage similarity of 80, and no dust filtering. **b** Structure of the plastid *D. carota* DcMP sequence. Open reading frames (ORFs) were detected using Open Reading Frame Finder (http://www.ncbi.nlm.nih.gov/gorf/gorf.html). The sequence was oriented according to 5′–3′ (indicated by arrows); ORF orientation is in opposite direction as related to other figures. Thick vertical blue lines indicate target site duplication (TSD). Thin red vertical lines indicate relative position of P1, P2, and P3 tnrV promoters. The red box indicates the region comprising partial sequence of cox1 gene. The scheme is drawn to scale

Palmer (1985) mentioned the maternal inheritance of plastid DNA, documented for most species by Tilney-Bassett (1978). Corriveau and Coleman (1988) developed a rapid cytological screen based on epifluorescence microscopy for maternal inheritance and examined 235 plant species from 80 angiosperm families. They detected putative plastid DNA in the generative and/or sperm cells of pollen from 43 species in 26 genera of 15 families, but not in the generative or sperm cells of pollen from the remaining 192 species (82%), strongly suggesting that they have only maternal inheritance. Their results corroborated most reports of maternal plastid inheritance, and suggested that biparental inheritance of plastids is rare, occurring in about 14% of flowering plant genera, scattered among 19% of the families examined. The carrot plastid genome follows a pattern of maternal inheritance (Vivek et al. 1999). Jansen and Ruhlman (2012) reviewed data on maternal inheritance of plastids in angiosperms and provided a similar figure (80%) for angiosperm species with maternal inheritance, the remaining 20% with biparental inheritance.

Wolfe et al. (1987) compared mutational rates among plant mitochondrial (mtDNA), plastid (cpDNA), and nuclear DNA (nDNA) sequences; and among plant and animal mitochondrial DNA sequences. He documented that (1) in contrast to mammals, where mtDNA evolves at least five times faster than nDNA, angiosperm mtDNA evolves at least five times slower than nDNA, (2) plant mtDNA undergoes much more frequent rearrangements and is larger and variable in size than mammalian mtDNA, (3) cpDNA evolves much slower than plant nDNA, and (4) DNA from the cpDNA IR region evolves much more slowly that the plant LSC or SSC regions. The relative structural conservatism and slower evolution rate of cpDNA in plants made it an ideal molecule for plant phylogenetic studies.

Early plastid phylogenetic studies were based partly on DNA restriction site procedures, but were largely replaced by massive data from next-generation DNA sequencing, stimulating the rapid accumulation of whole plastid DNA sequences. For example, Jansen and Ruhlman (2012) reported the public availability of 200 plastid genomes that as of June 2018 has grown to over 3000 (https://www.ncbi.nlm.nih.gov/genomes/GenomesGroup.cgi?taxid=2759&opt=plastid), allowing for finer comparisons of plastid DNA sequences. Raubeson and Jansen (2005) documented varying rates of change in different regions of the plastid genome, favoring phylogenetic studies at different taxonomic levels. Plastid DNA analyses (first DNA restriction site studies, and then DNA sequences from portions of the genome) dominated much of the molecular phylogenetic literature in the 1980s and 1990s. Jansen and Ruhlman (2012) documented additional lineages of both gymnosperms and angiosperms (the Campanulaceae) deviating from stability of plastid architecture, gene and intron content, and gene order across seed plants. They documented highly rearranged plastomes to exhibit three general phenomena: (1) highly accelerated rates of nucleotide substitutions, (2) an increase in the number of dispersed repeats, many of which are associated with rearranged endpoints, and (3) biparental plastid inheritance. They reviewed studies (e.g., Lilly et al. 2001) documenting deviations from the typical circular arrangement of the plastid molecule, to include multimeric circles or linear and branched structures.

The phylogenetic analysis of 81 plastid genes in 64 sequenced genomes by Jansen et al. (2007) allowed lineage-specific correlations between rates of nucleotide substitutions. They documented gene and intron content in plastids to be highly conserved among the early diverging angiosperms and basal eudicots, but found 62 independent gene and intron losses limited to the more derived monocot and eudicot clades. They showed that most angiosperm plastid genomes contain 113 different genes, 16 of which are duplicated in the inverted repeat, for a total of 129 genes. Intron content was shown to be highly conserved across angiosperms with most genomes containing 18 genes with introns. Like gene losses, intron losses were shown to be restricted to the more derived monocot and eudicot clades. Their fully resolved and strongly supported phylogenetic tree supported the genus *Amborella* as the earliest diverging lineage of flowering plants (now estimated to contain over 257,400 species classified into 52 orders and about 450 families; Judd et al. 2016), followed by the angiosperm orders Nymphaeales and Austrobaileyales, and provided strong support for a sister relationship between eudicots and monocots.

12.2 Plastid Structure in the Apiales (Apiaceae and the Sister Family Araliaceae)

Our literature survey of the in the Apiales (Table 12.1; data as of May 1, 2018) recovered 79 reports of published genomes in the Apiaceae and 33 reports (112 in total) in the Araliaceae. Like the Jansen et al. (2007) wider survey of the angiosperms, our survey of all 112 Apiales plastid genomes from these two families documents a single circular double-stranded DNA molecule, displaying the typical quadripartite structure of angiosperm plastid genomes, containing 111–114 nonduplicated genes. All plastid genomes are collinear, consistent with the rarity of recombination in plant plastomes (Palmer 1985). Total genome lengths varied from 146,512 in *Angelica nitida* to 171,083 in *Caucalis platycarpos*; with a large single-copy region from 83,553 in *Daucus crinitus* to 94,684 in *Pimpinella rhomboidea*; a small single-copy region ranging from 17,139 in *Crithmum maritimum* to 19,117 in *Schefflera delavayi*; and a pair of inverted repeats from 17,217 nt in *P. rhomboidea* to 27,993 in *C. maritimum*. Average CG contents range from 36.8% in *Eleutherococcus gracilistylus* to 38.1% in *Aralia undulata* and *Panax notoginseng*. The number of nonduplicated genes ranged from 111 in *Bupleurum falcatum* to 114 in many other species.

12.3 Plastid Structure in *Daucus* Sensu Lato

All reports of *Daucus* in its expanded sensu (sensu lato, Banasiak et al. 2016, see Chap. 2) likewise documented a typical chloroplast quadripartite circular genome consisting of a total length in nt varying from 155,441 in *Daucus involucratus* to 157,336 in *Daucus setulosus*; a large single-copy region from 83,553 in *D. crinitus* to 84,444 in *Rouya polygama*; a small single-copy region 17,314 in *R. polygama* to 17,887 in *Daucus tenuisectus*; and a pair of inverted repeats 26,924 nt in *Daucus bicolor* to 27,741 in *Daucus aureus*. Spooner et al. (2017) did not report average GC contents but they documented an inverse relationship between read coverage and GC content, most notably in the second half of the inverted repeat region, as seen in the coverage plots (Fig. 12.2). This observation is likely a reflection of the Illumina platform that introduces coverage bias in regions with high GC content (Ross et al. 2013). All reports documented 113 unique genes consisting of 80 protein-coding genes, 29 tRNA genes, and 4 RNA genes.

The inverted repeat junctions flanking the LSC were identical in all genotypes examined by Spooner et al. (2017), while those flanking the SSC were variable (Fig. 12.3). These variations form six distinct classes (A–F), with the out-group *Oenanthe virgata* (class F) having the largest fraction of the *ycf1* gene included in the inverted repeat, including a 9-nt insertion unique to this species. Relative to *Oenanthe*, class A consists of 15 accessions, which includes *D. carota*, and has a 326-nt contraction (reduction in the size of the inverted repeat); class B consisting of only *D. aureus* has the largest contraction, 422 nt; class C consisting of five accessions has a 318-nt contraction; class D consisting of 15 accessions has a 319-nt contraction; and class E consisting of only *C. platycarpos* has a 50-nt contraction. Relative to the plastid phylogeny of Spooner et al. (2017), there is a direct cladistic relationship of these inverted repeat junction classes with all accessions of *D. carota* and its immediate sister species *Pseudorlaya pumila* and *Rouya polygama* having class A; *D. aureus* class B; *D. muricatus*, *D. tenuisectus*, and *D. crinitus* class C; *D. conchitae*, *D. crinitus*, *D. glochidiatus*, *D. littoralis*, *D. pusillus*, *D. setulosus*, class D; out-group *Caucalis platycarpos* class E; and out-group *O. virgata* class F.

Table 12.1 Summary of genome statistics of fully sequenced plastids of members of the Apiaceae and sister family Araliaceae

Species	Reference	Total length in nucleotides	Large single copy	Small single copy	Percent average CG content	Inverted repeat	Number of unique (nonduplicated) genes
Apiaceae							
Anethum graveolens L.	NCBI: NC_029470	153,356					
Angelica acutiloba (Siebold & Zucc.) Kitag.	NCBI: NC_029391.1	147,074					
Angelica dahurica (Fisch.) Benth. & Hook.f.	NCBI: NC_029392	146,918					
Angelica decursiva (Miq.) Franch. & Sav.	Choi et al. (2016b)	146,719	93 256	17,497	37.56	17,983	113
Angelica gigas Nakai	Choi et al. (2016a)	146,916	93,118	17,582		18,108	113
Angelica gigas	NCBI: KX118044.1	152,185					
Angelica nitida H. Wolff	Deng et al. (2017)	146,512	93,298	18,068	37.48	17,573	113
Anthriscus cerefolium (L.) Hoffm	Downie and Jansen (2015)	154,719	84,774	17,551	37.4	26,197	
Arracacia xanthorrhiza Bancr.[a]	Alvarado et al. (2017)	143,989	49,169	17,439	37.48	31,370	106
Bupleurum boissieuanum H. Wolff	Wu et al. (2017)	156,108	86,007	17,495	37.7	26,303	112
Bupleurum falcatum L.	Shin et al. (2016)	155,989	85,912	17,517		26,280	111
Bupleurum latissimum Nakai	NCBI: NC_033346	155,621					
Carum carvi L.	NCBI: NC_029889.1	155,449					113
Caucalis platycarpos L.	Spooner et al. (2017)	171,083	85,042	17,553			113
Chuanminshen violaceum Sheh et Shan	Yuan et al. (2017)	154,529	84,171	17,800	37.8	26,279	112
Coriandrum sativum L.	NCBI: NC_029850	146,519					
Crithmum maritimum L.	Downie and Jansen (2015)	158,355	85,230	17,139	37.6	27,993	
Daucus aureus Desf.	Spooner et al. (2017)	156,984	83,655	17,846		27,741	113
Daucus bicolor Sm. in Sibth. and Sm	Spooner et al. (2017)	155,785; 155,833	84,282; 84,261	17,677; 17,677		26,924; 26,942	113; 113
Daucus capillifolius (Gilli) C. Arbizu	Spooner et al. (2017)	155,906	84,259	17,552		27,047	113
Daucus carota L. subsp. *carota*	Spooner et al. (2017)	155,676; 155,865; 155,870; 155,908; 155,909	84,102; 84,213; 84,250; 84,243; 84,243	17,503; 17,555; 17,527; 17,570; 17,571		27,035; 27,048; 27,046; 27,047; 27,047	113; 113; 113; 113; 113

(continued)

Table 12.1 (continued)

Species	Reference	Total length in nucleotides	Large single copy	Small single copy	Percent average CG content	Inverted repeat	Number of unique (nonduplicated) genes
Daucus carota subsp. *gummifer* (Syme) Hook.f.	Spooner et al. (2017)	155,857; 155,876; 155,883; 155,970	84,232; 84,257; 84,202; 84,323	17,528; 17,560; 17,594; 17,550		27,048; 27,028; 27,043; 27,048	113; 113; 113; 113
Daucus carota subsp. *maximus* (Desf.) Ball	Spooner et al. (2017)	155,870	84,250	17,527		27,046	113
Daucus carota subsp. *sativus*	Ruhlman et al. (2006)	155,911	84,243	17,571		27,048	113 (115)[b]
Daucus conchitae Greuter	Spooner et al. (2017)	155,835; 156,787; 156,821	84,227; 83,738; 83,735	17,676; 17,681; 17,682		26,966; 27,684; 27,702	113; 113; 113
Daucus crinitus Desf.	Spooner et al. (2017)	156,342; 156,388	83,553; 83,592	17,822; 17,829		27,483; 27,483	113; 113
Daucus glochidiatus (Labill.) Fisch., C. A. Mey. & Avé-Lall.	Spooner et al. (2017)	155,914	84,208	17,657		27,024	113
Daucus guttatus Sibth. and Sm.	Spooner et al. (2017)	157,194; 157,197	84,208; 84,231	17,657; 17,678		27,024; 27,644	113; 113
Daucus involucratus Sm.	Spooner et al. (2017)	155,441; 155,479	83,749; 83,738	17,717; 17,699		26,987; 27,021	113; 113
Daucus littoralis Sibth. and Sm.	Spooner et al. (2017)	156,923	83,940	17,698		27,642	113
Daucus muricatus L.	Spooner et al. (2017)	156,011; 156,052	83,905; 83,946	17,881; 17,881		27,112; 27,112	113; 113
Daucus pusillus Michx.	Spooner et al. (2017)	156,939; 157,032	84,191; 84,237	17,427; 17,451		27,667; 27,667	113; 113
Daucus setulosus Guss. ex DC.	Spooner et al. (2017)	157,292; 157,336	84,267; 84,311	17,681; 17,681		27,672 27,672	113; 113
Daucus syrticus Murb.	Spooner et al. (2017)	155,841; 155,898	84,208; 84,228	17,540; 17,585		27,046 27,042	113; 113
Daucus tenuisectus Coss. ex Batt.	Spooner et al. (2017)	156,931	83,615	17,887		27,714	113
Foeniculum vulgare Mill.	NCBI: NC_029469	153,628					
Glehnia littoralis	S-C. Lee et al. (2016b)	147,467	93 493	17,546		18,214	114
Glehnia littoralis	NCBI: KU866532	147,477					
Hansenia forbesii (H. Boissieu) Pimenov and Kljuykov	NCBI: NC_035054, NC_035056	159,287; 159,505					

(continued)

Table 12.1 (continued)

Species	Reference	Total length in nucleotides	Large single copy	Small single copy	Percent average CG content	Inverted repeat	Number of unique (nonduplicated) genes
Hansenia oviformis (R. H. Shan) Pimenov and Kljuykov	NCBI: NC_035055	157,292					
Hansenia weberbaueriana (Fedde ex H. Wolff) Pimenov and Kljuykov	NCBI: NC_035053	158,625					
Ledebouriella seseloides (Hoffm.) H. Wolff	H. O. Lee et al. (2016a)	147,880	93,222	17,324	37.5	18,667	113
Ligusticum tenuissimum (Nakai) Kitag	NCBI: NC_029394	158,500					
Notopterygium forrestii H. Wolff	Yang et al. (2017)	159,607	88,870	18,212	37.70	26,262	113
Notopterygium franchetii H. de Boissieu	Yang et al. (2017)	159,389	88,749	18,260	37.70	26,175	113
Notopterygium incisum C. C. Ting ex H. T. Chang	Yang et al. (2017)	158,684	88,260	18,232	37.70	26,096	113
Notopterygium oviforme R. H. Shan	Yang et al. (2017)	157,462	87,303	17,996	37.90	26,081	113
Oenanthe virgata Poir	Spooner et al. (2017)	154,218	84,411	17,163		26,445	
Ostericum koreanum Kitagawa	Choi et al. (2016c)	147,282	93,185	17,663	37.54	18,217	113
Pastinaca pimpinellifolia M. Bieb.	NCBI: NC_027450.1	149,758					
Petroselinum crispum (Mill.) Fuss	Downie and Jansen (2015)	152,890	86,116	17,508	37.8	24,633	
Peucedanum insolens Kitag.	NCBI: NC_033344	156,912					
Peucedanum japonicum Thunb.	NCBI: NC_034644	164,653					
Pimpinella rhomboidea var. *tenuiloba* Shan and Pu	Tan and Yu (2018)	146,655	94,684	17,537		17,217	113
Pleurospermum camtschaticum Hoffm.	NCBI: NC_033343.1	155,415					
Prangos trifida (Mill.) Herrnst. et Heyn	Samigullin et al. (2017)	153,510	86,481	17,445		24,792	113
Pseudorlaya pumila Grande	Spooner et al. (2017)	155,672	84,042	17,570		27,030	113
Pterygopleurum neurophyllum (Maxim.) Kitag.	NCBI: NC_033345.1	154,369					
Rouya polygama Coincy	Spooner et al. (2017)	155,864	84,444	17,314		27,053	113
Seseli montanum L.	Samigullin et al. (2016)	147,823	92,620	17,481	37.57	18,861	114
Tiedemannia filiformis subsp. *greenmannii* (Mathias and Constance) M. A. Feist and S. R. Downie	Downie and Jansen (2015)	154,737	84,535	17,140	37.3	26,506	

(continued)

Table 12.1 (continued)

Species	Reference	Total length in nucleotides	Large single copy	Small single copy	Percent average CG content	Inverted repeat	Number of unique (nonduplicated) genes
Araliaceae							
Aralia elata (Miq.) Seem.	Kim et al. (2017)	156,220					
Aralia undulata Hand.-Mazz.	Li et al. (2013)	156,333	86,028	18,089	38.1	26,108	114
Brassaiopsis hainla (Buch.-Ham.) Seem.	Li et al. (2013)	156,459	86,566	18,021	38.0	25,936	114
Dendropanax dentiger (Harms) Merr.	Wang et al. (2016)	156,687	86,680	18,247	38.0	25,880	114
Dendropanax morbifera H. Lev.	Kim et al. (2017)	156,366					
Eleutherococcus brachypus (Harms) Nakai	Zhang et al. (2018)	156,981	86,921	18,184		25,938	114
Eleutherococcus gracilistylus (W. W. Sm.) S. Y. Hu	Kim et al. (2016a)	156,770	86,729	18,175	36.8	25,938	113
Eleutherococcus senticosus (Rupr. & Maxim.) Maxim.	Yi et al. (2012)	156,768	86,755	18,153		25,930	
Eleutherococcus sessiliflorus (Rupr.& Maxim.) S.Y.Hu	Kim et al. (2017)	156,730					
Fatsia japonica (Thunb.) Decne. & Planch.	Chen et al. (2016)	155,613	86 487	17,866	37.91	25,929	114
Hydrocotyle sibthorpioides Lam.	Ge et al. (2017)	152,880	84,064	18,690		25,063	113
Hydrocotyle verticillata Thunb., non Turcz.	Downie and Jansen (2015)	153,207	84,352	18,739	37.6	25,058	
Kalopanax septemlobus (Thunb.) Koidz.	Li et al. (2013)	156,413	86,466	18,119	38.0	25,914	114
Metapanax delavayi (Franch.) J. Wen and Frodin	Li et al. (2013)	156,343	86,360	18,131	38.0	25,926	114
Panax bipinnatifidus Seem.	Manzanilla et al. (2018)	156,248					
Panax ginseng C. A. Mey.	Zhao et al. (2015)	156,354; 156,355	86,129; 86,130	18,007; 18,007		26,074; 26,074	114; 114
Panax ginseng	Kim et al. (2017)	156,248					
Panax japonicas C. A. Mey.	Kim et al. (2017)	156,188					
Panax notoginseng (Burk.) F. H. Chen	Dong et al. (2014)	156,387					
Panax notoginseng	Zhang et al. (2016)	156,324	86,082	18,032	38.1	26,105	114
Panax notoginseng	Kim et al. (2017)	156,466					
Panax quinquefolius L.	Han et al. (2016)	156,359	86,184	18,081	38.08	26,076	114
Panax quinquefolius	Kim et al. (2016b)	156,088					
Panax schin-seng T.Nees	Kim and Lee (2004)	156,318	86,106	18,070		26,071	114

(continued)

Table 12.1 (continued)

Species	Reference	Total length in nucleotides	Large single copy	Small single copy	Percent average CG content	Inverted repeat	Number of unique (nonduplicated) genes
Panax stipuleanatus H. T. Tsai and K. M. Feng	Manzanilla et al. (2018)	156,090					
Panax stipuleanatus	NCBI: NC_030598.1	156,064					
Panax vietnamensis Ha and Grushv.	Kim et al. (2017)	155,993					
Panax vietnamensis	Manzanilla et al. (2018)	156,022; 156,099					
Schefflera delavayi (Franch.) Harms	Li et al. (2013)	156,341	86,112	19,117	37.8	25,551	114
Schefflera octophylla (Lour.) Harms	Zong et al. (2016)	156,685	86,609	18,146	37.93	25,965	

[a]We report the numbers for *Arracacia xanthorrhiza* from Alvarado et al. (2017) but do not use them in our summaries in the text because of the atypical calculations in this paper

[b]Ruhlman et al. (2006) report 115 unique plastid genes, but Jansen et al. (2007) correct this to 113

The plastids of members of *D. carota* sensu lato have variable numbers of repeats (scanned for minimum length 30 nt) between 13 and 18, with a minimum size of 70 nt for *R. polygama* and a maximum size of 127 nt in *D. crinitus*. Twenty-five accessions share a maximum repeat size of 88, three accessions 106 nt, and two accessions 109 nt. Species in closely related clades share a larger number of repetitive sequences (Spooner et al. 2017).

12.4 Mitochondrial Structure and Function in Angiosperms

Mitochondrial DNA has the same basic role in plants as it does in other eukaryotes, encoding a small number of essential genes of the mitochondrial electron transfer chain. For the expression of these few genes, the mitochondrion has its own translation system that is also partially encoded by the mtDNA, including rRNAs, tRNAs, and a variable number of ribosomal proteins that vary across different species (Kubo and Newton 2008). A few proteins involved in the assembly of functional respiratory complexes are encoded by the plant mtDNA. However, all factors required for the maintenance of the mtDNA and the expression of its genes are encoded in the nucleus and imported from the cytosol, thus placing mtDNA replication, structural organization, and gene expression under nuclear control.

Although the number of mitochondrial genes varies little between species, the size of the mtDNA varies over more than a 100-fold, with land plant mitochondrial genomes by far the largest. Angiosperm mitogenomes are usually in the range of 200–700 kb, but can be as large as 11 Mb in *Silene conica* (Sloan et al. 2012). Although a few additional genes exist in plant mitogenomes, and several genes contain introns, these features do not contribute significantly to the large size or the size variation of plant mtDNA. Rather, most of the genome consists of noncoding sequences that are not conserved across species. Horizontal transfer seems to be responsible for the acquisition of exogenous

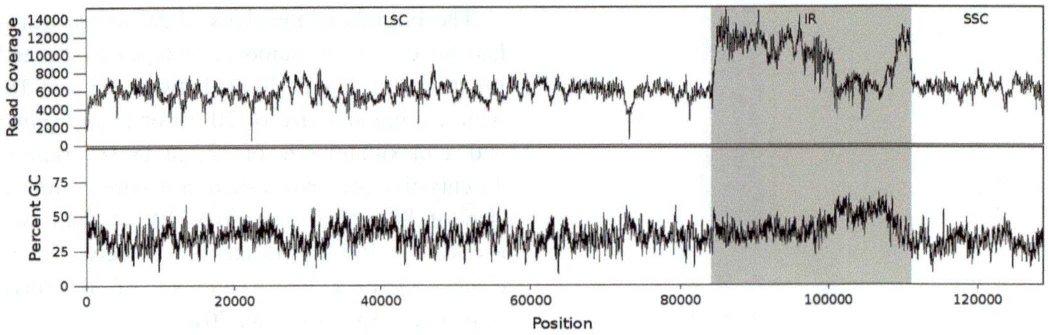

Fig. 12.2 Read coverage and percent GC plots spanning the plastid genome of *Daucus carota* subsp. *carota* PI 274297; inverted repeat regions highlighted in gray

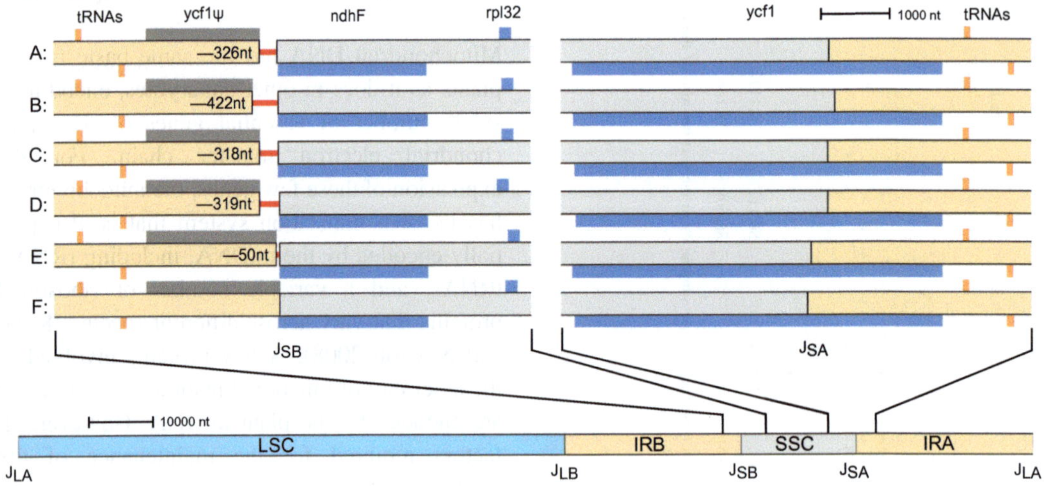

Fig. 12.3 Junctions of the inverted repeats and small single-copy plastid regions. Functional genes are represented in blue, tRNA in tan, and pseudogenes in gray. Numbers in the figure represent the number of nucleotides no longer present in inverted repeat B relative to *Oenanthe virgata*. A (*Daucus carota* NC_008325.1) is representative of 14 additional genotypes; B (*D. aureus* 319,403) is unique to this genotype; C (*D. crinitus* 652,413) is representative of four additional genotypes; D (*D. guttatus* 286,611) is representative of 14 additional genotypes; E (*Caucalis platycarpos* 649,446) and F (*O. virgata* Ames 30,293) are unique to these genotypes

sequences (Bergthorsson et al. 2003), and a fraction of plant mitogenomes can be recognized as derived from plastid, nuclear, or viral DNA. However, most noncoding sequences are of unknown origin.

The structure of angiosperm mitochondrial genomes is frequently characterized by repeat sequences (Gualberto et al. 2014). The number and the size of these repeats are important, as they influence the size of the genome, and they are the sites of intragenomic recombination, underlining evolutionary changes in mitochondrial genome organization and structural dynamism in vivo (Guo et al. 2017; Gupta et al. 2013). The repeats have often been classified as large repeats (>500 nucleotides), which can be involved in frequent homologous recombination; intermediate-size repeats (50–500 nucleotides), which are involved in infrequent ectopic homologous recombination; and small repeats (<50 nucleotides), which can promote illegitimate microhomology-mediated recombination (Arrieta-Montiel et al. 2009; Davila et al. 2011; Gualberto et al. 2014). Based on the very active recombination behavior of large repetitive sequences, early studies postulated that the entire genetic content of mtDNA could be assembled into a circular molecule, the so-called master circle, from which multiple subgenomic circular molecules are generated by intramolecular recombination across direct repeats. Although the repetitive sequences across species are not conserved, their organization and structure, which drive the recombination process, are conserved. Recent studies based on gel-based approaches or electron microscopy and quantitative sequence data from next-generation sequencing have indicated that circular and linear forms of mtDNA co-exist in vegetative tissue. Sequencing data also revealed the evolution of multichromosomal genomes associated with genome size expansion.

An economically important trait that can result from intraspecific variation promoted by recombination within mitogenomes is cytoplasmic male sterility (CMS)—the maternally transmitted inability of a plant to produce viable pollen. CMS is widespread in natural plant populations and is important for the evolution of gynodioecious species, in which females and hermaphrodites co-occur in populations (Dufay et al. 2007). In crop breeding, including in carrot it is an economically valuable trait used extensively for the production of hybrid seeds (see Chap. 3). It usually results from the expression of a chimeric gene created de novo by recombination processes, particularly microhomology-mediated recombination events, each of which involves just a few nucleotides of sequence identity. Multiple CMS phenotypes in carrot have been described and are used in breeding programs. A maternal mode of inheritance of the mitochondrial (mt)DNA has been observed in carrot CMS plants by several authors, and different genes/ORFs have been proposed to control this important trait (see Chap. 3).

Given the larger genome size relative to plastid, the diversity of repetitive sequences, and its dynamic organization, assembling mitochondrial genomes is challenging, and for this reason the number of mitochondrial genomes available is far lower than the plastomes.

12.5 Carrot Mitochondrial Genome, Structure, and Organization

In 2012, Iorizzo et al. (2012a) assembled and characterized the carrot mitochondrial genome, the first and still the only mitochondrial genome sequenced in the Apiaceae. With 281,132 nt, the carrot mitogenome is among the smallest mitochondrial genomes sequenced to date among the angiosperms and confirmed previous estimation (255,000 nt) made by Robison and Wolyn (2002) based on restriction digestion mapping. Although the genome could be assembled and represented as a master circle, Southern blot analysis confirmed the presence of two recombinant sub-circles. The overall GC content of carrot (45.4%) is comparable to other angiosperms (Alverson et al. 2011; Rodriguez-Moreno et al. 2011).

Annotation of the genome identified 44 protein-coding sequences and three ribosomal RNAs, which confirmed the previous report of Adams et al. (2002) based on Southern hybridization that surveyed mitochondrial gene presence or loss across 280 angiosperms. Truncated copies of *atp1* and *atp9* were detected, confirming observations previously reported by Bach et al. (2002). Considering a set of 51 mitochondrial conserved genes, the carrot mitogenome lack 7 genes (*sdh3*, *sdh4*, *rpl2*, *rps2*, *rps10*, *rps14*, and *rps19*), and three of them were identified in the carrot genome assembly. In addition to coding genes, the carrot mitogenome contains 18 tRNAs that recognize 15 amino acids and is missing tRNA genes for six amino acids, which are likely coded by the nuclear genome.

As expected, intergenic spacer regions represent the largest part of the genome, 224,526 nt (79.9%), with repetitive sequences occupying the majority of this space (49%). With 74 repeats ranging from 37 to 14,749 nt, the carrot mitochondrial genome has the lowest number of repeats among the sequenced plant mitochondrial genomes, which reflect its small genome size. All but one are dispersed repeats. Most of the repeats (about 90%) are between 20 and 202 nt in length accounting for just 2.0% of the total genome coverage. Nine large repeats ranging from 4220 to 14,749 nt account for 44.0% of the genome. The insertion of the large repeat 1, between repeat 2 and 3, forms a 35 kb super-repeat. After wild cabbage (Chang et al. 2011), this is the largest repeat region described in eudicot mitochondrial genomes to date. Other sequences in the intergenic spacer regions include additional open reading frames not associated with any conserved mt genes, and DNA of nuclear or plastid origin, derived from intracellular gene transfer (IGT) or possibly horizontal gene transfer (HGT), a prevalent and ongoing process in plant evolution.

12.6 Intracellular DNA Transfer in Angiosperms

While nuclear and mitochondrial genomes integrate foreign DNA via IGT and HGT, plastid genomes (plastomes) have resisted foreign DNA incorporation and only recently has IGT been uncovered in the plastomes of a few land plants. The emergence of contemporary genomics has dispelled traditional hypotheses of the sole evolution by vertical descent with modification. Drawing on phenotypic data, early investigators could not have predicted the impact of HGT on both the universality of the genetic code and diversity of organisms found on earth (Vetsigian et al. 2006). Although first recognized among eubacteria (Tatum and Lederberg 1947), HGT occurs across all domains of life and has shifted our views on the phylogeny of organisms from one of bifurcation to a more reticulate, web-like mode of evolution (Soucy et al. 2015).

Just as the sharing of DNA sequences among unrelated organisms has shaped their evolutionary history, so has the transfer of sequences among the genome-bearing compartments of individual cells shaped the evolution of eukaryotic species. Intracellular gene transfer, along with HGT, has played a pivotal role in the evolution of multicellularity and the oxygenation of earth's atmosphere, facilitating the evolution of plant and animal life (Timmis et al. 2004). The free-living, single-celled organisms that ultimately became mitochondria, and later plastids, of eukaryotic cells through endosymbiosis contained the necessary complement of genetic material for survival in the extracellular environment. Once housed within the host cell, much of that genetic material was transferred to the host nuclear genome. This massive transfer of DNA sequence fully integrated the processes of the organelles with those of the host nucleus.

Since the establishment of the cellular organelles, both mitochondrial and plastid genomes (mitogenomes and plastomes) of plants have continued to divest themselves of both coding and noncoding DNA. While mitogenomes exhibit more variation in overall size and retained gene content (Adams et al. 2002), most plastomes harbor a conserved set of coding sequences within a relatively stable size and configuration, with a small set of genes that tend to be transferred to the nucleus across the plant phylogeny (Jansen and Ruhlman 2012). The transfer of DNA sequence from both organelles to the nucleus is an ongoing process that has contributed to the evolution of the nuclear genome, regardless of whether those sequences were eventually purged from their original location or activated for their ancestral function elsewhere in the cell following nuclear transcription (Timmis et al. 2004). Likewise, plant mitogenomes contain extensive insertions of both plastid and nuclear DNA (nDNA), although, for the most part, these remain nonfunctional (Mower et al. 2012). Plastomes, however, appear to be recalcitrant to the incorporation of foreign DNA either by HGT or IGT, possibly because of the lack of an efficient DNA uptake system within plastids (Bock 2015; Richardson and Palmer 2007; Smith 2011).

Among the >3000 complete angiosperm plastomes now available in GenBank (https://www.ncbi.nlm.nih.gov/genbank/), just a few lineages have been recognized to contain DNA of nonplastome origin. Although a few studies explored putative plastome sequences with high identity to mtDNA, for the most part, the identity was due to the presence of sequences of plastid or nuclear origin in mitogenomes (Chumley et al. 2006; Ohtani et al. 2002).

The notion that land plant plastomes could incorporate foreign DNA sequences without biotechnological intervention was unheard of prior to 2009 (Goremykin et al. 2009). To date, legitimate cases of foreign DNA insertions into the plastome have been reported in four unrelated families/genus of angiosperms including *Daucus* (Iorizzo et al. 2012a), Apocynaceae (Straub et al. 2013), Bambusoideae (Ma et al. 2015), and *Anacardium* (Rabah et al. 2017). Identification of these rare events have been facilitated in part by the availability of complete mitogenome sequences. Given the wide distribution of these four families across four orders of land plants: Apiales (asterid II), Gentianales (asterid I), Sapindales (rosid II), and Poales (commelinid) combined with the lack of informative common features, suggested at least four independent events across all land plants, which likely occurred only once within each clade.

12.7 Inter-organelle DNA Transfer in the Apiaceae, a Story of First Discoveries

Goremykin et al. (2009), while analyzing the *Vitis vinifera* L. (grape) mitochondrial genome, detected two sequences of 74 and 126 nt which were similar to the carrot plastid genome (Ruhlman et al. 2006). The larger sequence has high similarity to the coding region of the mitochondrial cytochrome c oxidase subunit 1 gene (*cox1*), prompting the authors to suggest that its presence in the *Daucus* plastome might possibly represent a rare transfer of DNA from the mitochondrion into the plastid. These two sequences are contained within a large 1439-nt fragment of the *D. carota* inverted repeat at positions 99,309–100,747 and 139,407–140,845 (Ruhlman et al. 2006) that is a part of the 30rps12-trnV-GAC intergenic spacer region. This fragment, however, has no similarity to any other published plastid nucleotide region (Goremykin et al. 2009). Subsequently, Iorizzo et al. (2012a), in characterizing the entire carrot mitochondrial genome, verified the presence of this sequence in both plastid and mitochondrial genomes and designated this site as the *D. carota* mitochondrial-plastid (DcMP) region (Fig. 12.1 a). The DcMP sequence is 1452 nt-long in the carrot plastome and is present as three noncontiguous, rearranged sequences in the mitochondrial genome of *D. carota* (Iorizzo et al. 2012a). In the plastome, however, the DcMP sequence, or a large portion of it, is present only in *Daucus* (seven species) and its close relative *Cuminum* L. (cumin), both of Scandiceae subtribe Daucinae. Analysis of the plastid DcMP sequence identified three putative open reading frames (ORFs) with similarity to retrotransposon element domains (*gag* domain and reverse transcriptase) and a 6 nt direct repeat (CTTGAC), flanking the DcMP sequence, upstream of DcMP1, and downstream of DcMP4 (Fig. 12.1b) (Iorizzo et al. 2012b). These characteristics suggested that the DcMP might be a non-LTR retrotransposon and the direct repeats represent target site duplication (TSD) created because of the DcMP integration following its mobilization from a donor site localized in the mitochondrial genome. Overall, these two complementary studies demonstrated for the first time that DNA transfer from the mitochondrion to the plastid can occur in flowering plants and provided a hypothesis about its possible mode of integration.

Considering the stability of the plastid genome, it is legitimate to hypothesize that a mt-to-pt insertion within a phylogenetic clade is likely to have originated from a single event in a common ancestor, making this type of insertion useful to trace ancestry and genetic relationships within the Scandiceae tribe, which includes three subtribes Daucinae, Torilidinae, and Scandicinae. Analysis of 37 plastid genomes including members of the Daucinae and Torilidinae subtribes indicated that the DcMP region was detected in all 36 members of the Daucinae clade and in *C. platycarpos*, a member of the Torilidinae clade (Spooner et al. 2017). Comparative analysis of the DcMP region across the 37 plastid genomes revealed 21 structural variants (SVs) (insertions or deletions) (Fig. 12.4). Relative to the plastid phylogeny of Spooner et al. (2017), there is a direct cladistic relationship of these SVs with all accessions of *Daucus* and its immediate sister species

P. pumila, R. polygama, and C. platycarpos (Fig. 12.4). To expand the search for DcMP insertion within the Apiaceae, Downie and Jansen (2015) compared the plastomes of six Apiaceae species (*C. maritimum, D. carota, Hydrocotyle verticillata, Petroselinum crispum,* and *Tiedemannia filiformis* subsp. *greenmani)* including *Anthriscus cerefolium,* a member of the Scandicinae subtribe. Despite the observation that another putative insertion of mtDNA, unrelated to DcMP is present in the plastid genome of *P. crispum,* none of these six plastid genomes contain the DcMP sequence. Overall, these two studies indicated that the DcMP insertion is restricted to the Torilidinae subtribe (*C. platycarpos*) and Daucinae (36 species), which implies that within the Scandicinae tribe these two subtribes are genetically more closely related as compared with the Scandicinae subtribe where the insertion has not been detected. This hypothesis is supported by previous systematic and molecular marker work (Lee and Downie 2000; Lee et al. 2001) and confirms our hypothesis that detection of the DcMP sequence can be used as a marker to delineate relationships in this clade.

Sequence analysis of the DcMP regions detected in 36 species (Spooner et al. 2017) revealed other important aspects related to IGT in plants. Within the DcMP region, two large insertions were detected in the *C. platycarpos* plastid genome, named Cp MP5 (6663 nt) and Cp MP6 (360 nt). A large portion of the Cp MP5 sequence (KX832334 from 102,567 to 105,470) shares a high similarity (91% identity) with DCAR_022437, a nuclear gene located on carrot Chr6 annotated as an auxin response factor (ARF). The alignment covers seven of the 14 DCAR_022437 predicted exons, and none of its flanking nuclear sequences shares similarity with other plastid sequences (Fig. 12.5a). These findings represent the first evidence of a known nuclear sequence inserted in a plastid genome. Either the plastid ARF DNA sequence found in *C. platycarpos* could be part of the ancestral mitochondrial DcMP sequence, or it could have been transferred directly from the nucleus or mitochondrion into the plastid after the mt-to-pt DcMP insertion occurred. The mechanism of transfer of this nuclear DNA relative to the insertion of DcMP in the plastid genome is unknown. However, the sequence covering the DcMP and CpMP regions documented in *C. platycarpos* contains an intact *cox1* copy and fragments of *ARF* gene. Indeed, the Cp MP5 3′ end and Cp MP6 5′ end are contiguous to the pt-DcMP2 sequence and the carrot mt-Dc MP2 flanking sequences and cover the full length of the mitochondrial *cox1* gene (Fig. 12.5b). These findings indicate that direct insertion of nDNA into the plastome at the very same locus as mtDNA insertion is implausible compared with its insertion along with the mtDNA, as mitogenomes of land plants contain abundant foreign DNA from both IGT and HGT events (Knoop 2004; Alverson et al. 2010; Park et al. 2014). In particular, an *ARF* gene (*ARF17*) has been transferred to the mitogenome in several genera of Brassicaceae (Qiu et al. 2014).

In higher plants, horizontally transferred DNA is generally not functional in the recipient genome (Bock 2015; Richardson and Palmer 2007). In contrast, in carrot the DcMP sequence integrated three new functional promoters (P1, P2, and P3) located 105-, 41-, and 16-nt upstream of trnV, respectively, at the 3′—DcMP insertion junction. According to Manna et al. (1994), all three promoters are expressed in carrot cells and were responsible for the differential expression of trnV during embryogenesis. Assuming that all three promoters have a functional role, we expect their sequences to be conserved. Across all the samples harboring the pt-DcMP insertion, SVs resulted in the deletion of the P1 or P2 promoter sequences in at least one species (Spooner et al. 2017). In contrast, despite the observation that multiple independent insertion or deletion events occurred in the DcMP-4 region near the P3 promoter, its sequence is conserved across all accessions harboring the DcMP insertion (Fig. 12.4). Considering correct the hypothesis proposed by Manna et al. (1994) that the P3

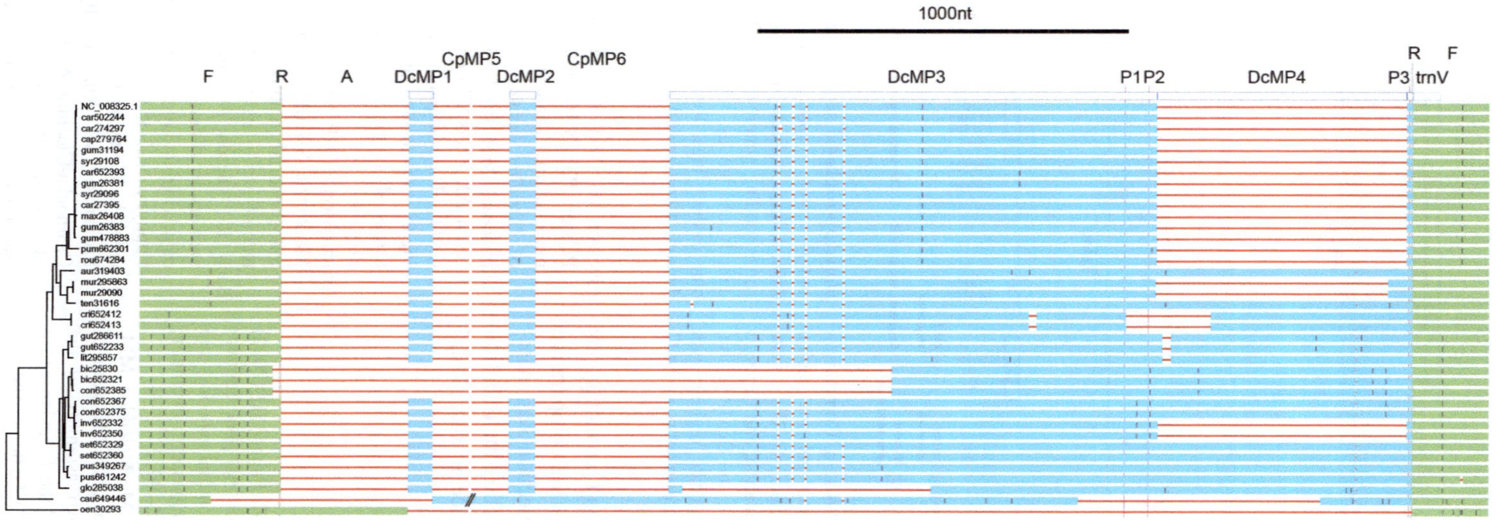

Fig. 12.4 Phylogenetic distribution and sequence comparison of plastid sequences spanning the mitochondrial-to-plastid (mt-to-pt) insertion designated as DcMP across all species included in Spooner et al. (2017). Green segments represent plastid sequence, and blue segments represent sequence of mitochondrial origin. The green region "F" designates conserved plastid sequences flanking the mt-to-pt insertion. The green region "A" designates a 339-nt region containing the ancestral promoter P4 and P5 (Tohdoh et al. 1981). DcMP1-2-3-4 (blue) designates the regions spanning the original mt-to-pt insertion described in Iorizzo et al. (2012a). CpMP5 and CpMP6 denote the two large insertions (6663 and 360 nt) identified Spooner et al. (2017) in *C. platycarpos*; trnV (green) represents the region coding for the trnV-GAC gene in the carrot plastid genome; P1, P2, and P3 indicate the location of the three putative promoters of the *D. carota* trnV (Manna et al. 1994). The vertical gray lines indicate the location of the 6-nt direct repeat flanking the DcMP insertion in *D. carota* and described in Iorizzo et al. (2012b). The double slash designates the masked portion of the 6663-nt insertion identified in *C. platycarpos*; this portion of DcMP5 insertion was masked to fit the figure in one panel. Vertical black lines indicate single-nucleotide polymorphisms (SNPs). Red lines indicate deletions identified based on the sequence alignment

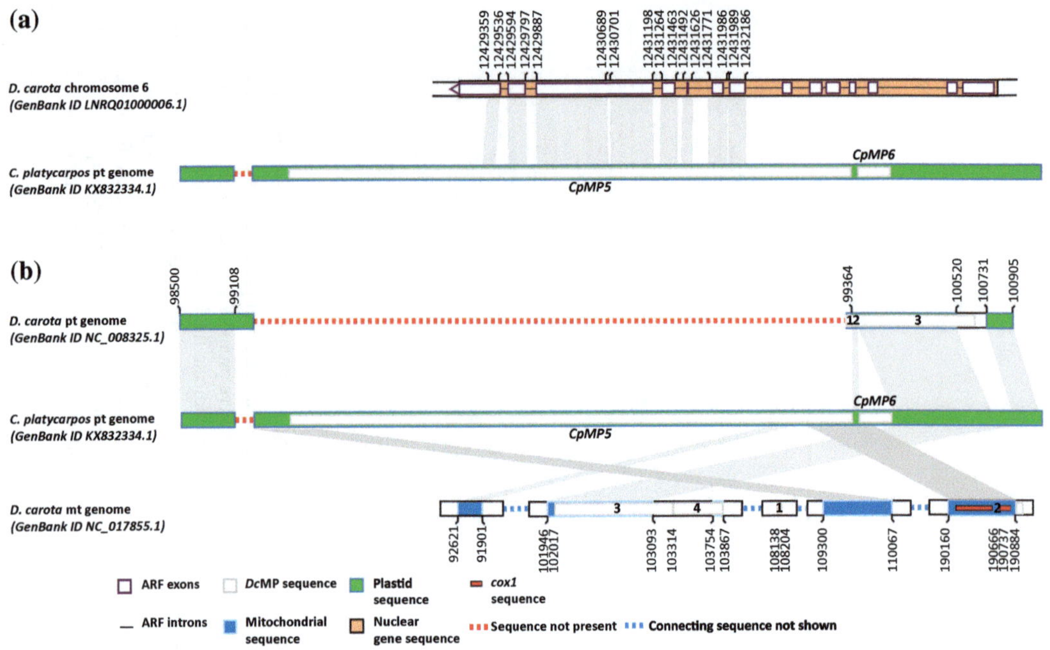

Fig. 12.5 DcMP comparative analysis. **a** Comparison between the *Daucus carota* nuclear genome region containing auxin response factor (ARF) gene DCAR_022437 in the antisense orientation, and *Caucalis platycarpos* plastid sequence spanning the DcMP region. Gray shading linking sequences indicate regions with >92% nucleotide similarity. **b** Comparison between the *C. platycarpos* plastid sequence spanning the DcMP region and *D. carota* plastid and mitochondrial genomes. Red dashed lines indicate deletions of the sequence in the corresponding genome. Mitochondrial sequences are not directly contiguous, which are represented by gaps and blue dashed lines. Regions labeled with single digits 1 through 4 correspond to DcMP regions 1 through 4

promoter plays a functional and advantageous role on the expression of trnV, the comparative studies suggest that natural selection has maintained its sequence intact promoting the retention of the ancestral DcMP sequence in the plastid genome after its first integration.

References

Adams KL, Qiu Y-L, Stoutemyer M, Palmer JD (2002) Punctuated evolution of mitochondrial gene content: high and variable rates of mitochondrial gene loss and transfer to the nucleus during angiosperm evolution. Proc Natl Acad Sci USA 99:9905–9912

Alvarado JS, López DH, Torres IM, Meléndez MM, Batista RA, Raxwal VK, Berríos Juan AN, Arun A (2017) Sequencing and de novo assembly of the complete chloroplast genome of the Peruvian carrot (*Arracacia xanthorrhiza* Bancroft). Genome Announc 5(7):e01519

Alverson AJ, Wei X, Rice DW, Stern DB, Barry K, Palmer JD (2010) Insights into the evolution of mitochondrial genome size from complete sequences of *Citrullus lanatus* and *Cucurbita pepo* (Cucurbitaceae). Mol Biol Evol 27:1436–1448

Alverson AJ, Zhuo S, Rice DW, Sloan DB, Palmer JD (2011) The mitochondrial genome of the legume *Vigna radiata* and the analysis of recombination across short mitochondrial repeats. PLoS ONE 6: e16404

Arrieta-Montiel MP, Shedge V, Davila J, Christensen AC, Mackenzie SA (2009) Diversity of the Arabidopsis mitochondrial genome occurs via nuclear-controlled recombination activity. Genetics 183:1261–1268

Bach IC, Olesen A, Simon PW (2002) PCR-based markers to differentiate the mitochondrial genomes of petaloid and male fertile carrot (*Daucus carota* L.). Euphytica 127:353–365

Banasiak Ł, Wojewódzka A, Baczyński J-P, Reduron M, Piwczński M, Kurzyna-Młynik R, Gutaker R, Czarnocka-Cieciura A, Kosmala-Grzechnik S, Spalik K (2016) Phylogeny of Apiaceae subtribe Daucinae and the taxonomic delineation of its genera. Taxon 65:563–585

Bergthorsson U, Adams KL, Thomason B, Palmer JD (2003) Widespread horizontal transfer of mitochondrial genes in flowering plants. Nature 424:197–201

Bock R (2015) Engineering plastid genomes: methods, tools, and applications in basic research and biotechnology. Annu Rev Plant Biol 66:211–241

Camacho C, Coulouris G, Avagyan V, Ma N, Papadopoulos J, Bealer K, Madden TL (2009) BLAST+: architecture and applications. BMC Bioinform 10:421

Chang S, Yang T, Du T, Huang Y, Chen J, Yan J, He J, Guan R (2011) Mitochondrial genome sequencing helps show the evolutionary mechanism of mitochondrial genome formation in *Brassica*. BMC Genom 12:497

Chen Q, Feng X, Li M, Yang B, Gao C, Zhang L, Tian J (2016) The complete chloroplast genome sequence of *Fatsia japonica* (Apiales: Araliaceae) and the phylogenetic analysis. Mitochondr DNA Part A 27:3050–3051

Choi SA, Kim Y, Kim K-Y, Kim JH, Seong RS (2016a) The complete chloroplast genome sequence of the medicinal plant, *Angelica gigas* (Apiaceae). Mitochondr DNA Part B 1:280–281

Choi SA, Kim YJ, Lee WK, Kim KY, Kim JH, Seong RS (2016b) The complete chloroplast genome of the medicinal plant *Angelica decursiva* (Apiaceae) in Peucedani Radix. Mitochondr DNA Part B 1:210–211

Choi SA, Lee WK, Kim Y, Kim KY, Kim JH, Seong RS (2016c) The complete chloroplast genome sequence of *Ostericum koreanum* (Apiaceae). Mitochondr DNA Part B 1:252–253

Chumley TW, Palmer JD, Mower JP, Fourcade HM, Calie PJ, Boore JL, Jansen RK (2006) The complete chloroplast genome sequence of *Pelargonium* × *hortorum*: organization and evolution of the largest and most highly rearranged chloroplast genome of land plants. Mol Biol Evol 23:2175–2190

Corriveau JL, Coleman AW (1988) Rapid screening method to detect potential biparental inheritance of plastid DNA and results for over 200 angiosperm species. Am J Bot 75:1443–1458

Davila JI, Arrieta-Montiel MP, Wamboldt Y, Cao J, Hagmann J, Shedge V, Xu Y-Z, Weigel D, Mackenzie SA (2011) Double-strand break repair processes drive evolution of the mitochondrial genome in Arabidopsis. BMC Biol 9:64

Deng Y-Q, Wen J, Yu Y, He X-J (2017) The complete chloroplast genome of *Angelica nitida*. Mitochondr DNA Part B 2:694–695

Dong W, Liu H, Xu C, Zuo Y, Chen Z, Zhou S (2014) A chloroplast genomic strategy for designing taxon specific DNA mini-barcodes: a case study on ginsengs. BMC Genet 15:138

Downie SR, Jansen RK (2015) A comparative analysis of whole plastid genomes from the Apiales: expansion and contraction of the inverted repeat, mitochondrial to plastid transfer of DNA, and identification of highly divergent noncoding regions. Syst Bot 40:336–351

Dufay M, Touzet P, Maurice S, Cuguen J (2007) Modelling the maintenance of male-fertile cytoplasm in a gynodioecious population. Heredity (Edinb) 99:349–356

Ge L, Shen L, Chen Q, Li X, Zhang L (2017) The complete chloroplast genome sequence of *Hydrocotyle sibthorpioides* (Apiales: Araliaceae). Mitochondr DNA Part B 2:29–30

Goremykin VV, Salamini F, Velasco R, Viola R (2009) Mitochondrial DNA of *Vitis vinifera* and the issue of rampant horizontal gene transfer. Mol Biol Evol 26:99–110

Gualberto JM, Mileshina D, Wallet C, Niazi AK, Weber-Lotfi F, Dietrich A (2014) The plant mitochondrial genome: dynamics and maintenance. Biochimie 100:107–120

Guo W, Zhu A, Fan W, Mower JP (2017) Complete mitochondrial genomes from the ferns *Ophioglossum californicum* and *Psilotum nudum* are highly repetitive with the largest organellar introns. New Phytol 213:391–403

Gupta R, Ryzhikov M, Koroleva O, Unciuleac M, Shuman S, Korolev S, Glickman MS (2013) A dual role for mycobacterial RecO in RecA-dependent homologous recombination and RecA-independent single-strand annealing. Nucleic Acids Res 41:2284–2295

Han Z-J, Li W, Liu Y, Gao L-Z (2016) The complete chloroplast genome of North American ginseng, *Panax quinquefolius*. Mitochondr DNA Part A 27:3496–3497

Iorizzo M, Grzebelus D, Senalik D, Szklarczyk M, Spooner D, Simon P (2012a) Against the traffic: the first evidence for mitochondrial DNA transfer into the plastid genome. Mobile Genet Elem 2:261–266

Iorizzo M, Senalik D, Szklarczyk M, Grzebelus D, Spooner D, Simon P (2012b) De novo assembly of the carrot mitochondrial genome using next generation sequencing of whole genomic DNA provides first evidence of DNA transfer into an angiosperm plastid genome. BMC Plant Biol 12:61

Jansen RK, Ruhlman TA (2012) Plastid genomes of seed plants. In: Bock R, Knoop V (eds) Genomics of chloroplasts and mitochondria. Springer, Netherlands, Dordrecht, pp 103–126

Jansen RK, Cai Z, Raubeson LA, Daniell H, dePamphilis CW, Leebens-Mack J, Müller KF, Guisinger-Bellian M, Haberle RC, Hansen AK, Chumley TW, Lee S-B, Peery R, McNeal JR, Kuehl JV, Boore JL (2007) Analysis of 81 genes from 64 plastid genomes resolves relationships in angiosperms and identifies genome-scale evolutionary patterns. Proc Natl Acad Sci USA 104:19369–19374

Judd WS, Campbell CS, Kellogg EA, Stevens PF, Donoghue MJ (2016) Plant systematics: a phylogenetic approach, 4th edn. Sinauer Associates, Sunderland

Kim K-J, Lee H-L (2004) Complete chloroplast genome sequences from Korean ginseng (*Panax schinseng*

Nees) and comparative analysis of sequence evolution among 17 vascular plants. DNA Res 11:247–261

Kim K, Lee J, Lee S-C, Kim N-H, Jang W, Kim S, Sung S, Lee J, Yang T-J (2016a) The complete chloroplast genome of *Eleutherococcus gracilistylus* (W.W.Sm.) S.Y.Hu (Araliaceae). Mitochondr DNA Part A 27:3741–3742

Kim K, Lee S-C, Lee J, Kim N-H, Jang W, Yang T-J (2016b) The complete chloroplast genome sequence of *Panax quinquefolius* (L.). Mitochondr DNA Part A 27:3033–3034

Kim K, Nguyen VB, Dong J, Wang Y, Park JY, Lee S-C, Yang T-J (2017) Evolution of the Araliaceae family inferred from complete chloroplast genomes and 45S nrDNAs of 10 *Panax*-related species. Sci Rep 7:4917

Knoop V (2004) The mitochondrial DNA of land plants: peculiarities in phylogenetic perspective. Curr Genet 46:123–139

Krzywinski M, Schein J, Birol İ, Connors J, Gascoyne R, Horsman D, Jones SJ, Marra MA (2009) Circos: an information aesthetic for comparative genomics. Genome Res 19:1639–1645

Kubo T, Newton KJ (2008) Angiosperm mitochondrial genomes and mutations. Mitochondrion 8:5–14

Lee BY, Downie SR (2000) Phylogenetic analysis of cpDNA restriction sites and rps16 intron sequences reveals relationships among Apiaceae tribes Caucalideae, Scandiceae and related taxa. Pl Syst Evol 221:35–60

Lee B-Y, Levin GA, Downie SR (2001) Relationships within the spiny-fruited Umbellifers (Scandiceae subtribes Daucinae and Torilidinae) as assessed by phylogenetic analysis of morphological characters. Syst Bot 26:622–642

Lee HO, Kim K, Lee S-C, Lee J, Lee J, Kim S, Yang T-J (2016a) The complete chloroplast genome sequence of *Ledebouriella seseloides* (Hoffm.) H. Wolff. Mitochondr DNA Part A 27:3498–3499

Lee S-C, Oh Lee H, Kim K, Kim S, Yang T-J (2016b) The complete chloroplast genome sequence of the medicinal plant *Glehnia littoralis* F.Schmidt ex Miq. (Apiaceae). Mitochondr DNA Part A 27:3674–3675

Li R, Ma P-F, Wen J, Yi T-S (2013) Complete sequencing of five Araliaceae chloroplast genomes and the phylogenetic implications. PLoS ONE 8:e78568

Lilly JW, Havey MJ, Jackson SA, Jiang J (2001) Cytogenomic analyses reveal the structural plasticity of the chloroplast genome in higher plants. Plant Cell 13:245–254

Ma PF, Zhang YX, Guo ZH, Li DZ (2015) Evidence for horizontal transfer of mitochondrial DNA to the plastid genome in a bamboo genus. Sci Rep 5:11608

Manna F, Massardo DR, Wolf K, Luccarini G, Carlomagno MS, Rivellini F, Alifano P, Del Giudice L (1994) A tRNA gene mapping within the chloroplast rDNA cluster is differentially expressed during the development of *Daucus carota*. Nucleic Acids Res 22:1712–1718

Manzanilla V, Kool A, Nguyen Nhat L, Van Nong H, Le Thi ThuH, de Boer HJ (2018) Phylogenomics and barcoding of *Panax*: toward the identification of ginseng species. BMC Evol Biol 18:44

Mower JP, Sloan DB, Alverson AJ (2012) Plant mitochondrial genome diversity: the genomics revolution. In: Wendel JF, Greilhuber J, Dolezel J, Leitch IJ (eds) Plant genome diversity, vol 1. Plant genomes, their residents, and their evolutionary dynamics. Springer, Vienna, pp 123–144

Ohtani K, Yamamoto H, Akimitsu K (2002) Sensitivity to *Alternaria alternata* toxin in citrus because of altered mitochondrial RNA processing. Proc Natl Acad Sci USA 99:2439–2444

Palmer JD (1985) Comparative organization of chloroplast genomes. Ann Rev Genet 19:325–354

Park S, Ruhlman TA, Sabir JS, Mutwakil MH, Baeshen MN, Sabir MJ, Baeshen NA, Jansen RK (2014) Complete sequences of organelle genomes from the medicinal plant *Rhazya stricta* (Apocynaceae) and contrasting patterns of mitochondrial genome evolution across asterids. BMC Genom 15:405

Qiu Y, Filipenko SJ, Darracq A, Adams KL (2014) Expression of a transferred nuclear gene in a mitochondrial genome. Curr Plant Biol 1:68–72

Rabah SO, Lee C, Hajrah NH, Makki RM, Alharby HF, Alhebshi AM, Sabir JSM, Jansen RK, Ruhlman TA (2017) Plastome sequencing of ten nonmodel crop species uncovers a large insertion of mitochondrial DNA in cashew. Plant Genome. https://doi.org/10.3835/plantgenome2017.03.0020

Raubeson LA, Jansen RK (2005) Chloroplast genomes of plants. In: Henry RJ (ed) Plant diversity and evolution: phenotypic variation in higher plants. CABI Publishing, Wallingford, UK, pp 45–68

Richardson AO, Palmer JD (2007) Horizontal gene transfer in plants. J Exp Bot 58:1–9

Robison MM, Wolyn DJ (2002) Complex organization of the mitochondrial genome of petaloid CMS carrot. Mol Genet Genom 268:232–239

Rodríguez-Moreno L, González VM, Benjak A, Martí MC, Puigdomènech P, Aranda MA, Garcia-Mas J (2011) Determination of the melon chloroplast and mitochondrial genome sequences reveals that the largest reported mitochondrial genome in plants contains a significant amount of DNA having a nuclear origin. BMC Genom 12:424

Ross MG, Russ C, Costello M, Hollinger A, Lennon NJ, Hegarty R, Nusbaum C, Jaffe DB (2013) Characterizing and measuring bias in sequence data. Genome Biol 14:R51

Ruhlman T, Lee S-B, Jansen RK, Hostetler JB, Tallon LJ, Town CD, Daniell H (2006) Complete plastid genome sequence of *Daucus carota*: implications for biotechnology and phylogeny of angiosperms. BMC Genom 7:222

Samigullin TH, Logacheva MD, Terenteva EI, Degtjareva GV, Vallejo-Roman CM (2016) Plastid genome of *Seseli montanum*: complete sequence and comparison with plastomes of other members of the Apiaceae family. Biochem (Moscow) 81:981–985

Samigullin TH, Logacheva MD, Degtjareva GV, Terentieva EI, Vallejo-Roman CM (2017) Complete plastid genome of critically endangered plant *Prangos trifida* (Apiaceae: Apioideae). Conserv Genet Res

Shin D-H, Lee J-H, Kang S-H, Ahn B-O, Kim C-K (2016) The complete chloroplast genome of the Hare's Ear root *Bupleurum falcatum*: its molecular features. Genes 7:20

Sloan DB, Alverson AJ, Chuckalovcak JP, Wu M, McCauley DE, Palmer JD, Taylor DR (2012) Rapid evolution of enormous, multichromosomal genomes in flowering plant mitochondria with exceptionally high mutation rates. PLoS Biol 10:e1001241

Smith DR (2011) Extending the limited transfer window hypothesis to inter-organelle DNA migration. Genome Biol Evol 3:743–748

Soucy SM, Huang J, Gogarten JP (2015) Horizontal gene transfer: building the web of life. Nat Rev Genet 16:472–482

Spooner DM, Ruess H, Iorizzo M, Senalik D, Simon P (2017) Entire plastid phylogeny of the carrot genus (*Daucus*, Apiaceae): concordance with nuclear data and mitochondrial and nuclear DNA insertions to the plastid. Am J Bot 104:296–312

Straub SC, Cronn RC, Edwards C, Fishbein M, Liston A (2013) Horizontal transfer of DNA from the mitochondrial to the plastid genome and its subsequent evolution in milkweeds (Apocynaceae). Genome Biol Evol 5:1872–1885

Tan J, Yu Y (2018) The complete chloroplast genome of *Pimpinella rhomboidea* var. *tenuiloba*. Mitochondr DNA Part B 3:101–102

Tatum EL, Lederberg J (1947) Gene recombination in the bacterium *Escherichia coli*. J Bacteriol 53:673–684

Tilney-Bassett RAE (1978) The inheritance and genetic behavior of plastids. In: Kirk JTO, Tilney-Bassett RAE (eds) The plastids. Elsevier/North Holland, Amsterdam, pp 251–324

Timmis JN, Ayliffe MA, Huang CY, Martin W (2004) Endosymbiotic gene transfer: organelle genomes forge eukaryotic chromosomes. Nat Rev Genet 5:123–135

Tohdoh N, Shinozaki K, Sugiura M (1981) Sequence of a putative promoter region for the rRNA genes of tobacco chloroplast DNA. Nucleic Acids Res 9:5399–5406

Vetsigian K, Woese C, Goldenfeld N (2006) Collective evolution and the genetic code. Proc Natl Acad Sci USA 103:10696–10701

Vivek BS, Ngo QA, Simon PW (1999) Evidence for maternal inheritance of the chloroplast genome in cultivated carrot (*Daucus* carota L. ssp. *sativus*). Theor Appl Genet 98:669–672

Wang L, Du X-J, Li X-F (2016) The complete chloroplast genome sequence of the evergreen plant *Dendropanax dentiger* (Araliaceae). Mitochondr DNA Part A 27:4193–4194

Wolfe KH, Li WH, Sharp PM (1987) Rates of nucleotide substitution vary greatly among plant mitochondrial, chloroplast, and nuclear DNAs. Proc Natl Acad Sci USA 84:9054–9058

Wu Y, Zhang T-Z, Qiu D-Y, Chai Q, Fan W-B, Li Z-H, Fang M-F (2017) Complete plastid genome of *Bupleurum boissieuanum*, an endemic herb plant in western China. Conserv Genet Resour. https://doi.org/10.1007/s12686-017-0890-2

Yang J, Yue M, Niu C, Ma X-F, Li Z-H (2017) Comparative analysis of the complete chloroplast genome of four endangered herbals of *Notopterygium*. Genes 8:124

Yi D-K, Lee H-L, Sun B-Y, Chung MY, Kim K-J (2012) The complete chloroplast DNA sequence of *Eleutherococcus senticosus* (Araliaceae); comparative evolutionary analyses with other three asterids. Mol Cells 33:497–508

Yuan C, Zhong W, Mou F, Gong Y, Pu D, Ji P, Huang H, Yang Z Zhang C (2017) The complete chloroplast genome sequence and phylogenetic analysis of *Chuanminshen* (*Chuanminshen* violaceum Sheh et Shan). Physiol Mol Biol Plants 23:35–41

Zhang D, Li W, Gao C, Liu Y, Gao L (2016) The complete plastid genome sequence of *Panax notoginseng*, a famous traditional Chinese medicinal plant of the family Araliaceae. Mitochondr DNA Part A 27:3438–3439

Zhang Y-J, Gao H, Chen Y, Guo F-X, Bai G, Guo Y-Y, Yan F, Wang E-J (2018) Characterization of the complete plastid genome sequence of *Eleutherococcus brachypus* (Araliaceae), an endangered shrub in China. Conserv Genet Resour. https://doi.org/10.1007/s12686-018-1012-5

Zhao Y, Yin J, Guo H, Zhang Y, Xiao W, Sun C, Wu J, Qu X, Yu J, Wang X, Xiao J (2015) The complete chloroplast genome provides insight into the evolution and polymorphism of *Panax ginseng*. Frontiers Plant Sci 5:696

Zong X, Song J, Lv J, Wang S (2016) The complete chloroplast genome sequence of *Schefflera octophylla*. Mitochondr DNA Part A 27:4685–4686

Carrot Genetics, Omics and Breeding Toolboxes

13

Hamed Bostan, Douglas Senalik, Philipp W. Simon and Massimo Iorizzo

Abstract

Today, researchers routinely generate and analyze large and complex omics, genetics and breeding datasets for both model and nonmodel crop species including carrot. This has resulted in the massive production and availability of omics data, which opened multiple challenges to store, organize and make those data available to the research and breeding communities. The value of these resources increases significantly when it is organized, annotated, effectively integrated with other data and made available to browse, query and analyze. In this chapter, we summarize the available omics, genetics and breeding resources for carrot and other *Daucus* species in different public and private databases. We also discuss the challenges for collecting, integrating and interpreting this data with a focus on the lack of dedicated, centralized and user-friendly bioinformatics platforms, breeding toolboxes and infrastructures for the carrot genome.

13.1 Introduction to "Omics" Data Resources

The last decade has witnessed a remarkable advancement of new technologies and high-throughput methods across all facets of plant biology (Suravajhala et al. 2016). This has enabled the study of molecular components and their interactions with a significant high resolution. Over time, these high-throughput techniques have become cost-effective and affordable for any crop, which has resulted in a massive production of "omics" data (Gomez-Cabrero et al. 2014). "Omics" resources are referring to the data collections produced by one of the biological sciences ending with "-omics" such as genomics, transcriptomics, proteomics, metabolomics, metagenomics and phenomics. The suffix "-ome" used in molecular biology reflects the concept of "a totality of some sort" which aims to identify, characterize and quantify all of the possible biological components of a specific molecular system involved in the structure, function and dynamics of a cell, tissue or organism (Keusch 2006; Vailati-Riboni et al. 2017). Such data not only enables a deeper investigation and annotation of

H. Bostan
Plants for Human Health Institute, North Carolina State University, 600 Laureate Way, 28081 Kannapolis, NC, USA

D. Senalik · P. W. Simon
USDA, Agricultural Research Service, Vegetable Crops Research Unit, Department of Horticulture, University of Wisconsin, 1575 Linden Dr, 53706 Madison, WI, USA

M. Iorizzo (✉)
Department of Horticultural Sciences, Plants for Human Health Institute, North Carolina State University, 600 Laureate Way, 28081 Kannapolis, NC, USA
e-mail: miorizz@ncsu.edu

P. Simon et al. (eds.), *The Carrot Genome*, Compendium of Plant Genomes,
https://doi.org/10.1007/978-3-030-03389-7_13

different molecular aspects from the target organisms, but may also allow the cross-annotation and characterization of similar aspects in closely related species (Arnold et al. 2006; Korf et al. 2001; Reed et al. 2006). Other than what are traditionally named omic resources described above, to date, scientists routinely generate and analyze larger and ever more complex genomic, genetic and breeding (GGB) datasets for both model and nonmodel crop species. Genetic and breeding datasets include genotyping and phenotyping data used in genome-wide association studies (GWAS) or quantitative trait loci (QTLs), germplasm information, validated markers to assist breeding (Acquaah 2009; Collard et al. 2005; Collard and Mackill 2007; He et al. 2014; Paran and Zamir 2003; Poland and Rife 2012; Stevens and Rick 1986; Varshney et al. 2009; Welsh 1981; Xu et al. 2012). Besides the value of these resources, appropriately collecting, storing, processing, the integration of such data (inter- and intra-collection) into databases is a great challenge (Cambiaghi et al. 2017; Gomez-Cabrero et al. 2014; Palsson and Zengler 2010).

In the last two decades, several databases have been developed to store and use these big datasets. Initially, large databases such as the National Center for Biotechnology Information (NCBI) (https://www.ncbi.nlm.nih.gov/), the DNA Data Bank of Japan (DDBJ) (https://www.ddbj.nig.ac.jp/index-e.html) or the European Nucleotide Archive (ENA) (https://www.ebi.ac.uk/ena) were developed to serve as a major storage of data. Since use of omics data has become a routine resource in the research community, specialized databases have been developed, and new functions were added to facilitate their application in genetic and genomic studies. NCBI is the primary database to store omics data; with the increasing amount and types of omic data generated, the database structure has expanded in several divisions each hosting a specific set of data (Coordinators 2013). NCBI divisions include dedicated databases and query interfaces to various data types, for example: (1) RefSeq: a curated nonredundant sequence database of genomes, transcripts and proteins (Pruitt et al. 2005); (2) SRA: sequence read archive (Leinonen et al. 2010); (3) GEO: gene expression omnibus database (Barrett et al. 2006); and (4) dbSNP: single-nucleotide polymorphism (SNP) database (Sherry et al. 2001). In the last ten years, with the increased application and integration of genomic, genetic and breeding resources, community or crop-based databases have been developed. For example, the SOL Genomics Network (SGN) (http://sgn.cornell.edu), released in 2005, is a comparative resource for the plants of the Solanaceae family, which includes important crop and model plants such as potato (*Solanum tuberosum*), eggplant (*Solanum melongena*), pepper (*Capsicum annuum*) and tomato (*Solanum lycopersicum*). The aim of SGN was to relate these species to one another using a comparative genomics approach and to tie them to the other dicots through the fully sequenced genome of Arabidopsis (*Arabidopsis thaliana*). The database houses map and marker data for Solanaceae species, expressed sequence tag (EST) collections with computationally derived unigene sets and phenotypic information for a mutagenized tomato population. Since then, several Solanaceae genomes have been released (e.g., tomato, potato, eggplant, pepper), and a large set of genetic and genomic tools to assist plant scientists and breeding programs have been developed, which required the development of new bioinformatics infrastructure to make those data available to browse, query and analyze. In the last ten years, crop and community-based databases that integrate genetic, genomic and breeding resources have become common, such as the Cacao Genome Database (CGD) (https://www.cacaogenomedb.org/main) which was released in 2008, the genetics and genomics database for Brassica plants (BRAD) (http://brassicadb.org/) released in 2011 (Cheng et al. 2011), the Genome Database of Rosaceae (GDR) (https://www.rosaceae.org/) (Jung et al. 2007), the Tobacco Genetics and Breeding (TGD) database (http://yancao.sdau.edu.cn/tgb)

released in 2013 and the genomic, genetic and breeding database for cotton research (CottonGen) (http://www.cottongen.org/) which was released in 2014 (Yu et al. 2013). These databases provide valuable genomic and genetic resources with relevant toolboxes to mine and investigate for research and breeding purposes.

To date, omics, genetics and breeding resources developed for carrot and other *Daucus*-related species include genomic (DNA) and transcriptome (RNA) data, genes and genome annotations, genetic maps, phenomic data (visually scored or analytical data such as metabolite profiles), different collections of molecular markers used for population genetics or phylogenetic studies and molecular markers associated with economically important traits detected through QTLs or GWAS. In this chapter, we review where these resources are stored, which are made available and we will provide a prospective to what resource are additionally needed. In this chapter, we will refer to any resources developed and available for *Daucus* species.

13.2 Accessing Carrot "Omics" Data

13.2.1 Genomic Data

Genomic or DNA sequences available for carrot and the *Daucus* species include nuclear and organelle genome assemblies and their corresponding un-assembled sequences and resequencing data. All of this data is stored and available through five databases, NCBI, DDBJ, ENA, CarrotDB (http://apiaceae.njau.edu.cn:8080/carrotdb/) and Phytozome (Goodstein et al. 2011) (https://phytozome.jgi.doe.gov/pz/portal.html) (Table 13.1). Among these databases, NCBI hosts the most comprehensive collection of carrot genomic data. The data hosted in DDBJ and ENA databases is also available in NCBI.

As presented in Chap. 11, currently there are two draft nuclear genomes for carrot (*Daucus carota* subsp. *sativus*) which are publicly available. The first carrot genome assembly of an inbred orange Nantes-type carrot named "DC-27" (*D. carota* subsp. *sativus L.*) was released in 2014 and sequenced using the Roche 454GS FLX$^+$ Titanium and Illumina HiSeq 2000 sequencing technologies (Xu et al. 2014). In total, 4.7 M 454 shotgun reads and 120 M pair-end Illumina reads, corresponding to a 32× depth of coverage, were generated for this project. The assembly spanned approximately 371.6 Mb and contained 185,376 contig or scaffold sequences. These assembled sequences are available through the CarrotDB platform, while the un-assembled sequences are not publicly available.

In 2016, a high-quality chromosome-scale genome assembly of an orange Nantes-type double haploid carrot, DH1, was sequenced using Illumina sequencing technology (Iorizzo et al. 2016). In total 147.2 Gb of data, approximately 186× coverage of the genome was generated by sequencing eight pair-end libraries (three pair-end libraries with insert sizes of 170, 280, 800 nt and five mate-pair libraries with an estimated insert size of 2, 5, 10, 20 and 40 kb). This new chromosome-scale genome assembly (DCARv2) covered approximately 421.5 Mb of the genome sequence and included 4826 contigs, scaffolds and super-scaffolds. Over 362 Mb was assembled and included in 60 super-scaffolds anchored to nine pseudomolecules/chromosomes. All of the sequencing data associated with the DH1 genome project is available under NCBI umbrella bio-project accession PRJNA285926. The assembled genome sequences can be accessed through NCBI accession LNRQ00000000. The pair-end (PE) and mate-pair data used for the genome assembly process can be accessed using NCBI bio-project accession PRJNA268187.

Pair-end, BAC-end sequences and 454 sequences were also produced and used for the DH1 genome assembly. A collection of 29,875 PE BAC-end sequences with an estimated insert size of 150 ± 70 kb were used to join the assembly scaffolds into super-scaffolds. These BAC-end sequences are not available through any public database. Pair-end data from a 454 library of DH1 was also sequenced using a GS-FLX

Table 13.1 List of databases hosting different types of sequences for carrot and other *Daucus* relatives

Database	Sequences						Annotation				
	Un-assembled DNA	Assembled DNA	Un-assembled RNA	Assembled RNA	Small and noncoding RNA	Protein	Gene	Protein	Small and noncoding	GO	InterPro
NCBI	x	x	x	x	x	x	x	x	x		
ENA	x	x	x	x	x		x		x		
DDBJ	x			x	x		x				
Phytozome		x	x	x		x	x	x		x	x
CarrotDB		x		x		x	x			x	x
RoBuST	x	x	x	x			x				
PDB						x		x			
KEGG							x	x			

Genetics			Remarks	Remarks
SNP	Phenotypic/germplasm	Pathways	Publicly available	Platform accessibility
x			Open	Accessible
			Open	Accessible
			Open	Accessible
			Password protected	Accessible
	x		Open/password protected	Partially accessible
	x	x	Open	Not accessible
			Open	Accessible
		x	Open	Accessible

NCBI: National Center for Biotechnology Information; ENA: European Nucleotide Archive; DDBJ: DNA Databank of Japan; PDB: Protein Data Bank; KEGG: Kyoto Encyclopedia of Genes and Genomes

platform, generating about 0.23 Gb (0.5 M reads) of data with an insert size of 8 kb. This data was used to verify the quality of the nuclear genome assembly and to develop a de novo assembly of the organelle genomes. The 454 raw data is available through NCBI SRA accession SRX1135252. Resequencing of 35 carrot accessions (NCBI bio-samples SAMN03766317–SAMN03766351) is available through NCBI bio-project accession PRJNA291976. Whole genome resequencing of these 35 *Daucus* accessions was carried out using Illumina sequencing technology which generated 5.2–29 Gb nucleotides of sequence with an average of 106 Gb at a median depth of 14×.

Other genomic data for *Daucus* species includes 38 assembled organelle genomes (Table 13.2). Two are mitochondrial genomes (Iorizzo et al. 2016, 2012), and 36 are plastid genomes (Iorizzo et al. 2012; Ruhlman et al. 2006; Spooner et al. 2017). The first mitochondrial genome was assembled using sequencing data from a 454 and Illumina whole genome library generated from a male fertile plant of USDA carrot inbred line B493B. Five 454 shotgun sequence sets totaling ~3.8 Gb were used for the initial assembly. In addition, ~50.5 M Illumina reads (single-end 100 nt) were used to correct homopolymer ambiguity. The final mitochondrial genome assembly spans 281,132 nt and was presented as a circular conformation. The raw Illumina and 454 sequences are not available through any public database, but the assembled mitochondrial genome is available through the NCBI organelle genome as well as the ENA database (accession number JQ248574). The second mitochondrial genome was assembled using sequences from the 280 nt, 2 and 5 kb Illumina libraries described above for the DH1 genome project (SRA accessions SRX1135259, SRX1135263 and SRX1135266). This assembly resulted in a single linear molecule of 244,980 nt which is accessible through the NCBI under accession number NC_017855.

The first plastid genome for carrot was released in 2006 (Ruhlman et al. 2006). The 155,911 nt organelle genome (NCBI GenBank accession DQ898156) was assembled using 1231 high-quality reads with an average length of 808 nt, generated using Sanger sequencing technology. The DNA was extracted from the leaves of a "Half Long" carrot. The second plastid genome for carrot (NCBI GenBank accession NC_008325.1) was released in 2016 (Iorizzo et al. 2016). The resulting circular assembly was 155,848 nt and was assembled using Roche 454 reads accessible through NCBI accession SRX1135252. In another study, 34 plastid genomes from different *Daucus* accessions were assembled and released (Spooner et al. 2017). All examined materials are wild taxa except for one cultivated accession. All accessions were obtained from the United States National Plant Germplasm System. Pair-end (100 nt) sequencing was performed on either an Illumina HiSeq 2000 sequencer and v3 SBS chemistry, or HiSeq 2500 sequencer and Rapid v1 SBS chemistry. All the un-assembled and assembled sequences from this study are available through NCBI SRA and GenBank databases (Table 13.2).

In addition to the whole genome sequencing data described above, a large set of DNA sequences for *Daucus* species were produced through genotyping-by-sequencing (GBS), a reduced representation library sequencing method (see Chap. 7). Although six studies have utilized GBS in *Daucus* species to identify SNP markers for phylogenetic, linkage map construction and marker-trait association analysis (Ellison et al. 2017, 2018; Iorizzo et al. 2016; Keilwagen et al. 2017; Macko-Podgórni et al. 2017), none of these sequences are publicly available.

Results from data mining analysis, such as gene prediction or annotation, of the genomic resources described above are organized in dedicated platforms that will be discussed in detail in Sect. 13.3.

13.2.2 Transcriptomics and Proteomics

RNA or protein sequences provide knowledge about gene structure and their expression profile (Bostan and Chiusano 2015). Development of a

Table 13.2 Sequenced organelle genomes of *Daucus* species

Organelle	Daucus species	NCBI accession (bio-sample)	GenBank accession	Reference
Plastid	*D. aureus* Desf.	SAMN03766348	KX832313	Spooner et al. (2017)
	D. bicolor Sm. in Sibth.	SAMN05713676	KX832323	Spooner et al. (2017)
	D. bicolor	SAMN05713677	KX832324	Spooner et al. (2017)
	D. capillifolius (Gilli) C. Arbizu	SAMN03766340	KX832302	Spooner et al. 2017)
	D. carota L. subsp. *carota*	SAMN03766335	KX832299	Spooner et al. (2017)
	D. carota subsp. *carota*	SAMN03766343	KX832307	Spooner et al. (2017)
	D. carota subsp. *carota*	SAMN03766342	KX832308	Spooner et al. (2017)
	D. carota subsp. *carota*	SAMN03766339	KX832300	Spooner et al. (2017)
	D. carota subsp. *gummifer* (Syme) Hook. f.	SAMN03766344	KX832305	Spooner et al. (2017)
	D. carota subsp. *gummifer*	SAMN03766345	KX832304	Spooner et al. (2017)
	D. carota subsp. *gummifer*	SAMN03766351	KX832306	Spooner et al. (2017)
	D. carota subsp. *gummifer*	SAMN03766341	KX832301	Spooner et al. (2017)
	D. carota subsp. *maximus* (Desf.) Ball	SAMN03766350	KX832303	Spooner et al. (2017)
	D. carota L. subsp. *sativus* (Hoffm.) Schübl. and G. Martens	–	NC_008325.1	Spooner et al. (2017)
	D. conchitae Greuter	SAMN05713682	KX832329	Spooner et al. (2017)
	D. conchitae	SAMN05713683	KX832330	Spooner et al. (2017)
	D. conchitae	SAMN05713684	KX832331	Spooner et al. (2017)
	D. crinitus Desf.	SAMN05713670	KX832316	Spooner et al. (2017)
	D. crinitus	SAMN05713671	KX832317	Spooner et al. (2017)
	D. glochidiatus (Labill.) Fisch., C. A. Mey. and Avé-Lall.	SAMN05713675	KX832317	Spooner et al. (2017)
	D. guttatus Sibth. and Sm.	SAMN05713673	KX832319	Spooner et al. (2017)
	D. guttatus	SAMN03766349	KX832320	Spooner et al. (2017)
	D. involucratus Sm.	SAMN05713685	KX832332	Spooner et al. (2017)
	D. involucratus	SAMN05713686	KX832333	Spooner et al. (2017)
	D. littoralis Sibth. and Sm.	SAMN05713674	KX832321	Spooner et al. (2017)
	D. muricatus L.	SAMN05713668	KX832314	Spooner et al. (2017)
	D. muricatus	SAMN05713669	KX832315	Spooner et al. (2017)
	D. pusillus Michx.	SAMN05713680	KX832327	Spooner et al. (2017)
	D. pusillus	SAMN05713681	KX832328	Spooner et al. (2017)
	D. syrticus Murb.	SAMN03766346	KX832309	Spooner et al. (2017)
	D. syrticus	SAMN03766347	KX832310	Spooner et al. (2017)
	D. setulosus Guss. ex DC.	SAMN05713678	KX832325	Spooner et al. (2017)
	D. setulosus	SAMN05713679	KX832326	Spooner et al. (2017)
	D. tenuisectus Coss. ex Batt.	SAMN05713672	KX832318	Spooner et al. (2017)
	D. carota	–	NC_008325.1	Ruhlman et al. (2006)
	D. carota	–	CM004358.1	Iorizzo et al. (2016)
Mitochondrion	*D. carota*	–	NC_017855.1	Iorizzo et al. (2012)
	D. carota	–	JQ248574	Iorizzo et al. (2016)

diverse collection of RNA sequences representing multiple tissues, developmental stages, growing conditions such as biotic or abiotic stress, and genetic backgrounds is a key resource to catalog genes that are expressed and enable the study of genes that are involved in economically important phenotypes. Early RNA sequencing data for plants included expressed sequence tags (ESTs) which were obtained from one-shot sequencing of a cloned cDNA. The cDNAs used for EST generation were typically individual clones from a cDNA library that were sequenced using Sanger technology. The resulting sequence was a relatively low-quality fragment, and its length was limited to approximately 500–800 nucleotides. ESTs have been extensively used in plant genomics to study gene structure and their expression. However, the techniques used to generate ESTs were low throughput and not cost-effective, compared to next-generation sequencing (NGS) technologies. The advent of NGS has provided a cost-effective strategy to sequence the whole transcriptome without cloning, enabling the development of diverse sets of transcriptome data. This technological advancement represents the foundation to characterize the gene repertoire of several crop genomes.

To date, most of the transcriptome and protein sequences available for *Daucus* species were developed from cultivated carrot (*D. carota* subsp. *sativus*) and largely represent root tissue. Transcriptomic and proteomic data available for *Daucus* species includes ESTs, short NGS reads or RNA-seq, assembled sequences, predicted mRNA and protein sequences and sequenced protein structure. This data is available through the ENA, DDBJ, RCSB Protein Data Bank (PDB) (https://www.rcsb.org/) and the following NCBI divisions: GenBank database (https://www.ncbi.nlm.nih.gov/genbank/); expressed sequence tags database (dbEST, https://www.ncbi.nlm.nih.gov/dbEST/); Gene Expression Omnibus (GEO, https://www.ncbi.nlm.nih.gov/geo/); Sequence Read Archive (SRA, https://www.ncbi.nlm.nih.gov/sra).

To date, the largest collection of ESTs for carrot includes 18,044 sequences (Iorizzo et al. 2016). Sanger sequencing for the ESTs was carried out using two normalized cDNA libraries constructed from pooled root and leaf RNA samples collected from a carrot inbred line, B493. The two libraries included RNA fragments with size ranging from 1.0 to 2.0 kb and above 2 kb. The collection is accessible under NCBI accession numbers JG753039–JG771082. Other *D. carota* ESTs available in NCBI include 93 sequences described in four publications (Lin et al. 1996; Park et al. 2006; Tanaka et al. 2009; Zhao et al. 2000) and two unpublished projects titled: "*Daucus carota cold responsive cDNA subtraction library Daucus carota cDNA similar to Fasicated protein, mRNA sequence*" and "*Differential display analysis of genetically transformed carrot roots colonised and uncolonised by the arbuscular mycorrhizal fungus Gigaspora margarita.*"

Only one microarray transcriptomic dataset for carrot is available in NCBI GEO database. Bio-project PRJNA215221 (GEO accession GSE49873) includes a GEO dataset representing raw and processed data from six samples (two tissues with threes biological replicates) from cultivated carrots. In this study, a cross-species hybridization (CSH) approach was used to evaluate whole transcriptome changes during carotenoid accumulation in the storage root (Bowman 2012).

As of December 2018, the NCBI SRA includes 95 ($\simeq$444 Gb) short read (RNA-seq) collections that are stored under 10 bio-projects for different carrot accessions that were published from 2011 to 2018 (Table 13.3).

Bio-project PRJNA80035 includes six sets of pair-end and three sets of single-end sequences, sequenced by Illumina Genome Analyzer II (Iorizzo et al. 2011). The RNA was extracted from root and leaves of inbred lines B6274 and B7262 as well as the pool of F_4 B493xQAL recombinant inbred lines (RILs). The assembled sequences were integrated with the 18,044 ESTs and were used to annotate carrot genes, identify

SNPs or insertion/deletion (indel) and develop ESSR, KASPar markers (see Chap. 7). Bio-project PRJEB3356 includes 11 sets of pair-end sequences from four *Daucus* genetic stocks (6 cultivated carrots and 5 wild carrots) sequenced by Illumina Genome Analyzer IIx (Rong et al. 2014). All RNA samples are representing root tissue. These sequences were used for SNP discovery, phylogenetic analysis and differential expression analysis to investigate carrot domestication. Bio-project PRJNA291977 includes 20 sets of pair-end sequences, sequenced by Illumina HiSeq 2000 and representing different DH1 tissues (Iorizzo et al. 2016). The RNA was extracted from seeds, leaves (under physiological/stressed conditions in three developmental stages), root (physiological and stressed combined), callus, fibrous roots, xylem, phloem, hypocotyl, petiole, whole flowers (two developmental stages), and bracts and buds. This study provided a comprehensive transcriptome analysis for the carrot genome. These sequences were used for gene model prediction and assembly quality verification (Iorizzo et al. 2016). Bio-project PRJNA350691 includes 29 sets of pair-end sequences, sequenced by Illumina HiSeq 2000 sequencer. The plant material was extracted from the root tissue of plants from a F_4 population that was derived from a cross between USDA carrot inbred line B493, an orange-colored root line and Queen Anne's Lace (QAL), a wild-type white-rooted carrot from the USA (Ellison et al. 2017). The sequences were used for quantitative transcriptome analysis to study the gene expression profile associated with the *Y2* gene, which conditions β-carotene accumulation in carrot. Bio-project PRJNA482951 includes 18 pair-end sets sequenced by Illumina HiSeq 4000 sequencer (Machaj et al. 2018). The RNA was extracted from root tissue collected from a cultivated orange-rooted breeding line, 2874B, and a wild accession of *D. carota* subsp. *commutatus*, JKI-W232/07. Root tissue was collected at 3 developmental stages, 55 days after sowing, 110 days after sowing (developing roots) and 165 days after sowing (mature roots). Bio-project PRJNA484382 included nine sets of single-end sequences sequenced by Illumina HiSeq 2500 sequencer (Iorizzo et al. 2018). For this experiment, RNA was extracted from root and petiole tissue representing three genetic backgrounds of cultivated carrots with different pigmentations (purple and green petioles, purple and orange roots). The data was used to study the regulatory mechanisms controlling anthocyanin accumulation in carrot root and petioles.

In addition, there are four bio-projects that have not been published. Bio-project PRJEB4558 includes two sets of pair-end sequences from cultivated carrot cv. Hapa-ochon is sequenced by Illumina HiSeq 2000 sequencer. The collection includes sequences from a cDNA library for random sequencing of the whole mRNA from a carrot taproot. Bio-project PRJNA391808 includes two sets of single-end sequences, sequenced by Illumina HiSeq 2000 sequencer and represents leaf and root tissue of a "Kuroda" cultivar. Bio-project PRJNA401383 includes two sets of single-end sequences sequenced by Illumina NextSeq 550 from two carrot cultivars with a different color/pigmentation; bio-project PRJNA413468 includes two sets of single-end sequences sequenced by Illumina HiSeq 2500 and includes RNA extracted from a pool of leaves and roots from an unspecified cultivated carrot.

Besides RNA sequencing data, other transcriptome data available for carrot includes mRNA sequence structure that is based on in silico gene model prediction analysis. To date, three gene model predictions are available for carrot. The first collection is from Xu et al. (2014) that includes 78,935 predicted genes. The sequences of these predicted genes are available through CarrotDB under the "*Download*" section. Another collection, DCARv2, is available from Iorizzo et al. (2016). This collection includes 30,113 mRNA sequences and the corresponding protein translation and can be downloaded through Phytozome under the "*Bulk download*" section. Another collection of predicted genes for the DH1 genome was independently developed by NCBI and is available through the NCBI RefSeq division. This collection includes 44,485 genes or isoform models and their associated proteins and can be downloaded through NCBI nucleotide and protein databases. All three of these collections include

Table 13.3 Summary of *Daucus* species short read data available in NCBI SRA

Strategy	Source	Bio-project accession	Run accessions	Year	*Daucus* species	Bases sequenced
WGS	Genomic	PRJNA268187	SRR2148143, SRR2148141, SRR2148137, SRR2148135, SRR2148133, SRR2148131, SRR2148129, SRR2148122, SRR2147675, SRR2147674, SRR2147673	2016	*Daucus carota* subsp. *sativus*	94,367,094,902
		PRJNA291976	SRR2147184, SRR2147183, SRR2147181, SRR2147180, SRR2147153, SRR2147152, SRR2147151, SRR2147149, SRR2147124, SRR2147123, SRR2147122, SRR2147121, SRR2147120, SRR2146951, SRR2146950, SRR2146949, SRR2146948, SRR2146946, SRR2146945, SRR2146944, SRR2146943, SRR2146942, SRR2146941, SRR2146940, SRR2146939, SRR2146938, SRR2146937, SRR2146936, SRR2146935, SRR2146934, SRR2146932, SRR2146927, SRR2146926, SRR2146925, SRR2146923	2016	*Daucus carota* subsp. *capillifolius*	17,375,538,630
					Daucus carota subsp. *carota*	102,239,071,988
					Daucus carota subsp. *gummifer*	59,152,225,890
					Daucus carota subsp. *sativus*	127,072,406,500
Total						544,205,875,834

(continued)

Table 13.3 (continued)

Strategy	Source	Bio-project accession	Run accessions	Year	*Daucus* species	Bases sequenced
RNA-Seq	Transcriptomic	PRJNA80035	SRR187763, SRR187762, SRR187761, SRR187760, SRR187759, SRR187758, SRR187757, SRR187756, SRR187755	2011	*Daucus carota* subsp. *sativus*	8,898,949,596
		PRJEB3356	ERR185934, ERR185937, ERR185938, ERR185929, ERR185935, ERR185932, ERR185931, ERR185936, ERR185933, ERR185930, ERR185939,	2013	*Daucus carota*	7,017,995,400
		PRJEB4558	ERR338560, ERR338561	2015	*Daucus carota*	5,713,695,038
		PRJNA291977	SRR2148990, SRR2148999, SRR2148997, SRR2148991, SRR2148994, SRR2148989, SRR2148987, SRR2148996, SRR2148993, SRR2148998, SRR2148986, SRR2148985, SRR2148984, SRR2148988, SRR2148983, SRR2148981, SRR2148992, SRR2148979, SRR2148982, SRR2148980,	2016	*Daucus carota* subsp. *sativus*	115,911,111,692

(continued)

Table 13.3 (continued)

Strategy	Source	Bio-project accession	Run accessions	Year	*Daucus* species	Bases sequenced
		PRJNA350691	PRJNA350691, SRR2148990, SRR2148999, SRR2148997, SRR2148991, SRR2148994, SRR2148989, SRR2148987, SRR2148996, SRR2148993, SRR2148998, SRR2148986, SRR2148985, SRR2148984, SRR2148988, SRR2148983, SRR2148981, SRR2148992, SRR2148979, SRR2148982, SRR2148980,	2017	*Daucus carota*	46,621,845,253
		PRJNA391808	SRR5829255, SRR5829254	2017	*Daucus carota*	4,726,675,800
		PRJNA401383	SRR6007613, SRR6007614	2018	*Daucus carota*	3,385,315,272
		PRJNA413468	SRR6144211, SRR6144212	2018	*Daucus carota*	1,773,335,350
		PRJNA482951	SRR7601367, SRR7601366, SRR7601369, SRR7601364, SRR7601359, SRR7601358, SRR7601368, SRR7601371, SRR7601370, SRR7601357, SRR7601356, SRR7601363, SRR7601373, SRR7601372, SRR7601365, SRR7601362, SRR7601361, SRR7601360,	2018	*Daucus carota* L.	66,339,131,590
		PRJNA484382	SRR7641993, SRR7641992, SRR7641991, SRR7641990, SRR7641989, SRR7641988, SRR7641987,	2018	*Daucus carota* L.	183,887,654,750

(continued)

Table 13.3 (continued)

Strategy	Source	Bio-project accession	Run accessions	Year	*Daucus* species	Bases sequenced
			SRR7641986, SRR7641985, SRR7641984, SRR7641983, SRR7641982, SRR7641981, SRR7641980, SRR7641979, SRR7641978, SRR7641977, SRR7641976, SRR7641975, SRR7641974, SRR7641973, SRR7641972, SRR7641971, SRR7641970, SRR7641969, SRR7641968, SRR7641967, SRR7641966, SRR7641965, SRR7641964, SRR7641963, SRR7641962, SRR7641961, SRR7641960, SRR7641959, SRR7641958, SRR7641957, SRR7641956, SRR7641955, SRR7641954, SRR7641953, SRR7641952, SRR7641951, SRR7641950, SRR7641949, SRR7641948, SRR7641947, SRR7641946, SRR7641945, SRR7641944, SRR7641943, SRR7641942, SRR7641941, SRR7641940, SRR7641939, SRR7641938, SRR7641937, SRR7641936, SRR7641935, SRR7641934, SRR7641933,			

(continued)

Table 13.3 (continued)

Strategy	Source	Bio-project accession	Run accessions	Year	*Daucus* species	Bases sequenced
			SRR7641932, SRR7641931, SRR7641930, SRR7641929, SRR7641928, SRR7641927, SRR7641926, SRR7641925, SRR7641924, SRR7641923, SRR7641922, SRR7641921, SRR7641920, SRR7641919, SRR7641918, SRR7641917, SRR7641916, SRR7641915, SRR7641914, SRR7641913, SRR7641912, SRR7641911, SRR7641910, SRR7641909, SRR7641908, SRR7641907, SRR7641906, SRR7641905, SRR7641904			
Total						444,275,709,741
Amplicon	Metagenomics	PRJEB6729	ERR578896, ERR578895, ERR578894, ERR578893, ERR578892, ERR578891, ERR578812, ERR578811, ERR578810, ERR578809, ERR578808, ERR578807, ERR578728, ERR578727, ERR578726, ERR578725, ERR578724, ERR578723	2014	*Daucus carota* subsp. *sativus*	1,069,951,560
		PRJNA397712	SRR5930483	2017	*Daucus carota*	996,187
Total						1,070,051,178
Final Total						989,551,636,753

the functional annotation that will be discussed further in the "Annotations" section of this chapter. In addition, there are 185 manually annotated (reviewed) proteins reported in UniProtKB/Swiss-Prot (https://www.uniprot.org/uniprot/?query=reviewed:yes), four proteins with 3D structure in RCSB PDB and some other proteins reported in NCBI for carrot that were derived from different experimental researches. However, the majority of the proteins available for the carrot genome are in silico predicted proteins from the two carrot genome releases.

13.2.3 Annotations

The first collection of gene prediction and annotation for the carrot genome were provided through the CarrotDB publication (Xu et al. 2014). The in silico annotation of the predicted genes was performed using Blast2GO with gene ontology and orthologous analyses. The annotation file can be downloaded from CarrotDB under "*Annotation*" section of the "*Download*" page. With the second sequenced carrot genome v2 (Iorizzo et al. 2016), several genome and gene annotation tracks became available. The publication included the repeat annotation and masking of the genome sequence, the gene and protein model prediction (DCARv2) and their in silico functional annotation, and the in-depth functional annotation of different major gene families such as transcription factors, resistance genes and putative candidate genes involved in the flavonoid and anthocyanin biosynthetic pathways. The predicted genes were also annotated for the associated Pfam, KOG, GO, best hit on *A. thaliana* genes and some other major accessory annotations. In addition, it also included the prediction and annotation of some noncoding RNAs as well (Iorizzo et al. 2016). All the data mentioned can be downloaded from Phytozome v12 (Goodstein et al. 2011) under "*Bulk data*" section of the "*Daucus carota v2.0*" genome. The gene models, structures and some annotation tracks for each gene can also be visualized within the JBrowse section of the Phytozome platform for the "*Daucus carota v2.0*" genome. Later, the RefSeq gene model prediction and annotation of the carrot v2 genome were provided by the NCBI RefSeq pipeline. This data can be downloaded from NCBI database under the RefSeq division. The gene models, structures and some annotation tracks can be visualized and accessed through the NCBI genome browser associated with the carrot v2 genome, by clicking on the "*Graphics*" button for each queried RefSeq gene.

The availability of such information provided a unique platform for the in-depth characterization and annotation of the carrot genome and its functionality. For example, KEGG (Kanehisa and Goto 2000) is a reference platform for understanding high-level functions and utilities of the biological system; it takes advantage of the published gene/protein sequences and structures from completely sequenced genomes and integrates them with chemical and systemic functional information of the cell, organism and ecosystem, when available. This integrational analysis is mainly conducted using similarity searches versus functionally characterized and annotated genes/proteins from model organisms which is sometimes followed with some manual curation processes (Kanehisa and Goto 2000) (https://www.genome.jp/kegg/kegg1.html). A simple query of "carrot" in KEGG resulted in the involvement of carrot genes or a homology in 135 KEGG pathways and 131 KEGG modules (https://www.genome.jp/dbget-bin/www_bget?gn:T05350). The collection provides the catalog of several carrot genes/proteins involved in the KEGG reference enzymatic pathways/steps with their accessory annotation (e.g., sequences, Pfam, orthologs) and references (publications, reference genomes, major database accessions, etc.) when available. Such integrated and processed information is a valuable resource to investigate and decipher the carrot genome functionality.

13.2.4 Other Genetics, Phenomics or Breeding Resources

As presented in detail in Chap. 7, a large number of molecular markers and other genetic resources are available for carrot and other related species.

Table 13.4 Summary of sequence data available for *D. carota* genetic markers

Marker	Sequence source/ genotyping assay	Markers	Accession	Database	Year	Reference
SSR	GSSR: enriched SSR/PCR	156	FJ816111–FJ816266	DDBJ/EMBL/ GenBank	2011	Cavagnaro et al. (2011)
SSR	BSSR: BAC-end sequences/PCR	144	FJ147759–FJ149613	DDBJ/EMBL/ GenBank	2009, 2011	Cavagnaro et al. (2009, 2011)
SNP	GBS-SNP: genomic/GBS	78,850	PRJNA348698	NCBI dbSNP	2011	Ellison et al. (2017)

SSR: single sequence repeat; SNP: single-nucleotide polymorphism

Thousands of SSRs, SNPs and other types of molecular markers have been developed and used in carrot genetic studies. Over 21 linkage maps have been developed and include over 70 QTLs or simply inherited phenotypic markers linked with economically important traits. These studies used multiple phenotypic data including visual scores (e.g., for nematode resistance, root color, fertile/sterile flower types) and metabolite profile (e.g., carotenoid or anthocyanin content and type). Despite the large number of resources/data developed and reported in published manuscripts, the majority of those resources/data are not available and accessible through publicly available databases. As presented in Table 13.4, 300 sequences containing SSR (GSSRs and BSSRs) motifs and 78,850 SNPs are available through open-source databases. The information associated with this data is limited to the target sequence containing the polymorphisms and does not include any other information such as polymorphism rate, number of alleles, their position in the genome or any linkage map, or their association with economically important traits, which diminishes their value for possible application in other genetic studies. In some cases, data was made available through supplementary files associated with publications. For example, a set of 4000 sequences and primers that were used to validate the KASPar assay were made available through a supplementary file associated with the manuscript (Iorizzo et al. 2013). None of the linkage maps and QTLs or markers associated with economically important traits are available through any database making it very difficult to reuse the data by scientists that were not directly involved in the study. Over 23 manuscripts/ studies used analytical approaches such as NMR, GC-MS, HLPC, and LC-MS to characterize the metabolite profile of nutritionally related compounds in carrot (mainly carotenoids, anthocyanin, terpenoid) and identify molecular markers or candidate genes associated with them (Hampel et al. 2005; Ma et al. 2018; Simpson et al. 2016; Wang et al. 2015). Besides the methods and results summarized in those studies, we were not able to find any trace of such data in any of the reference metabolomics databases. In addition, we were not able to find any other metabolomics or epigenomics data in open access databases for carrot or other members of the *Daucus* genus.

Overall, this highlights that the number of available bioinformatics resources to store and integrate the available data for carrot is quite limited. In addition, the genetic or breeding toolboxes available to appropriately query and extract the data are partial or nonexistent. In the following section, we review some major bioinformatics platforms which integrate "omics" data for plants, and report the state-of-the-art bioinformatics resources for carrot.

13.3 Crop or Community-Based Databases

As indicated briefly before, some major platforms such as NCBI, ENA, DDBJ and RCSB PDB collect and store the generic data (e.g., raw sequences, assembled or/and predicted sequences, some annotation tracks, publication info) from different bio-projects. These platforms accept raw or processed data (with specific pre-defined formats to keep the databases homogeneous) and store them into dedicated partitions/divisions based on the nature of the data (e.g., NCBI dbEST for EST sequences, NCBI SRA for short NGS reads). These databases represent an important resource for long-term storage of the data and to make the data publicly available. Nonetheless, these platforms lack a lot of features and analytical toolboxes that a platform dedicated to a specific crop or groups of plants could offer to the research and breeding community. Indeed, several dedicated platforms were developed and published to fulfill this gap for specific genomes, families or group of genomes. For example, the Ensembl Plants Genome Browser (Fernández and Birney 2010) and Phytozome (Goodstein et al. 2011) provide NGS data for many different plant species, in the form of short read tracks mapped onto the corresponding reference genome sequence, and are visually accessible from the associated genome browser interface. Phytozome also provides a VISTA plot of some major or closely related species to the target genome of interest for further comparative genomics analysis. Some databases are specialized with a toolbox that facilitates access to omics, genetics and breeding data. The TIGR Rice Annotation Project (Ouyang et al. 2007) which includes data from RNA-seq collections offers tools for gene-based investigations and for genome-based views of their mapped distribution. SoyBase (Grant et al. 2009), the USDA-ARS soybean genetic database, also includes RNA-seq expression data from their collection including different libraries, offering querying capability, and comparative and clustering tools for the analyses. SGN (Fernandez-Pozo et al. 2014; Mueller et al. 2005) is a dedicated platform for exploring and visualizing "omics" data from some of the Solanaceae family members (tomato, potato, pepper, etc.) and provides access to the tomato genome browser with the possibility of visualizing different transcriptomics and expression data in the form of short reads aligned to the tomato reference genome. It also provides access to the SNP and structural variant data detected from resequencing of 150 tomato genomes (a part of the 150 tomato genome resequencing project) within a dedicated genome browser. The Potato Genomic Resource (Spud DB) (Hirsch et al. 2014) also provides potato RNA-seq data in the form of raw short read sequence files and of FPKM normalized results as defined by the Cufflinks pipeline (Trapnell et al. 2012) which is analogous to the RPKM normalization (Mortazavi et al. 2008) in Excel format, as well as visualization of expression tracks by library in a potato genome browser. The maize gene atlas (Sekhon et al. 2013) also provides a consistent RNA-seq collection of 18 tissues representing five organs. This dataset, together with microarray data from 60 unique spatially and temporally separated tissues from 11 maize organs, offers a comprehensive collection for understanding the transcriptome during maize development. NexGenEx (Bostan and Chiusano 2015) is designed to enable the exploration of NGS-based collections. The NexGenEx-Tom platform includes NGS expression data from three different tomato genotypes, their functional and GO annotation with their cross-reference to the AmiGO/GO Database (Carbon et al. 2008) with several enhanced tools to conduct DEG, GO enrichment and cluster analysis of a specific gene set of interest. The platform is also enriched with a genome browser which visualizes different EST sequence collections and unigenes from 20 Solanaceae species.

Some databases and platforms also include advanced toolboxes for breeding purposes. GDR is a central repository for curated and integrated genetics and genomics data of Rosaceae, an economically important family which includes apple, cherry, peach, pear, raspberry, rose and strawberry (Jung et al. 2007). GDR contains all publicly available Rosaceae ESTs and their assembled unigenes, the genetically anchored peach physical map, Rosaceae genetic maps and comprehensively annotated ESTs, markers and traits. Most of the published maps can be viewed through dedicated map viewer combining genetic, transcriptome and physical mapping information in an integrative view. Tracks such as ESTs, BACs, markers and traits can be queried, and the search result sites are linked to the mapping visualization tools. GDR also provides online analysis tools such as a NCBI BLAST server for the GDR datasets, a sequence assembly server and microsatellite and primer detection tools. Other annotations include putative function, microsatellites, single-nucleotide polymorphisms and anchored map position, when available. CottonGen is another curated and integrated platform providing access to publicly available genomic, genetic and breeding data for cotton (Yu et al. 2013). CottonGen provides enhanced tools for easier data sharing, mining, visualization and data retrieval of cotton research data. It contains annotated whole genome sequences, unigenes from expressed sequence tags (ESTs), markers, trait loci, genetic maps, genes, taxonomy, germplasm, publications and communication resources for the cotton community. Annotated whole genome sequences of *Gossypium raimondii* are available with aligned genetic markers and transcripts. Most of the published cotton genetic maps can be visualized and compared, and are searchable via map search tools. Other query interfaces also exist for markers, QTLs, germplasm, publications and trait evaluation data. CottonGen also provides online analysis tools such as NCBI BLAST for the CottonGen datasets. CGD developed for Cacao genome, TGD developed for tobacco genome, BRAD database developed for the Brassica plants (Cheng et al. 2011) and SGN dedicated to Solanaceae family also includes some breeding toolboxes and genetics datasets for similar purposes. Such platforms can help researchers and breeders to have ease of access to preprocessed data and integrated resources for specific genomes or genome families of interest.

As summarized above, most of the genetics and genomics resources developed for carrot are available through NCBI. Though, some data can be downloaded or visualized through other databases. Phytozome is one of the reference databases that includes carrot DCARv2 release (Iorizzo et al. 2016). This platform provides a keyword query interface, a BLAST search on the genome and a section for bulk data download of the genome sequence and its annotation tracks. Phytozome also provides a JBrowse genome browser visualizing the gene prediction and annotation, and assembly gaps across the genome. It also includes several tracks of BLASTx from some major collections (such as basal dicots, eudicots and embryophytes) and some model plant genomes (e.g., *A. thaliana*, rice). It also includes the alignment of some ESTs and assembled ESTs, and the VISTA plot of several model and nonmodel genome organisms for cross-annotation and comparative genomics. Other databases such as Ensembl Plants and NCBI also provide the carrot DCARv2 genome and its annotation in the form of raw data or a genome browser visualizing its annotation across the reference genome. In the last ten years, two crop or community-based platforms have been developed to organize carrot data. RoBuST database was a bioinformatics platform dedicated to root and bulb vegetables (RBV) (Bhasi et al. 2010) which included carrot. The database was a part of an effort that was initiated to collect and organize genomic information useful for RVB researchers. In the release of this platform (Bhasi et al. 2010), it is reported that the database included genomics data for 294 Alliaceae and 816 Apiaceae plant species. Genomic resources included 3663 genes, 5959 RNAs, 22,723 ESTs and 11,438 regulatory sequence elements and their functional annotations. The RoBuST database was enriched with graphical tools for visualization and analysis of sequence data, with query interfaces to access the traits, biosynthetic pathways, genetic linkage maps and molecular taxonomy data associated with Alliaceae and Apiaceae plants (Bhasi et al. 2010). Unfortunately, this database is no longer accessible.

CarrotDB was the first effort to develop a dedicated platform to host the carrot DC-27 genome (Xu et al. 2014) and other related genetic and genomic resources. As presented in the original manuscript, the platform was enriched with a genome browser visualizing the predicted gene and gene fragments across the draft carrot genome. A GBrowse track reported the best BLAST matches to *A. thaliana* protein sequences. Such functions are not available in the current version of the CarrotDB. The accessible version includes five subpages of: (1) NCBI Blast service, (2) genome map, (3) transcription factors, (4) germplasm and (5) download. However, many parts of the database are not functional, such as the genome map, BLAST page and the bioinformatics tools available in the subpages 1 and 2. The transcription factor page includes a table summarizing the number of carrot genes annotated in multiple transcription factor families. The sequences associated with those gene families cannot be downloaded. The germplasm page includes pictures and general information such as the name, color of cortex, phloem and xylem parts of taproots, and origin of 45 carrot accessions. The download page provides assembled genome sequence and gene sequences and an in silico functional annotation. However, all data is related to the DC-27 genome and does not provide access to the newer DH1 chromosome-scale genome assembly (Iorizzo et al. 2016). Despite the great efforts to develop a platform dedicated to carrot, this platform provides limited access and tools to carrot researchers and breeders.

13.4 Future Perspectives

Currently, the carrot research community is lacking a comprehensive bioinformatics platform and resources to enable the deep investigation and efficient integration of multiple levels of "omics" data which could be utilized for breeding efforts. This can be due but not limited to the fact that not enough or no data is available from specific "omics" levels for carrot. For example, the carrot genome severely lacks good-quality coverage of transcriptomics data from different tissues and developmental stages under normal conditions and during abiotic and biotic stress treatments. To further model and decipher the mechanisms underlying the biological processes in this plant species, similar data from transcriptomics, epigenomics, proteomics, metabolomics as well as its genetic properties is needed. To the best of our knowledge, the carrot genome does not have a single data collection from any of the before mentioned treatments/stages for all of the "omics" data levels. Such data and platforms in biomolecular research are extremely relevant. This highlights the need for focused efforts that are required to understand carrot omics and its functionality. In contrast, driven by the need and desire to apply marker-assisted breeding for carrot, the number of molecular markers used for linkage map construction, marker-trait association analysis and phenotyping data is rapidly increasing. To integrate this data, the carrot genome consortium is setting up a dedicated bioinformatics platform called carrot omics (https://carrotomics.org) to store and integrate "omics" and phenotypic data resources obtained from different carrot accessions produced within this project and will include information for Apiaceae family members. The database will provide access to various genomics and genetics resources by using integrative and efficient user-friendly query interfaces and toolboxes, and a centralized data download of genomic sequences, genes and annotations. The consortium is also undertaking another effort to improve the carrot gene prediction quality aided by long-read transcript sequencing methods. This provides a valuable full-length transcript collection representing different gene isoforms from several tissues of the carrot DH1 genome. This database will provide carrot and Apiaceae researchers and breeders a centralized repository of integrated information which will expedite carrot research and breeding efforts.

References

Acquaah G (2009) Principles of plant genetics and breeding. Wiley, London

Arnold K, Bordoli L, Kopp J, Schwede T (2006) The SWISS-MODEL workspace: a web-based environment for protein structure homology modelling. Bioinformatics 22:195–201

Barrett T, Troup DB, Wilhite SE, Ledoux P, Rudnev D, Evangelista C, Kim IF, Soboleva A, Tomashevsky M, Edgar R (2006) NCBI GEO: mining tens of millions of expression profiles—database and tools update. Nucleic Acids Res 35:D760–D765

Bhasi A, Senalik D, Simon PW, Kumar B, Manikandan V, Philip P, Senapathy P (2010) RoBuST: an integrated genomics resource for the root and bulb crop families Apiaceae and Alliaceae. BMC Plant Biol 10:161

Bostan H, Chiusano ML (2015) NexGenEx-Tom: a gene expression platform to investigate the functionalities of the tomato genome. BMC Plant Biol 15:48

Bowman MJ (2012) Gene expression and genetic analysis of carotenoid pigment accumulation in carrot (Daucus carota L.). Ph.D. thesis, University of Wisconsin–Madison

Cambiaghi A, Ferrario M, Masseroli M (2017) Analysis of metabolomic data: tools, current strategies and future challenges for omics data integration. Brief Bioinform 18:498–510

Carbon S, Ireland A, Mungall CJ, Shu S, Marshall B, Lewis S, Hub A, Group WPW (2008) AmiGO: online access to ontology and annotation data. Bioinformatics 25:288–289

Cavagnaro PF, Chung S-M, Manin S, Yildiz M, Ali A, Alessandro MS, Iorizzo M, Senalik DA, Simon PW (2011) Microsatellite isolation and marker development in carrot—genomic distribution, linkage mapping, genetic diversity analysis and marker transferability across Apiaceae. BMC Genom 12:386

Cavagnaro PF, Chung S-M, Szklarczyk M, Grzebelus D, Senalik D, Atkins AE, Simon PW (2009) Characterization of a deep-coverage carrot (*Daucus carota* L.) BAC library and initial analysis of BAC-end sequences. Mol Genet Genomics 281:273–288

Cheng F, Liu S, Wu J, Fang L, Sun S, Liu B, Li P, Hua W, Wang X (2011) BRAD, the genetics and genomics database for Brassica plants. BMC Plant Biol 11:136

Collard B, Jahufer M, Brouwer J, Pang E (2005) An introduction to markers, quantitative trait loci (QTL) mapping and marker-assisted selection for crop improvement: the basic concepts. Euphytica 142:169–196

Collard BC, Mackill DJ (2007) Marker-assisted selection: an approach for precision plant breeding in the twenty-first century. Philos Trans R Soc B Biol Sci 363:557–572

Coordinators NR (2013) Database resources of the national center for biotechnology information. Nucleic Acids Res 41:D8

Ellison S, Senalik D, Bostan H, Iorizzo M, Simon P (2017) Fine mapping, transcriptome analysis, and marker development for Y2, the gene that conditions beta-carotene accumulation in carrot (*Daucus carota* L.). G3-Genes Genom Genet, 117.043067

Ellison SL, Luby CH, Corak KE, Coe KM, Senalik D, Iorizzo M, Goldman IL, Simon PW, Dawson JC (2018) Carotenoid presence is associated with the or gene in domesticated carrot. Genetics 210:1497–1508

Fernandez-Pozo N, Menda N, Edwards JD, Saha S, Tecle IY, Strickler SR, Bombarely A, Fisher-York T, Pujar A, Foerster H (2014) The sol genomics network (SGN)—from genotype to phenotype to breeding. Nucleic Acids Res 43:D1036–D1041

Fernández XM, Birney E (2010) Ensembl genome browser. In: Vogel and motulsky's human genetics. Springer, Berlin, pp 923–939

Gomez-Cabrero D, Abugessaisa I, Maier D, Teschendorff A, Merkenschlager M, Gisel A, Ballestar E, Bongcam-Rudloff E, Conesa A, Tegnér J (2014) Data integration in the era of omics: current and future challenges. BioMed Central 8(Suppl 2):1

Goodstein DM, Shu S, Howson R, Neupane R, Hayes RD, Fazo J, Mitros T, Dirks W, Hellsten U, Putnam N (2011) Phytozome: a comparative platform for green plant genomics. Nucleic Acids Res 40: D1178–D1186

Grant D, Nelson RT, Cannon SB, Shoemaker RC (2009) SoyBase, the USDA-ARS soybean genetics and genomics database. Nucleic Acids Res, gkp798

Hampel D, Mosandl A, Wüst M (2005) Biosynthesis of mono-and sesquiterpenes in carrot roots and leaves (*Daucus carota* L.): metabolic cross talk of cytosolic mevalonate and plastidial methylerythritol phosphate pathways. Phytochemistry 66:305–311

He J, Zhao X, Laroche A, Lu Z-X, Liu H, Li Z (2014) Genotyping-by-sequencing (GBS), an ultimate marker-assisted selection (MAS) tool to accelerate plant breeding. Front Plant Sci 5:484

Hirsch CD, Hamilton JP, Childs KL, Cepela J, Crisovan E, Vaillancourt B, Hirsch CN, Habermann M, Neal B, Buell CR (2014) Spud DB: a resource for mining sequences, genotypes, and phenotypes to accelerate potato breeding. Plant Genome 7(1)

Iorizzo M, Cavagnaro P, Bostan H, Zhao Y, Zhang J, Simon PW (2018) A Cluster of MYB transcription factors regulate anthocyanin biosynthesis in carrot (*Daucus carota* L.) root and petiole. Front Plant Sci 9:1927

Iorizzo M, Senalik DA, Ellison SL, Grzebelus D, Cavagnaro PF, Allender C, Brunet J, Spooner DM, Van Deynze A, Simon PW (2013) Genetic structure and domestication of carrot (Daucus carota subsp. sativus) (Apiaceae). Am J Bot 100:930–938

Iorizzo M, Ellison S, Senalik D, Zeng P, Satapoomin P, Huang J, Bowman M, Iovene M, Sanseverino W, Cavagnaro P et al (2016) A high-quality carrot genome assembly provides new insights into carotenoid accumulation and asterid genome evolution. Nat Genet 48:657

Iorizzo M, Senalik D, Szklarczyk M, Grzebelus D, Spooner D, Simon P (2012) De novo assembly of the carrot mitochondrial genome using next generation sequencing of whole genomic DNA provides first evidence of DNA transfer into an angiosperm plastid genome. BMC Plant Biol 12:61

Iorizzo M, Senalik DA, Grzebelus D, Bowman M, Cavagnaro PF, Matvienko M, Ashrafi H, Van Deynze A, Simon PW (2011) De novo assembly and characterization of the carrot transcriptome reveals novel genes, new markers, and genetic diversity. BMC Genom 12:389

Jung S, Staton M, Lee T, Blenda A, Svancara R, Abbott A, Main D (2007) GDR (genome database for Rosaceae): integrated web-database for Rosaceae genomics and genetics data. Nucleic Acids Res 36: D1034–D1040

Kanehisa M, Goto S (2000) KEGG: kyoto encyclopedia of genes and genomes. Nucleic Acids Res 28:27–30

Keilwagen J, Lehnert H, Berner T, Budahn H, Nothnagel T, Ulrich D, Dunemann F (2017) The terpene synthase gene family of carrot (*Daucus carota* L.): identification of QTLs and candidate genes associated with terpenoid volatile compounds. Front Plant Sci 8:1930

Keusch GT (2006) What do *-omics* mean for the science and policy of the nutritional sciences? Am J Clin Nutr 83:520S–522S

Korf I, Flicek P, Duan D, Brent MR (2001) Integrating genomic homology into gene structure prediction. Bioinformatics 17:S140–S148

Leinonen R, Sugawara H, Shumway M (2010) The sequence read archive. Nucleic Acids Res 39:D19–D21

Lin X, Hwang G-JH, Zimmerman JL (1996) Isolation and characterization of a diverse set of genes from carrot somatic embryos. Plant Physiol 112:1365–1374

Ma J, Li J, Xu Z, Wang F, Xiong A (2018) Transcriptome profiling of genes involving in carotenoid biosynthesis and accumulation between leaf and root of carrot (*Daucus carota* L.). Acta Biochim Biophys Sin 50:481–490

Machaj G, Bostan H, Macko-Podgórni A, Iorizzo M, Grzebelus D (2018) Comparative transcriptomics of root development in wild and cultivated carrots. Genes 9:431

Macko-Podgórni A, Machaj G, Stelmach K, Senalik D, Grzebelus E, Iorizzo M, Simon PW, Grzebelus D (2017) Characterization of a genomic region under selection in cultivated carrot *(Daucus carota* subsp. *sativus*) reveals a candidate domestication gene. Front Plant Sci 8:12

Mortazavi A, Williams BA, McCue K, Schaeffer L, Wold B (2008) Mapping and quantifying mammalian transcriptomes by RNA-Seq. Nat Methods 5:621–628

Mueller LA, Solow TH, Taylor N, Skwarecki B, Buels R, Binns J, Lin C, Wright MH, Ahrens R, Wang Y (2005) The SOL genomics network. a comparative resource for Solanaceae biology and beyond. Plant Physiol 138:1310–1317

Ouyang S, Zhu W, Hamilton J, Lin H, Campbell M, Childs K, Thibaud-Nissen F, Malek RL, Lee Y, Zheng L (2007) The TIGR rice genome annotation resource: improvements and new features. Nucleic Acids Res 35:D883–D887

Palsson B, Zengler K (2010) The challenges of integrating multi-omic data sets. Nat Chem Biol 6:787

Paran I, Zamir D (2003) Quantitative traits in plants: beyond the QTL. Trends Genet 19:303–306

Park J-S, Kim IS, Cho MS, Park S, Park SG (2006) Identification of differentially expressed genes involved in spine formation on seeds of *Daucus carota* L. (carrot), using annealing control primer (ACP) system. J Plant Biol 49:133–140

Poland JA, Rife TW (2012) Genotyping-by-sequencing for plant breeding and genetics. Plant Genome 5:92–102

Pruitt KD, Tatusova T, Maglott DR (2005) NCBI reference sequence (RefSeq): a curated non-redundant sequence database of genomes, transcripts and proteins. Nucleic Acids Res 33:D501–D504

Reed JL, Famili I, Thiele I, Palsson BO (2006) Towards multidimensional genome annotation. Nat Rev Genet 7:130

Rong J, Lammers Y, Strasburg JL, Schidlo NS, Ariyurek Y, De Jong TJ, Klinkhamer PG, Smulders MJ, Vrieling K (2014) New insights into domestication of carrot from root transcriptome analyses. BMC Genom 15:895

Ruhlman T, Lee SB, Jansen RK, Hostetler JB, Tallon LJ, Town CD, Daniell H (2006) Complete plastid genome sequence of *Daucus carota*: implications for biotechnology and phylogeny of angiosperms. BMC Genom 7:222

Sekhon RS, Briskine R, Hirsch CN, Myers CL, Springer NM, Buell CR, de Leon N, Kaeppler SM (2013) Maize gene atlas developed by RNA sequencing and comparative evaluation of transcriptomes based on RNA sequencing and microarrays. PLoS ONE 8:e61005

Sherry ST, Ward M-H, Kholodov M, Baker J, Phan L, Smigielski EM, Sirotkin K (2001) dbSNP: the NCBI database of genetic variation. Nucleic Acids Res 29:308–311

Simpson K, Quiroz LF, Rodriguez-Concepción M, Stange CR (2016) Differential contribution of the first two enzymes of the MEP pathway to the supply of metabolic precursors for carotenoid and chlorophyll biosynthesis in carrot (*Daucus carota*). Front Plant Sci 7:1344

Spooner DM, Ruess H, Iorizzo M, Senalik D, Simon P (2017) Entire plastid phylogeny of the carrot genus (Daucus, Apiaceae): concordance with nuclear data and mitochondrial and nuclear DNA insertions to the plastid. Am J Bot 104:296–312

Stevens MA, Rick CM (1986) Genetics and breeding. In: The tomato crop. Springer, Berlin, pp 35–109

Suravajhala P, Kogelman LJ, Kadarmideen HN (2016) Multi-omic data integration and analysis using systems genomics approaches: methods and applications in animal production, health and welfare. Genet Sel Evol 48:38

Tanaka M, Kikuchi A, Kamada H (2009) Isolation of putative embryo-specific genes using stress induction of carrot [*Daucus carota*] somatic embryos. Breed Sci 59:37–46 (Japan)

Trapnell C, Roberts A, Goff L, Pertea G, Kim D, Kelley DR, Pimentel H, Salzberg SL, Rinn JL, Pachter L (2012) Differential gene and transcript expression analysis of RNA-seq experiments with TopHat and Cufflinks. Nat Protoc 7:562–578

Vailati-Riboni M, Palombo V, Loor JJ (2017) What are omics sciences? In: Ametaj BN (ed) Periparturient diseases of dairy cows: a systems biology approach. Springer, Cham, pp 1–7
Varshney RK, Nayak SN, May GD, Jackson SA (2009) Next-generation sequencing technologies and their implications for crop genetics and breeding. Trends Biotechnol 27:522–530
Wang G-L, Xu Z-S, Wang F, Li M-Y, Tan G-F, Xiong A-S (2015) Regulation of ascorbic acid biosynthesis and recycling during root development in carrot (*Daucus carota* L.). Plant Physiol Biochem 94:10–18
Welsh JR (1981) Fundamentals of plant genetics and breeding. Wiley, London
Xu Y, Lu Y, Xie C, Gao S, Wan J, Prasanna BM (2012) Whole-genome strategies for marker-assisted plant breeding. Mol Breed 29:833–854
Xu ZS, Tan HW, Wang F, Hou XL, Xiong AS (2014) CarrotDB: a genomic and transcriptomic database for carrot. Database (Oxford) 2014
Yu J, Jung S, Cheng C-H, Ficklin SP, Lee T, Zheng P, Jones D, Percy RG, Main D (2013) CottonGen: a genomics, genetics and breeding database for cotton research. Nucleic Acids Res 42:D1229–D1236
Zhao D, Chen D, Yang C, Zhai Z (2000) Sequence analysis of keratin-like proteins and cloning of intermediate filament-like cDNA from higher plant cells. Sci China Ser C Life Sci 43:265–271

Carrot Carotenoid Genetics and Genomics

14

Philipp W. Simon, Emmanuel Geoffriau, Shelby Ellison and Massimo Iorizzo

Abstract

Carotenoids are essential for photosynthesis, and they are the ultimate source of all dietary vitamin A. They account for the striking diversity of orange, yellow, and red carrot storage root color, and this likely contributes to the fact that carotenoids are the most extensively studied class of compounds in carrot, where their biosynthesis and accumulation have been evaluated across diverse genetic backgrounds and environments. Many genes in the 2-C-methyl-D-erythritol-4-phosphate pathway (MEP) and carotenoid biosynthetic pathways have been identified and characterized in carrot, and genes in those pathways are expressed in carrot roots of all colors, including white carrots which contain at most trace amounts of carotenoids. The active functioning of genes in the carotenoid pathway in carrot roots of all colors should be expected since pathway products serve as precursors for hormones important in plant development. 1-Deoxy-d-xylulose-5-phosphate synthase (*DXS*) in the MEP pathway and the phytoene synthase and lycopene β-cyclase (*PSY*, *LCYB*) genes in the carotenoid pathway provide some level of overall regulation or modulation of these respective pathways, and these genes are incrementally upregulated in carrots with higher carotenoid content but variation in their expression does not account for the diverse content and composition of carotenoids in different colors of carrots. In contrast, genetic polymorphism in the *Y* and Y_2 genes accounts for much for the variation in carotenoids accumulated in white, yellow, and orange carrots, and with the sequencing of the carrot genome, the genetic basis for these genes is becoming revealed. A candidate for the *Y* gene, *DCAR_032551,* is not a member of either the MEP or carotenoid biosynthesis pathway but rather a regulator of photosystem development and carotenoid storage. A clear candidate for the Y_2 gene has not been identified, but no carotenoid biosynthetic gene was found in the genomic region defined by fine mapping of Y_2. The *Or* gene, which regulates chromoplast development in other crops, was also recently associated with the presence of carotenoids in carrot. The discovery of genes outside the carotenoid biosynthetic pathway that contributes to carotenoid colors of carrots is but one exciting consequence of sequencing the carrot genome.

P. W. Simon (✉) · S. Ellison
USDA–Agricultural Research Service, Vegetable Crops Research Unit, Department of Horticulture, University of Wisconsin, 1575 Linden Dr., Madison, WI 53706, USA
e-mail: Philipp.Simon@ars.usda.gov

E. Geoffriau
Agrocampus Ouest, Institut de Recherche en Horticulture et Semences – UMR 1345, SFR 4207 QUASAV, Angers, France

M. Iorizzo
Department of Horticultural Sciences, North Carolina State University, Plants for Human Health Institute, 600 Laureate Way, Kannapolis, NC 28081, USA

P. Simon et al. (eds.), *The Carrot Genome*, Compendium of Plant Genomes,
https://doi.org/10.1007/978-3-030-03389-7_14

14.1 Carotenoids in Carrots

Carotenoids are 40-carbon isoprenoids including non-oxygenated carotenes and oxygenated xanthophylls that are widely distributed in nature, being found in all green plants as well as in animals, bacteria, and fungi. Carotenoids are synthesized in plants where they play an essential role in growth and development as accessory pigments contributing to light-harvesting in photosynthesis and imparting protection from photooxidative damage (Cazzonelli and Pogson 2010; DellaPenna and Pogson 2006; Lu and Li 2008; Nisar et al. 2015). Derivatives of the carotenoid pathway also play a critical role in plant development, flavor, and mycorrhizal associations (Cazzonelli and Pogson 2010; DellaPenna and Pogson 2006). The color of carotenoids is also important to non-photosynthetic tissues in plants since they attract animals which pollinate and disperse seeds (e.g., Simkin et al. 2004 and Walter et al. 2010). Animals, including humans, realize important nutritional benefits from carotenoids, but they rely on diet as their source since they are unable to synthesize carotenoids de novo, with very rare exceptions. Several dietary carotenoids, including α-carotene, β-carotene, and β-cryptoxanthin, can be converted to vitamin A, and these provitamin A carotenoids are the ultimate source of all vitamin A, which is essential for sustaining immune function, vision, growth, and reproduction. These and other dietary carotenoids, including lycopene and lutein, also provide additional health benefits, contributing to reduced risk of cancers, osteoporosis, macular degeneration, and cardiovascular disease (Tanumihardjo 2012).

The content and compositional profile of carotenoids vary widely among tissues and organs of a plant during development, between comparable tissues and organs of different plants, and between the same tissues or organs among genotypes or cultivars of the same plant species. The bright colors that carotenoids impart, which act as attractants to animals, may have also attracted the attention of humans during the domestication of many crops. The regulatory mechanisms underlying the flux of metabolites into, through, and out of the biosynthetic, degradation, and sequestration pathways and cellular compartments involved in carotenoid accumulation are variable and numerous, and they often include variation in aspects of photomorphogenesis and plastid development (Llorente et al. 2017; Lu and Li 2008; Sun et al. 2018; Yuan et al. 2015). The biological bases of carotenoid accumulation in carrots are beginning to be understood.

Wild carrots, or Queen Anne's Lace, and white cultivated carrots contain only trace amounts of carotenoids, while most cultivated carrots contain carotenoids. The major color classes of carrots attributed to carotenoids are yellow, orange, and red where lutein, α- and β-carotene, and lycopene, respectively, are the primary carotenoids that account for those colors (Arscott and Tanumihardjo 2010). In addition to lutein, yellow carrots also can contain small amounts of zeaxanthin and α- and β-carotene (Alasalvar et al. 2001; Arscott and Tanumihardjo 2010; Grassmann et al. 2007; Nicolle et al. 2004; Surles et al. 2004), while orange carrots can also contain small amounts of phytoene, lutein, ζ-carotene, and lycopene in addition to α- and β-carotene (Alasalvar et al. 2001; Arscott and Tanumihardjo 2010; Grassmann et al. 2007; Nicolle et al. 2004; Simon and Wolff 1987; Surles et al. 2004). Red carrots usually contain some α- and β-carotene, and lutein, in addition to lycopene (Arscott and Tanumihardjo 2010; Grassmann et al. 2007). Orange carrots are unusual among dietary sources of provitamin A carotenoids in that α-carotene can account for a much larger fraction of their total carotenoids—from 13 to 40% of their total carotenoid content, with higher percentages in carrot roots with higher total carotenoid content (Santos and Simon 2006; Simon and Wolff 1987). Carrots have been estimated to provide 67% of the α-carotene in the US diet (Simon et al. 2009). The leaves of carrot plants with orange roots also have a higher α-carotene-to-β-carotene ratio than do leaves of plants with yellow roots (Arango et al. 2014; Perrin et al. 2016; Wang et al. 2014). Total carotenoid content can reach 500 ppm fresh weight basis in dark orange cultivars

(Simon et al. 1989), and this capacity to accumulate carotenoids is associated with the development of 'carotenoid bodies' and crystal formation in carrot root chromoplasts (Baranska et al. 2006; Ben-Shaul and Klein 1965; Fuentes et al. 2012; Kim et al. 2010; Li et al. 2016; Maass et al. 2009; Sun et al. 2018).

Carotenoids have been a distinguishing characteristic of carrots since their first definitive mention as a root crop around 1100 years in Central Asia where the colors of carrots were noted to be either yellow or purple (Banga 1957, 1963; Simon 2000). It is interesting to note that wild carrots have never been reported to contain more than a trace of either carotenoids or anthocyanins. This indicates that there clearly had been a period of carrot domestication ongoing before the records of 1100 years ago, as wild carrots were apparently selected for color as well as other domestication traits (also see Chap. 5). Beyond the lutein found in yellow carrots, variation in the accumulation of several other carotenoids has also played a major role in the history of carrot, with orange carrots first appearing in southern Europe around 1500 (Stolarczyk and Janick 2011) and red carrots in Asia in 1700s (Rubatzky et al. 1999; Simon 2000). Orange has been the predominant color of carrots since relatively soon after their initial appearance, and hundreds of orange carrot cultivars were developed in Europe since the 1600s (Banga 1963). Orange carrots predominate worldwide today, but sizable markets of red carrots are also grown in much of Asia.

14.2 Carotenoid Biosynthesis in Carrot

Carotenoid isoprenoids are synthesized in plastids by the 2-C-methyl-D-erythritol-4-phosphate pathway (MEP) where pyruvate and glyceraldehyde 3-phosphate are metabolized to isopentenyl diphosphate (IPP), and in the cytosol by the mevalonic acid pathway (MVA) which converts acetyl-CoA to IPP and geranylgeranyl diphosphate (GGPP). GGPPs then form phytoene, the first committed step of the carotenoid pathway (Fig. 14.1). Most carotenoid precursors are produced by the MEP pathway (Rodriguez-Concepcion 2010).

Forty-four genes in the isoprenoid biosynthetic pathway (Iorizzo et al. 2016) and 24 genes in the carotenoid biosynthetic pathway in carrot have been identified (Iorizzo et al. 2016; Just et al. 2007; Stange Klein and Rodriguez-Concepcion 2015) with multiple paralogues in a number of pathway genes, suggesting that different paralogues evolved a specialized function in different types of plastids, tissue types or developmental stages, environmental conditions or pathway cross-talk mechanisms (Iorizzo et al. 2016; Rodriguez-Concepcion and Stange 2013; Simpson et al. 2016b). The large diversity of genes in the carotenoid pathway in carrot likely accounts for the variability in gene expression behavior reported in several studies and different phenotypes. Most of the pathway genes are expressed in storage roots of all colors, including white cultivated and wild carrots (Bowman et al. 2014; Clotault et al. 2008; Just et al. 2007; Ma et al. 2017; Perrin et al. 2016, 2017a; Wang et al. 2014). This might be expected since pathway products serve as precursors for important compounds in plant development, including the hormones abscisic acid and strigolactones (Walter et al. 2010; Walter and Strack 2011).

Carotenoid content increases during root development in orange, yellow, and red roots (Clotault et al. 2008; Fuentes et al. 2012; Hansen 1945; Wang et al. 2014) and in carrot leaves (Wang et al. 2014; Perrin et al. 2016). The relative expression of carotenoid genes also increases during development, but increases in gene expression are usually many-fold less than increases in pigment accumulation (Bowman et al. 2014; Clotault et al. 2008; Fuentes et al. 2012; Ma et al. 2017; Stange et al. 2008; Wang et al. 2014). For example, compared to young plants, Lycopene β-cyclase (*DcLcyb1*) expression rose 25-fold in mature leaves and 14-fold in mature roots (Moreno et al. 2013). To further evaluate the role of *DcLcyb1* in that same study, overexpression in transgenic carrots increased total carotenoid content of leaves and roots, and the expression of *DcPsy1*, *DcPsy2*, and *DcLcyb2*

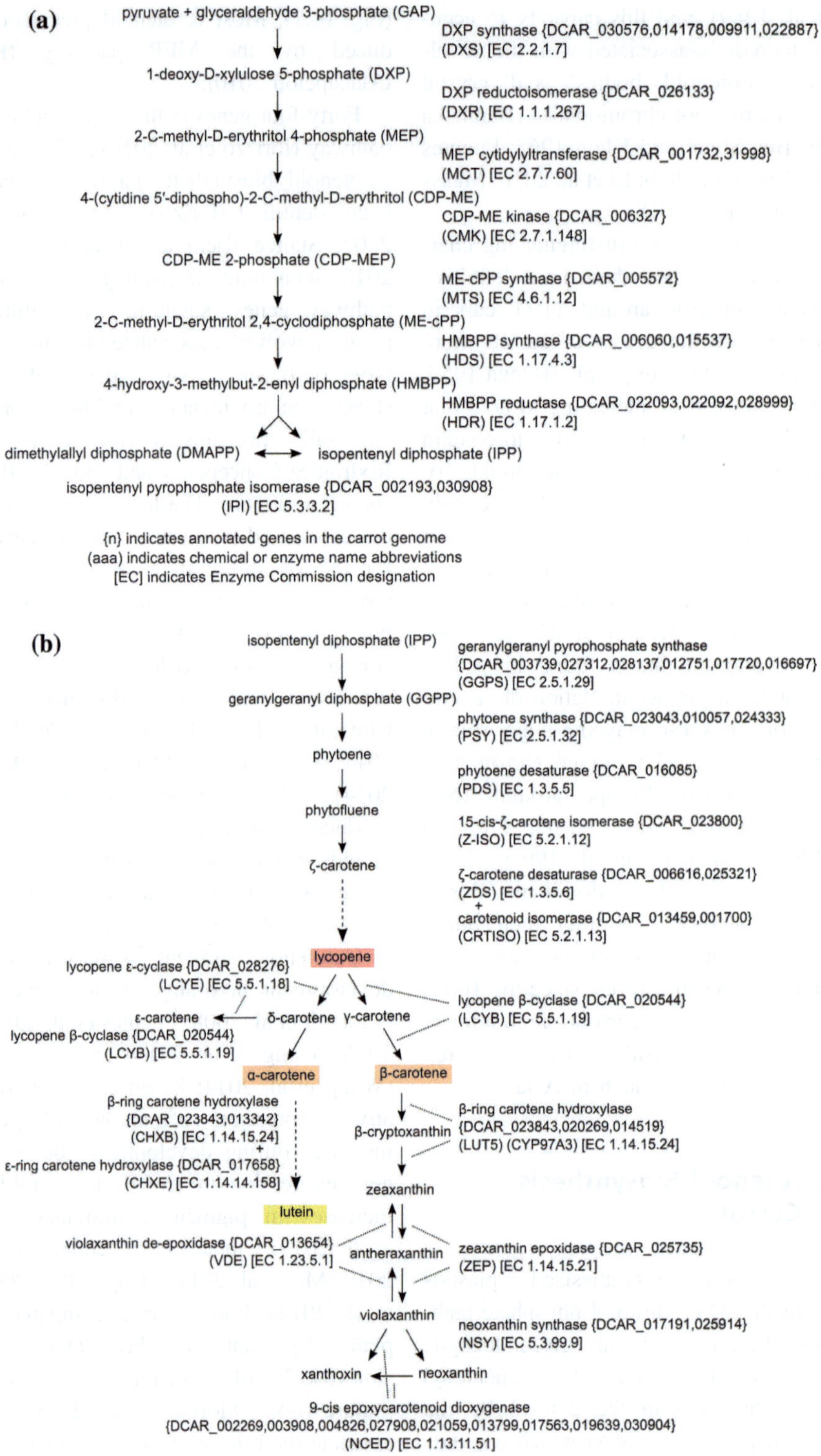

Fig. 14.1 MEP (**a**) and carotenoid (**b**) pathways. Enzyme names, carrot locus tags (in curly brackets), abbreviations (in parentheses), and Enzyme Commission numbers (in square brackets) are included

were also upregulated, while post-transcriptional silencing of *DcLcyb1* demonstrated its essential role in contributing to β-carotene accumulation in leaves and roots and lutein in leaves. Interestingly, transgenic tobacco bearing this same carrot gene increased not only carotenoid content, but also increased plant biomass and stimulated early flowering (Moreno et al. 2016).

In another study involving transgenics to investigate the role of specific genes in carrot carotenoid metabolism, Arango et al. (2014) overexpressed carotene hydroxylase *CYP97A3* in orange carrots to better understand why orange carrots contain high storage root and leaf α-carotene content, relative to other crops. They observed that α-carotene content of leaves of orange carrots was over tenfold higher in carrots with orange roots than in leaves of wild and cultivated carrots with white roots. Overexpression of *CYP97A3* in transgenic orange carrots resulted in a lower α-carotene content of leaves similar to that in untransformed wild carrot. Root carotenoid levels were also significantly reduced in orange transformed carrots, and PSY protein levels were reduced even though *PSY* expression was not. This suggested a feedback system in carotenoid metabolism.

One dramatic change in growing environment that does significantly alter carrot carotenoid accumulation and gene expression is the exposure of growing storage roots to light. Interestingly, when carrot storage roots are exposed to light during development, chloroplasts develop instead of chromoplasts and the pattern of carotenoid profile and accumulation of the root becomes more like that of leaf tissue, instead of the typical chromoplast development in underground roots (Fuentes et al. 2012; Stange et al. 2008; Stange Klein and Rodriguez-Concepcion 2015). This process is similar to the developmental pattern of gene expression in the process of de-etiolation in seedlings grown in the dark and then exposed to light (Rodriguez-Concepcion and Stange 2013). Carrot root morphology also changed with exposure to light, and both carotenoid content and transcript levels for most genes in the pathway were reduced. Similarly, suboptimal environments for light and temperature reduced the carotenoid content in both leaves and roots (Perrin et al. 2016). This reduced accumulation was explained by regulation at the transcriptional level for all tested carotenoid genes in leaves and for phytoene desaturase (*PDS*) and zeaxanthin epoxidase (*ZEP*) genes in roots. However, in environments involving combined stresses (*Alternaria dauci* infection and water restriction), variation of carotenoid transcript levels did not explain the differences in carotenoid content (Perrin et al. 2017b), suggesting the involvement of other regulatory mechanisms outside of the pathway.

Based on the wide variation for carotenoid composition and color intensity observed among diverse genetic stocks and cultivars of carrots, variation in carotenoid gene expression in carrots of varying color has been evaluated in several studies. Qualitative variation in the expression of pathway genes does not vary in a pattern to account for the very large and diverse differences in carotenoid composition among the range of storage root colors, as noted earlier, but several trends in gene expression do follow patterns for carotenoid accumulation. In studies evaluating carotenoid pathway gene expression in white and orange carrots, quantitative variation in *PSY1* and *PSY2* expression was consistently two- to fourfold higher in orange carrots compared to white carrots (Bowman et al. 2014; Clotault et al. 2008; Wang et al. 2014). Clotault et al. (2008) also reported that lycopene ε-cyclase (*LCYE*) and ζ-carotene desaturase (*ZDS*) expression were higher in yellow carrots than orange or white carrots, while Ma et al. (2017) observed that the expression levels of genes involved in xanthophyll formation in yellow cultivars were higher than in orange cultivars. Perrin et al. (2017a) found that carotenoid gene expression varies in a pattern similar to carotenoid accumulation in phloem tissue for plants grown in control conditions.

Among genes for those enzymes in the MEP and MVA pathways, 1-deoxy-d-xylulose-5-phosphate (DXP) synthase 1 (*DXS1*) was the only gene upregulated in a pattern following carrot root carotenoid content (Iorizzo et al. 2016). Several studies in Arabidopsis identified *DXS* as playing a regulatory role in isoprenoid biosynthesis (reviewed by Rodriguez-Concepcion and Boronat 2015), and Simpson et al. (2016a) found Arabidopsis *DXS* expressed in transgenic carrots to be the rate-limiting enzyme in carotenoid production, so this was not surprising. The latter study also observed increased *PSY* transcript as part of the *DXS* regulatory cascade, reflecting the important role of *PSY* in carotenoid metabolism (Cazzonelli and Pogson 2010; Li et al. 2016; Lu and Li 2008; Nisar et al. 2015; Welsch et al. 2000; Yuan et al. 2015), and mirroring observations of upregulated *PSY* in comparisons with orange versus white carrot roots (Bowman et al. 2014; Wang et al. 2014).

Sequence variation for carotenoid genes has also been evaluated across the diverse range of root colors in several studies. The geographic structure of carrot diversity based on the nucleotide variation of carotenoid genes demonstrated an important role for color in carrot domestication (Clotault et al. 2010). Clotault et al. (2012) evaluated sequence variation for seven carotenoid genes in 46 carrots varying in root color from diverse global origins and observed evidence for varying levels of selection for *PDS* and IPP isomerase (*IPI*), which are upstream in the pathway, relative to those later in the pathway. They also observed evidence for balancing selection for carotene isomerase (*CRTISO*), *LCYB1*, and *LCYE* genes, closer to lycopene in the pathway, and sequence variation for *LCYB1* differed between color groups, suggesting it was selected during domestication, but they noted no pattern of sequence variation pointing to candidate genes in the carotenoid pathway to account for color variation. Soufflet-Freslon et al. (2013) found a signature for balancing selection for *CRTISO* polymorphism during carrot breeding history that was associated with root color in a globally diverse collection of white, yellow, orange, red, and purple carrots, regardless of geographic origin. Jourdan et al. (2015) performed an association mapping study with 67 geographically and phenotypically diverse carrot cultivars evaluating 17 carotenoid genes, and they observed associations between α-carotene content and plastid terminal oxidase (*PTOX*) and *CRTISO* polymorphism and between total carotenoid content and β-carotene content with *ZEP*, *PDS*, and *CRTISO* polymorphism. Since *ZEP* and *PDS* are on the same chromosomes as the *Y* and Y_2 genes, respectively, genetic linkage (Just et al. 2007) may have contributed to those associations.

14.3 Carrot Carotenoid Genetics

Carrot color attributed to carotenoids was the first trait noted by early carrot breeders to indicate that outcrossing with wild carrots had occurred in orange cultivated carrots (Vilmorin 1859). Vilmorin also noted that white color was dominant to orange. In studies by Borthwick and Emsweller 75 years later involving intercrosses between yellow and white cultivated carrots, white color was observed to be dominant to yellow (Borthwick and Emsweller 1933; Emsweller et al. 1935). Emsweller et al. (1935) and Lamprecht and Svensson (1950) both attributed single gene control of yellow storage root color over orange with nearly complete dominance. Katsumata et al. (1966) reported orange root color to be dominant to red in orange × red crosses.

In an extensive series of studies from the late 1960s to early 1980s, W. H. Gabelman and his students identified and named several genes controlling carrot storage root color. Laferriere and Gabelman (1968) and Imam and Gabelman (1968) found in multiple populations that a single dominant gene controls white root color over yellow and that segregation ratios in white × orange crosses yield white, yellow, and orange progeny in F_2, F_3, and backcross populations that fit 2–3 gene patterns of inheritance, depending on the population. A monogenic pattern of inheritance controlling yellow root color over orange was observed, as had been observed in earlier studies. Kust (1970) named the gene controlling white color over yellow characterized by Laferriere and Gabelman (1968) *Y*, and the one or two additional dominant genes in white × orange crosses that reduce xylem color Y_1 and Y_2. He also named two genes enhancing phloem color, *O* and *IO*. Buishand and Gabelman (1979) confirmed these results in additional segregating populations and expanded observations to more detailed descriptions of orange pigmentation in phloem and xylem in orange × white crosses. In nearly all cases, the dominant alleles in progeny of white × orange, white × yellow, and yellow × orange crosses reduced the accumulation of carotenoids.

In addition to white × orange and white × yellow crosses, Umiel and Gabelman (1972) observed orange root color to be dominant to red as had been observed by Katsumata et al. (1966). In red × orange F_2 and backcross populations, they observed two genes that they named *A* and *L*, conditioning lycopene and α-carotene accumulation in progeny. Buishand and Gabelman (1980) observed segregation patterns reflecting three major genes segregating in red × yellow crosses: Y_2, inhibiting the synthesis of carotenoids, *L*, stimulating lycopene synthesis, and A_1, with action similar to either *O* or *IO* described by Kust (1970).

Studies by Gabelman and his students established several valuable basic genetic principles of carrot color attributable to carotenoids. Intercrosses were not performed between progeny of different crosses to test for allelism or genetic complementation (i.e., whether genes identified in different studies were allelic or different genes), so the process of naming alleles was based solely on phenotype. Research has continued on the *Y* and Y_2 genes, based on descriptions of their phenotypes, but subsequent characterization of *L, O,* and *IO* has not been reported. The A_1 mutant controlling α-carotene accumulation has likely been characterized with the discovery, using transgenic carrots and described earlier in this chapter, that carotene hydroxylase *CYP97A3* in orange carrots is defective, resulting in an increased content of α-carotene relative to β-carotene (Arango et al. 2014). In that study, they also evaluated the genome sequence of *CYP97A3* in a large association panel of diverse carrots, and they discovered a frameshift mutation in that gene that only occurs in orange carrots, thus revealing the genetic and molecular basis of the high α-carotene content that has long been observed in orange carrots.

Beyond the numerous studies characterizing the biochemical and molecular basis of the range of typical colors attributable to carotenoids established during carrot domestication, Goldman and Breitbach (1996) identified and characterized a newly discovered, naturally occurring mutant in orange carrots that was conditioned by a recessive gene designated *rp*. This gene reduces storage root α- and β-carotene content by over 90% and α-tocopherol content by 25–43%, while elevating phytoene content (Koch and Goldman 2005). Leaves are chlorotic and even white early in the development of *rprp* plants, but by the sixth leaf, they are a typical green color. The reduced plant vigor associated with the *rp* mutation is unique among carrot genes controlling carotenoid color. It has been suggested that the *rp* mutant represses the carotenoid pathway (Goldman and Breitbach 1996), but this has not yet been examined. Given the unique nature of *rp*, further characterization may provide unique insights into carrot carotenoid metabolism.

Simon (1992) established carotenoid mapping populations to evaluate linkage of Y_2 to genes conditioning sugar and anthocyanin accumulation and confirmed monogenic inheritance for Y_2 in three orange × yellow crosses. This gene was designated as Y_2 since the observed phenotype best fit the earlier descriptions of the phenotype by Buishand and Gabelman (1979) for this trait (Fig. 14.2). To facilitate phenotyping of carotenoid accumulation genes in breeding programs, an AFLP fragment linked to Y_2 was converted to a PCR-based marker for that gene in a yellow × orange cross (Bradeen and Simon 1998). This was soon followed by a more general mapping study, also using AFLPs, where the Y_2 gene controlling orange color and carotenoid content was first placed on a genetic map (Vivek and Simon 1999). In these studies, both root color based on visual scores and carotene content based on HPLC analysis were mapped.

To place more genes controlling carrot carotenoid accumulation on the carrot genetic map, Santos and Simon (2002) used AFLP markers to

White	Yellow	Pale Orange	Orange
Y_Y_2	$yyY_{2}\_$	$Y\_y_2y_2$	yyy_2y_2

Fig. 14.2 Phenotypes and genotypes for white, yellow, pale orange, and orange carrots (left to right) from the B493 × QAL mapping population segregating for the *Y* and *Y2* genes

map the content of individual carotenoids quantified by HPLC in two unrelated mapping populations. They reported that eight QTLs for orange color were associated with α-carotene content and three with β-carotene content. One of the QTLs for β-carotene content was in common to both populations. They also noted that for several QTLs as many as four carotenoids (phytoene, lycopene, β-carotene, and α-carotene) were all associated with the same QTL. Before that point, carrot carotenoid genetic analysis for color usually assumed that genes contributing to carotenoid accumulation were genes in the biosynthetic pathway, as has been observed in several other crops. The clustering noted in this study suggested that carrot carotenoid color genes may not be pathway genes, but rather that they regulate carotenoid accumulation (Santos et al. 2005).

Santos and Simon (2006) also reported the broad-sense heritability values for individual carotenoids and total carotenoid content in the two populations they studied. One of the mapping populations utilized was a population derived from a cross between a wild, white-rooted carrot, referred to as QAL, and B493, a dark orange cultivated carrot, so progeny of this cross were segregating for both the *Y* and Y_2 genes, in addition to quantitative loci that contributed to variation in total carotenoid content. The other mapping population was a cross between an orange carrot with an average carotene content ('Brasilia') and a high-carotene, dark orange-colored carrot (HCM), so all progeny were yyy_2y_2, but segregating for quantitative loci that contributed to the roughly fivefold difference in carotenoid content between the two parents. The heritability values for total carotenoid content ranged from 0.89 to 0.98 for the B493 × QAL population, while it ranged from 0.38 to 0.45 for the 'Brasilia' × HCM population, indicating the large effects that the *Y* and Y_2 genes conferred to progeny in the B494 × QAL population. Variation in carotenoid content for the same orange carrot genetic stocks can vary twofold across environments (Perrin et al. 2016, 2017b; Simon and Wolff 1987), which provides some rationale for the lower heritability observed in the 'Brasilia' × HCM population.

To evaluate the relationship between carotenoid biosynthetic genes and carotenoid color genes in carrot, Just et al. (2007, 2009) mapped 22 putative genes coding for carotenoid enzymes, based on the hypothesis that they may be candidate genes for carotenoid color. This study utilized a population derived from a cross between a wild, white-rooted carrot, referred to as QAL, and B493, a dark orange cultivated carrot—one of the same populations used by Santos and Simon (2002). In that population, the segregation pattern for orange, yellow, and white root color fits a two-gene model. These two genes included the Y_2 gene mapped by Bradeen and Simon (1998) on linkage group 5 and a second gene on linkage group 2, which was called *Y* since its phenotype best fit the description of the *Y* gene first described by members of the Gabelman group. The *Y* gene was linked to ε-ring carotene hydroxylase (*CHXE*), 9-*cis*-epoxycarotenoid dioxygenase 2 (*NCED2*), and more distantly to *PDS*; and the Y_2 gene was linked to *ZEP* and *ZDS*. These were considered to be positional candidates since linkages were not close (Just et al. 2009).

The sequencing of the carrot genome (Iorizzo et al. 2016) advanced a candidate for the *Y* gene. Fine mapping of orange (yyy_2y_2) versus pale orange (YYy_2y_2) root color (Fig. 14.2) segregating as a monogenic trait in a population derived from the B493 × QAL cross by Just (2004), and of yellow (yyY_2Y_2) and white (YYY_2Y_2) root color in an unrelated population, revealed in this study that both traits mapped to the same 75-kb region of chromosome 5 which Just et al. (2009) had identified as *Y*. While the monogenic control of yellow (yyY_2Y_2) versus white (YYY_2Y_2) color was expected, the pale orange phenotype had not been associated with the YYy_2y_2 genotype with certainty before this study since it had not been clearly distinguished from orange (yyy_2y_2) in earlier phenotyping. Based on differential expression analysis and sequence polymorphisms, a candidate gene (*DCAR_032551*) controlling the *Y* locus was identified (Iorizzo et al. 2016). This gene was upregulated in the *yy* segregant in these populations and was found to be co-expressed with two genes involved in the isoprenoid pathway, *DXS1* and *LCYE*, and several other genes involved in photosynthetic system activation and function, plastid biogenesis, and chlorophyll metabolism—an unexpected finding in non-photosynthetic root tissue. The homolog to the *Y* gene candidate in Arabidopsis, *pseudo-etiolation in light*, has an etiolated phenotype and interacts with genes directly involved in the regulation of light response/photomorphogenesis (Ichikawa et al. 2006). Interestingly, *DXS1* expression is induced by light (Cordoba et al. 2011; Kim et al. 2005), and it catalyzes the biosynthesis of carotenoid precursors in photosynthetic metabolism (Estévez et al. 2001; Saladié et al. 2014). Based on this information, Iorizzo et al. (2016) hypothesized that a recessive allele (*yy*) lifts the repression of photomorphogenic development typically found in etiolated roots, which then induces the over-expression of *DXS1* and thus carotenoid biosynthesis. Awaiting validation, this model would explain the light-induced changes observed in orange carrot roots exposed to light described earlier (Fuentes et al. 2012; Stange et al. 2008; Stange Klein and Rodriguez-Concepcion 2015).

In addition to the *Y* gene, the Y_2 gene was also segregating in the B493 × QAL cross evaluated in earlier studies (Just et al. 2009). Using a combination of fine mapping with transcriptome analysis in a population derived from this cross homozygous for *y*, but segregating at the Y_2 locus, Y_2 was mapped to a 650-kb region that included 72 predicted genes (Ellison et al. 2017). Transcriptome analysis was performed at 40 and 80 days after planting, and several genes in the carotenoid pathway were differentially expressed genes in orange (yyy_2y_2) but not yellow (yyY_2Y_2) roots. These included *PSY1, PSY3,* geranylgeranyl diphosphate synthase 1 (*GPPS1*), *LUTEIN DEFICIENT 5* (*LUT5*) carotenoid cleavage dioxygenase 1 (*CCD1*), neoxanthin synthase 1 (*NSY1*), and two cytochrome genes, but none of these genes were in the 650-kb region. The only MEP or carotenoid pathway gene in that region was DXP reductoisomerase (*DXR*) (Iorizzo et al. 2016), but it was not differentially expressed in comparisons between orange and yellow carrots. Within the fine-mapped region, 17 genes were differentially expressed and of these only four were differentially expressed at both 40 and 80 days. Of those four, only one had lower expression in orange compared to yellow roots, as would be expected for a recessive trait—*Protein DEHYDRATION-INDUCED 19 homolog 5 (Di19)* (*DCAR_026175*). Members of the Arabidopsis *Di19* gene family can function in an ABA-independent fashion and are regulated by other abiotic stimuli such as *AtDi19-7*, which has been implicated in regulating light signaling and responses (Milla et al. 2006). Consequently, altered expression of *Di19* could potentially influence the coordinated production of chlorophyll and carotenoids that occurs during photomorphogenesis. The relatively large number of genes and candidates in the 650-kb region indicates the need for additional evaluation to confirm the y_2 candidate with more certainty.

Another gene controlling the accumulation of carotenoids in carrot roots was recently identified in association analysis of 154 wild and 520 cultivated carrots from geographically diverse global growing regions and included the full range of carotenoid colors. This collection of carrots was evaluated to assess genomic signatures of domestication based upon GBS (genotyping by sequencing) (Ellison et al. 2018). An association between a 143-kb genomic region of chromosome 3 with carotene presence was identified, and this region contained no MEP or carotenoid genes, but it did include the *Or* gene. *Or* is important for chromoplast development which in turn provides a sink for the accumulation of carotenoids in cauliflower, sweet potato, and

Fig. 14.3 Phenotypes for orange, light orange, and yellow carrots (left to right) from a mapping population segregating for the *Or* gene but fixed homozygous recessive at the *y* and *y2* loci

Arabidopsis (Li et al. 2016; Lu and Li 2008; Sun et al. 2018; Yuan et al. 2015), and a similar function is suggested for carrot. It is interesting to note that allelic variation for the *Or* gene observed in this study was more common in cultivated carrots from Central Asia, the center of diversity for carrots (Iorizzo et al. 2013), than it was for cultivated carrots from Europe. In an orange carrot background (yyy_2y_2) the wild-type allele for *Or* (Or_w) when homozygous, conditions yellow color, heterozygotes are light orange, and the storage roots of $Or_cOr_cyyy_2y_2$ plants are yellow, where Or_c is the cultivated allele for *Or* (Fig. 14.3). This suggests that *Or* was fixed for the Or_c allele in the development of European carrots, but variation at the *Or* gene apparently played a role early in carrot domestication in Central Asia. Evaluations of gene expression and phenotypes associated with *Or* allelic variation are in progress.

14.4 Future Perspectives on Carrot Carotenoid Genetics and Genomics

Domesticated carrot storage roots can accumulate large quantities of diverse carotenoids. A complex pattern of genetic analysis, gene expression, and metabolic interaction is beginning to be documented in the examination of the carotenoid biosynthetic pathway and carotenoid accumulation in carrot storage roots. Within any of the diverse root color categories—yellow, orange, and red—the relationships between genetic variation for pathway genes, gene expression, and carotenoid accumulation throughout development present a relatively straightforward series of metabolic events. But when comparing those root colors attributable to carotenoid content, it has been challenging to understand how the variation observed in the carotenoid pathway can account for the diverse composition and quantities of carotenoids that we observe in orange, yellow, and red carrots. Variation not only in the carotenoid biosynthetic pathway but also in photomorphogenesis and plastid development contribute to that variation in carrot color. The sequencing of the carrot genome has provided surprising insights into genome function involving regulatory genes outside of the carotenoid biosynthetic pathway that we are just beginning to understand.

References

Alasalvar C, Grigor JM, Zhang D et al (2001) Comparison of volatiles, phenolics, sugars, antioxidant vitamins, and sensory quality of different colored carrot varieties. J Agric Food Chem 49:1410–1416

Arango J, Jourdan M, Geoffriau E et al (2014) Carotene hydroxylase activity determines the levels of both alpha-carotene and total carotenoids in orange carrots. Plant Cell 26:2223–2233

Arscott SA, Tanumihardjo SA (2010) Carrots of many colors provide basic nutrition and bioavailable phytochemicals acting as a functional food. Compr Rev Food Sci Food Saf 9:223–239

Banga O (1957) Origin of the European cultivated carrot. Euphytica 6:54–63

Banga O (1963) Main types of the western carotene carrot and their origin. W.E.J. Tjeenk Willink, Zwolle, The Netherlands

Baranska M, Baranski R, Schulz H, Nothnagel T (2006) Tissue-specific accumulation of carotenoids in carrot roots. Planta 224:1028–1037

Ben-Shaul Y, Klein S (1965) Development and structure of carotene bodies in carrot roots. Bot Gaz 126:79–85

Borthwick HA, Emsweller SL (1933) Carrot breeding experiments. Proc Am Soc Hortic Sci 30:531–533

Bowman MJ, Willis DK, Simon PW (2014) Transcript abundance of phytoene synthase 1 and phytoene synthase 2 is association with natural variation of storage root carotenoid pigmentation in carrot. J Am Soc Hortic Sci 139:63–68

Bradeen JM, Simon PW (1998) Conversion of an AFLP fragment linked to the carrot Y_2 locus to a simple, codominant PCR-based marker form. Theor Appl Genet 97:960–967

Buishand JG, Gabelman WH (1979) Investigations on the inheritance of color and carotenoid content in phloem and xylem of carrot roots (*Daucus carota* L.). Euphytica 28:611–632

Buishand JG, Gabelman WH (1980) Studies on the inheritance of root color and carotenoid content in red × yellow and red × white crosses of carrot, *Daucus carota* L. Euphytica 29:241–260

Cazzonelli CI, Pogson BJ (2010) Source to sink: regulation of carotenoid biosynthesis in plants. Trends Plant Sci 15:266–274

Clotault J, Peltier D, Berruyer R et al (2008) Expression of carotenoid biosynthesis genes during carrot root development. J Exp Bot 59:3563–3573

Clotault J, Geoffriau E, Lionneton E, Briard M, Peltier D (2010) Carotenoid biosynthesis genes provide evidence of geographical subdivision and extensive linkage disequilibrium in the carrot. Theor Appl Genet 121:659–672

Clotault J, Peltier D, Soufflet-Freslon V et al (2012) Differential selection on carotenoid biosynthesis genes as a function of gene position in the metabolic pathway: a study on the carrot and dicots. PLoS ONE 7:e38724

Cordoba E et al (2011) Functional characterization of the three genes encoding 1-deoxy-D-xylulose 5-phosphate synthase in maize. J Exp Bot 62:2023–2038

DellaPenna D, Pogson BJ (2006) Vitamin synthesis in plants: Tocopherols and carotenoids. Annu Rev Plant Biol 57:711–738

Ellison S, Senalik D, Bostan H, Iorizzo M, Simon P (2017) Fine mapping, transcriptome analysis, and marker development for *Y2*, the gene that conditions beta-carotene accumulation in carrot (*Daucus carota* L). G3: Genes, Genomes, Genet 7:2665–2675

Ellison SL, Luby CH, Corak K, Coe K et al (2018) Carotenoid presence is associated with the *Or* gene in domesticated carrot. Genetics 210:1–12

Emsweller SL, Burrell PC, Borthwich HA (1935) Studies on the inheritance of color in carrots. Proc Am Soc Hortic Sci 33:508–511

Estévez JM, Cantero A, Reindl A, Reichler S, León P (2001) 1-Deoxy-d-xylulose-5-phosphate synthase, a limiting enzyme for plastidic isoprenoid biosynthesis in plants. J Biol Chem 276:22901–22909

Fuentes P, Pizarro L, Moreno JC, Handford M, Rodriguez-Concepcion M, Stange C (2012) Light-dependent changes in plastid differentiation influence carotenoid gene expression and accumulation in carrot roots. Plant Mol Biol 79:47–59

Goldman IL, Breitbach DN (1996) Inheritance of a recessive character controlling reduced carotenoid pigmentation in carrot (*Daucus carota* L.). J Hered 87:380–382

Grassmann J, Schnitzler WH, Habegger R (2007) Evaluation of different coloured carrot cultivars on antioxidative capacity based on their carotenoid and phenolic contents. Int J Food Sci Nutr 58:603–611

Hansen E (1945) Variations in carotene content of carrots. Proc Am Soc Hortic Sci 46:355–358

Ichikawa T et al (2006) The FOX hunting system: an alternative gain-of-function gene hunting technique. Plant J 48:974–985

Imam MK, Gabelman WH (1968) Inheritance of carotenoids in carrots, *Daucus carota*, L. Proc Am Soc Hortic Sci 93:419–428

Iorizzo M, Senalik DA, Ellison SL et al (2013) Genetic structure and domestication of carrot (*Daucus carota* L. subsp. *sativus* L.) (Apiaceae). Am J Bot 100:930–938

Iorizzo M, Ellison S, Senalik D, Zeng P, Satapoomin P, Huang J et al (2016) A high-quality carrot genome assembly provides new insights into carotenoid accumulation and asterid genome evolution. Nat Genet 48 (6):657–666

Jourdan M, Gagne S, Dubois-Laurent C et al (2015) Carotenoid content and root color of cultivated carrot: a candidate-gene association study using an original broad unstructured population. PLoS ONE 10: e0116674

Just BJ (2004) Genetic mapping of carotenoid pathway structural genes and major gene QTLs for carotenoid accumulation in wild and domesticated carrot (*Daucus carota* L.). Dissertation, University of Wisconsin-Madison

Just BJ, Santos CAF, Fonseca MEN et al (2007) Carotenoid biosynthesis structural genes in carrot (*Daucus carota*): isolation, sequence-characterization, single nucleotide polymorphism (SNP) markers and genome mapping. Theor Appl Genet 114:693–704

Just BJ, Santos CA, Yandell BS, Simon PW (2009) Major QTL for carrot color are positionally associated with carotenoid biosynthetic genes and interact epistatically in a domesticated × wild carrot cross. Theor Appl Genet 119:1155–1169

Katsumata HH, Yasui H, Matsue Y, Hamazaki K (1966) Studies on the premature bolting and carotene, lycopene, content in carrot. Bul Hort Res Stn Japan, Ser D 4:107–129

Kim BR, Kim SU, Chang YJ (2005) Differential expression of three 1-deoxy-D-xylulose-5-phosphate synthase genes in rice. Biotechnol Lett 27:997–1001

Kim JE, Rensing KH, Douglas CJ, Cheng KM (2010) Chromoplasts ultrastructure and estimated carotene content in root secondary phloem of different carrot varieties. Planta 231:549–558

Koch T, Goldman IL (2005) Relationship of carotenoids and tocopherols in a sample of carrot root-color accessions and carrot germplasm carrying *Rp* and *rp* alleles. J Agric Food Chem 53:325–331

Kust AF (1970) Inheritance and differential formation of color and associated pigments in xylem and phloem of carrot, *Daucus carota*, L. Dissertation, University of Wisconsin-Madison

Laferriere L, Gabelman WH (1968) Inheritance of color, total carotenoids, alpha-carotene, and beta-carotene in carrots, *Daucus carota* L. Proc Am Soc Hortic Sci 93:408–418

Lamprecht H, Svensson V (1950) The carotene content of carrots and its relation to various factors. Agr Hort Genet 8:74–108

Li L, Yuan H, Zeng Y, Xu Q (2016) Plastids and carotenoid accumulation. In: Stange C (ed) Carotenoids in nature. Subcellular biochemistry, vol 79. Springer, Cham, pp 273–293

Llorente B, Martinez-Garcia JF, Stange C, Rodriguez-Concepcion M (2017) Illuminating colors: regulation of carotenoid biosynthesis and accumulation by light. Curr Opin Plant Biol 37:49–55

Lu S, Li L (2008) Carotenoid metabolism: biosynthesis, regulation, and beyond. J Integr Plant Biol 50:778–785

Ma J, Xu Z, Tan G, Wang F, Xiong A (2017) Distinct transcription profile of genes involved in carotenoid biosynthesis among six different color carrot (*Daucus carota* L.) cultivars. Acta Biochim Biophysica Sinica 49:817–826

Maass D, Arango J, Wust F, Beyer P, Welsch R (2009) Carotenoid crystal formation in Arabidopsis and carrot roots caused by increased phytoene synthase protein levels. PLoS ONE 4:e6373

Milla MAR, Townsend J, Chang I, Cushman JC (2006) The Arabidopsis *AtDi19* gene family encodes a novel type of Cys2/His2 Zinc-finger protein implicated in ABA-independent dehydration, high-salinity stress and light signaling pathways. Plant Mol Biol 61:13

Moreno JC, Pizarro L, Fuentes P et al (2013) Levels of lycopene beta-cyclase 1 modulate carotenoid gene expression and accumulation in *Daucus carota.* PLoS ONE 8:e58144

Moreno JC, Cerda A, Simpson K et al (2016) Increased *Nicotiana tabacum* fitness through positive regulation of carotenoid, gibberellin and chlorophyll pathways promoted by *Daucus carota* lycopene β-cyclase (*Dclcyb1*) expression. J Exp Bot 67:2325–2338

Nicolle C, Simon G, Rock E, Amouroux P, Rémésy C (2004) Genetic variability influences carotenoid, vitamin, phenolic, and mineral content in white, yellow, purple, orange, and dark-orange carrot cultivars. J Am Soc Hortic Sci 129:523–529

Nisar N, Li L, Lu S, Khin NC, Pogson BJ (2015) Carotenoid metabolism in plants. Mol Plant 8:68–82

Perrin F, Brahem M, Dubois-Laurent C et al (2016) Differential pigment accumulation in carrot leaves and roots during two growing periods. J Agric Food Chem 64:906–912

Perrin F, Hartmann L, Dubois-Laurent C et al (2017a) Carotenoid gene expression explains the difference of carotenoid accumulation in carrot root tissues. Planta 245:737–747

Perrin F, Dubois-Laurent C, Gibon Y, Citerne S et al (2017b) Combined *Alternaria dauci* infection and water stresses impact carotenoid content of carrot leaves and roots. Environ Exp Bot 143:125–134

Rodriguez-Concepcion M (2010) Supply of precursors for carotenoid biosynthesis in plants. Arch Biochem Biophys 504:118–122

Rodriguez-Concepcion M, Boronat A (2015) Breaking new ground in the regulation of the early steps of plant isoprenoid biosynthesis. Curr Opin Plant Biol 25:17–22

Rodriguez-Concepcion M, Stange C (2013) Biosynthesis of carotenoids in carrot: an underground story comes to light. Arch Biochem Biophys 539:110–116

Rubatzky VE, Quiros CF, Simon PW (1999) Carrots and related vegetable Umbelliferae. CABI, New York

Saladié M, Wright LP, Garcia-Mas J et al (2014) The 2-C-methylerythritol 4-phosphate pathway in melon is regulated by specialized isoforms for the first and last steps. J Exp Bot 65:5077–5092

Santos C, Simon PW (2002) QTL analyses reveal clustered loci for accumulation of major provitamin A carotenes and lycopene in carrot roots. Mol General Genet 268:122–129

Santos CAF, Senalik D, Simon PW (2005) Path analysis suggests phytoene accumulation is the key step limiting the carotenoid pathway in white carrot roots. Genet Mol Biol 28:287–293

Santos C, Simon P (2006) Heritabilities and minimum gene number estimates of carrot carotenoids. Euphytica 151:79–86

Simkin AJ, Schwartz SH, Auldridge M et al (2004) The tomato carotenoid cleavage dioxygenase 1 genes contribute to the formation of the flavor volatiles beta-ionone, pseudoionone, and geranylacetone. Plant J 40:882–892

Simon PW, Wolff XY (1987) Carotenes in typical and dark orange carrots. J Agric Food Chem 35:1017–1022

Simon PW (1992) Inheritance and expression of purple and yellow storage root color in carrot. J Hered 87:63–66

Simon PW, Wolff XY, Peterson CE et al (1989) High Carotene Mass carrot population. HortScience 24:174

Simon PW (2000) Domestication, historical development, and modern breeding of carrot. Plant Breed Rev 19:157–190

Simon PW, Pollak LM, Clevidence BA et al (2009) Plant breeding for human nutritional quality. Plant Breed Rev 31:325–392

Simpson K, Quiroz LF, Rodriguez-Concepción M, Stange C (2016a) Differential contribution of the first two enzymes of the MEP pathway to the supply of metabolic precursors for carotenoid and chlorophyll biosynthesis in carrot (*Daucus carota*). Front Plant Sci 7:1344

Simpson K, Cerda A, Stange C (2016b) Carotenoid biosynthesis in *Daucus carota.* In: Stange C (ed) Carotenoids in nature. Subcellular biochemistry, vol 79. Springer, Cham, pp 199–217

Soufflet-Freslon V, Jourdan M, Clotault J et al (2013) Functional gene polymorphism to reveal species history: the case of the CRTISO gene in cultivated carrots. PLoS ONE 8(8):e70801

Stange C, Fuentes P, Handford M, Pizarro L (2008) *Daucus carota* as a novel model to evaluate the effect of light on carotenogenic gene expression. Biol Res 41:289–301

Stange Klein C, Rodriguez-Concepcion M (2015) Carotenoids in carrots. In: Chen C (ed) Pigments in fruits and vegetables. Springer, New York, pp 217–228

Stolarczyk J, Janick J (2011) Carrot: History and iconography. Chron Hortic 51:13–18

Sun T, Yuan H, Cao H, Yazdani M, Tadmor Y, Li L (2018) Carotenoid metabolism in plants: the role of plastids. Mol Plant 11:58–74

Surles RL, Weng N, Simon PW, Tanumihardjo SA (2004) Carotenoid profiles and consumer sensory evaluation of specialty carrots (*Daucus carota* L.) of various colors. J Agric Food Chem 52:3417–3421

Tanumihardjo S (ed) (2012) Carotenoids and human health. Springer, New York

Umiel N, Gabelman WH (1972) Inheritance of root color and carotenoid synthesis in carrot, *Daucus carota*, L.: Orange vs. red. J Am Soc Hort Sci 97:453–460

Vilmorin M (1859) L'hérédité dans les végétaux. In: Vilmorin M (ed) Notice sur l'amelioration des plantes par la semis. Librairie Agricole, Paris, France, pp 5–29

Vivek BS, Simon PW (1999) Linkage relationships among molecular markers and storage root traits of carrot (*Daucus carota* L. ssp *sativus*). Theor Appl Genet 99:58–64

Walter M, Floss D, Strack D (2010) Apocarotenoids: hormones, mycorrhizal metabolites and aroma volatiles. Planta 232:1–17

Walter MH, Strack D (2011) Carotenoids and their cleavage products: biosynthesis and functions. Nat Prod Rep 28:663–692

Welsch R, Beyer P, Hugueney P, Kleinig H, von Lintig J (2000) Regulation and activation of phytoene synthase, a key enzyme in carotenoid biosynthesis, during photomorphogenesis. Planta 211:846–854

Wang H, Ou CG, Zhuang FY, Ma ZG (2014) The dual role of phytoene synthase genes in carotenogenesis in carrot roots and leaves. Mol Breed 34:2065–2079

Yuan H, Zhang J, Nageswaran D, Li L (2015) Carotenoid metabolism and regulation in horticultural crops. Hortic Res 2:15036

Carrot Anthocyanin Diversity, Genetics, and Genomics

15

Pablo F. Cavagnaro and Massimo Iorizzo

Abstract

Purple carrots (*Daucus carota* ssp. *sativus* var. *atrorubens* Alef.) accumulate anthocyanins in their roots, petioles, and other plant parts. These flavonoid pigments represent an excellent dietary source of antioxidant and anti-inflammatory agents. In addition, carrot anthocyanins are also used as food dyes. Compositional variation in carrot root, mainly with regard to the content of acylated (AA) and non-acylated anthocyanins (NAA), strongly influences the bioavailability and chemical stability of these pigments, therefore conditioning their potential use as nutraceutical agents or as food colorants. In this context, genetic diversity analysis for root anthocyanin composition is relevant for selecting materials for either purpose. Also, knowledge on the genetic basis underlying anthocyanin biosynthesis and modification is expected to aid in the development of new varieties with high nutraceutical or for extracting food dyes. In the last decades, germplasm collections have been characterized for anthocyanin content and composition. Various simply inherited traits for root and petiole anthocyanin pigmentation and acylation, including P_1, P_3 and *Raa1*, and QTL for root anthocyanins, have been described and mapped to two regions of chromosome 3, in different genetic backgrounds. Recent advances in high-throughput sequencing and bioinformatic analyses have facilitated the discovery of candidate regulatory genes for root and petiole pigmentation associated with the P_3 region in chromosome 3, as well as structural genes involved in anthocyanin glycosylation and acylation. In this chapter, we reviewed recent advances in diversity, genetic, and genomic studies related to carrot anthocyanin pigmentation.

Abbreviations

AA	Acylated anthocyanins
AC	Antioxidant capacity
NAA	Non-acylated anthocyanins

P. F. Cavagnaro (✉)
National Scientific and Technical Research Council (CONICET), National Institute of Agricultural Technology (INTA) E.E.A. La Consulta, San Carlos, Mendoza, Argentina
e-mail: cavagnaro.pablo@inta.gob.ar

P. F. Cavagnaro
Faculty of Agricultural Sciences, National University of Cuyo, Mendoza, Argentina

M. Iorizzo
Department of Horticultural Sciences, Plants for Human Health Institute, North Carolina State University, 600 Laureate Way, Kannapolis, NC 28081, USA

P. Simon et al. (eds.), *The Carrot Genome*, Compendium of Plant Genomes,
https://doi.org/10.1007/978-3-030-03389-7_15

15.1 Introduction

Anthocyanins are nearly ubiquitous in the plant kingdom. These water-soluble flavonoid compounds confer purple, red, and blue pigmentation to several organs and tissues of numerous plants species (Harborne and Williams 2000). In the plant, these pigments serve various roles, including attraction of animals and insects for seed dispersal and pollination, protection against ultraviolet light, and amelioration of different abiotic and biotic stresses, such as drought, salinity, wounding, cold temperatures, and phytopathogen attacks (reviewed by Shirley 1996).

As dietary components, anthocyanins possess various health benefits, mainly due to their antioxidant and anti-inflammatory properties. Because oxidative stress and inflammation are considered root causes of many chronic diseases (reviewed by He and Giusti 2010), the consumption of anthocyanin-rich foods (mainly present in fruits and vegetables) contributes to the prevention and improvement of various health conditions. Thus, the consumption of these pigments has been associated with reduced risk of cardiovascular disease (Bell and Gochenaur 2006), improved glucose regulation (Jayaprakasam et al. 2005), prevention of autoimmune arthritis (Min et al. 2015), decrease risk of some types of cancer (Lin et al. 2017), as well as aiding in the prevention of cognitive decline and neurological disorders (Joseph et al. 2005).

Anthocyanins are synthesized via the phenylpropanoid pathway, a late branch of the shikimic acid pathway (Herrmann and Weaver 1999). A schematic representation of the anthocyanin pathway is presented in Fig. 15.1. The chemical structure of these pigments consists of a core chromophore (anthocyanidin) that varies in the number and positions of hydroxyl and methyl groups. Although, to date, a total of eighteen different anthocyanidins have been identified, only six of them are ubiquitously distributed in higher plants: pelargonidin, cyanidin, delphinidin, peonidin, petunidin, and malvidin (Davies et al. 2017). When anthocyanidins are bound to sugar moieties, they are known as anthocyanins. In addition, these sugar moieties may be acylated by a range of aromatic or aliphatic acids. The different combinations of anthocyanidins, sugars, and organic acids result in a huge molecular diversity array for these pigments, with over 600 naturally occurring anthocyanins reported to date (Andersen and Jordheim 2006). The extent of anthocyanin glycosylation and acylation has a significant effect on their chemical stability, bioavailability, and biological activities (reviewed by Prior and Wu 2006). While glycosylation confers increased stability and water solubility, the acylation of the sugar residues significantly increases stability (Mazza et al. 2004). Thus, because acylated anthocyanins (AA) are chemically more stable than their non-acylated counterparts, the formers are more suitable and preferred for their use as natural food dyes. Conversely, non-acylated anthocyanins (NAA) are significantly more bioavailable (~fourfolds) than their acylated counterparts from the same species (Charron et al. 2009; Kurilich et al. 2005). The latter is a relevant aspect when it comes to breeding for nutritional value, as the potential of a dietary component to provide health benefits is largely affected by its bioavailability.

15.2 Carrot Anthocyanins in Human Health

A number of health-related studies have used purple carrots as a source of anthocyanins. Raw and microwave-cooked purple carrots were administered to human subjects in clinical feeding trials, and anthocyanin recovery rate in blood plasma and urine was assessed (Kurilich et al. 2005). It was found that the recovery of NAA was 8–10 and 11–14 times higher than the recovery of AA in plasma and urine samples, respectively, whereas cooking further increased the recovery of NAA but not AA. Because the study of Kurilich et al. (2005) was conducted with whole carrots, the effects of the plant matrix and of anthocyanins chemical structure on the pigments bioavailability could not be separated. Thus, in order to circumvent matrix effects, a similar feeding trial using purple carrot juice, instead of whole carrots,

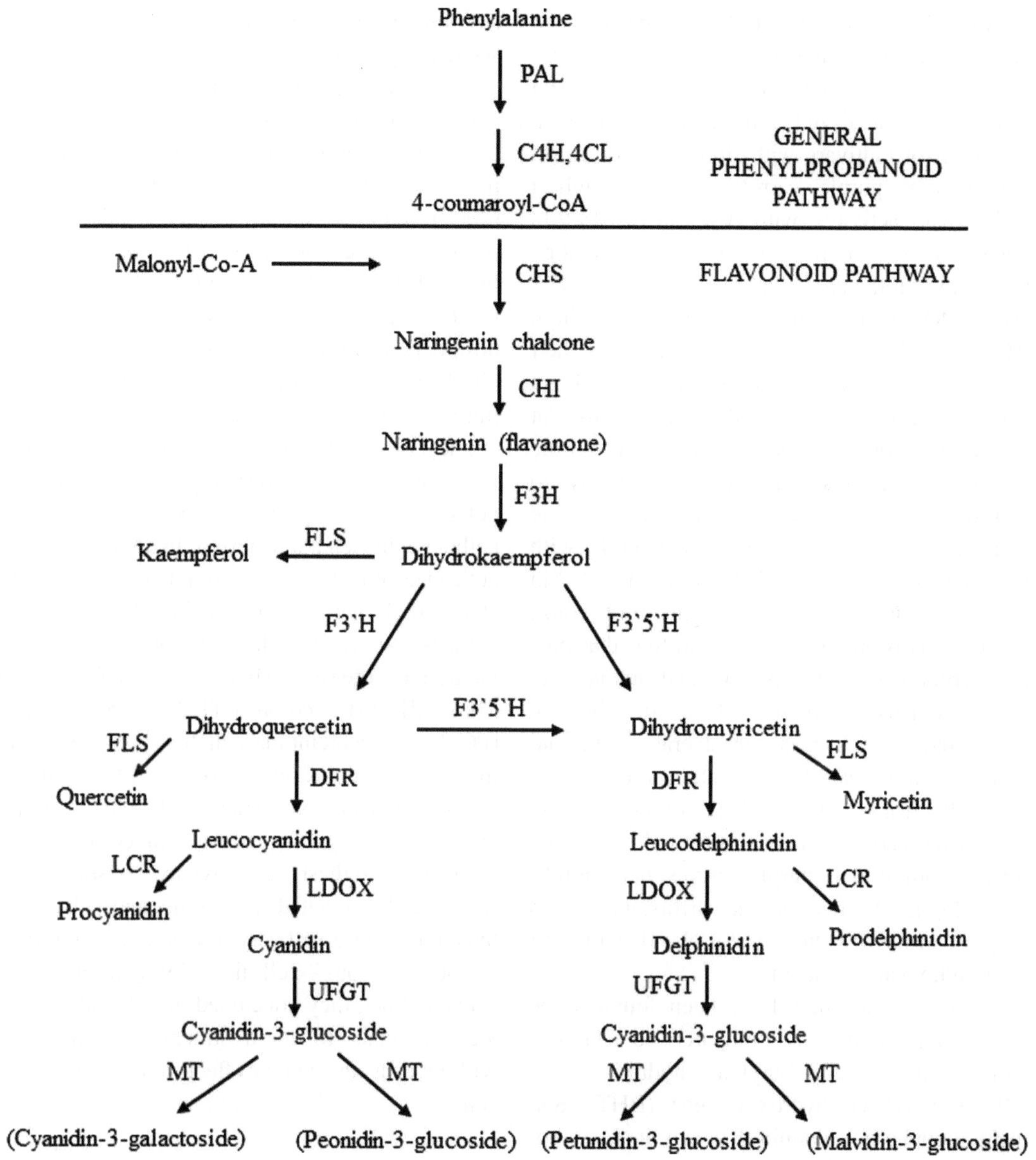

Fig. 15.1 Schematic representation of the anthocyanin biosynthesis pathway. PAL, phenylalanine ammonia-lyase; C4H, cinnamate 4-hydroxylase; 4CL, 4-coumarate: CoA ligase; CHS, chalcone synthase, CHI, chalcone isomerase; F3H, flavanone 3-hydroxylase; FLS, flavonol synthase; F3′H, flavonoid 3′-hydroxylase; F3′5′H, flavonoid 3′,5′-hydroxylase; DFR, dihydroflavonol 4-reductase; LDOX, leucoanthocyanidin dioxygenase; UFGT, UDP-glucose: flavonoid 3-*O*-glucosyltransferase; MT, methyltransferase. Modified from Holton and Cornish (1995)

was conducted, revealing recovery rates of NAA fourfold higher those of AA (Charron et al. 2009). Because anthocyanins were consumed as juice, the higher bioavailability of NAA versus AA was attributed to their chemical structure and not to interactions with the plant matrix. These results using carrot anthocyanins coincide with those reported for anthocyanins from steamed red cabbage, finding that NAA was fourfold more bioavailable than AA (Charron et al. 2007).

Antioxidant capacity (AC) of purple and non-purple carrots and its association with pigment composition has been investigated. Sun et al. (2009) analyzed antioxidant activities in seven carrot cultivars with different root color, finding that purple carrots had the highest antioxidant activity, with values 3.6–28-fold higher than activities in other root colors, depending on the analytical method used (ABTS and DPPH) and the cultivars compared. The high AC of purple carrots was associated with their high concentration of phenolic compounds in general, and in particular with anthocyanins, but not with carotenoids concentration. In purple carrots, carotenoids contributed minimally to their overall AC, representing less than 3% of the total AC (Sun et al. 2009). Coincidently with these results, Leja et al. (2013) evaluated AC in 35 carrot cultivars with different root color, including two purple carrots, reporting that purple carrots had the highest AC, and this activity was associated with their level of phenolic compounds which was, on average, ninefold higher than in carrots with other root colors. Similarly, Algarra et al. (2014) evaluated AC in two purple and one orange carrot cultivars from Spain, finding that purple carrots had significantly higher AC than orange carrots, and AC in purple carrots was directly correlated with their total anthocyanin content.

Carrot anthocyanins have been found to be potent antioxidants as compared to classical antioxidants, such as butylated hydroxyanisole (BHA), butylated hydroxytoluene (BHT), and alpha-tocopherol (vitamin E), and it was proposed that these pigments might prevent lipids auto-oxidation and peroxidation in biological systems (Narayan et al. 1999). More recently, Olejnik et al. (2016) reported reduced DNA damage due to oxidative stress in colon mucosa cells treated with purple carrot extract, indicating that anthocyanin-rich extracts from purple carrots confer protection to colonic cells against oxidative stress. In line with these results, extracts of purple carrots demonstrated strong AC and reduced level of inflammation markers in a human intestinal cell line, suggesting that carrot anthocyanins can ameliorate oxidative stress-mediated intestinal inflammatory responses (Zhang et al. 2016).

In addition to their antioxidant and anti-inflammatory properties, carrot anthocyanins have also been found to exert antiproliferative properties against some cancer types. Netzel et al. (2007) reported antiproliferative activity of a purple carrot extract on two human cancer cell lines (HT-29 colorectal adenocarcinoma and HL-60 promyelocytic leukaemia) in a dose-dependent manner. Similarly, Jing et al. (2008) found inhibition of colon cancer cell proliferation when exposed to anthocyanin-rich extracts of carrot and other vegetables and fruits, with carrot extracts presenting the second most potent antiproliferative effects among the anthocyanin-rich plants tested. In addition, potent anticancer effects were reported by Sevimli-Gur et al. (2013), who tested the effect of purple carrot extracts on various human cancer cell lines including breast (MCF-7, SK-BR-3, and MDA-MB-231), colon (HT-29), and prostate (PC-3) adenocarcinoma cell lines, as well as in mouse neuroblastomas (neuro-2A), finding dose-dependent cytotoxicity of carrot anthocyanins against all investigated cancer cell lines, with the highest cytotoxicity observed in 'neuro-2A' associated with brain cancer. Because the authors observed very little toxicity in normal —not cancerous—cell lines by natural purple carrot extract, they concluded that the latter is an ideal candidate for the treatment of brain cancer without causing negative effects to normal healthy cells.

15.3 Carrot Acylated Anthocyanins as Natural Food Dyes and Non-acylated Anthocyanins for Increased Bioavailability

Vegetable species containing high levels of AA, such as purple carrots (Kammerer et al. 2004) and cabbage (Charron et al. 2007), have potential as natural food dyes due to their higher chemical stability, as compared to NAA. Chemical stability is an important requisite for any food colorant, in order to avoid oxidation or decomposition, to be able to maintain the desired color in the food. In general, natural pigments are more prone to degradation than synthetic dyes. The latter (synthetic food dyes) are chemically stable under a broad range of temperature, pH, light intensity, and oxygen concentration, making them suitable for a wide range of food conservation conditions. However, there is increasing public concern regarding potential health problems associated with the consumption of synthetic dyes, including allergic reactions (Koutsogeorgopoulou et al. 1998), behavioral and neurological adverse effects (McCann et al. 2007), and potential carcinogenesis (Dees et al. 1997). Thus, because of these health issues, alternative colorants from natural sources are desired.

Purple carrots have, in general, a higher proportion of AA than NAA (discussed below). However, substantial intra-specific variation for AA:NAA ratio, as well as for total AA concentration in the root, has been found among purple carrot lines (Kammerer et al. 2004), suggesting that not all carrots genetic stocks are equally suitable for food dye production. On the other hand, data from previous studies of Kurilich et al. (2005) and Charron et al. (2009) suggest that carrots with high levels of NAA will have increased bioavailability and—therefore—increased functional value, both desirable quality traits for fresh consumption carrots. In this context, it appears important to characterize purple carrot germplasm for their root anthocyanin composition, in order to identify and select materials for either purpose. Also, advances in genetic research aiming at elucidating the genetic factors conditioning pigment biosynthesis and modification (i.e., glycosylation and acylation) will provide knowledge for the development of new cultivars suitable for each purpose.

15.4 Anthocyanin Composition and Diversity in Purple Carrots

Broad genetic variation for anthocyanin concentration and pigment distribution in carrot root tissues can be found in the purple carrot germplasm (Fig. 15.2). Montilla et al. (2011) analyzed four commercial cultivars and reported a range of 1.5–17.7 mg/100 g fresh weight (fw). Kammerer et al. (2004) evaluated 15 accessions, reporting a range of 45–17,400 mg/kg dry weight (dw), which corresponds to ~0.5–191 mg/100 g fw (considering a dry matter content of 11%), whereas Algarra et al. (2014) found 93.4–126.4 mg/100 g fw in two cultivars from Spain. Very recently, Bannoud et al. (2018) examined 26 accessions from diverse geographic origins and phenotypes, including open-pollinated (OP) and hybrid cultivars, and found root total anthocyanin levels in a range of 1–229 mg/100 g fw. Total anthocyanin content was significantly and positively correlated ($r = 0.85$) with total phenolic content (evaluated by spectrophotometry), suggesting that anthocyanins account for a large proportion of the phenolic compounds present in purple carrot roots. Noteworthy, the anthocyanin content of the accessions, relative to each other, can be roughly predicted visually by the color intensity and coverage of the root tissues with purple color. Thus, accessions with dark purple phloem and xylem tissues tend to have the highest anthocyanin concentration, whereas accessions pigmented only in the outermost phloem cell layers had the lowest pigment content. Thus, visual examinations can be useful for rapid selection of materials based on their root total pigment content.

Anthocyanin composition varies among carrot genotypes. The main anthocyanins in purple carrot roots are cyanidin glycosides (Algarra et al. 2014; Bannoud et al. 2018; Kammerer et al. 2004; Montilla et al. 2011), although traces of pelargonidin, petunidin, and peonidin have been reported in some genetic backgrounds (Algarra et al. 2014; Kammerer et al. 2003; Montilla et al. 2011). Among the cyanidin glycosides, five major compounds, two non-acylated, and three acylated are commonly found in purple carrots (Table 15.1). The percentage of AA relative to the total anthocyanin content found across different studies varied from 49.6 to 99% (Algarra et al. 2014; Kammerer et al. 2004; Montilla et al. 2011; Netzel et al. 2007). In terms of absolute

Fig. 15.2 Examples of the extent of phenotypic variation for root anthocyanin pigmentation in the purple carrot germplasm

Table 15.1 Carrot cyanidin derivatives with approximate HPLC retention times and molecular masses

Compound	Abbreviation	RT	MW
Cy-3-(2″-xylose-6-glucose-galactoside)	Cy3XGG	14.0	743
Cy-3-(2″-xylose-galactoside)	Cy3XG	15.1	581
Cy-3-(2″-xylose-6″-sinapoyl-glucose-galactoside)	Cy3XSGG	15.4	949
Cy-3-(2″-xylose-6″-feruloyl-glucose-galactoside	Cy3XFGG	16.0	919
Cy-3-(2″-xylose-6″-(4-coumaroyl) glucose-galactoside)	Cy3XCGG	16.4	889

RT is retention time (min) for the chromatographic procedure described by Kurilich et al. (2005). MW is molecular weight

content, carrot accessions with up to ~155 mg/100 g fw of AA and ~36 mg/100 g fw of NAA have been reported [a dry matter content of 11% was considered for calculations for expressing the original data (expressed as mg/kg dw) on a fresh weight basis] (Kammerer et al. 2004).

The content and relative proportion of individual anthocyanin pigments varies across accessions. Cyanidin glycosides acylated with ferulic (Cy3XFGG), sinapic (Cy3XSGG), and coumaric acid (Cy3XCGG) are, in that order, the most abundant pigments found in purple carrot roots. According to Kammerer et al. (2004), Cy3XFGG was the most abundant anthocyanin in 13 of the 15 accessions evaluated, representing in these materials 42.5–83.8% of the total anthocyanin content, with Cy3XSGG being the predominant pigment in the remaining two accessions and accounting for 42–51% of the total anthocyanins. Similarly, Montilla et al. (2011) reported Cy3XFGG as the predominant pigment in 3 of the 4 commercial cultivars analyzed by them, with Cy3XSGG being the major pigment in the other cultivar. Among the non-acylated pigments, Cy3XG is generally at a higher concentration than Cy3XGG (Kammerer et al. 2004; Montilla et al. 2011).

Altogether, these data indicate that there is sufficient genetic variability in the purple carrot germplasm for anthocyanin concentration and composition, allowing for selection of materials with high AA content for food dye production, and relatively high NAA content for increased nutritional value in, for example, fresh consumption carrots.

15.5 Inheritance of Anthocyanin Pigmentation and Mapping of Simply Inherited Traits (P_1, P_2, P_3, and *Raa1*)

In the last decades, substantial progress toward understanding the genetics underlying anthocyanin pigmentation in carrot has been made. A simply inherited gene, P_1, controlling purple pigmentation in the carrot root was first described by Simon (1996), by means of segregation analysis in F_1, F_2, F_3, and backcross (BC) populations derived from crosses between purple and non-purple-rooted plants. In addition to P_1, a simply inherited dominant locus conditioning purple pigmentation in the nodes, called P_2, was also described in the same study, with P_1 and P_2 being linked and separated by approximately 36 cM (Simon, 1996). P_1 was genetically mapped to chromosome 3 by Vivek and Simon (1999) and Yildiz et al. (2013) in the 'B7262' genetic background, which corresponds to a Turkish purple-rooted carrot with green petioles. More recently, P_3, a dominant locus conditioning purple pigmentation in the roots and petioles of P9547 and PI652188, two carrot lines with purple roots and petioles from Turkey and China, respectively, was described and mapped to chromosome 3 (Cavagnaro et al. 2014). Comparative linkage mapping using segregating populations developed from crosses using B7262, P9547, and PI652188 as the purple-rooted source progenitors, revealed that P_1 and P_3 correspond to different loci that map to chromosome 3 at more than 30 cM apart (Cavagnaro et al. 2014) (Fig. 15.3).

Very recently, the segregation for root and petiole pigmentation was investigated and mapped in an F_2 family of a different genetic background derived from the purple-rooted source BP85682 of Syrian origin, along with segregation and mapping analyses for these traits in advanced generations (F_3, F_5) of the mapping populations used previously by Cavagnaro et al. (2014) (Iorizzo et al. 2019). Root and petiole pigmentation fully co-segregated and revealed a 3:1 purple:non-purple ratio in the F_2, consistent with a single dominant gene model. By means of linkage analysis using common markers between this and the other populations with previously known genetic backgrounds, it was revealed that this locus in the Syrian background BP85682 corresponds to P_3. Thus, P_3 controls root and petiole pigmentation in all of the purple carrot genetic backgrounds evaluated to date, except B7262, in which 'root pigmentation' is conditioned by the P_1 locus.

In addition to P_1, P_2, and P_3, which condition the presence or absence of pigmentation in the root, nodes, and petioles, respectively, a simply inherited trait conditioning the relative content (%) of AA versus NAA, with high % of AA being dominant over low % AA (i.e., high % NAA), was discovered and mapped in chromosome 3, separated from P_3 at 17.9 cM. This gene was termed *Raa1* (for 'root anthocyanin acylation').

15.6 Quantitative Trait Loci (QTL) Mapping

Quantitative trait loci (QTL) analysis for individual root anthocyanin pigments and for total root anthocyanins ('*RTPE,*' for 'root total pigment estimate') was performed in the P9547 background (Cavagnaro et al. 2014). Fifteen significant QTL for *RTPE* and four individual

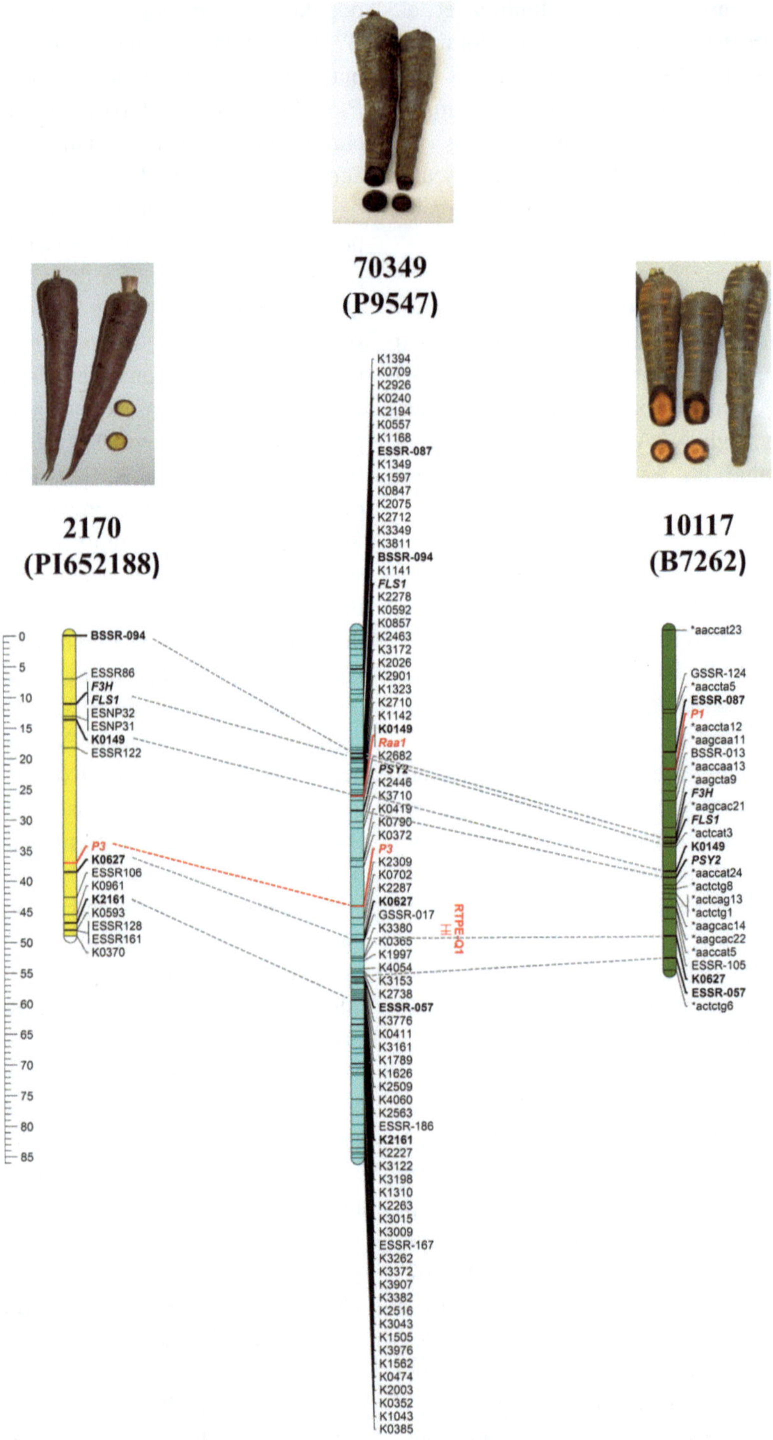

Fig. 15.3 Comparative mapping of loci on chromosome 3 controlling anthocyanin pigmentation in three carrot genetic backgrounds. The anthocyanin simply inherited loci P_1, P_3, and *Raa1* (root anthocyanin acylation), and the QTL conditioning root total anthocyanin content root total pigment estimate (*RTPE*), are indicated in red and italic font. The F_2 mapping populations and the respective purple-rooted progenitors (in parenthesis) are indicated above each linkage group. Root images correspond to the ultimate purple-rooted source used for developing the F_2s. All linkage groups. Modified from Cavagnaro et al. (2014)

root pigments were detected and mapped across five carrot chromosomes (Chr. 1, Chr. 2, Chr. 3, Chr. 6, and Chr. 8). Eight of the QTL with largest effects (26.6–73.3% variation explained) co-localized to two regions of Chr. 3. In one of these regions, a QTL for *RTPE* explaining 50.5% of the variation (*RTPE-Q1*) and major QTL for four individual root anthocyanins co-localized with P_3 (Fig. 15.4a), further confirming that this region conditions root and petiole pigmentation in the P9547 background (Cavagnaro et al. 2014). Very recently, Iorizzo et al. (2019) performed high-resolution QTL mapping in this region using a larger size (N = 421) of the same population used previously by Cavagnaro et al. (2014) (N = 187), revealing the same five major QTL (Fig. 15.4b). The larger phenotypic and genotypic data set used in this study allowed the construction of a linkage map with better resolution for *RTPE-Q1* and the other anthocyanin QTL, as indicated by the substantially smaller map region as delimited by the QTL confidence intervals—obtained in the new map. In the study of Cavagnaro et al. (2014), these 5 QTL spanned a 12 cM region (Fig. 15.4a), whereas in the new map they spanned a 6.3 cM region, with co-localized QTL for RTPE and three root anthocyanins (Cy3XG, Cy3XSGG, Cy3XFGG) within a 3 cM region (Fig. 15.4b).

The other region in Chr. 3 with co-localized QTL spanned a 3.6 cM map region (from 24.1–27.7 cM), as defined by the QTL confidence

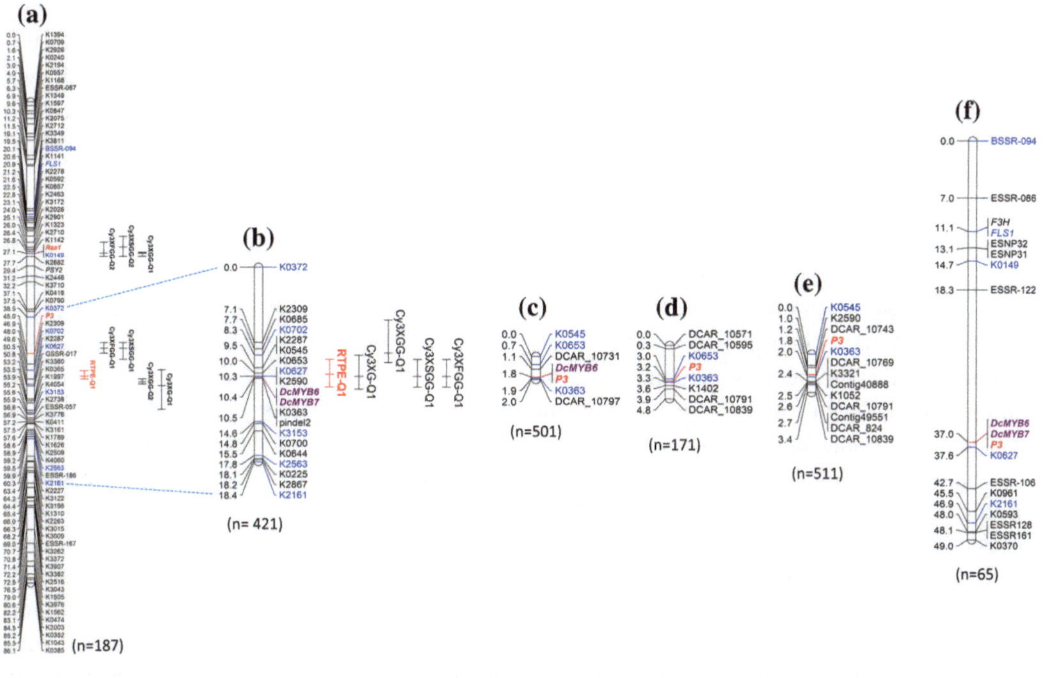

Fig. 15.4 Genetic mapping of anthocyanin pigment traits in chromosome 3 of different carrot populations. P_3, *Raa1*, and QTL for 'root total pigment estimate' (RTPE) and root anthocyanin pigments (Cy3XG, Cy3XGG, Cy3XSGG, Cy3XFGG) mapped in population 70,349 by Cavagnaro et al. (2014) (**a**). High-resolution mapping in the *RTPE-Q1* region using a larger 70,349 population (**b**). The connecting dotted lines indicate the flanking markers of the map region further analyzed in the study of Iorizzo et al. (2019). In (**a**) and (**b**), bars to the right of the linkage groups represent support intervals of the QTL. P_3, *Raa1*, and *RTPE-Q1* explaining 52.3% of the variation are denoted in red. The P_3 locus, conditioning root and leaf pigmentation, was mapped in populations 95,710 (**c**), 5394 (**d**), 5723 (**e**), and 2170 (**f**). Population size is indicated in parenthesis under each linkage group. Markers in blue denote common markers across different maps. Two anthocyanin candidate MYB genes, *DcMYB6* and D*cMYB7*, are denoted in bold purple letters. Modified from Iorizzo et al. (2019)

intervals, and harbored overlapping QTL for two acylated anthocyanins, Cy3XSGG and Cy3XFGG, and the non-acylated pigment Cy3XGG (Fig. 15.4a). These three QTL co-localized with *Raa1*. The Cy3XGG QTL had the highest statistical support (LOD = 104.7), the largest phenotypic effect (73.3%), and the shortest map distance covered by its confidence interval (0.7 cM) of all 15 mapped QTL. Based on examination of the structure of the five cyanidin glycosides present in carrot roots (Table 15.1), together with available information on anthocyanin biochemistry in carrot (Gläßgen and Seitz 1992; Rose et al. 1996), and other species (He et al. 2010), the authors suggested that Cy3XGG is the most likely substrate for acylation, presumably by an acyltransferase, to produce the acylated pigments Cy3XCGG, Cy3XFGG, and Cy3XSGG. In agreement with the proposed model, they found a strong negative correlation between the non-acylated Cy3XGG and the total content of acylated anthocyanins ($r = -0.99$, $p < 0.001$), strongly suggesting that acylation of Cy3XGG causes this shift from non-acylated to acylated anthocyanin forms. In addition, a clear bimodal distribution was found for Cy3XGG in the F_2, displaying a segregation ratio compatible with a simply inherited dominant trait. Altogether, these data and the co-localization Cy3XGG and *Raa1* suggest that indeed a single dominant gene (*Raa1*) controls 'high' versus 'low' content of acylated anthocyanins in carrot roots.

Because acylation of carrot anthocyanins influences bioavailability (Charron et al. 2009; Kurilich et al. 2005) and pigment stability (Mazza et al. 2004; Prior and Wu 2006), further characterization of *Raa1* is of interest. Ongoing efforts towards the structural and functional characterization of this gene may have a positive impact in carrot breeding programs aiming at developing carrot cultivars with higher nutraceutical value (e.g., with increased content of bioavailable non-acylated anthocyanins) as well as carrots for the production of chemically stable acylated pigments for the food industry.

15.7 Candidate Genes for the Control of Anthocyanin Biosynthesis and Modification

In the last decade, research efforts towards the identification of candidate genes involved in anthocyanin biosynthesis, its regulation and modification (i.e., glycosylation and acylation) have been undertaken, although the most fruitful initiatives have been reported very recently, concomitantly with the completion and publication of the carrot genome sequence (Iorizzo et al. 2016). With the release of the carrot genome, a curated annotation of structural genes involved in the biosynthesis of flavonoids and anthocyanins identified 97 genes. These included 20 genes involved in biosynthesis of anthocyanidins and 76 UDP-glycosyltransferases (UDPG-like) and one methyltransferase (OMT) which are involved in anthocyanin glycosylation and methylation. Compared with some other crops like grapevine and *Arabidopsis*, the carrot genome lacks the anthocyanidin reductase (ANR) gene. The ANR enzyme catalyzes the first committed step of the proanthocyanidin (PA) pathway, converting cyanidin to epicatechin. As a result of whole genome duplications, several flavonoid/anthocyanin genes are duplicated. For example, three copies of the phenylalanine ammonia-lyase (*PAL1*, *PAL3*, and *PAL4*) were retained after three whole genome duplications (WGD). Although the role of these duplicated genes in carrot is still unknown, they may have specialized in the regulation and the expression of the pathway in a specific tissue or under specific environmental conditions (e.g., abiotic stresses such as cold temperatures or UV light). Efforts towards the identification of candidate genes controlling anthocyanin biosynthesis and their modifications, including glycosylation and acylation, have been made.

A first attempt to identify candidate genes for P_1, by means of linkage analysis of structural [*PAL3*, flavanone 3-hydroxylase (*F3H*), flavonol synthase (*FLS1*), UDP-glucose: flavonoid 3-*O*-

glucosyltransferase (*UFGT*), and leucoanthocyanidin dioxygenase (*LDOX2*)] and regulatory (*DcEFR1*, *DcMYB3*, and *DcMYB5*) anthocyanin biosynthesis genes in a population segregating for P_1, was reported by Yildiz et al. (2013). In their study, the position of these genes did not coincide with P_1, despite the fact that two genes (*F3H* and *FLS1*) were linked to the trait locus, suggesting that none of the genes evaluated is a candidate for P_1. In the same study, expression analysis of six anthocyanin structural genes in purple and non-purple carrots revealed higher expression level for *F3H*, *LDOX2*, *PAL3*, chalcone synthase (*CHS1*), and dihydroflavonol 4-reductase (*DFR1*) in solid purple carrots as compared to purple–orange and orange carrots, suggesting a coordinated regulatory control of anthocyanin expression in the carrot root. However, because none of these genes co-localized with P_1 in the linkage map, their candidacy for this trait must be ruled out.

Similar results were obtained by Xu et al. (2014), who analyzed expression levels of 13 anthocyanin structural genes [*PAL1*, PAL3/PAL4, CHS1, *CHS2/CHS9*, *F3H1*, F3′H1, DFR1, LDOX1/LDOX2, chalcone isomerase 1 (CHI1), cinnamate 4-hydroxylase 1 and 2 (*C4AH1*, *C4AH2*), and 4-coumaroyl-coenzyme A ligase 1 and 2 (4CL1, 4CL2)] in purple and non-purple roots of nine carrot cultivars and reported significant upregulation for nine of these genes (PAL3/PAL4, *C4AH1*, 4CL1, CHS1, CHI1, F3H1, F3′H1, DFR1, LDOX1/LDOX2) in purple-rooted carrots as compared to non-purple ones. The expression of these genes was associated with anthocyanin concentration in carrot root. Together, data from this study suggest that these structural genes are involved in anthocyanin biosynthesis in purple carrot roots. The concomitant expression of numerous structural genes in purple carrot roots suggests a coordinated transcriptional regulation for these genes. The position of these genes in the carrot genome assembly was investigated revealing that only one of them, F3H1, was located in Chr. 3, although in a position (7,964,484–7,965,154) unrelated to the positions of P_1, P_3/*RTPE-Q1*, and *Raa1*, indicating that none of the structural genes evaluated in this study are candidates for these traits.

As described earlier, anthocyanin glycosylation and acylation play important roles in the stability and, therefore, accumulation of these pigments in the carrot root. Such pigment modifications occur in the last steps of anthocyanin biosynthesis. Chen et al. (2016) cloned and characterized a UDP-glucose: sinapic acid glucosyltransferase (USAGT) gene, called *DcUSAGT1* (GenBank accession number: KT595241), from purple carrot roots. USAGT catalyzes the transfer the glucose moiety to the carboxyl group of sinapic acid, thereby forming the ester bond between the carboxyl-C and the C1 of glucose (1-*O*-sinapoyl-glucose). 1-*O*-sinapoyl-glucose can serve as an acyl donor in the acylation of anthocyanins to generate cyanidin 3-xylosyl (sinapoylglucosyl) galactoside (Cy3XSGG). Sequence homology analysis of *DcUSAGT1*, as compared against the NCBI database, revealed that this gene belongs to the glycosyltransferase-B-type superfamily and showed highest sequence homology to UDP-glucose: glycosyltransferases from *Medicago truncatula*, *Clitoria ternatia*, and *Vitis vinifera*. The DcUSAGT1 protein was purified from purple carrot roots, and various parameters associated with enzyme activity were characterized, confirming the theoretical role of this enzyme in the transformation of sinapic acid to produce 1-*O*-sinapoyl-glucose. Gene expression analysis of *DcUSAGT1* in the roots of three purple and three non-purple (orange) carrot cultivars revealed significant upregulation of this gene in all the purple-rooted cultivars. These data suggest that *DcUSAGT1* may indeed play a role in the glycosylation of sinapic acid to produce 1-*O*-sinapoyl-glucose, which can then serve as a sinapoyl donor to produce Cy3XSGG. However, because this enzyme uses, specifically, sinapic acid as a substrate, and not other hydroxycinnamic acids such as coumaric and ferulic acids, its contribution to the overall production of acylated pigments may be limited considering that cyanidin glycosides acylated with coumaric acid (Cy3XCGG) are the most abundant anthocyanin pigments in the root of most purple carrots (Kammerer et al. 2004; Montilla et al. 2011). Nonetheless, the activity of *DcUSAGT1*, and perhaps also of other carrot USAGT enzymes not yet identified, may be important in particular genetic backgrounds displaying high content of Cy3XSGG (Kammerer et al. 2004; Montilla et al. 2011). Noteworthy, *DcUSAGT1* is located in Chr. 9 (position 1,585,289–1,586,523) and therefore is unrelated to P_1, P_3, and *Raa1*. It is possible that regulatory genes in the region of P_1 or P_3 may regulate the coordinated expression of *DcUSAGT1* and other structural anthocyanin biosynthetic genes distributed elsewhere in the genome.

A gene involved in anthocyanin glycosylation was cloned from purple carrot roots and further described (Xu et al. 2016). This gene, called *DcUCGalT1* (accession number AB103471), is a galactosyltransferase responsible for glycosylation of cyanidin with galactose. Specifically, this enzyme catalyzes the first step of cyanidin glycosylation by transferring of the galactosyl moiety from UDP-galactose to cyanidin, to produce cyanidin-3-*O*-galactoside. Heterologous expression of *DcUCGalT1* in *Escherichia coli* BL21 (DE3) followed by analysis of the enzyme activity with other anthocyanin aglycones and flavonoid substrates revealed much lower galactosylation activity for peonidin and pelargonidin, kaempferol and quercetin, as compared to cyanidin. However, when different glycosyl donors were tested for glycosylating cyanidin, it was found that DcUCGalT1 accepted only UDP-galactose, but not UDP-glucose or UDP-xylose. These results indicate that DcUCGalT1 is highly specific for galactosylation of cyanidin. Comparative analysis of gene expression of *DcUCGalT1* in purple and non-purple roots of nine carrot cultivars revealed upregulation in all purple roots, as compared to non-purple ones, and the expression level of this gene was positively correlated with root anthocyanin concentration. Altogether, these data indicate that *DcUCGalT1* is involved in the glycosylation of cyanidin with galactose and suggest that such galactosylation improves anthocyanin stability and accumulation in the root. A BLAST search of *DcUCGalT1* in the carrot genome assembly revealed highest sequence homology with DCAR_009912 which is located in Chr. 3 at a position (12,350,188–12,351,646) unrelated to P_1, P_3, *Raa1*, and *RTPE-Q1*, suggesting that *DcUCGalT1* is not the key gene responsible for carrot root and petiole pigmentation.

Recently, a MYB transcription factor, called *DcMYB6*, associated with anthocyanin biosynthesis in purple carrot roots was cloned described (Xu et al. 2017). Phylogenetic analysis of this gene along with other flavonoid-related MYB transcription factors from other species clustered *DcMYB6* into a clade with MYBs involved in anthocyanin biosynthesis. Expression analysis in purple and non-purple roots of nine carrot cultivars (three purple and six non-purple) revealed significant upregulation of *DcMYB6* in the purple-rooted carrots, as compared to the non-purple ones. Heterologous overexpression of *DcMYB6* in *Arabidopsis thaliana* resulted in enhanced accumulation of anthocyanins in vegetative and reproductive tissues, concomitantly with an upregulation of *Arabidopsis* structural anthocyanin genes, suggesting transcriptional regulation of the latter by *DcMYB6*. Together, these results suggest that *DcMYB6* is involved in the regulation of anthocyanin synthesis in purple carrot roots.

Very recently, a study was conducted to identify candidate genes for the *RTPE-Q1*/P_3 region in chromosome 3 conditioning root and petiole pigmentation (Iorizzo et al. 2019). High-resolution mapping of *RTPE-Q1* and four other QTL for root anthocyanins was performed by expanding the QTL analysis reported by Cavagnaro et al. (2014) to a larger population size (Fig. 15.4a, b). In addition, fine mapping of P_3 in four populations from three different genetic backgrounds was performed (Fig. 15.4c–f). Considering these four individual maps, the fine mapping of *RTPE-Q1*/P_3 was done in a total of 1669 individuals, and allowed to delimit the confidence interval of *RTPE-Q1* to a 2.6 cM region, and the smallest map interval containing P_3 (as defined by the closest flanking markers of the trait locus) to a region of 0.3–0.8 cM. Analysis of the genotypic scores for the SNP markers in the map region containing *RTPE-Q1* and P_3, and their

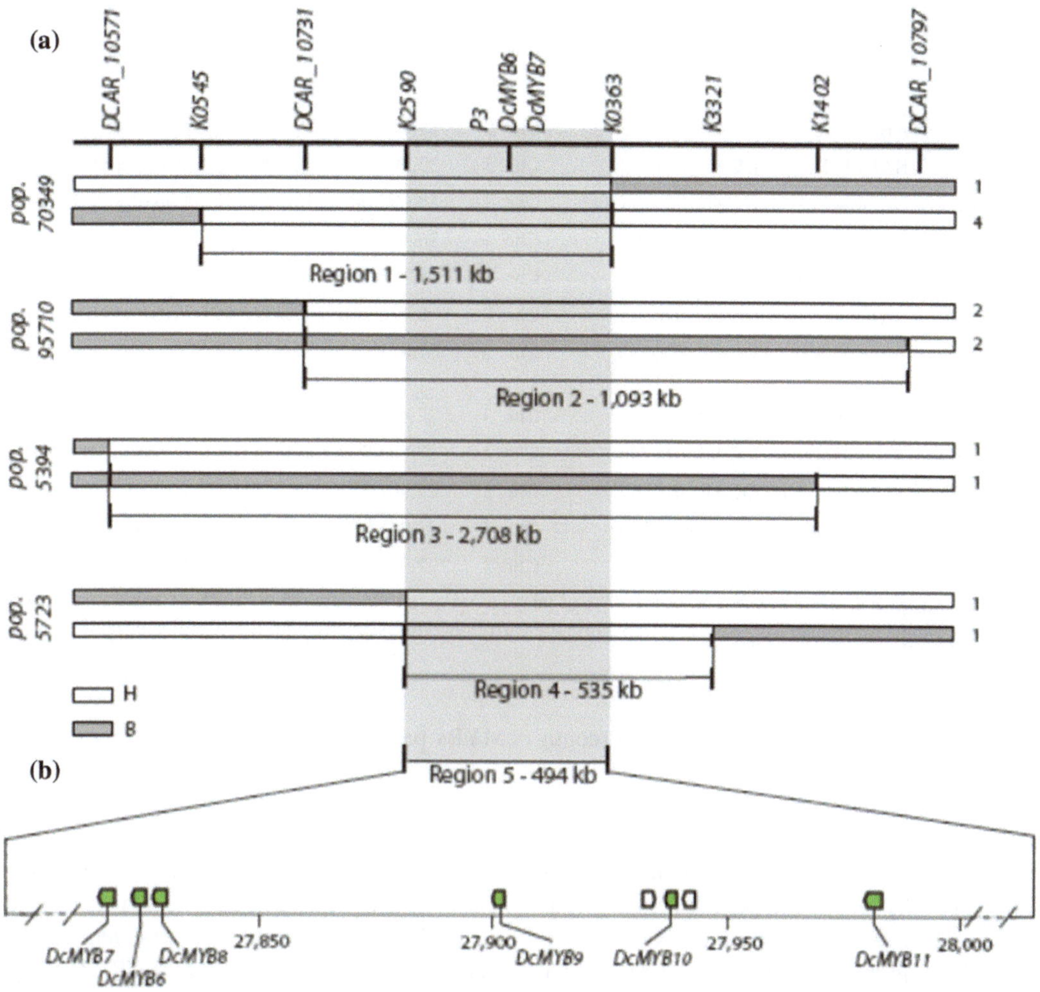

Fig. 15.5 Scheme of the fine mapping approach and identification of candidate genes in the *RTPE-Q1*/P_3 region associated with anthocyanin pigmentation in the carrot root and petioles. A: haplotypes delimiting the genomic regions controlling the *RTPE-Q1* QTL in population 70,349 (region 1), and the P_3 locus in populations 95,710, 5394 and 5723 (region 2–4). White bars indicate the heterozygous haplotypes (H = *Aa*), and gray bars indicate the homozygous recessive haplotypes (B = *aa*). Region 5 represents the genomic sequence delimited by the nearest markers flanking *RTPE-Q1* and the P_3 locus across regions 1–4. Numbers on the right side of each bar represent the number of recombinant genotypes for each haplotype. B: schematic representation of carrot chromosome 3 containing regions 1–5 and the six anthocyanin-related MYBs (*DcMYB6-DcMYB11*) denoted in green boxes. The scheme was drawn to scale. Modified from Iorizzo et al. (2019)

correspondence in the genome assembly, allowed the identification of linkage blocks harboring both traits, spanning a 1511 kb region for *RTPE-Q1*, and a 535–2708 kb region for P_3 (depending on the population) (Fig. 15.5a). These regions were further analyzed for the identification of candidate genes for P_3 and *RTPE-Q1*.

Analysis of the coordinates of the annotated genes involved in the flavonoid or anthocyanin biosynthetic pathways revealed the absence of anthocyanin structural genes in these regions. Also, none of the previously reported transcription factors (TFs) associated with anthocyanin biosynthesis was found within the *RTPE-Q1*/P_3 region (Table 15.2).

A closer examination at *DcMYB6*, described by Xu et al. (2017), revealed that this gene had not been included in the carrot genome assembly, but

Table 15.2 MYB transcription factors associated with carrot anthocyanin pigmentation described to date

Gene ID	Accession no. at NCBI	Gene ID in the carrot genome assembly	Chr. #	Genome coordinates		Reference
				Start	End	
DcMYB1	AB218778.1	DCAR_030745	Chr. 9	29,370,465	29,370,597	Maeda et al. (2005)
DcMYB2	BAF49441.1	DCAR_011083	Chr. 3	32,098,199	32,100,603	Wako et al. (2010)
DcMYB5	BAF49445.1	DCAR_024737	Chr. 7	18,692,086	18,693,681	Wako et al. (2010)
DcMYB4	BAF49444.1	DCAR_015002	Chr. 4	16,426,186	16,428,272	Wako et al. (2010)
DcMYB3-2	BAF49443.1	DCAR_028315	Chr. 8	8,778,266	8,779,336	Wako et al. (2010)
DcMYB3-1	BAF49442.1	DCAR_028315	Chr. 8	8,778,266	8,779,336	Wako et al. (2010)
*DcMYB6**	ARD08872.1	DCAR_000385	Chr. 3	27,831,723	27,833,545	Xu et al. (2017), Iorizzo et al. (2019)
DcMYB7	Not submitted	DCAR_010745	Chr. 3	27,816,911	27,819,103	Iorizzo et al. (2019)
DcMYB8	Not submitted	DCAR_010746	Chr. 3	27,824,309	27,826,050	Iorizzo et al. (2019)
DcMYB9	Not submitted	DCAR_010747	Chr. 3	27,901,372	27,903,024	Iorizzo et al. (2019)
DcMYB10	Not submitted	DCAR_010749	Chr. 3	27,938,999	27,939,453	Iorizzo et al. (2019)
DcMYB11	Not submitted	DCAR_010751	Chr. 3	27,980,959	27,982,962	Iorizzo et al. (2019)

**DcMYB6* was first described by Xu et al. (2017) and later incorporated manually [this gene was not included in the published genome assembly (Iorizzo et al. 2016)] into the carrot genome assembly with the coordinates above (Iorizzo et al. 2019)

it was found in a small unassembled contig (C10735702). This contig was further analyzed, and it was included in the genome assembly with position 27,830,908–27,834,114, which is within the *RTPE-Q1/P*$_3$ region. In addition, *DcMYB6* was genetically mapped in populations segregating for *RTPE-Q1* and *P*$_3$, finding that this gene mapped within the *RTPE-Q1* QTL support interval (Fig. 15.4 B), and fully co-segregated (100% linked) with *P*$_3$ (Fig. 15.4c, f), indicating that this *DcMYB6* is a candidate gene for *RTPE-Q1/P*$_3$.

In addition to defining the genetic and physical position of *DcMYB6* in the region of *RTPE-Q1* and *P3*, and therefore considering this TF as a putative candidate gene, a comprehensive analysis of the genes in the *RTPE-Q1/P*$_3$ region, including gene prediction, orthologous, and phylogenetic analyses, was performed to identify other anthocyanin-related TFs. As result, fourteen carrot TFs (7 MYB-HB-like and 7 bHLH) were found in one or more of the genomic regions associated with *RTPE-Q1* and *P*$_3$, and eight of them (6 MYB-HB-like and 2 bHLH) were present in all five of these regions (Fig. 15.5b). On the basis of orthology and phylogenetic analysis with MYB and bHLH TFs from other species, five MYBs present in all the genomic regions associated with *RTPE-Q1/P*$_3$ and showing high homology with other functionally characterized MYBs involved in the regulation of anthocyanin biosynthesis were selected as candidate genes for *RTPE-Q1/P*$_3$. These genes belong to the R2-R3-MYB family, and they were denominated *DcMYB6* (DCAR_000385), *DcMYB7* (DCAR_010745), *DcMYB8* (DCAR_010746), *DcMYB9* (DCAR_010747), *DcMYB10* (DCAR_010749), and *DcMYB11* (DCAR_010751) (Table 15.2).

Comparative transcriptome (RNA-Seq) and gene expression (qRT-PCR) analyses for these five candidate genes in purple versus non-purple roots, as well as in purple versus non-purple petioles, revealed that *DcMYB7* was the only gene upregulated in all purple tissues from both root and petioles, while *DcMYB11* was exclusively upregulated in all the purple petiole tissues. The gene expression pattern of *DcMYB6* across the different comparisons was, altogether, not positively correlated with anthocyanin pigmentation in neither root nor petioles. This gene was upregulated in purple roots of one of the genetic backgrounds used (population 5394), but it was downregulated in purple roots of population 95,710 and in purple petioles of population 5732 and 95,710. These data suggest a genotype-dependent activity for *DcMYB6*.

Altogether, these results strongly suggest *DcMYB7* as a key gene controlling anthocyanin pigmentation in the carrot root, whereas *DcMYB11* specifically regulates or co-regulates (with *DcMYB7*) petiole pigmentation. These data also indicate that *DcMYB6*, previously described by Xu et al. (2017), is not a key gene controlling anthocyanin pigmentation in either tissue, at least for the carrot genetic backgrounds used in the study of Iorizzo et al. (2019).

Very recently, a study using comparative transcriptome (RNA-Seq) analysis among purple and non-purple tissue samples from carrot calli and carrot tap roots from three purple carrot cultivars was performed with the aim of identifying MYB, bHLH, and WD40 genes that may function as positive or negative regulators in the carrot anthocyanin biosynthesis pathway (Kodama et al. 2018). Among all the comparison performed, a total of 104 MYB, bHLH, and WD40 genes were differentially expressed. Using anthocyanin content measured from a subset of calli and tissue samples, the expression of 32 genes (out of the 104 genes) was shown to be significantly correlated with anthocyanin content. Expression patterns of these genes were compared across the three purple carrot cultivars and, for one of the cultivars, at three different time points, and 11 of the 32 genes were found to be consistently up- or downregulated in a purple color-specific manner. The differential expression of ten of these genes was confirmed by means of qRT-PCR analysis. These results suggest that these TFs may be involved in regulating carrot anthocyanin biosynthesis.

The eleven genes that were consistently differentially expressed in purple versus non-purple tissue comparisons and which expression levels were correlated with anthocyanin content included six MYBs, four bHLH, and one WD40 transcription factors. These genes are located throughout seven of the nine carrot chromosomes, with only one of them located in Chr. 3. The latter gene is a MYB TF with LOC number 108213488, which the authors referred to as *DcMYB6*, the MYB TF previously described by Xu et al. (2017). However, a closer examination at this gene revealed that LOC108213488 corresponds to the *DcMYB7* gene recently described by Iorizzo et al. (2019), not *DcMYB6*. Both *DcMYB6* and *DcMYB7* are located within the cluster of MYBs that is in the P_3/*RTPE* region of Chr. 3, and they seem to regulate anthocyanin pigmentation in the carrot root in different genetic backgrounds, whereas *DcMYB7* is also involved in petiole pigmentation (Iorizzo et al. 2019; Xu et al. 2017). Interestingly, in the study of Kodama et al. (2018), *DcMYB7* was consistently upregulated in cultivars 'CH5544' and 'superblack,' but not in 'night bird,' suggesting a genotype-dependent activity for the gene. The rest of the TFs were located in other chromosomes and therefore are not candidates for P_1 or P_3. One can speculate that the identified bHLH and WD40 genes showing differential expression and correlation with anthocyanin content may act together with *DcMYB7*, conforming the MYB-bHLH-WD40 complex involved in transcriptional regulation of structural flavonoid/anthocyanin genes, as reported in other species (Zhang et al. 2014).

15.8 Chromosome Organization of the Carrot R2-R3-MYBs into a Gene Cluster

The six R2-R3-MYB genes described in the study of Iorizzo et al. (2019) are arranged in a small cluster of genes within a ~166 kb region of chromosome 3 (Fig. 15.5b). Genome-wide analyses of R2-R3-MYB family members have revealed that these transcription factors (and other MYB subfamilies) are commonly found in gene clusters in the genomes of many plant species, including *Arabidopsis* (Stracke et al. 2001), soybean (Du et al. 2012a), maize (Du et al. 2012b), poplar (Wilkins et al. 2009), and grapevine (Matus et al. 2008). Tandem and segmental duplication events have been hypothesized as the main driving mechanism for the expansion and chromosomal organization into clusters of MYB gene families (Feller et al. 2011). In agreement with this hypothesis, in carrot, tandem, and segmental duplications represented the main mode of duplication for MYB-HB-like TFs which include R2-R3-MYB genes (Iorizzo et al. 2016), suggesting that these same evolutionary forces may have also shaped the genome organization and diversity of the R2-R3-MYB gene family in carrot.

References

Algarra M, Fernandes A, Mateus N et al (2014) Anthocyanin profile and antioxidant capacity of black carrots (*Daucus carota* L. ssp. *sativus* var. *atrorubens* Alef.) from Cuevas Bajas, Spain. J Food Compos Anal 33:71–76

Andersen OM, Jordheim M (2006) The Anthocyanins. In: Andersen OM, Markham KR (eds) Flavonoids chemistry, biochemistry and applications. CRC Press, Taylor & Francis, Boca Raton, pp 471–551

Bannoud F, Da Peña Hamparsomián J, Insani M et al (2018) Assessment of genetic diversity for root anthocyanin composition and phenolic content in purple carrots. In: 2nd international symposium on carrot and other Apiaceae, Krakow, Poland. Abstracts book, pp 45–46

Bell DR, Gochenaur K (2006) Direct vasoactive and vasoprotective properties of anthocyanin-rich extracts. J Appl Physiol 100:1164–1170

Cavagnaro PF, Iorizzo M, Yildiz M et al (2014) A gene-derived SNP-based high resolution linkage map of carrot including the location of QTL conditioning root and leaf anthocyanin pigmentation. BMC Genom 15:1118

Charron CS, Clevidence BA, Britz SJ et al (2007) Effect of dose size on bioavailability of acylated and nonacylated anthocyanins from red cabbage (*Brassica oleracea* L. var. *capitata*). J Agric Food Chem 55:5354–5362

Charron CS, Kurilich AC, Clevidence BA et al (2009) Bioavailability of anthocyanins from purple carrot juice: effects of acylation and plant matrix. J Agric Food Chem 57:1226–1230

Chen YY, Xu ZS, Xiong AS (2016) Identification and characterization of *DcUSAGT1*, a UDP-glucose: sinapic acid glucosyltransferase from purple carrot taproots. PLoS ONE 11:e0154938

Davies KM, Schwinn KE, Gould KS (2017) Anthocyanins. In: Thomas B, Murray BG, Murphy DJ (eds) Encyclopedia of applied plant sciences, vol 2, 2nd edn. Academic Press, Elsevier, Oxford, pp 355–363

Dees C, Askari M, Garret S (1997) Estrogenic and DNA-damaging activity of Red No. 3 in human breast cancer cells. Environ Health Perspect 105:625–632

Du H, Feng B-R, Yang S-S et al (2012a) The R2R3-MYB transcription factor gene family in maize. PLoS ONE 7:e37463

Du H, Yang SS, Liang Z et al (2012b) Genome-wide analysis of the MYB transcription factor superfamily in soybean. BMC Plant Biol 12:106

Feller A, Machemer K, Braun EL et al (2011) Evolutionary and comparative analysis of MYB and bHLH plant transcription factors. Plant J 66:94–116

Gläßgen W, Seitz H (1992) Acylation of anthocyanins with hydroxycinnamic acids via 1-*O*-acylglucosides by protein preparations from cell cultures of *Daucus carota* L. Planta 186:582–585

Harborne JB, Williams CA (2000) Advances in flavonoid research since 1992. Phytochemistry 55:481–504

He J, Giusti MM (2010) Anthocyanins: natural colorants with health-promoting properties. Annu Rev Food Sci Technol 1:163–187

He F, Mu L, Yan GL et al (2010) Biosynthesis of anthocyanins and their regulation in colored grapes. Molecules 15:9057–9091

Herrmann KM, Weaver LM (1999) The Shikimate Pathway. Annu Rev Plant Physiol Plant Mol Biol 50:473–503

Holton TA, Cornish EC (1995) Genetics and biochemistry of anthocyanin biosynthesis. Plant Cell 7:1071–1083

Iorizzo M, Ellison S, Senalik D et al (2016) A high-quality carrot genome assembly provides new insights into carotenoid accumulation and asterid genome evolution. Nat Genet 48:657–666

Iorizzo M, Cavagnaro PF, Bostan H et al (2019) A cluster of MYB transcription factors regulates anthocyanin biosynthesis in carrot (*Daucus carota* L.) root and petiole. Front Plant Sci 9:1927

Jayaprakasam B, Vareed SK, Olson LK et al (2005) Insulin secretion by bioactive anthocyanins and anthocyanidins present in fruits. J Agric Food Chem 53:28–31

Jing P, Bomser JA, Schwartz SJ et al (2008) Structure-function relationships of anthocyanins from various anthocyanin-rich extracts on the inhibition of colon cancer cell growth. J Agric Food Chem 56:9391–9398

Joseph JA, Shukitt-Hale B, Casadesus G (2005) Reversing the deleterious effects of aging on neuronal communication and behavior: beneficial properties of fruit polyphenolic compounds. Am J Clin Nutr 81:313S–316S

Kammerer D, Carle R, Schieber A (2003) Detection of peonidin and pelargonidin glycosides in black carrots (*Daucus carota* ssp. *sativus* var. *atrorubens* Alef.) by high-performance liquid chromatography/electrospray ionization mass spectrometry. Rapid Commun Mass Spectrom 17:2407–2412

Kammerer D, Carle R, Schieber A (2004) Quantification of anthocyanins in black carrot extracts (*Daucus carrota* ssp. *sativus* var. *atrorubens* Alef.) and evaluation of their colour properties. Eur Food Res Technol 219:479–486

Kodama M, Brinch-Pedersen H, Sharma S et al (2018) Identification of transcription factor genes involved in anthocyanin biosynthesis in carrot (*Daucus carota* L.) using RNA-Seq. BMC Genom 19:811

Koutsogeorgopoulou L, Maravelias C, Methenitou G et al (1998) Immunological aspects of the common food colorants, amaranth and tartrazine. Vet Hum Toxicol 40:1–4

Kurilich AC, Clevidence BA, Britz SJ et al (2005) Plasma and urine responses are lower for acylated versus nonacylated anthocyanins from raw and cooked purple carrots. J Agric Food Chem 53:6537–6542

Leja M, Kamińska I, Kramer M et al (2013) The content of phenolic compounds and radical scavenging activity varies with carrot origin and root color. Plant Foods Hum Nutr 68:163–170

Lin BW, Gong CC, Song HF et al (2017) Effects of anthocyanins on the prevention and treatment of cancer. Br J Pharmacol 174:1226–1243

Maeda K, Kimura S, Demura T (2005) *DcMYB1* acts as a transcriptional activator of the carrot phenylalanine ammonia-lyase gene (*DcPAL1*) in response to elicitor treatment, UV-B irradiation and the dilution effect. Plant Mol Biol 59:739–752

Matus JT, Aquea F, Arce-Johnson P (2008) Analysis of the grape MYB R2R3 subfamily reveals expanded wine quality-related clades and conserved gene structure organization across *Vitis* and *Arabidopsis* genomes. BMC Plant Biol 8:83

Mazza G, Cacace JE, Kay CD (2004) Methods of analysis for anthocyanins in plants and biological fluids. J AOAC Int 87:129–145

McCann D, Barrett A, Cooper A et al (2007) Food additives and hyperactive behaviour in 3-year-old and 8/9-year-old children in the community: a randomised, double-blinded, placebo-controlled trial. Lancet 370:1560–1567

Min HK, Kim S-M, Baek S-Y et al (2015) Anthocyanin extracted from black soybean seed coats prevents autoimmune arthritis by suppressing the development of Th17 cells and synthesis of proinflammatory cytokines by such cells, via inhibition of NF-κB. PLoS ONE 10:e0138201

Montilla EC, Arzaba MR, Hillebrand S et al (2011) Anthocyanin composition of black carrot (*Daucus carota* ssp. *sativus* var. *atrorubens* Alef.) cultivars antonina, beta sweet, deep purple, and purple haze. J Agric Food Chem 59:3385–3390

Narayan MS, Naidu KA, Ravishankar GA et al (1999) Antioxidant effect of anthocyanin on enzymatic and non-enzymatic lipid peroxidation. Prostaglandins Leukot Essent Fatty Acids 60:1–4

Netzel M, Netzel G, Kammerer DR et al (2007) Cancer cell antiproliferation activity and metabolism of black carrot anthocyanins. Innov Food Sci Emerg 8:365–372

Olejnik A, Rychlik J, Kidoń M et al (2016) Antioxidant effects of gastrointestinal digested purple carrot extract on the human cells of colonic mucosa. Food Chem 190:1069–1077

Prior RL, Wu X (2006) Anthocyanins: Structural characteristics that result in unique metabolic patterns and biological activities. Free Radic Res 40:1014–1028

Rose A, Gläßgen W, Hopp W et al (1996) Purification and characterization of glycosyltransferases involved in anthocyanin biosynthesis in cell suspension cultures of *Daucus carota* L. Planta 198:397–403

Sevimli-Gur C, Cetin B, Akay S et al (2013) Extracts from black carrot tissue culture as potent anticancer agents. Plant Foods Hum Nutr 68:293–298

Shirley BW (1996) Flavonoid biosynthesis: "new" functions for an "old" pathway. Trends Plant Sci 1:377–382

Simon PW (1996) Inheritance and expression of purple and yellow storage root color in carrot. J Hered 87:63–66

Stracke R, Werber M, Weisshaar B (2001) The R2R3-MYB gene family in *Arabidopsis thaliana.* Curr Opin Plant Biol 4:447–456

Sun T, Simon PW, Tanumihardjo SA (2009) Antioxidant phytochemicals and antioxidant capacity of biofortified carrots (*Daucus carota* L.) of various colors. J Agric Food Chem 57:4142–4147

Vivek BS, Simon PW (1999) Linkage relationships among molecular markers and storage root traits of carrot (*Daucus carota* L. ssp. *sativus*). Theor Appl Genet 99:58–64

Wako T, Kimura S, Chikagawa Y et al (2010) Characterization of MYB proteins as transcriptional regulatory factors for carrot phenylalanine ammonia-lyase gene (*DcPAL3*). Plant Biotechnol 27:131–139

Wilkins O, Nahal H, Foong J et al (2009) Expansion and diversification of the *Populus* R2R3-MYB family of transcription factors. Plant Physiol 149:981–993

Xu ZS, Huang Y, Wang F (2014) Transcript profiling of structural genes involved in cyanidin-based anthocyanin biosynthesis between purple and non-purple carrot (*Daucus carota* L.) cultivars reveals distinct patterns. BMC Plant Biol 14:262

Xu ZS, Ma J, Wang F, Ma HY, Wang QX, Xiong AS (2016) Identification and characterization of *DcUCGalT1*, a galactosyltransferase responsible for anthocyanin galactosylation in purple carrot (*Daucus carota* L.) taproots. Sci Rep. 6:27356

Xu Z-S, Feng K, Que F et al (2017) A MYB transcription factor, *DcMYB6*, is involved in regulating anthocyanin biosynthesis in purple carrot taproots. Sci Rep 7:45324

Yildiz M, Willis DK, Cavagnaro PF et al (2013) Expression and mapping of anthocyanin biosynthesis genes in carrot. Theor Appl Genet 126:1689–1702

Zhang Y, Butelli E, Martin C (2014) Engineering anthocyanin biosynthesis in plants. Curr Opin Plant Biol 19:81–90

Zhang H, Liu R, Tsao R (2016) Anthocyanin-rich phenolic extracts of purple root vegetables inhibit pro-inflammatory cytokines induced by H_2O_2 and enhance antioxidant enzyme activities in Caco-2 cells. J Funct Foods 22:363–375

Carrot Volatile Terpene Metabolism: Terpene Diversity and Biosynthetic Genes

16

Mwafaq Ibdah, Andrew Muchlinski, Mossab Yahyaa, Bhagwat Nawade and Dorothea Tholl

Abstract

Carrot is considered one of the leading horticultural crops in the world in terms of its nutritional value, health benefits, and unique flavor based on its high content of carotenoids and volatile aroma compounds. Terpenes such as monoterpenes and sesquiterpenes represent some of the predominant volatile compounds that contribute to carrot aroma and flavor. Variation of terpene composition based on genotypic differences or environmental factors has significant effects on taste perception by consumers and, therefore, is a critical quality attribute for carrot breeders and growers. Surprisingly, little is known about the biosynthesis of volatile terpenes in carrots and the various enzymes involved in their formation. In this chapter, we provide an overview of volatile terpene sampling and terpene diversity in different *D. carota* genotypes. Facilitated by the recent elucidation of the carrot genome, we further describe and discuss latest findings of the function of genes and enzymes in the terpene synthase family involved in the biosynthesis of carrot terpene volatiles.

Abbreviations

DMAPP	Dimethylallyl diphosphate
FPP	Farnesyl diphosphate
GC/MS	Gas chromatography-mass spectrometry
GGPP	Geranylgeranyl diphosphate
GPP	Geranyl diphosphate
HS	Headspace
IDS	Isoprenyl diphosphate synthases
MEP	Methylerythritol phosphate
MVA	Mevalonate
SPME	Solid-phase micro-extraction
TPS	Terpene synthase
VOC	Volatile organic compound

16.1 Introduction

The flavor of carrots is determined predominantly by compounds in the diverse class of terpenes. Various publications have estimated that the number of distinct terpene compounds in plants is in the score of tens of thousands, and likely many more have not yet been described (Chen et al. 2011). In all living organisms,

M. Ibdah (✉) · M. Yahyaa · B. Nawade
Newe Ya'ar Research Center, Agriculture Research Organization, Post Office Box 1021, 30095 Ramat Yishay, Israel
e-mail: mwafaq@volcani.agri.gov.il

A. Muchlinski · D. Tholl (✉)
Department of Biological Sciences, Virginia Polytechnic Institute and State University, 409 Latham Hall, 220 Ag Quad Lane, Blacksburg, VA 24061, USA
e-mail: tholl@vt.edu

P. Simon et al. (eds.), *The Carrot Genome*, Compendium of Plant Genomes,
https://doi.org/10.1007/978-3-030-03389-7_16

terpenes have roles as primary metabolites that include functions as hormones, components of electron transfer systems, determinants of membrane fluidity, or agents of protein modification (Pichersky and Raguso 2016). As specialized or secondary metabolites, terpenes represent important tools in the various interactions of plants with the environment by functioning, for instance, in the attraction of beneficial organisms or as a defense against pests and pathogens (Tholl 2006). Moreover, and more relevant to this chapter, terpenes are important constituents of the aroma and flavor of fruits, rhizomes, or roots such as tomato, ginger, or carrot, respectively (Ekundayo et al. 1988; El Hadi et al. 2013; Keilwagen et al. 2017; Kreutzmann et al. 2008; Mamede et al. 2017; Marais 2017; Pang et al. 2017).

Despite their large structural diversity, all plant terpenes are made from the 5-carbon (C5) precursor isopentenyl diphosphate (IPP) and its isomer dimethylallyl diphosphate (DMAPP), both of which are derived from two alternative pathways, the mevalonate (MVA) pathway in the cytosol (and ER and peroxisomes) or the methylerythritol phosphate (MEP) pathway in plastids (Fig. 16.1). Combination of the C5 precursors leads to the formation of 10-carbon, 15-carbon, and 20-carbon *trans-* or *cis*-prenyl diphosphate intermediates including geranyl

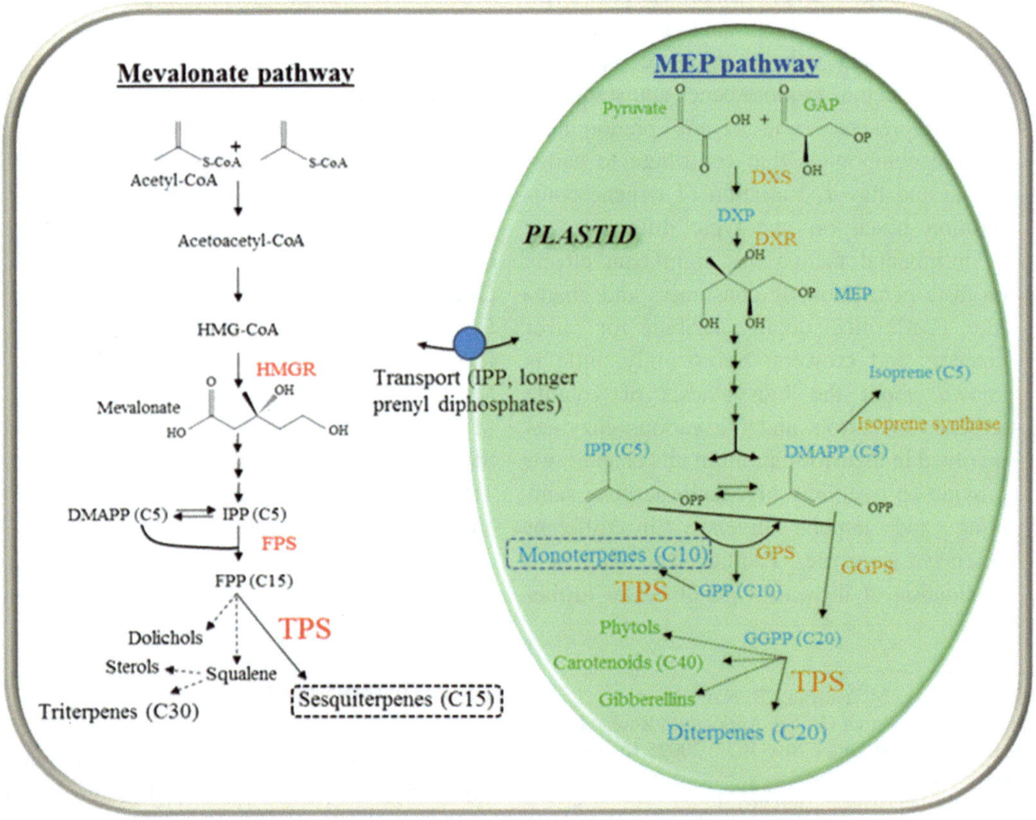

Fig. 16.1 Terpene biosynthetic pathways in the plant cell. MEP: 2-C-methyl-d-erythritol-4-phosphate; HMG-CoA: 3-hydroxyl-3-methyl-glutaryl CoA; IPP: isopentenyl diphosphate; DMAPP: dimethylallyl diphosphate; GPP: geranyl diphosphate; FPP: farnesyl diphosphate; GGPP: geranylgeranyl diphosphate; GPS: geranyl diphosphate synthase; FPS: farnesyl diphosphate synthase; GGPS: geranylgeranyl diphosphate synthase; TPS: terpene synthase

diphosphate (GPP), neryl diphosphate (NPP), farnesyl diphosphate (FPP), and geranylgeranyl diphosphate (GGPP). These intermediates are converted into monoterpenes (C10), sesquiterpenes (C15), and diterpenes (C20) by the activity of terpene synthase (TPS) enzymes (Fig. 16.1) (Tholl 2006). GGPP also functions as a precursor in the carotenoid biosynthetic pathway by condensation of two GGPP units to C40-phytoene (Cunningham and Gantt 1998). TPSs are encoded by large gene families and have the ability to produce multiple terpene products from a single substrate, which results in the formation of mixtures of structurally diverse compounds (Chen et al. 2011; Degenhardt et al. 2009) that contribute to the specific aroma and flavor characteristics of plant tissues. The formation of TPS products is largely regulated at the level of *TPS* gene transcription (Hong et al. 2012; Tholl and Lee 2011).

Carrots produce a large number of different volatile monoterpenes and sesquiterpenes in leaf, root, and seed tissues (Habegger and Schnitzler 2000; Yahyaa et al. 2015b, 2016). In carrot roots, terpenes are primarily synthesized in the upper part of the root and in an interconnected network of oil ducts located in the phloem (Senalik and Simon 1986). Labeling experiments with stable isotope precursors also indicated de novo biosynthesis of terpenes in the xylem (Hampel et al. 2005). Despite the substantial accumulation of terpenes in carrot tissues and their important role in carrot aroma and flavor, little attention has been paid to their biosynthesis and the various TPS enzymes involved in their formation. However, the recent elucidation of the carrot genome and available transcriptome assemblies have facilitated the identification of the carrot TPS gene family and resulted in the first functional characterization of carrot TPS enzymes. In this chapter, we briefly describe methods used in the sampling and analysis of volatile terpenes and other compounds in carrot roots. We further give an overview of the types of volatile terpenes produced in carrot roots and fruits (seeds) and present recent findings on the genes and proteins involved in their biosynthesis.

16.2 Volatile Terpenes in *Daucus carota*

16.2.1 Volatile Analysis

Carrots produce complex mixtures of volatile organic compounds (VOCs) that largely consist of terpenes but also include short-chain aldehydes, alcohols, hydroxyketones, and others (Kreutzmann et al. 2008; Rosenfeld et al. 2002; Senalik and Simon 1986; Yahyaa et al. 2015b). To achieve a complete qualitative and quantitative analysis of all VOCs, different sampling and analytical techniques may be combined. Heat-dependent distillation–extraction methods are less favored due to the formation of artifacts based on compound oxidation or decomposition (Rausch 2009). As a sensitive sampling method less prone to such artifacts, headspace (HS) sampling of VOCs emitted by carrot tissues has been applied. For example, Alasalvar et al. (1999) used statistic HS sampling coupled online to gas chromatography-mass spectrometry (GC-MS) to study the volatile compositions of raw, stored, and cooked carrot resulting in the identification of 35 VOCs.

A popular and simple HS sampling technique used in recent years for the analysis of food aroma compounds including those in carrot is solid-phase micro-extraction (SPME) (Alasalvar et al. 1999; Fukuda et al. 2013; Soria et al. 2008). SPME is an affordable, solvent-free method that requires little or no manipulation/preparation of samples and it substantially shortens the time of analysis. Soria et al. (2008) tested different adsorbent coatings of SPME fibers and examined other factors such as sample amount, matrix modification by water and sodium chloride addition, agitation of sample matrix, equilibrium time, and extraction time and temperature to ensure efficient extraction and high recoveries of volatiles from dehydrated carrots. Recently, Yahyaa et al. (2015b, 2016) examined the volatile composition of freshly harvested tissue of colored carrot cultivars and wild carrot fruits using auto-HS-SPME-GC-MS. More than 41 compounds were detected in roots, while more than 200 VOCs were found in wild carrot fruits. Fukuda et al. (2013) performed headspace sorptive extraction (HSSE), a technique related to SPME, to correlate volatile terpenoid profiles with carrot sensory attributes. This technique used a twister stir bar coated with polydimethylsiloxane (PDMS) to adsorb VOCs in the headspace of diced carrot tissue prior to thermal desorption and GC-MS analysis.

Despite the obvious advantages of applying HS sampling techniques, SPME fiber of twister-dependent methods still bear challenges for quantitative analysis. Therefore, mostly semi-quantitative data (e.g., percentage of total volatile composition) have been reported (Soria et al. 2008). To determine absolute amounts of VOCs, a thorough calibration of the headspace sampling method using authentic standards is required. To circumvent these challenges, extraction of ground tissue with an organic solvent such as hexane can be performed although less abundant compounds may be lost in this process due to sample concentration. Quantitative analysis is then performed by GC-flame ionization detection (FID) or by GC-MS depending on calibration with authentic standards.

16.2.2 Terpene Diversity in *D. carota*

Terpene VOCs occur ubiquitously in the Apiaceae family including the genus *Daucus*. The volatiles are constituents of an essential oil, which is present in roots, leaves, umbels, and seeds. The composition of the essential oils is highly species-specific (Alasalvar et al. 2001; Maxia et al. 2009) and can be of significance in chemotaxonomy, e.g., geranyl acetate and carotol are characteristic for the genus *Daucus*. In most studied species, monoterpene hydrocarbons dominate and make up 25–75% of all volatile compounds. They are usually followed by sesquiterpene hydrocarbons while oxygenated terpenes do not usually exceed 10% (Alasalvar et al. 1999; Kjeldsen et al. 2001; Simon et al. 1982; Yahyaa et al. 2015b, 2016). Besides their variation at the species level, volatile terpenes also vary highly in roots, leaves, flowers, and fruits of various *D. carota* subspecies (Dib et al. 2010; Fukuda et al. 2013; Yahyaa et al. 2015b, 2016). The most common terpene compounds that have been reported are α-pinene, sabinene, myrcene, limonene, γ-terpinene, terpinolene, β-caryophyllene, and γ-bisabolene (Fig. 16.2). Variations of terpene profiles are further observed under different growth conditions and depend on the physiological/developmental state (Alasalvar et al. 1999, 2001; Yahyaa et al. 2015b).

Considerable research has been done to identify terpene VOCs in roots of *D. carota* and define their contribution to carrot aroma and flavor (Alasalvar et al. 2001; Dib et al. 2010; Fukuda et al. 2013; Staniszewska and Kula 2001; Yahyaa et al. 2015b, 2016). Simon (1982) measured volatile terpene levels in different segregating carrot populations and found terpinolene, (*E*)-β-caryophyllene, and (*E*)-γ-bisabolene to be the most abundant terpenes. Furthermore, Alasalvar et al. (2001) and Ulrich et al. (2015) examined four different colored carrots, orange, purple with orange core, yellow, and white, for the content of their volatiles. The levels of the most predominant terpenes with highest concentrations in fresh carrot roots of different color are summarized in Table 16.1. Similar analyses were performed by Yahyaa et al. (2015b) on roots of four different colored carrot cultivars grown in Israel: The yellow carrot cultivar ('Yellowstone') had higher levels of terpenes than the other genotypes, orange 'Nairobi' and 'Rothild,' purple 'Purple Haze,' and white 'Crème de Lite' (see Table 16.1). A principal component analysis (PCA) for these carrot cultivars found that mono- and sesquiterpene profiles varied widely among the different genotypes (Fig. 16.3). These findings of genotypic differences are supported by other previous analyses of diverse carrot varieties (Alasalvar et al. 1999; Kjeldsen et al. 2001).

In addition to carrot line or cultivar, factors such as maturity, storage, processing, and climate can influence concentrations of carrot volatiles. The effect of climate (temperature) on volatile terpenes of carrot roots was investigated by

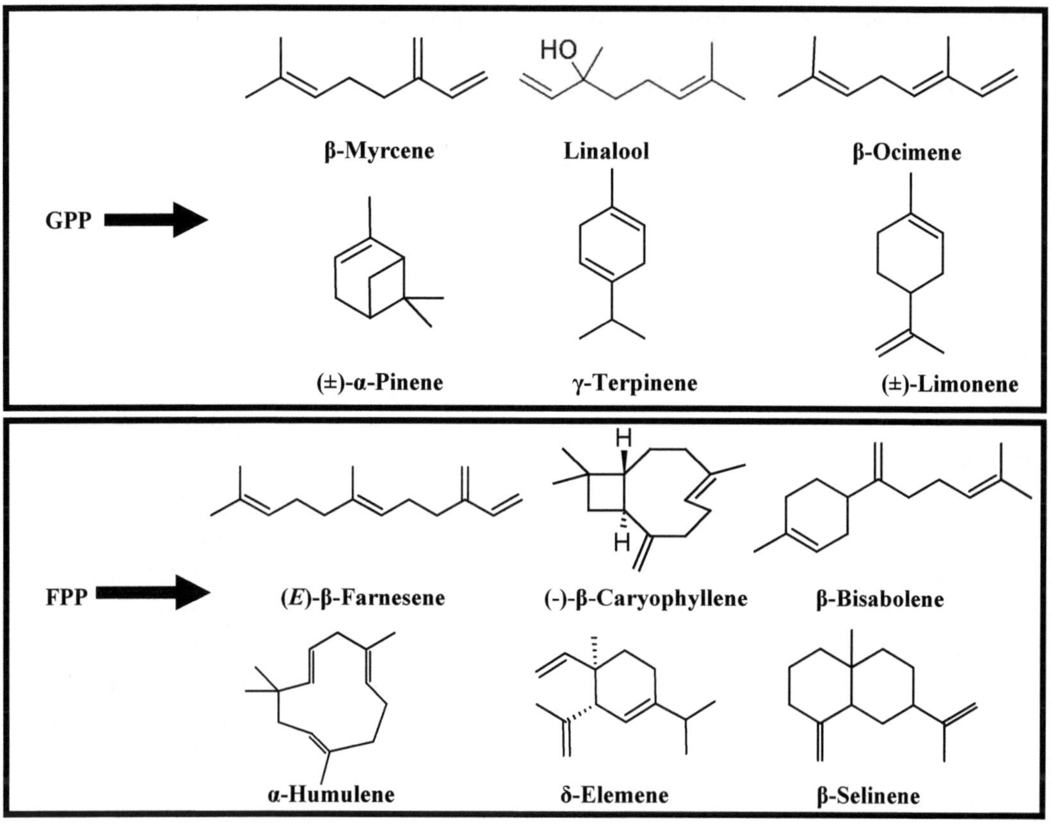

Fig. 16.2 Representative terpenes biosynthesized by *D. carota*. Monoterpenes and sesquiterpenes are derived from the prenyl diphosphate substrates, geranyl diphosphate (GPP), and farnesyl diphosphate (FPP), respectively, and are produced in the roots of carrots

Rosenfeld et al. (2002), who found that higher temperatures (18 and 21 °C) led to a higher content of terpene volatiles. The study also found significant differences in the concentrations of terpenes except for bornyl acetate, *β*-farnesene, and α-humulene, when the plant density was 'high' as compared to 'normal'.

Besides the importance of terpene-rich oils in carrot root aroma, essential oils derived from *D. carota* fruit (seed) are of commercial interest in cosmetic products and aroma therapy. Therefore, we will list a few studies that exemplify efforts to identify terpene constituents and their diversity in seed oils of cultivated and wild *D. carota* accessions. Benecke et al. (1987) studied the accumulation of VOCs in the fruit of *D. carota* ssp. *sativus*, and they found that the main constitutes of terpenes were geranyl acetate and terpinyl acetate. Similar findings were observed by Kilibarda et al. (1996). More recently, Flamini et al. (2014) reported the isolation and identification of 70 terpene VOCs from the essential oil of nine different commercial varieties of *D. carota* ssp. *sativus* fruit. This study identified α-pinene, sabinene, myrcene, *p*-cymene, and limonene as the predominant terpene compounds.

Volatile terpenes in carrot fruit/seed of wild accessions were also studied (Chizzola 2010; Grzebelus et al. 2011; Rokbeni et al. 2013; Yahyaa et al. 2016). In general, carrot seeds contain more volatile terpenes than leaves and roots and their amount increases with ripening. Highest levels were reported for the terpenes *β*-bisabolene and elemicin, the content of which increases two- to fourfold in comparison to other

Table 16.1 Abundance of predominant monoterpenes and sesquiterpenes in fresh roots of different colored carrot cultivars

Compound	Orange	Purple	Yellow	White	Reference
α-Pinene	−	−	−	+	Ulrich et al. (2015)
	−	−	−	−	Yahyaa et al. (2015b)
	−	−	−	+	Alasalvar et al. (2001)
Sabinene	−	−	−	+	Ulrich et al. (2015)
	−	−	+	−	Yahyaa et al. (2015b)
	−	−	+	+	Alasalvar et al. (2001)
β-Pinene	−	−	−	+	Ulrich et al. (2015)
	−	−	−	−	Yahyaa et al. (2015b)
	−	−	−	+	Alasalvar et al. (2001)
Myrcene	−	−	−	+	Ulrich et al. (2015)
	−	−	+	−	Yahyaa et al. (2015b)
	+	+	+	+	Alasalvar et al. (2001)
α-Phellandrene	−	−	−	+	Ulrich et al. (2015)
	−	−	−	−	Yahyaa et al. (2015b)
	−	−	−	+	Alasalvar et al. (2001)
Limonene	−	−	−	+	Ulrich et al. (2015)
	−	−	−	−	Yahyaa et al. (2015b)
	−	−	−	+	Alasalvar et al. (2001)
γ-Terpinene	−	−	−	−	Ulrich et al. (2015)
	+	+	+	−	Yahyaa et al. (2015b)
	+	−	−	+	Alasalvar et al. (2001)
Terpinolene	+	+	+	+	Ulrich et al. (2015)
	+	−	+	+	Yahyaa et al. (2015b)
	+	+	+	+	Alasalvar et al. (2001)
β-Caryophyllene	+	+	+	+	Ulrich et al. (2015)
	+	+	+	+	Yahyaa et al. (2015b)
	+	+	+	+	Alasalvar et al. (2001)
γ-bisabolene	−	−	−	−	Ulrich et al. (2015)
	−	−	−	−	Yahyaa et al. (2015b)
	+	+	+	+	Alasalvar et al. (2001)

Terpene compounds in high (+) or low (−) concentrations

VOCs in ripe umbels of *D. carota* subsp. *carota*. For *D. carota* subsp. *halophilus*, only the content of elemicin increases significantly. On the other hand, the monoterpenes limonene, myrcene, α-pinene, and terpine-4-ol were all found at elevated levels in the umbel at all flowering stages (Grzebelus et al. 2011).

In another study, Mockute and Nivinskiene (2004) examined the composition of the essential oil from seeds of wild growing *D. carota* ssp. *carota* in Lithuania. Here, a total of 61 VOC terpene compounds were identified with sabinene, α-pinene, terpine-4-ol, limonene, and γ-terpinene as major terpenes. Several other

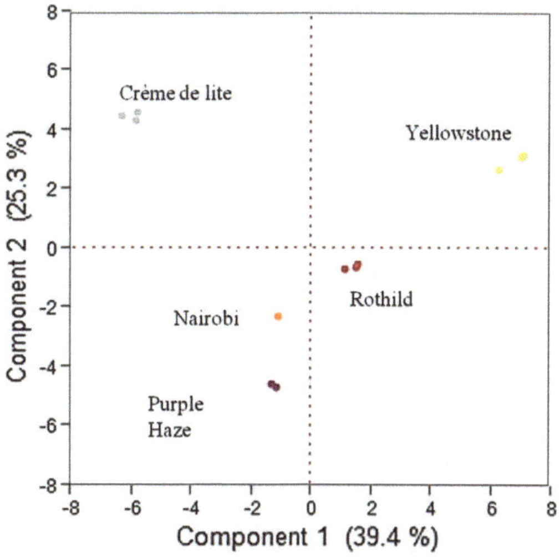

Fig. 16.3 PCA of 41 volatile compounds in five differently colored carrot cultivars: orange 'Nairobi', orange 'Rothild', purple 'Purple Haze', yellow 'Yellowstone', and white 'Creme de Lite'. All analyses were performed using three biological replicates. From Yahyaa et al. (2015b)

studies of fruit (seed) oil compositions from wild accessions of *D. carota* or other *Daucus* species demonstrate similarities and differences with those listed above and thereby clearly indicate plasticity and diversification in seed terpene profiles dependent on genotype and environmental conditions (Aćimović et al. 2016; Chizzola 2010; Mansour et al. 2004; Rokbeni et al. 2013).

16.3 Identification and Functional Characterization of Terpene Biosynthetic Genes in *D. carota*

16.3.1 Genes Involved in the Early Steps of Terpenoid Biosynthesis in *D. carota*

Previously, little attention has been paid to genes and enzymes involved in the early steps in terpene biosynthesis of *D. carota*. However, available genome resources now allow a more detailed identification of genes of the core MEP and MVA pathways. For instance, a family of four putative 1-deoxy-D-xylulose-5-phosphate synthases (DXS) (type I DCAR_030576, type II DCAR_009911 and DCAR_014178, and type III DCAR_022887) and one putative 1-deoxy-D-xylulose 5-phosphate reductoisomerase (DXR) involved in catalyzing the first two steps of the MEP pathway were recently annotated (Iorizzo et al. 2016). Of the four DXS-type genes, only the type I DXS was correlated in its expression with high carotenoid content. Several other genes in the plastidial MEP pathway and downstream steps leading to carotenoid biosynthesis are represented by more than a single gene model (Iorizzo et al. 2016). By contrast, a single copy of a geranyl diphosphate synthase (GPS, DCAR_019378) and a geranylgeranyl diphosphate synthase-related gene (GGR) were annotated, both of which are likely to be involved in the formation of GPP in monoterpene biosynthesis. Work by Hampel et al. (2005) using deuterated precursors previously demonstrated that monoterpenes are biosynthesized exclusively via the MEP pathway, whereas both MVA and the MEP pathways contribute to the synthesis of sesquiterpenes. Further analysis of the genes involved in both pathways and the downstream reactions leading to diphosphate intermediates should provide more detailed insight in the spatial and temporal regulation of these central steps in carrot terpene metabolism.

16.3.2 Terpene Synthases

16.3.2.1 Plant TPS Gene Families

Terpene synthases (TPSs) represent the key enzyme catalysts responsible for the formation of monoterpenes, sesquiterpenes, and diterpenes (Bohlmann et al. 1998; Tholl 2006; Tholl and Lee 2011). Their diphosphate substrates are synthesized by *trans-* or *cis-*isoprenyl diphosphate synthases (IDSs) in conjugation reactions of IPP and DMAPP. TPS-catalyzed reactions are initiated by ionization of the substrate and formation of a carbocation either through a divalent cation-dependent subtraction of the diphosphate group (class I TPSs) or by protonation of the substrate (class II TPSs). The carbocation intermediate then undergoes further conversions, which often leads to the formation of several products by a single promiscuous enzyme (Degenhardt et al. 2009). The primary enzymatic products are in most cases acyclic or cyclic hydrocarbons, which often are converted in secondary enzymatic reactions into an even larger and structurally diverse number of terpene end products.

TPS enzymes may accept just a single substrate or convert two or more substrates (e.g., GPP and FPP) in vitro. Subcellular localization, however, typically determines the type of TPS reaction in vivo. TPS enzymes that are targeted to plastids are likely to synthesize monoterpenes or diterpenes from the plastidial pools of GPP and GGPP, respectively, whereas TPS enzymes in the cytosol are considered to convert FPP to sesquiterpenes. An exception to this general rule are plastid-targeted TPSs that are capable of converting (Z,Z)-FPP produced in plastids and thereby function as sesquiterpene synthases (Akhtar et al. 2013; Sallaud et al. 2009).

The plant TPS superfamily is divided into several subfamilies (TPS-a-h), some of which are unique to angiosperm species (clades a, b, and g), gymnosperm species (clade d), or lower land plants (clade h) (Chen et al. 2011). Class I sesqi-TPSs of angiosperms are typically represented by the TPS-a subfamily but also occur in the TPS e/f and g clades. Angiosperm class I mono-TPSs generally cluster in the TPS-b and g families, while the TPS-c family largely comprises class II and bifunctional class I/II-type diterpene synthases and the TPS e/f family contains class I diterpene synthases (Chen et al. 2011). Species-specific clades of TPS genes often emerge by gene duplication and neofunctionalization resulting in metabolic plasticity and allowing selective adaptation to different environments. In diterpene biosynthesis, TPS genes are frequently arranged in clusters with other biosynthetic genes leading to a coordinated tissue- and time-specific regulation of entire pathways (Zi et al. 2014).

Elucidation and comparison of structural domains of mono-TPSs, sesqui-TPSs, and di-TPSs in angiosperms and gymnosperms have predicted an emergence of these proteins from class I/class II bifunctional diterpene synthases such as the copalyl diphosphate (CPP)/kaurene synthase in the moss *Physcomitrella patens* (Chen et al. 2011). These bifunctional enzymes consist of three domains: (1) a class I α-domain that carries the highly conserved aspartate-rich motif, DDxxD, and a less conserved NSE/DTE motif, both of which are involved in binding of the phosphorylated substrate; (2) a class II γ- and (3) β-domain, the latter of which carries a conserved DxDD motif required for protonation-initiated carbocation formation (Cao et al. 2010; Chen et al. 2011). During the evolution of monofunctional TPSs, class I or class II functions were lost and protein sizes reduced to a functional α- domain and residual β-domain in several class I enzyme subfamilies. Thus, depending on the domain structure and the presence of transit peptides, sizes of TPS proteins vary between approximately 50–100 kDa. Recently, additional families of TPSs specific to lower land plants have been identified. As shown in *Selaginella* and many other lower land plants, these TPSs seem to have evolved from ancestral microbial TPSs, while they are absent in gymnosperms and angiosperms (Jia et al. 2016).

16.3.2.2 The Carrot TPS Family

In the past decade, access to genome sequences has facilitated the characterization of TPS gene families from several plant species including crops and trees such as grape (*Vitis vinifera*), tomato (*Solanum lycopersicum*), apple (*Malus domestica*), poplar (*Populus trichocarpa*), *Eucalyptus*, and switchgrass (Aubourg et al. 2002; Falara et al. 2011; Külheim et al. 2015; Martin et al. 2010; Matsumoto et al. 2005; Nieuwenhuizen et al. 2013; Paterson et al. 2009; Pelot et al. 2018; Tuskan et al. 2006). The genome of *Eucalyptus* carries one of the largest TPS gene families with a total of more than 100 putative TPS genes in *E. grandis* and *E. globulus* (Külheim et al. 2015). By contrast, only 10 TPS genes are currently known from apple (Nieuwenhuizen et al. 2013).

Characterization of TPS gene and enzyme activities in *D. carota* was previously neglected despite the obvious importance of terpenes for carrot aroma and flavor attributes and hence consumer acceptance. With the available high-quality carrot genome assembly (Iorizzo et al. 2016), more precise associations can now be established between tissue-specific terpene profiles and enzyme function. Genomic and transcriptomic analysis of the orange Nantes-type carrot (line DH1) led Iorizzo et al. (2016) to predict 36 potentially functional TPS genes (Iorizzo et al. 2016). However, a recent analysis by Keilwagen et al. (2017) expanded this number to 65 full-length putative TPS genes. In this study, the authors performed homology-based gene prediction using TPS gene models from five reference species. The large size of the carrot TPS gene family is not unexpected given the presence of a large number of terpenes in carrot tissues.

The identified TPS gene models were associated with all TPS subfamilies except for the gymnosperm-specific type-d family and the type-h family of lower land plants (Fig. 16.4) (Keilwagen et al. 2017). Most carrot TPS genes distribute between the type-a and b families indicating a substantial expansion of these clades in class I mono-TPSs and sesqui-TPSs in correlation with the dominance of monoterpenes and sesquiterpenes in carrot volatile mixtures. The large number (32) of genes in the TPS-b subfamily characteristic for mono-TPSs is noteworthy since in other species clade expansions more typically occur in TPS-a subfamilies (e.g., in *A. thaliana*) (Aubourg et al. 2002). Most proteins in the type-b clade carry plastidial transit peptides suggesting functions as mono-TPSs. Moreover, most members of the type- and b families carry the RRX8W motif characteristic of the synthesis of cyclic products (Chen et al. 2011). Other putative mono-TPS gene candidates were found in the TPS-g subfamily, while only very few putative di-TPS gene models, most likely including CPP synthase and kaurene synthase involved in gibberellin biosynthesis, are present in the type-c and e/f families (Fig. 16.4) (Keilwagen et al. 2017). The limited presence of di-TPSs in the carrot genome is in stark contrast to the abundance of these genes and their expression in other species, especially in grasses such as switchgrass (Pelot et al. 2018). Most genes in the TPS subfamilies a, b, and g contain an expected number of 7 exons, whereas this number is typically higher (12–15) in the genes of the subfamilies TPS-c, TPS-e, and TPS-f. The predicted protein sizes range between 550 amino acids for putative sesqui-TPSs and more than 580 amino acids for predicted mono-TPSs.

The identified carrot TPS genes were found to be distributed over all nine carrot chromosomes. However, substantial clustering of TPS genes was found on chromosomes 1, 3, 4, and 9 suggesting multiple events of gene duplications (Keilwagen et al. 2017). A better understanding of the expression of the TPS genes can be deducted from 20 transcript profiles generated by Iorizzo et al. (2016) (Fig. 16.5). These transcriptome sets of different tissues including seeds, hypocotyl, leaves, flowers, and roots, and stress conditions indicate clearly distinct tissue-specific and stress-induced expression patterns (Fig. 16.5). Interestingly, transcripts of a majority of TPS genes accumulate in flower and leaf tissues, and only about a dozen of genes are expressed at higher levels in the hypocotyl and roots. These findings suggest substantial constitutive defenses based on terpene-rich essential oils in carrot aboveground tissues.

To define distinct loci responsible for the synthesis of major volatile terpene constituents in carrot, Keilwagen et al. (2017) performed a combination of terpene metabolite profiling, genotyping-by-sequencing (GBS), and genome-wide association study (GWAS). By using a panel of 85 carrot cultivars and accessions, four genomic regions on chromosomes 2, 4, and 8 were associated with the monoterpene γ-terpinene, sabinene, terpineol, and bornyl acetate. The gene *TPS03* was found in a QTL associated with bornyl acetate, and *TPS29* was located on a QTL associated with γ-terpinene. The other two QTLs for sabinene and terpineol were found to be located in the same region of chromosome 4, and both were associated with a gene cluster comprising *DcTPS04*, *DcTPS26*,

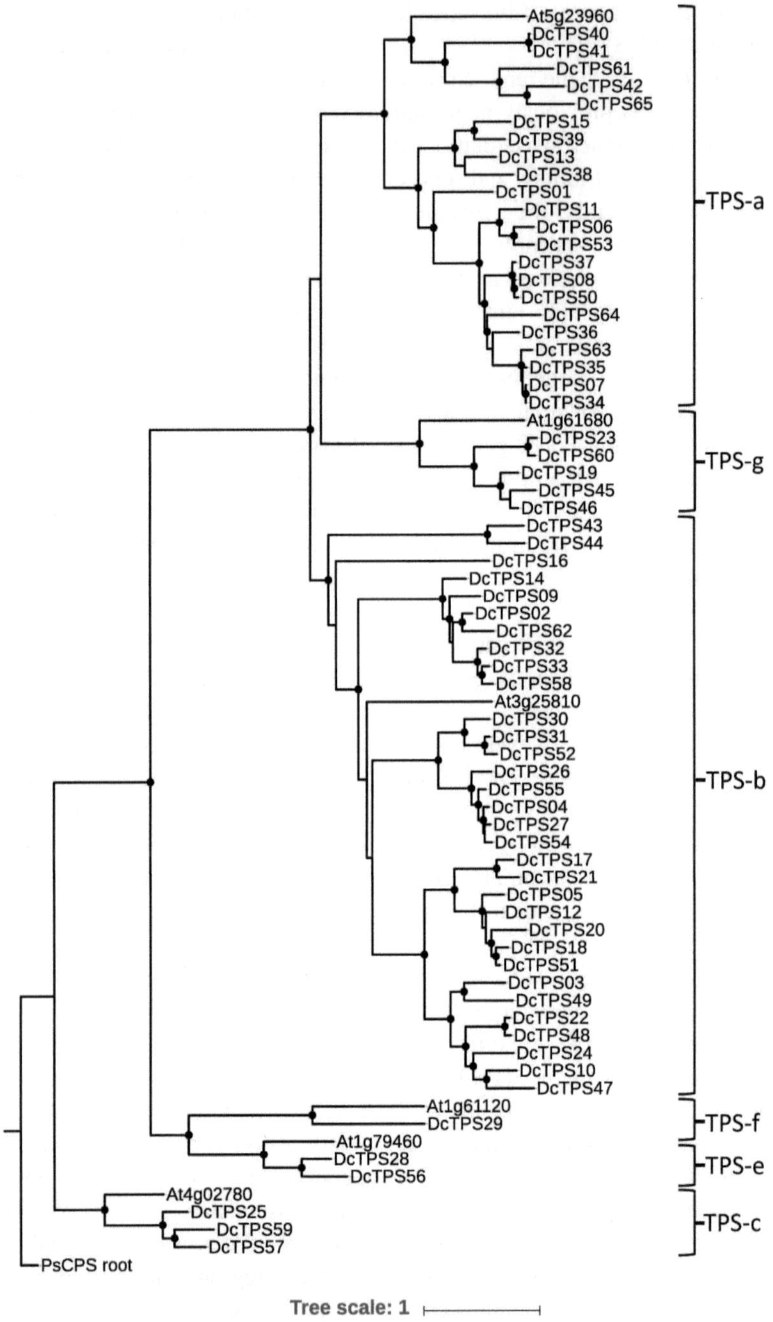

Fig. 16.4 Maximum-likelihood phylogenetic tree of predicted TPSs in *Daucus carota* (line DH1). Shaded circles indicate bootstrap support of >80% where bootstrap replicates = 500. The tree was rooted with the gymnosperm *ent*-CPP synthase from *Picea sitchensis* (PsCPS). *Arabidopsis* TPSs representative for each subfamily have been included

DcTPS27, *DcTPS54,* and *DcTPS55*. While transcriptome profiles suggested a predominant expression of these genes in aboveground tissues (except *TPS03*) (Fig. 16.5), Keilwagen et al. (2017) observed transcripts of these genes also in the roots of different cultivars. Several other

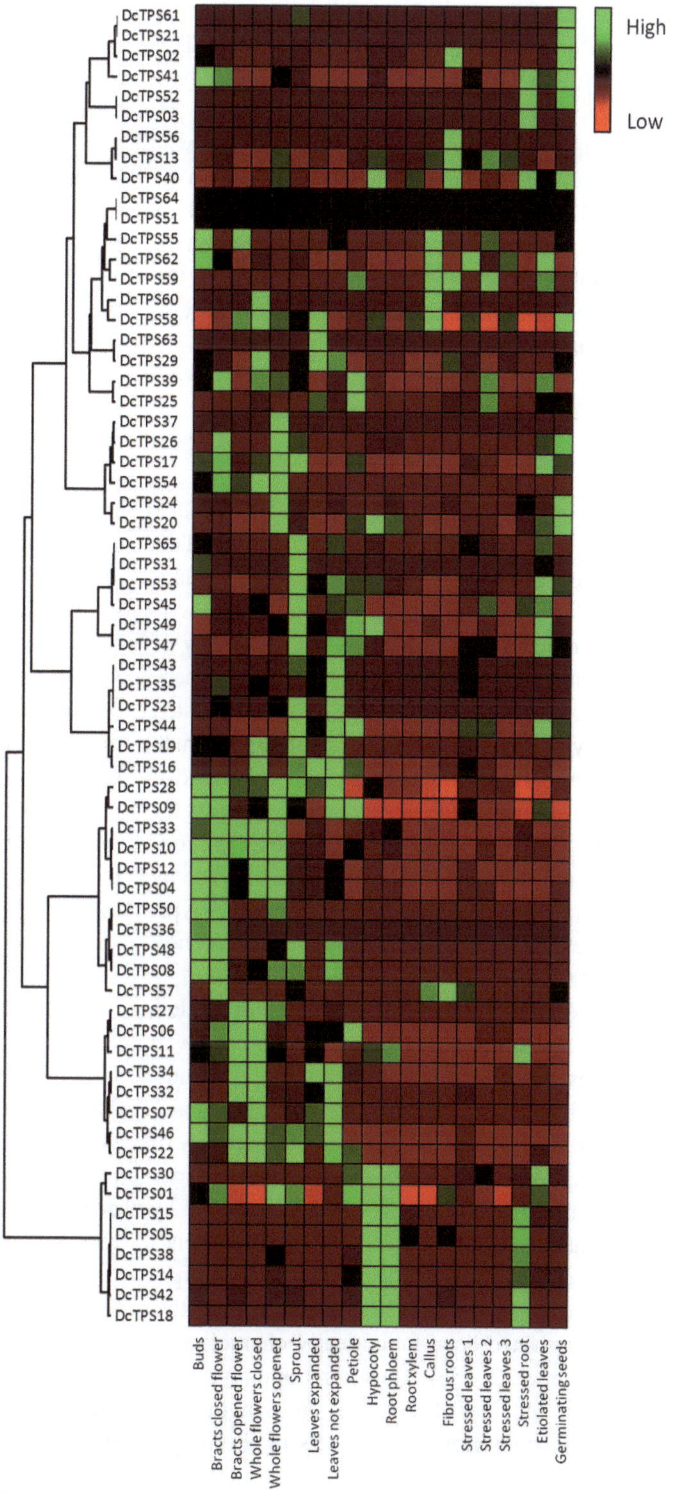

Fig. 16.5 Hierarchical cluster analysis of TPS gene expression profiles across carrot tissues. Heat map comparing relative transcript abundance for all TPS gene candidates in FPKM (fragments per kilobase of transcript per million mapped reads) across 20 tissue-specific gene expression data sets from Iorizzo et al. (2016)

QTLs were identified for terpenes but did not contain any TPS genes. However, some QTLs were found to contain cytochrome P450 monooxygenases, which may be involved in oxygenation reaction of TPS products (Keilwagen et al. 2017). Functional characterization of the TPS genes (see Sect. 16.3.2.3) positioned on the identified regions is currently performed in the Tholl Laboratory to further corroborate the associations of distinct enzyme activities with individual terpene metabolites.

16.3.2.3 Functional Characterization of Carrot Terpene Synthases

To establish distinct correlations between terpene profiles or individual compounds and the TPS gene loci responsible for their synthesis, functional characterization of individual enzymes is usually required. Since the product specificity of a TPS enzyme is not predictable from its primary structure, recombinant proteins have to be tested for enzyme activity using different diphosphate substrates. Truncation of proteins to remove plastidial transit peptides may be necessary for optimal expression and activity. Volatile products are either sampled in the assay headspace by SPME or extracted with organic solvent (typically hexane) and analyzed by GC-MS (Tholl et al. 2005; Yahyaa et al. 2015a, b, 2016).

In the large TPS family of *D. carota,* preference for functional characterization may be given to genes with highest transcript levels in individual tissues (Fig. 16.5), since transcript abundance often correlates with the in vivo activity of TPS enzymes (Tholl 2006). Moreover, priority should be given to the TPS loci identified by QTL mapping and GWAS, since these genes have been associated with common monoterpenes (Keilwagen et al. 2017). To date, only two TPS genes have been functionally characterized (Yahyaa et al. 2015b). These first attempts were based on the identification of full-length TPS cDNAs in transcriptome data sets of the purple cultivar B7267 and an orange cultivar (B6274) (Iorizzo et al. 2011). Functional characterization of *DcTPS01* in the TPS-a subfamily (Fig. 16.4) revealed that this enzyme makes exclusively (*E*)-β-caryophyllene and α-humulene in vitro. *TPS01* shows highest transcript abundance in roots and open flowers (Fig. 16.5) and is most likely responsible for the in vivo formation of (*E*)-*β*-caryophyllene as one of the most abundant sesquiterpenes in carrot root tissue responsible for spicy and woody aroma attributes (Kjeldsen et al. 2003). The second gene, *DcTPS02*, of the TPS-b subfamily (Fig. 16.4) was found to encode a protein, which converts GPP into geraniol and *β*-myrcene in vitro. *β*-Myrcene, a common monoterpene in carrot roots, contributes together with sabinene to the so-called carrot-top aroma (Kjeldsen et al. 2003).

Further characterization of other TPS genes will reveal the extent at which individual terpene compounds are catalyzed by single TPS enzymes or made together with several other products by promiscuous enzymes. For instance, proteins encoded by the closely related genes *DcTPS04*, *DcTPS26*, *DcTPS27*, *DcTPS54,* and *DcTPS55* positioned in a gene cluster on a QTL on chromosome 4 may produce similar overlapping blends of compounds. Furthermore, polymorphisms of TPS genes among different cultivars might influence product specificity depending on the effects of residue changes on protein topography and enzymatic reaction mechanisms.

16.4 Conclusions

Identification and characterization of the carrot TPS gene family have been facilitated by recent genome and transcriptome assemblies. These efforts will allow making associations of TPS gene loci with individual terpenes and their corresponding aroma attributes although more than a single gene might be involved in the same terpene trait. Deeper insight into gene-terpene correlations will benefit the breeding of high-quality carrot cultivars with improved aroma and taste characteristics. In this context, it will be important to examine changes in terpene profiles and TPS gene expression upon different stress or under post-harvest conditions. For example, elevated temperature has been shown to enhance terpene levels in carrot roots leading to an undesired harsh taste (Rosenfeld et al. 2002). Terpene products and genes largely responsible for these taste attributes could be identified and eliminated using RNAi and probably gene editing approaches (Klimek-Chodacka et al. 2018). Overall, the TPS gene family will provide a useful genetic toolbox for future modifications and improvements of terpene-based traits in carrot cultivars.

Acknowledgements This work was supported by grant IS-4745-14R from the US-Israel Binational Agricultural Research and Development Fund (to M.I. and D.T.).

References

Aćimović M, Stanković J, Cvetković M, Ignjatov M, NIKOLIĆ L (2016) Chemical characterization of essential oil from seeds of wild and cultivated carrots from Serbia. Bot Serb 40:44–60

Akhtar TA, Matsuba Y, Schauvinhold I, Yu G, Lees HA, Klein SE, Pichersky E (2013) The tomato cis-prenyltransferase gene family. Plant J 73:640–652

Alasalvar C, Grigor J, Quantick P (1999) Method for the static headspace analysis of carrot volatiles. Food Chem 65:391–397

Alasalvar C, Grigor JM, Zhang D, Quantick PC, Shahidi F (2001) Comparison of volatiles, phenolics, sugars, antioxidant vitamins, and sensory quality of different colored carrot varieties. J Agric Food Chem 49:1410–1416

Aubourg S, Lecharny A, Bohlmann J (2002) Genomic analysis of the terpenoid synthase (AtTPS) gene family of *Arabidopsis thaliana*. Mol Genet Genomics 267:730–745

Benecke R, Reichold K, Kessel M, Schmidt W (1987) Essential oil content and composition of the seeds of variuos cultivars of *Daucus carota* L. ssp. sativus (Hoffm.) Arcang. Pharmazie 42:256–259

Bohlmann J, Meyer-Gauen G, Croteau R (1998) Plant terpenoid synthases: molecular biology and phylogenetic analysis. PNAS 95:4126–4133

Cao R, Zhang Y, Mann FM, Huang C, Mukkamala D, Hudock MP, Mead ME, Prisic S, Wang K, Lin FY (2010) Diterpene cyclases and the nature of the isoprene fold. Proteins 78:2417–2432

Chen F, Tholl D, Bohlmann J, Pichersky E (2011) The family of terpene synthases in plants: amid-size family of genes for specialized metabolism that is highly diversified throughout the kingdom. Plant J 66:212–229

Chizzola R (2010) Composition of the essential oil from *Daucus carota* ssp. carota growing wild in Vienna. J Essent Oil Bearing Plants 13:12–19

Cunningham FXJ, Gantt E (1998) Genes and enzymes of carotenoid biosynthesis in plants. Annu Rev Plant Physiol Plant Mol Biol 49:557–583

Degenhardt J, Köllner TG, Gershenzon J (2009) Monoterpene and sesquiterpene synthases and the origin of terpene skeletal diversity in plants. Phytochemistry 70:1621–1637

Dib MEA, Djabou N, Desjobert J-M, Allali H, Tabti B, Muselli A, Costa J (2010) Characterization of volatile compounds of *Daucus crinitus* Desf. Headspace Solid Phase Microextraction as alternative technique to Hydrodistillation. Chem Cent J 4:16

Ekundayo O, Laakso I, Hiltunen R (1988) Composition of ginger (*Zingiber officinale Roscoe*) volatile oils from Nigeria. Flavour Frag J 3:85–90

El Hadi MAM, Zhang F-J, Wu F-F, Zhou C-H, Tao J (2013) Advances in fruit aroma volatile research. Molecules 18:8200–8229

Falara V, Akhtar TA, Nguyen TTH, Spyropoulou EA, Bleeker PM, Schauvinhold I, Matsuba Y, Bonini ME, Schilmiller AL, Last RL, Schuurink RC, Pichersky E (2011) The tomato terpene synthase gene family. Plant Physiol 157:770–789

Flamini G, Cosimi E, Cioni PL, Molfetta I, Braca A (2014) Essential oil composition of *Daucus carota* ssp. major (Pastinocello Carrot) and nine different commercial varieties of *Daucus carota* ssp. sativus fruits. Chem Biodivers 11:1022–1033

Fukuda T, Okazaki K, Shinano T (2013) Aroma characteristic and volatile profiling of carrot varieties and quantitative role of terpenoid compounds for carrot sensory attributes. J Food Sci 78:S1800–S1806

Grzebelus D, Baranski R, Spalik K, Allender C, Simon PW (2011) *Daucus*. In: Kole Ch (ed) Wild crop relatives: genomic and breeding resources. Vegetables. Springer-Berlin, Berlin, Heidelberg, pp 91–113

Habegger R, Schnitzler WH (2000) Aroma compounds in essential oils of carrots (*Daucus carota* L. ssp. *sativus*) leaves in comparison with roots. J Appl Bot 74:220–223

Hampel D, Mosandl A, Wust M (2005) Biosynthesis of mono- and sesquiterpenes in carrot roots and leaves (*Daucus carota* L.): metabolic cross talk of cytosolic mevalonate and plastidial methylerythritol phosphate pathways. Phytochemistry 66:305–311

Hong GJ, Xue XY, Mao YB, Wang LJ, Chen XY (2012) Arabidopsis MYC2 interacts with DELLA proteins in regulating sesquiterpene synthase gene expression. Plant Cell 24:2635–2648

Iorizzo M, Senalik DA, Grzebelus D, Bowman M, Cavagnaro PF, Matvienko M, Ashrafi H, Van Deynze A, Simon PW (2011) De novo assembly and characterization of the carrot transcriptome reveals novel genes, new markers, and genetic diversity. BMC Genom 12:389

Iorizzo M, Ellison S, Senalik D, Zeng P, Satapoomin P, Huang J, Bowman M, Iovene M, Sanseverino W, Cavagnaro P (2016) A high-quality carrot genome assembly provides new insights into carotenoid accumulation and asterid genome evolution. Nat Genet 48:657

Jia Q, Li G, Köllner TG, Fu J, Chen X, Xiong W, Crandall-Stotler BJ, Bowman JL, Weston DJ, Zhang Y (2016) Microbial-type terpene synthase genes occur widely in nonseed land plants, but not in seed plants. PNAS 113:12328–12333

Keilwagen J, Lehnert H, Berner T, Budahn H, Nothnagel T, Ulrich D, Dunemann F (2017) The terpene synthase gene family of carrot (*Daucus carota* L.): identification of QTLs and candidate genes associated with terpenoid volatile compounds. Front Plant Sci 8

Kilibarda V, Nanusević N, Dogović N, Ivanić R, Savin K (1996) Content of the essential oil of the carrot and its antibacterial activity. Die Pharmazie 51:777–778

Kjeldsen F, Christensen LP, Edelenbos M (2001) Quantitative analysis of aroma compounds in carrot (*Daucus carota* L.) cultivars by capillary gas chromatography using large-volume injection technique. J Agric Food Chem 49:4342–4348

Kjeldsen F, Christensen LP, Edelenbos M (2003) Changes in volatile compounds of carrots (*Daucus carota* L.) during refrigerated and frozen storage. J Agric Food Chem 51:5400–5407

Klimek-Chodacka M, Oleszkiewicz T, Lowder LG, Qi Y, Baranski R (2018) Efficient CRISPR/Cas9-based genome editing in carrot cells. Plant Cell Rep 37:575–586

Kreutzmann S, Thybo AK, Edelenbos M, Christensen LP (2008) The role of volatile compounds on aroma and flavour perception in coloured raw carrot genotypes. Int J Food Sci Technol 43:1619–1627

Külheim C, Padovan A, Hefer C, Krause ST, Köllner TG, Myburg AA, Degenhardt J, Foley WJ (2015) The *Eucalyptus* terpene synthase gene family. BMC Genom 16:450

Mamede AM, Soares AG, Oliveira EJ, Farah A (2017) Volatile composition of sweet passion fruit (*Passiflora alata* Curtis). J Chem

Mansour E-SS, Maatooq GT, Khalil AT, Marwan E-SM, Sallam AA (2004) Essential oil of *Daucus glaber* Forssk. Z Naturforsch C 59:373–378

Marais J (2017) Terpenes in the aroma of grapes and wines: a review. S Afr J Enol Vitic 4:49–58

Martin DM, Aubourg S, Schouwey MB, Daviet L, Schalk M, Toub O, Lund ST, Bohlmann J (2010) Functional annotation, genome organization and phylogeny of the grapevine (*Vitis vinifera*) terpene synthase gene family based on genome assembly, FLcDNA cloning, and enzyme assays. BMC Plant Biol 10:226

Matsumoto T, Wu J, Kanamori H, Katayose Y (2005) The map-based sequence of the rice genome. Nature 436:793

Maxia A, Marongiu B, Piras A, Porcedda S, Tuveri E, Gonçalves MJ, Cavaleiro C, Salgueiro L (2009) Chemical characterization and biological activity of essential oils from *Daucus carota* L. subsp. *carota* growing wild on the Mediterranean coast and on the Atlantic coast. Fitoterapia 80:57–61

Mockute D, Nivinskiene O (2004) The sabinene chemotype of essential oil of seeds of *Daucus carota* L. ssp. *carota* growing wild in Lithuania. J Essent Oil Res 16:277–281

Nieuwenhuizen NJ, Green SA, Chen X, Bailleul EJ, Matich AJ, Wang MY, Atkinson RG (2013) Functional genomics reveals that a compact terpene synthase gene family can account for terpene volatile production in apple. Plant Physiol 161:787–804

Pang X, Cao J, Wang D, Qiu J, Kong F (2017) Identification of ginger (*Zingiber officinale Roscoe*) volatiles and localization of aroma-active constituents by GC–olfactometry. J Agric Food Chem 65:4140–4145

Paterson AH, Bowers JE, Bruggmann R et al (2009) The sorghum bicolor genome and the diversification of grasses. Nature 457(7229):551–556

Pelot KA, Chen R, Hagelthorn DM, Young CA, Addison JB, Muchlinski A, Tholl D, Zerbe P (2018) Functional diversity of diterpene synthases in the biofuel crop switchgrass. Plant Physiol 178:54–71

Pichersky E, Raguso RA (2016) Why do plants produce so many terpenoid compounds? New Phytol 220:692–702

Rausch T (2009) Influence of extrusion parameters and recipe compounds on flavor formation and its quantification. Doctoral Thesis at the Technische Universität Berlin, Fakultät III, Prozesswissenschaften

Rokbeni N, M'rabet Y, Dziri S, Chaabane H, Jemli M, Fernandez X, Boulila A (2013) Variation of the chemical composition and antimicrobial activity of the essential oils of natural populations of Tunisian *Daucus carota* L. (Apiaceae). Chem Biodivers 10:2278–2290

Rosenfeld HJ, Knut S, Dalen KS, Haffner K (2002) The growth and development of carrot roots. Gartenbauwissenschaft 67:11–16

Sallaud C, Rontein D, Onillon S, Jabès F, Duffé P, Giacalone C, Thoraval S, Escoffier C, Herbette G, Leonhardt N (2009) A novel pathway for sesquiterpene biosynthesis from Z, Z-farnesyl pyrophosphate in the wild tomato *Solanum habrochaites*. Plant Cell 21:301–317

Senalik DA, Simon PW (1986) Relationship between oil ducts and volatile terpenoid content in carrot roots. Am J Bot 73:60–63

Simon PW (1982) Genetic variation for volatile terpenoids in roots of carrot, *Daucus carota*, backcrosses and F2 generations. Phytochemistry 21:875–879

Simon PW, Peterson CE, Lindsay RC (1982) Genotype, soil, and climate effects on sensory and objective components of carrot flavor. J Am Soc Hort Sci 107:644–648

Soria AC, Sanz J, Villamiel M (2008) Analysis of volatiles in dehydrated carrot samples by solid-phase microextraction followed by GC-MS. J Sep Sci 31:3548–3555

Staniszewska M, Kula J (2001) Composition of the essential oil from wild carrot umbels (*Daucus carota* L. ssp. *carota*) growing in Poland. J Essent Oil Res 13:439–441

Tholl D (2006) Terpene synthases and the regulation, diversity and biological roles of terpene metabolism. Curr Opin Plant Biol 9:297–304

Tholl D, Lee S (2011) Terpene specialized metabolism in *Arabidopsis thaliana*. Arabidopsis Book 9:e0143. https://doi.org/10.1199/tab.0143

Tholl D, Chen F, Petri J, Gershenzon J, Pichersky E (2005) Two sesquiterpene synthases are responsible for the complex mixture of sesquiterpenes emitted from *Arabidopsis* flowers. Plant J 42:757–771

Tuskan GA, Difazio S, Jansson S, Bohlmann J, Grigoriev I, Hellsten U, Putnam N, Ralph S, Rombauts S, Salamov A (2006) The genome of black cottonwood, *Populus trichocarpa* (Torr. & Gray). Science 313:1596–1604

Ulrich D, Nothnagel T, Schulz H (2015) Influence of cultivar and harvest year on the volatile profiles of leaves and roots of carrots (*Daucus carota* spp. *sativus* Hoffm.). J Agric Food Chem 63:3348–3356

Yahyaa M, Matsuba Y, Brandt W, Doron-Faigenboim A, Bar E, McClain A, Davidovich-Rikanati R, Lewinsohn E, Pichersky E, Ibdah M (2015a) Identification, functional characterization, and evolution of terpene synthases from a basal dicot. Plant Physiol 169:1683–1697

Yahyaa M, Tholl D, Cormier G, Jensen R, Simon PW, Ibdah M (2015b) Identification and characterization of terpene synthases potentially involved in the formation of volatile terpenes in carrot (*Daucus carota* L.) roots. J Agric Food Chem 63:4870–4878

Yahyaa M, Ibdah M, Marzouk S, Ibdah M (2016) Profiling of the terpene metabolome in carrot fruits of wild (*Daucus carota* L. ssp. *carota*) accessions and characterization of a geraniol synthase. J Agric Food Chem 66:2378–2386

Zi J, Mafu S, Peters RJ (2014) To gibberellins and beyond! Surveying the evolution of (di) terpenoid metabolism. Annu Rev Plant 65:259–286

Genetics and Genomics of Carrot Sugars and Polyacetylenes

17

Pablo F. Cavagnaro

Abstract

Carrot root carbohydrates, composed mainly of reducing and non-reducing free sugars, influence flavor, total dissolved solids and dry mater content, all quality traits for fresh-market and processing carrots. In the last decades, important advances have been made in biochemistry, physiology and genetics of carrot sugar metabolism. Several enzymes involved in sucrose metabolism and their corresponding genes have been isolated and functionally characterized, increasing our understanding of their individual roles and of their interactions in complex regulatory systems that influence major plant physiological processes, including partitioning of photo-assimilates, plant growth and storage of different sugar types in the carrot taproot. Polyacetylenes represent a large group of non-volatile lipid compounds produced primarily by members of the Apiaceae family. The major carrot polyacetylenes have been extensively studied with regard to their analytical identification and elucidation of their chemical structures as well as their biological activities, which have revealed numerous health-promoting properties for these compounds. Very recently, with the publication of the carrot genome sequence and related genomic and transcriptomic sequence resources, key genes and enzymes involved in the biosynthesis of carrot polyacetylenes were discovered. In this chapter, advances in genetics and genomics of carrot sugars and polyacetylenes were reviewed and discussed.

17.1 Introduction

Carrot root carbohydrates, composed mainly of reducing and non-reducing free sugars (mainly glucose, fructose and sucrose) and other more complex carbohydrates have been studied since nearly a century. Sugar content strongly influences carrot flavor (e.g., sugar level is associated with the sweetness perception), a major quality trait associated with consumer's preferences regarding organoleptic grade. Sugar content in the taproot is a major contributor of the total dissolved solids (TDS) and dry matter contents (the three variables are strongly and positively correlated), and all three traits influence the quality of carrots for fresh market and

P. F. Cavagnaro (✉)
National Scientific and Technical Research Council (CONICET), Buenos Aires, Argentina
e-mail: cavagnaro.pablo@inta.gob.ar

P. F. Cavagnaro
La Consulta Experimental Station, National Institute of Agricultural Technology (INTA), Ex Ruta 40 Km 96, La Consulta 5567, Mendoza, Argentina

P. F. Cavagnaro
Horticulture Institute, Faculty of Agricultural Sciences, National University of Cuyo, Almirante Brown 500, Lujan de Cuyo M5528AHB, Mendoza, Argentina

P. Simon et al. (eds.), *The Carrot Genome*, Compendium of Plant Genomes,
https://doi.org/10.1007/978-3-030-03389-7_17

processing. The positive association between sugar levels and carrot quality raised the attention of carrot breeders and researchers and oriented early studies which mainly focused on elucidating the type and content of sugars in the carrot root and its compositional variation during root development in the field and in post-harvest storage, as well as its variation due to genotype, growing conditions (e.g., soil and climate) and locations. Since the mid-80s to date, with the advent of molecular biology techniques and—more recently—high-throughput sequencing technologies becoming of widespread use among plant biologists, important advances have been made in biochemistry, physiology and genetics of carrot sugar metabolism. Thus, in the last decades, a number of enzymes involved in carrot sucrose metabolism have been isolated and biochemically characterized at various levels, including pH optima, solubility, substrate specificity, subcellular localization and enzymatic activity in different tissues and development stages of the carrot plant, as well as in response to biotic and abiotic stresses. In addition, cDNAs and genomic clones of the genes coding for these enzymes, along with their functional characterization, have been described. Also, a genetic mutation in an invertase gene conditioning the predominant sugar type stored in carrot roots was characterized in detail.

Polyacetylenes are a large group of non-volatile lipid compounds that comprise at least two, usually conjugated, triple carbon–carbon bonds. They are primarily produced by members of the Apiaceae and Araliaceae families, with carrot being the major dietary source of polyacetylenes. Some polyacetylenes, such as falcarinol-type polyacetylenes, are bioactive phytochemicals with proven antifungal, antibacterial, allergenic, anti-inflammatory, anti-platelet aggregatory and in vivo and in vitro anticancer properties. In addition, these compounds are largely responsible for the bitter unpleasant taste perceived in some carrots.

Early studies concerning polyacetylenes in carrot and other Apiaceae focused on their analytical identification and characterization of their chemical structure using different procedures (Raman spectroscopy, HPLC, gas chromatography, NMR, mass spectrometry). Also, characterization of the compositional diversity of polyacetylenes among carrot genotypes and among Apiaceae species was investigated. Variations due to cultivation, storage and processing conditions were also studied. In addition, their biological activities, mainly with regard to their potential human health benefits, as dietary constituents, were extensively studied. The association of some specific polyacetylenes with the carrot bitter off-taste was also investigated. Very recently, with the publication of the carrot genome sequence and related genomic and transcriptomic sequence resources, key genes involved in the biosynthesis of carrot polyacetylenes were discovered and described.

The following sections review and discus major advances in genetics and genomics concerning carrot sugars and polyacetylene metabolism.

17.2 Carrot Carbohydrates

17.2.1 Sugars in the Carrot Root and Their Relation with Flavor and Quality

Sugars and volatile terpenoids in the root are the two major components of carrot flavor. Glucose, fructose and sucrose account for more than 95% of the free sugars and 40–60% of the stored carbohydrates in the carrot root (Platenius 1934; Rygg 1945; Alabran and Mabrouk 1973). The reducing sugars glucose and fructose are generally present in equimolar amounts (Rygg 1945; Simon et al. 1980a). The ratio of sucrose to reducing sugars increases with root maturity, but decreases after harvest and during cold storage (Hasselbring 1927; Platenius 1934; Werner 1941; Phan and Hsu 1973; Phan et al. 1973).

Total sugar content is a major component of the total dissolved solids (TDS) and dry matter contents in the carrot root, accounting for 60–75% of the TDS and 50–70% of the dry weight, with the three variables being positively and strongly correlated (Carlton and Peterson 1963).

Thus, in the latter study, total sugar quantity and TDS had correlation coefficients (*r*) of 0.75–0.95, whereas TDS and dry matter had *r* values of 0.85–0.95. These three correlated variables influence the quality of carrots for fresh market and processing.

Significant genetic variation in total sugar content as well as in 'reducing sugar/total sugar' ratios can be found in the carrot germplasm, as indicated by results from studies that used analytical methods (high-performance liquid chromatography, HPLC) to estimate glucose, fructose and sucrose and sensory evaluations that measured several flavor parameters including sweetness, which is influenced by sugar levels (Simon et al. 1980a, b, c). In addition to genotype, which was the main factor influencing carrot total and reducing sugar levels, sweetness and preference, environmental factors such as soil, climate and location can also influence sugar content and flavor (Simon et al. 1980b, c; Freeman and Simon 1983).

In a study by Simon et al. (1980c), which analyzed three carrot lines grown under three different soil and climate conditions, total sugar content varied moderately within a range of 45.8–72.8 mg/g fresh weight (fw), whereas striking differences were found for percentage of reducing sugar (RS), relative to the total sugar content, reporting a range of variation of 8–66.3% RS. Similar results were found in another study by Simon et al. (1980a) which examined sugar levels in four carrot lines grown in three different locations, finding an overall range of variation for total sugars, including all genotypes and locations, of 38.2–81.7 mg/g fw and 0.2–70.6% RS. Similarly, Freeman and Simon (1983) reported a range of 40–80 mg/g fw for total sugar content and 10–78% RS among six carrot inbred lines classified as either 'high RS' (51–78% RS) or 'low RS' (10–12% RS), which were then intercrossed to study inheritance of this trait. In line with the results from these studies, considerable genetic variation for type and amount of free sugars among carrot genotypes was also reported by Carlton and Peterson (1963) and Bajaj et al. (1980). Conversely, a lower range of variation was found in another study that evaluated four carrot cultivars of different root colors for various nutritional and flavor components, including sugars, reporting a range of 50.4–54.7 mg/g fw for total sugars and 23.6–61.1% RS (Alasalvar et al. 2001).

More recently, a large carrot germplasm collection, composed of 118 accessions from Europe, mostly from the Warwick Genetic Resources Unit (United Kingdom), was characterized on the basis of their root sugar and carotenoids levels (Baranski et al. 2012). It was found that total sugar content varied from 51 to 136 mg/g fw, with half of the accessions having 86–106 mg/g fw. No significant differences were found for sugar content among accessions from different geographical origins or different root colors. However, European and American accessions had, on average, about 18% more total sugars than Asian accessions, and this trend was observed for advanced cultivars as well as for landraces, and even when only orange-rooted accessions were considered. The non-reducing sugar (NRS) to reducing sugar (RS) ratio ranged from 1.1 to 9.6, with a mean of 2.5, which is equivalent to 10–48.1% RS (mean: 28.6% RS). The vast majority of the accessions (105 out of 118) had a NRS/RS ratio > 1.5 (equivalent to 40% RS). Results from this study indicate that there is broad variation for sugar content and even broader variation (more than ninefolds) for non-reducing/reducing sugar ratio in the carrot germplasm.

Altogether, these studies revealed the existence of significant variation for total sugar content and striking differences in the ratio of NRS to RS contents. These data are relevant from a breeding point of view, due to the influence of this trait on various carrot quality and flavor parameters. For example, carrots low in RS content are necessary for ensuring the quality of deep-fried carrot chips, a specialty snack product, whereas high RS may be desirable for improving flavor of fresh-market carrots, since the average sweetness of glucose plus fructose is 20% greater than that of sucrose (Shallenberger and Birch 1975). The influence of root sugar type and content on various carrot quality parameters, as demonstrated in some of the studies described above, encouraged research on the genetics underlying this trait, as such knowledge would facilitate the development of better tasting and improved-quality carrots.

17.2.2 Inheritance and Genetic Mapping of Rs, a Monogenic Trait Conditioning Root Sugar Type

The genetic control of reducing sugar (RS) to non-reducing sugar (NRS) balance was first investigated using crosses derived from high and low RS carrot inbred lines (Freeman and Simon 1983). The parental lines with 'high RS' had 51–78% reducing sugars, whereas inbreds with 'low RS' had 10–12% RS. Thus, the parental lines and their respective F_1, F_2, F_3 and backcross progenies were analyzed for percentage of RS (glucose, fructose and sucrose contents were estimated by HPLC analysis). Segregation analysis revealed high RS:low RS ratios that were consistent with a simply inherited locus for the genetic control of this trait, with high RS being dominant over low RS. These segregation data, indicating a single dominant gene for the control of %RS, were consistent across the 20 F_1, F_2 and BC populations initially evaluated and were further confirmed in F_3 families. This monogenic trait conditioning storage root sugar type was called *Rs*.

Exploiting the fact that TDS and sugar contents are correlated variables (Carlton and Peterson 1963), phenotypic recurrent selection for these traits was performed during five cycles, and heritability estimates were calculated after the fifth selection cycle (Stommel and Simon 1989). To this end, divergent recurrent selection was applied on four carrot populations with high or low percentage of TDS with high or low levels of reducing sugars. Effective selection for both TDS and RS was achieved, as indicated by significant gains or reductions in TDS and RS levels over five selection cycles. Increases in TDS content of 22.4–28.2% were obtained in the 'high TDS' populations, whereas reduction in TDS levels by 15.9–21.9% was obtained in the 'low TDS' populations. Successful selection for total sugars and sugar type was also achieved. After five cycles of selection, total sugar levels in the 'high TDS' populations were 2–2.5 times higher than in the 'low TDS' populations, demonstrating that TDS estimates are a useful selection criterion for total sugar level. Within the 'high TDS' and 'low TDS' populations, selection for 'high percentage of RS' and 'low percentage of RS' was very effective, as indicated by the fact that undetectable quantities of RS were found in the 'low RS' samples, whereas the 'high RS' samples represented 31 and 71% of the total sugars in the 'low TDS' and 'high TDS' populations, respectively.

Significant and moderate realized heritability estimates were obtained for TDS from the four populations analyzed, values of which ranged from 0.40 to 0.45, indicating that a significant genetic component exists for TDS content in carrot.

Inheritance analysis of *Rs*, P_1 and Y_2 (conditioning purple and yellow root colors, respectively) in 33 F_2 and backcross populations segregating for these traits revealed absence of linkage among these monogenic loci (Simon 1996). *Rs* was later genetically mapped in an F_2 population derived from the cross of B9304, an orange-rooted inbred presenting high percentage of RS and homozygote dominant for *Rs* (*RsRs*), and YC7262, a purple-rooted inbred with low percentage of RS and homozygote recessive for *Rs* (*rsrs*) (Vivek and Simon 1999). In addition to *Rs*, this population was also segregated for P_1 and Y_2. These 3 trait loci were mapped to different linkage groups (LG), confirming previous results of Simon (1996), indicating absence of linkage among them. *Rs* was mapped to the end of linkage group C, a largely unsaturated LG composed of *Rs* and 11 anonymous markers (10 AFLPs and 1 SAMPL) covering a total map distance of 103.9 cM. Because none of the markers in this LG had sufficient sequence information to perform a BLAST search against the carrot genome assembly, the correspondence of this LG with the carrot chromosomes could not be established.

17.2.3 Carrot Sucrose Metabolism

Sucrose metabolism determines the type and content of sugars in the carrot root. Key enzymes in storage organs that regulate sucrose metabolism are sucrose phosphate synthase (EC 2.4.1.14), sucrose phosphatase (EC 3.1.3.24), sucrose synthase (EC 2.4.1.13) and invertase (or β-fructofuranosidase; EC 3.2.1.26). Of these, sucrose phosphate synthase and sucrose phosphatase are involved in sucrose synthesis, while sucrose synthase and invertase are involved in sucrose cleavage (Copeland 1990; Huber and Huber 1996; Tang et al. 1999; Sturm and Tang 1999). Sucrose synthase is a glycosyltransferase that converts sucrose in the presence of uridine diphosphate (UDP) into UDP-glucose and fructose, whereas invertase is a hydrolase which cleaves sucrose into glucose and fructose.

Sucrose is the major end product of leaf photosynthesis and is the major sugar transported in the phloem. However, sucrose must be first cleaved into hexoses—by invertase or sucrose synthase—in order to be used for most metabolic processes. Invertases are present in most plant tissues and catalyze the breakdown of the disaccharide sucrose into fructose and glucose. Isoforms of invertases are characterized and classified according to pH optima (acid, neutral and alkaline), subcellular locations (vacuole or cell wall) and solubility properties (soluble or insoluble), and they have been described in several plants including carrot (Stommel and Simon 1990; Sturm and Chrispeels 1990; Unger et al. 1992; Sturm et al. 1995; Sturm 1996). Acid invertase has a pH optimum between 3 and 5, whereas neutral and alkaline invertases have pH optima of about 6.8 and 8, respectively (Lee and Sturm 1996). Acid invertases either accumulate as a soluble protein in the matrix of the vacuole (called 'vacuolar' or 'soluble' invertase) or is ionically bound to the cell wall (called 'extracellular' or 'insoluble' invertase) (Unger et al. 1994; Laurière et al. 1988). It has been proposed that acid invertases are involved in phloem unloading and in the control of sugar type in storage organs (Klann et al. 1993).

A number of enzymes involved in carrot sucrose metabolism have been isolated and biochemically characterized at various levels, including pH optima, solubility, substrate specificity, subcellular localization and enzymatic activity in different tissues and development stages of the carrot plant, as well as in response to biotic and abiotic stresses (reviewed by Sturm 1996). In addition, cDNAs and genomic clones of the genes coding for these enzymes, along with their functional characterization, have been reported. A summary of the main finding for the carrot sucrose-cleaving enzymes and their corresponding genes is as follows.

17.2.4 Genes Involved in Carrot Sucrose Metabolism

Sucrose synthase has been isolated from carrot roots and characterized at the biochemical level, and cDNA clones encoding sucrose synthase were isolated and sequenced (Sebkova et al. 1995). Simple hybridization patterns in DNA gel blots and the fact that independent cDNA clones had identical DNA sequence suggested that only one sucrose synthase gene existed. However, subsequent analysis of genomic clones for carrot sucrose synthase revealed two different genes, denominated *Susy*Dc1* and *Susy*Dc2* (Sturm et al. 1999a). Very recently, three sucrose synthase genes, namely *DcSus1*, *DcSus2* and *DcSus3*, were identified by Liu et al (2018) by means of analysis of a carrot transcriptomic and genomic database (Xu et al. 2014) combined with cDNA cloning. As is shown in Table 17.1, presenting information on the carrot sucrose cleavage genes reported to date, the cDNA cloned by Sebkova et al. (Acc. X75332) corresponds to *Susy*Dc1* (Sturm et al. 1999a) and *DcSus1* (Liu et al. 2018), whereas *Susy*Dc2*, *DcSus2* and *DcSus3* correspond to different genes (i.e., different DCARs) in the carrot genome assembly (Iorizzo et al. 2016).

Carrot contains several invertases: a major form of the cell wall enzyme and two forms of vacuolar invertase (isoenzymes I and II). The different isoforms and the most abundant form of each cellular compartment were purified and described (Unger et al. 1992; Laurière et al. 1988). Partial sequences and specific antibodies led to the isolation of their cDNA clones (Sturm and Chrispeels 1990; Unger et al. 1994), and the corresponding genomic clones were later isolated and sequenced (Ramloch-Lorenz et al. 1993; Lorenz et al. 1995; Sturm 1996) (Table 17.1). Comparisons were made—at the biochemical and DNA and polypeptide sequence levels—between the vacuolar and the two cell wall invertases (Unger et al. 1992, 1994; Laurière et al. 1988). Among the main findings, it was reported that (i) the three isoenzymes are encoded by different genes and do not originate from differential splicing, as is the case of *Saccharomyces cerevisiae*; (ii) striking differences in isoelectric points were found for vacuolar (pH 3.8 and 5.7, for isozyme I and II, respectively) and cell wall invertases (pH 9.9); and (iii) vacuolar—but not cell wall—invertases contain a signal peptide for vacuolar targeting.

Analysis of the specific functions of different isoforms of sucrose synthase and invertase may provide a better understanding of their roles in carrot sucrose metabolism, translocation and storage. To this end, gene expression analysis of the carrot sucrose—cleaving enzymes (sucrose synthase, cell wall invertase and the vacuolar invertase isozymes I and II) in source and sink organs (leaf lamina, petioles and roots) of developing carrot plants was performed (ten time points, from sowing to 20-week-old plants, were analyzed) (Sturm et al. 1995). Different and individual expression patterns were revealed for the genes analyzed, suggesting a different function for each enzyme. Only plants with primary roots contained high and comparable levels of transcripts for cell wall invertase in leaf lamina, petioles and roots, indicating that the expression of this gene is development specific, but not organ specific. Conversely, the expression of the vacuolar invertase isoenzymes I and II seemed to be under spatial and temporal control. High levels of transcripts for both isoenzymes were only found in roots, with the highest level of isoenzyme I transcripts found in young primary roots (1–8-week-old plants) and isoenzyme II in developing taproots (10–20-week-old plants). Transcripts for sucrose synthase were found in all developing plant organs, but transcript levels varied during the plant growth, with markedly high levels in young leaves and in roots at the transition from primary to secondary roots.

The fact that only transcripts of sucrose synthase and vacuolar invertase isozyme II were detected in developing storage roots (i.e., during the filling phase of the root) suggests their involvement in sucrose partitioning. Conversely, transcripts of isozyme I and cell wall invertase were not detected in developing taproots, suggesting that they are not involved in this process. In order to further examine this hypothesis, the sink/source balance of carrot plants was altered by removing the leaves in field-growing plants (12-week old), thereby making the storage root become a strong source organ and the newly emerging leaves sink organs, and expression patterns of the same genes were analyzed. It was found that only the expression of vacuolar isoenzyme II and sucrose synthase markedly changed, becoming strongly upregulated in the newly emerging leaves (sink organ), supporting the involvement of these two genes in sucrose partitioning. Transcripts of cell wall invertase and vacuolar invertase isoenzyme I were not detected throughout the experiment, providing further evidence that these enzymes do not participate in this process.

Based on these results, the authors proposed a model for sucrose partitioning in which phloem unloading is driven by sucrose utilization in the cytoplasm and storage in the vacuole of taproot cells. According to this model, in developing taproots, sucrose is mainly utilized in the cytoplasm by sucrose synthase to produce UDP-glucose, which is utilized for growth and development (utilization sink), whereas at later stages of taproot development, storage of sugars is the main cellular activity (storage sink). For the latter purpose, sucrose is transported into the vacuole where it is subsequently cleaved by a vacuolar invertase (most likely isozyme II), and

Table 17.1 Carrot genes encoding sucrose-cleaving enzymes described to date

Gene ID	Accession number (NCBI)	Description (GenBank-NCBI)	Type of sequence	Gene ID in the carrot genome assembly*	Annotation in carrot genome	Chr.	Genome coordinates		ID checked§	Reference
							Start	End		
*Inv*Dc1*	X69321.1	*D. carota Inv*Dc1* gene (cell wall invertase, insoluble)	genomic	DCAR_013407	hypoth. prot.	4	30,421,649	30,423,613	yes	Sturm and Chrispeels (1990); Lorenz et al. (1995); Sturm (1996)
				DCAR_013406	hypoth. prot.	4	30,446,925	30,449,363	yes	
*Inv*Dc2*	X78424	*D. carota Inv*Dc2* gene (cell wall invertase, insoluble)	genomic	DCAR_020738	hypoth. prot.	6	28,751,059	28,753,544	yes	Lorenz et al. (1995); Sturm (1996)
*Inv*Dc3*	X78423.1	*D. carota Inv*Dc3* gene (cell wall invertase, insoluble)	genomic	DCAR_023361	hypoth. prot.	7	768,904	771,296	yes	Lorenz et al. (1995); Sturm (1996)
*Inv*Dc4*	Y18707	*D. carota Inv*Dc4* gene (vacuolar acid-soluble invertase isozyme I)	genomic	DCAR_018105	hypoth. prot.	5	26,561,530	26,569,454	yes	Sturm (1996)
*Inv*Dc5*	Y18706.1	*D. carota Inv*Dc5* gene (vacuolar acid-soluble invertase isozyme II)	genomic	DCAR_008466	hypoth. prot.	2	42,305,870	42,309,690	yes	Sturm (1996); Yau and Simon (2003); Yau et al. (2005)
	X75353.1	*D. carota* soluble acid invertase isoenzyme I	cDNA	DCAR_018105	hypoth. prot.	5	26,561,530	26,569,454	yes	Unger et al. (1994)
	X75352.1	*D. carota* soluble acid beta-fructofuranosidase isoenzyme I	cDNA	DCAR_018106	hypoth. prot.	5	26,561,530	26,569,454	yes	Unger et al. (1994)
	Y16262	*D. carota* neutral invertase	cDNA	DCAR_002523	hypoth. prot.	1	30,135,362	30,140,332	yes	Sturm et al. (1999b)
	X75332	*D.carota* (Nantaise) sucrose synthase	cDNA	DCAR_028527	hypoth. prot.	8	6,163,760	6,167,727	yes	Sebkova et al. (1995); Sturm (1996)

(continued)

Table 17.1 (continued)

Gene ID	Accession number (NCBI)	Description (GenBank-NCBI)	Type of sequence	Gene ID in the carrot genome assembly*	Annotation in carrot genome	Chr.	Genome coordinates		ID checked§	Reference
							Start	End		
*Susy*Dc1*	Y16090	*D. carota Susy*Dc1* gene (sucrose synthase)	genomic	DCAR_028527	hypoth. prot.	8	6,163,760	6,167,727	yes	Sturm et al. (1999a)
*Susy*Dc2*	Y16091	*D. carota Susy*Dc2* gene (sucrose synthase)	genomic	DCAR_024777	hypoth. prot.	7	19,289,899	19,294,038	yes	Sturm et al. (1999a)
				DCAR_024776	hypoth. prot.	7	19,273,726	19,277,413	yes	
DcSus1	n.a.	*D. carota* sucrose synthase 1†	cDNA	DCAR_028527	hypoth. prot.	8	6,163,760	6,167,727	yes	Liu et al. (2018)
DcSus2	n.a.	*D. carota* sucrose synthase 2†	cDNA	DCAR_002857	hypoth. prot.	1	33,487,640	33,502,693	yes	Liu et al. (2018)
				DCAR_002856	hypoth. prot.	1	33,480,847	33,486,390	yes	
DcSus3	n.a.	*D. carota* sucrose synthase 3†	cDNA	DCAR_010214	hypoth. prot.	3	16,510,683	16,514,823	yes	Liu et al. (2018)

* Refers to the DCAR with highest sequence homology to each sucrose-cleaving enzyme gene. n.a. sequence not available in GenBank (NCBI)

† Description from reference paper

§Because no precise annotation was available for the DCARs (annotated as hypothetical protein), their sequences were BLASTed against the GenBank (NCBI) database to confirm their sequence homology to sucrose-cleaving enzymes from other species. Genes *Inv*Dc1*, *Susy*Dc2* and *DcSus2* had highest sequence homology with two contiguous (presumably duplicated) genes each

this cleavage maintains the sucrose concentration gradient and thereby participates in driving the flow of sucrose from leaves into roots.

A follow-up study by Tang et al. (1999) was undertaken to further unravel the functions of cell wall and vacuolar invertases in carrot. For this purpose, antisense technique was used to generate transgenic carrot plants with reduced activity for these two enzymes. Compared with control plants, the dry weight leaf-to-root ratio of cell wall invertase antisense plants was shifted from 1:3 to 17:1. Plants expressing antisense mRNA for vacuolar invertase had more leaves and smaller roots than control plants (this was also observed for cell wall invertase antisense lines), suggesting less partitioning of photo-assimilates to the root. In antisense lines of both cell wall and vacuolar invertases, the carbohydrate content of leaves was elevated, and that of roots was reduced. Together, these data suggest that acid invertases play an important role in early plant development, most likely via control of sugar composition and metabolic fluxes. Later in plant development, both isoenzymes seem to have important functions in sucrose partitioning.

A similar study by the same group (Tang and Sturm 1999) further investigated the role of carrot sucrose synthase. Transgenic carrot plants with reduced sucrose synthase activity were obtained by antisense technique which knocked down the main form of carrot sucrose synthase. It was found that, in transgenic lines, sucrose synthase activity was reduced in the taproot but not in leaves. In the sink organs, sucrose utilization was markedly decreased and higher levels of sucrose but lower levels of UDP-glucose, glucose, fructose, starch and cellulose were found. In addition, antisense plants developed atypical phenotypes, presenting markedly smaller leaves and roots than in 'controls,' and plant size was correlated with sucrose synthase activity. Also, in most of the antisense lines, the leaf-to-root dry weight ratios were not altered, suggesting that sucrose synthase in carrot strongly influences plant growth rather than sucrose partitioning. The authors concluded that, in contrast to the vacuolar acid invertases, which are critical for partitioning of photo-assimilates between source leaves and taproots (Tang et al. 1999), sucrose synthase's main role seems to be sucrose-cleaving activity for feeding sucrose into metabolism and thereby influence plant growth.

Further investigations on the role of sucrose synthase genes in carrot sugar metabolism were reported. Expression patterns of *Susy*Dc1* and *Susy*Dc2* (Table 17.1) in different organs and organ sections at different developmental stages were examined by Sturm et al. (1999a). It was found that *Susy*Dc2* was exclusively expressed in flowers, whereas transcripts of *Susy*Dc1* were found in stems, in roots at different developmental stages and in flower buds, flowers and maturing seeds, with the highest expression levels found in strong utilization sinks for sucrose, such as the growing stems and the tip of taproots.

Very recently, transcript profiling of sucrose synthase genes in different tissues and developmental stages of four carrot cultivars were investigated (Liu et al. 2018). To this end, three sucrose synthase genes (*DcSus1*, *DcSus2* and *DcSus3*) were cloned from carrot roots, and their expression levels were analyzed in three tissues (root, leaf lamina and leaf petioles) at five plant developmental stages which were characterized by their root sucrose and soluble sugar contents. It was found that root sucrose and sugar content increased throughout root development and that enzymatic activity of sucrose synthase followed an inverse trend to sucrose content, decreasing in activity as the root develops [a negative correlation between sucrose synthase activity and sucrose concentration was reported ($r = -0.628$, $p < 0.01$)]. Concomitantly with increasing root sucrose levels and decreasing sucrose synthase activity during root development, *DcSus1*, *DcSus2* and *DcSus3* genes showed downregulation during root growth, suggesting that the expression of these genes conditions sucrose synthase activity and, consequently, sucrose levels in the carrot root. The inverse association between the expression levels of the three sucrose synthase genes and root development coincides with previous results from Sturm et al. (1995), indicating higher expression of a sucrose synthase gene in young leaves and roots, but little or no expression in 14- to 20-week-old roots. Interestingly, gene expression of the three sucrose synthases of Liu et al. (2018) followed an opposite trend in the leaf blade and petioles than that observed in roots (i.e., in leaf tissues, gene expression increased with plant growth), and gene expression preferences for specific leaf tissues were noted for blade (*DcSus2*, *DcSus3*) and petioles (*DcSus1*). Together, results from this study suggest that sucrose synthase genes condition soluble sugar and sucrose content in the carrot root. However, because no discrimination was made between reducing and non-reducing sugars, one cannot speculate on the *Rs* genotype of the cultivars used nor on the influence of sucrose synthase activity on the type of sugars accumulated in the root.

Altogether, the studies described above have contributed extensively to our understanding on the role of sucrose-cleaving enzymes in sucrose metabolism in carrot and plants in general.

17.2.5 A Mutation in the Acid-Soluble Invertase Isozyme II Gene Conditioning Sugar Type in the Carrot Root Is a Candidate for Rs

Yau and Simon (2003) examined expression of carrot invertase and sucrose synthase genes (by RT-PCR) in roots and leaves of *Rs/Rs* and *rs/rs* near-isogenic lines, characterized by storing their roots predominantly in reducing sugars (glucose and fructose) or fructose, respectively. Of the four carrot genes analyzed (3 invertases and 1 sucrose synthase), the acid-soluble invertase isozyme I, the cell wall invertase and the sucrose synthase, revealed no differences in expression between the *Rs/Rs* and *rs/rs* lines, whereas the acid-soluble isozyme II gene was only expressed in the *Rs/Rs* line (no transcripts of this gene were detected in the *rs/rs* line) in 18-week-old roots. In line with these results, invertase enzyme activity was less in all tissues of *rs/rs* plants than *Rs/Rs* plants, whereas activities of other enzymes (sucrose synthase and sucrose phosphate synthase) were comparable in *Rs/Rs* and *rs/rs* tissues. Comparative structural analysis of the genomic DNA sequence of the acid-soluble invertase isozyme II gene in *Rs/Rs* and *rs/rs* plants revealed a 2.5 kb insert in the *rs* allele. The insertion occurred in the first and largest intron near the 5' end of the gene. This insertion was later further characterized to reveal that it was caused by a transposable element of the *PIF/Harbinger*-like family, which was called *DcMaster1* (Grzebelus et al. 2006). Altogether, these results suggest that this insertion of a 2.5 kb DNA fragment accounts for the failure of acid invertase isozyme II transcription in *rs/rs* roots and, consequently, little to no acid-soluble invertase activity. The fact that a little invertase activity still remained after the knockdown of the acid-soluble invertase isozyme II gene was attributed to the activity of acid-soluble invertase isozyme I.

Based on these results, demonstrating no acid-soluble invertase isozyme II transcripts in roots of *rs/rs* while abundant transcripts occurred in *Rs/Rs*, and the fact that this is the only invertase enzyme found to be well expressed during carrot taproot development (Sturm 1996), the authors proposed that the carrot acid-soluble invertase isozyme II gene is the likely candidate for the *Rs* locus.

The discovery of a 2.5 kb insertion in the *rs* allele of the carrot acid-soluble invertase isozyme II gene facilitated the development of a codominant PCR-based marker for performing marker-assisted evaluation and selection for carrot root sugar type (Yau et al. 2005). In this study, a total of 1176 plants from 7 F_2 families derived from 12 diverse carrot inbreds and their selected F_3 and F_4 progenies were evaluated for PCR amplification products developed to evaluate *Rs* and *rs* alleles and for the predominant sugar stored in carrot roots. In all cases, plants scored as *rs/rs*—based on maker genotyping—had storage roots which stored predominantly sucrose, while *Rs/rs* and *Rs/Rs* plants all stored predominantly reducing sugars (glucose + fructose). Marker-assisted selection was successful in predicting the type of sugar stored in 11 F_3 families and three F_4 families. The markers and sugar type fit expected 1:2:1 or 3:1 segregation ratios in all cases. Complete co-segregation (100%) of the markers and the trait locus was found among plants in the F_2 and F_3 families, further confirming that the insertion found in the acid-soluble invertase isozyme II gene, indeed, conditions the *Rs* trait locus. A BLAST search of the carrot acid-soluble invertase isozyme II gene (*Inv*Dc5*) in the carrot genome assembly revealed highest sequence homology with DCAR_008466, which is located in Chr. 2 at position 42,305,870-42,309,690. Although DCAR_008466 was annotated in the carrot genome assembly as a 'hypothetical protein,' its BLAST search against the NCBI database revealed high sequence matches with acid invertases from carrot and other species.

17.3 Carrot Polyacetylenes

17.3.1 Biosynthesis of Polyacetylenes

The most abundant polyacetylenes found in Apiaceae plants, including carrot, are C_{17}-polyynes of the falcarinol type, like falcarinol, falcarindiol and falcarindiol-3-acetate (Fig. 17.1, compounds 1-3) (Lund 1992; Dawid et al. 2015). Although extensive research has been performed concerning the analytical and biochemical identification and characterization of falcarinol-type polyacetylenes, as well as their putative biological functions (both topics are discussed below), far less knowledge exists about the enzymes—and their corresponding genes—involved in the biosynthesis of these compounds. In addition, very little is known regarding the genetic architecture underlying falcarinol-type polyacetylene production in plants, thereby hindering progress in breeding for this trait.

Data from metabolic studies point out the important role of crepenynic and dehydrocrepenynic acids as precursors of polyacetylenes that normally occur in various plant families, including Apiaceae (Minto and Blacklock 2008). Based on the structure of falcarinol-type polyacetylenes, it has been hypothesized that these compounds are derived from ubiquitous fatty acids. Indeed, indirect evidence from biochemical (Bohlman 1988) and radiochemical tracer studies (Barley et al. 1988), along with the discovery of pathway intermediates (Jones et al. 1966; Kawazu et al. 1973), implicates a diversion of flux away from linoleic acid biosynthesis as the entry point into falcarinol-type polyacetylene biosynthesis (reviewed by Minto and Blacklock 2008). The final steps of linolenic biosynthesis are the conversion of oleic acid to linoleic acid, mediated by fatty acid desaturase 2 (FAD2), and linoleic acid to linolenic acid, catalyzed by another fatty acid desaturase (FAD3). Some plant species contain 'divergent' forms of FAD2 that, instead of or in addition to converting oleic acid to linoleic acid, catalyze the installation of unusual in-chain functional groups such as hydroxyl groups, epoxy groups, conjugated double bonds or carbon–carbon triple bonds into the acyl chain (Badami and Patil 1980) and thus divert flux from linolenic production into the accumulation of unusual fatty acids, such as crepenynic and dehydrocrepenynic acids, which are thought to be intermediate precursors to the biosynthesis of falcarinol-type polyacetylenes. Based on the structural comparisons between dehydrocrepenynic acid and the simplest C_{17}-falcarinol-type polyacetylene (falcarinol), it seems likely that additional modifications of the former compound are still needed, possibly mediated by a Δ14 acetylenase, an ω-3 hydroxylase and a ω-desaturase. A schematic representation of the proposed pathway is shown in Fig. 17.2.

17.3.2 Polyacetylenes in Carrot and Other Apiaceae

Polyacetylenes are primarily produced in members of the Apiaceae family, although other plant taxa may present these compounds, usually at lower concentrations. Apiaceae plants known to produce polyacetylenes in their edible parts include carrot (*Daucus carota* L.), celery (*Apium graveolens* L.), parsley (*Petroselinum crispum* Mill.), parsnip (*Pastinaca sativa* L.), fennel (*Foeniculum vulgare* Mill.) and caraway (*Carum carvi* L.). Falcarinol-type polyacetylenes, like

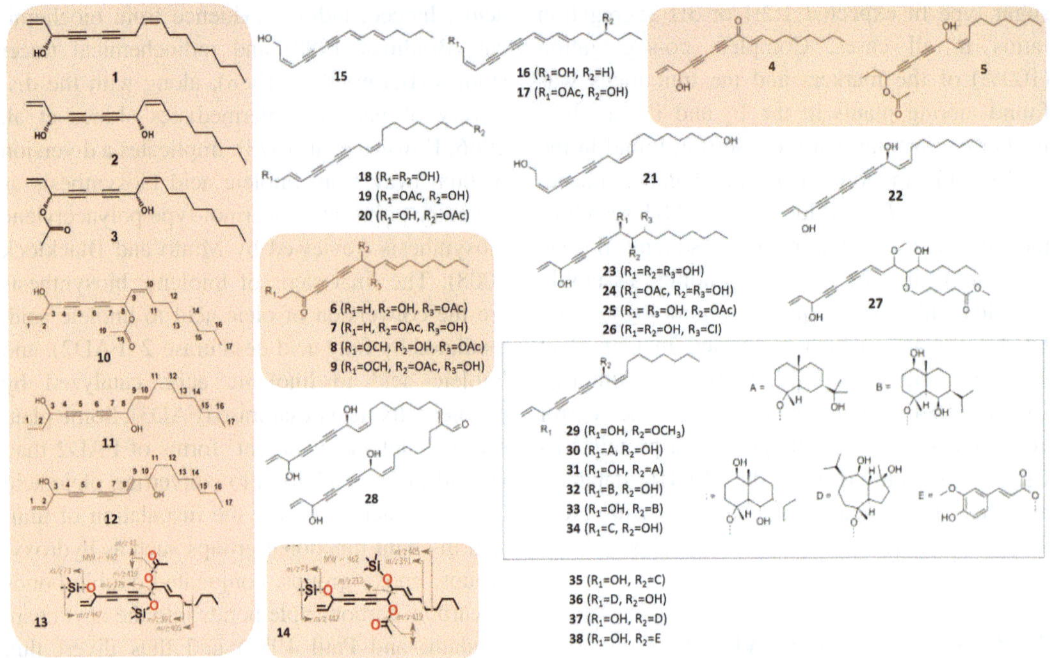

Fig. 17.1 Polyacetylenic structures found in carrot and other Apiaceae. The most abundant carrot polyacetylenes are falcarinol (**1**), falcarindiol (**2**) and falcarindiol-3-acetate (**3**). Other polyacetylenes identified in carrot roots are isofalcarinolone (**4**), 1,2-dihydrofalcarindiol-3-acetate (**5**), (*E*)-falcarindiolone-9-acetate (**6**), (*E*)-falcarindiolone-8-acetate (**7**), (*E*)-1-methoxy-falcarindiolone-9-acetate (**8**), (*E*)-1-methoxy-falcarindiolone-8-acetate (**9**), falcarindiol-8-acetate (**10**), 1,2-dihydrofalcarindiol (**11**), panaxydiol (**12**), falcarintriol-8-acetate (**13**) and falcarintriol-9-acetate (**14**). Polyacetylenes identified in carrot (1-14) are highlighted. Modified from Schmiech et al. (2009), Negri (2015) and Busta et al. (2018)

falcarinol, falcarindiol and falcarindiol-3-acetate (Fig. 17.1) have been indicated as important contributors to the bitter off-taste of fresh carrots, carrot puree and carrot juice (Czepa and Hofmann 2003, 2004). Due to their relevance for carrot flavor and nutritional value, several studies have focused on developing analytical methods for the identification and quantitation of these compounds (Metzger and Barnes 2009; Pferschy-Wenzig et al. 2009; Roman et al. 2011; Killeen et al. 2013).

The content and composition of polyacetylenes vary greatly within the carrot plant, with highest total polyacetylene content observed in the root periderm, followed by the root phloem and leaf lamina, with falcarindiol content being highest in the periderm and phloem tissues (Fig. 17.3) (Garrod and Lewis 1979; Baranska and Schulz 2005; Baranska et al. 2005; Busta et al. 2018). These data suggest that peeling fresh carrots decreases total polyacetylene and falcarindiol contents, as it has been previously reported (Garrod and Lewis 1979). Interestingly, variation in polyacetylene content and tissue distribution in the storage root were found between a dark orange carrot cultivar (HCM) and some wild carrots, namely *Daucus carota* subspecies *maritimus*, *D. carota* ssp. *gummifer*, *D. carota* ssp. *commutatus* and *D. halophilus*, as indicated by Raman spectroscopy analysis of root sections in these taxa (Baranska et al. 2005). Lower content of total polyacetylenes was found in the roots of cultivated carrots, as compared to those of wild carrots belonging to these *D. carota* subspecies. In addition, in the cultivated orange carrot root, the maximum polyacetylene content was found in the phloem tissue, close to the vascular cambium, with falcarinol being the

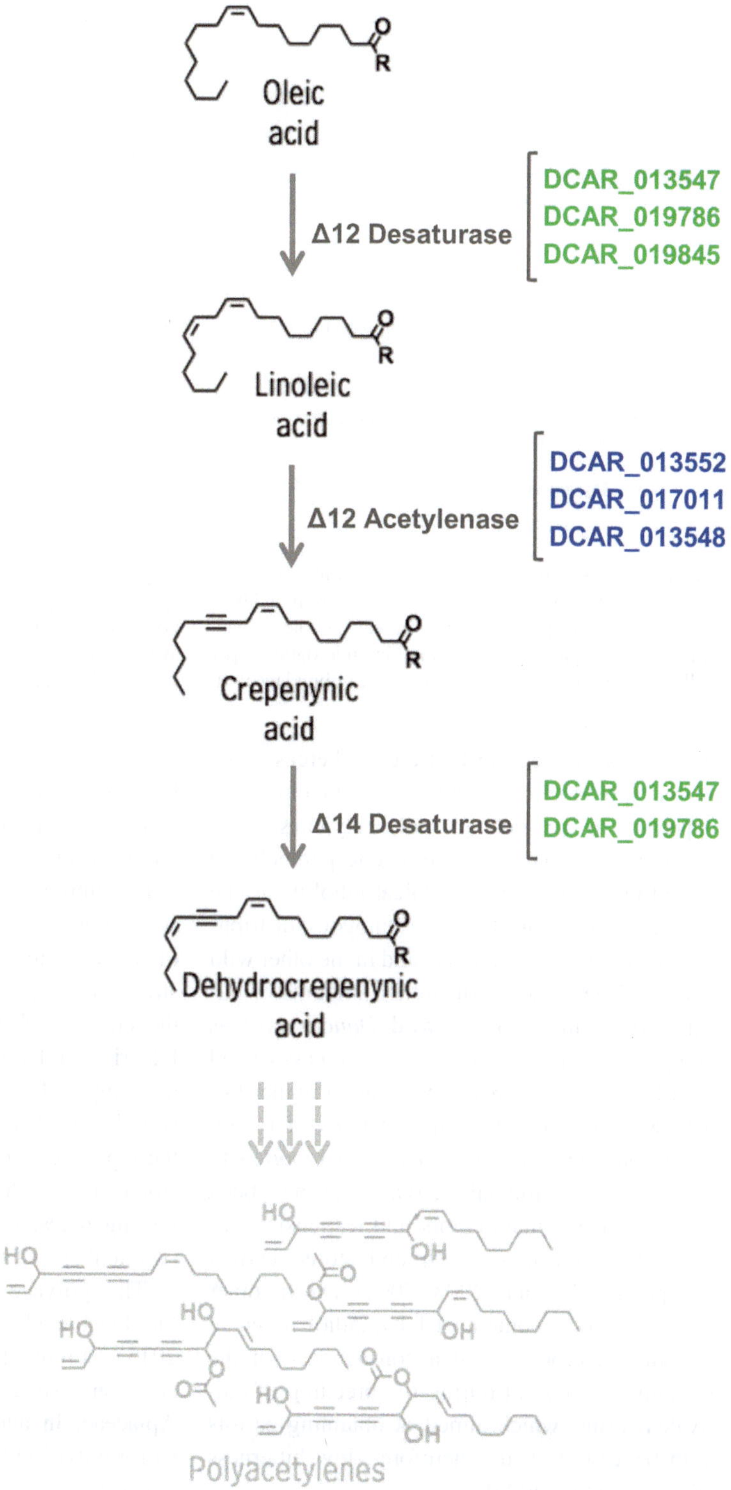

Fig. 17.2 Possible biosynthetic pathway of falcarinol-type C17-polyacetylenes in higher plants like carrots (adapted from Minto and Blacklock). In color font are indicated functionally characterized fatty acid desaturase (in green) and acetylenase (in blue) genes from carrot

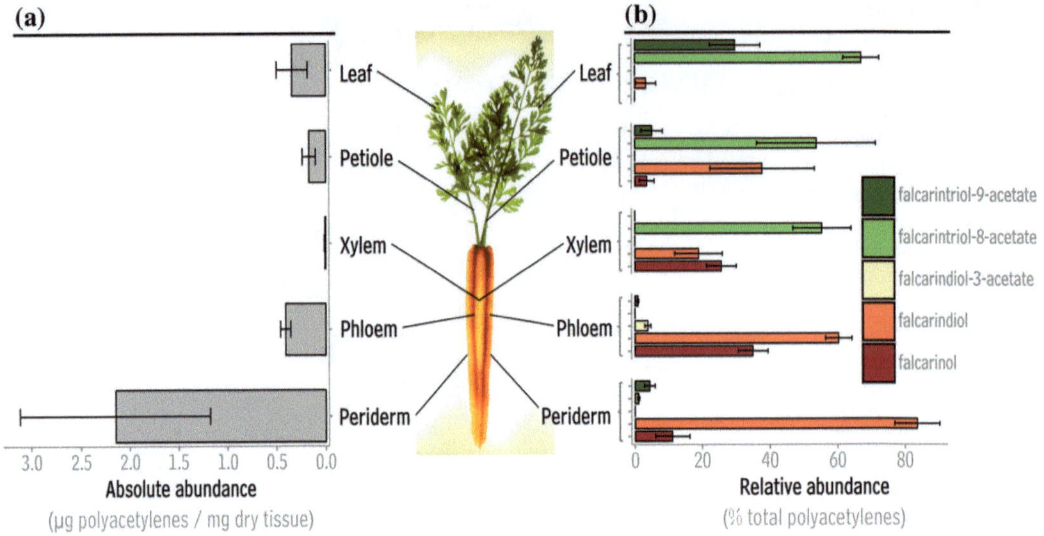

Fig. 17.3 Polyacetylene content and composition in aerial and subterranean tissues of *Daucus carota*. **a** Absolute abundance of polyacetylenes in leaf, petiole, xylem, phloem and epidermis is given in micrograms per milligram of dry tissue. **b** Relative abundance of polyacetylenes in leaf, petiole, xylem, phloem and epidermis is given as percent of total polyacetylenes. For both **a** and **b**, bar lengths and error bars represent the average and standard deviation of six independent samples. Reproduced from Busta et al. (2018)

major polyacetylene in this area, whereas in *D. carota* spp. *maritimus,* the whole phloem tissue appeared to be rich in polyacetylenes, with observed maxima in the pericyclic parenchyma close to the periderm, with falcarindiol being the predominant polyacetylene. Analogous distribution of polyacetylenes was found in the other wild *Daucus.* The presence of higher amounts of falcarindiol in the roots of wild *Daucus* taxa as compared to cultivated carrot has been associated with known disease resistances and a higher bitterness found in some wild relatives of carrot, given that this compound has been reported to have strong antifungal activity and has been indicated as the main compound responsible for the bitter off-taste of fresh and stored carrots (Czepa and Hofmann 2003, 2004). On the other hand, the lower content of falcarindiol found in cultivated carrot may be a consequence of the selection carried out during the breeding of cultivated forms, which aimed at obtaining carrots with better taste and—therefore—low bitterness (Baranska et al. 2005).

Genotype (cultivar), root size, location and year of cultivation, water stress due to drought or waterlogging, storage, industry processing and harvesting dates all influence polyacetylene content in carrot roots (Lund and White 1990; Kidmose et al. 2004; Kjellenberg et al. 2010). For example, small roots within a cultivar and storage at 1 °C for 4 months had higher content of polyacetylenes (Kidmose et al. 2004). Water stress by either drought or waterlogging reduced the content of the major carrot polyacetylenes falcarinol, falcarindiol and falcarindiol-3-acetate, as compared to control (not stressed) samples (Lund and White 1990). The concentration of these polyacetylenes varied significantly between stored and fresh carrots, as well as due to harvesting dates, genotype and year of cultivation (Kjellenberg et al. 2010).

The polyacetylenes found in several plant families have been described (reviewed by Negri 2015). Figure 17.1 depicts the main polyacetylenic structures found in carrot and other Apiaceae. In addition to the three predominant carrot polyacetylenes (falcarinol, falcarindiol and falcarindiol-3-acetate; Fig. 17.1, compounds 1-3), other C_{17}-polyynes with similar structure have been identified in carrot roots (compounds

4–14), with two of them being recently discovered (compounds 13–14) (Schmiech et al. 2009; Busta et al. 2018).

In addition to carrot polyacetylenes, several other polyacetylenic compounds have been isolated from other Apiaceae species (Fig. 17.1, compounds 15–38). Among them, four C_9-polyynes were isolated from the underground parts of *Selinum tenuifolium* Wall. (Chauhan et al. 2012), and five polyacetylenic β-D-glucopyranosyl glycosides were isolated from *Mediasia macrophylla* (Regel ex Schmalh.) Pimenov., a traditional medicinal plant from Uzbekistan (Kurimoto et al. 2010). From the poisonous plant *Bupleurum longiradiatum* Turcz., three compounds were isolated: a C_{15}-polyyne (15), a C_{18}-polyacetylene diol (21) and a C_{16}-acetylated polyyne (17) (Huang et al. 2009). From a close relative species of the latter, called *Bupleurum acutifolium* Boiss, wildly grown in Portugal and Spain, the alcohol (16), the diol (18), and two mono-acetate (19, 20) polyacetylenes were isolated (Barrero et al. 1999). A methyl ether derivative of falcarindiol (29) was isolated from celery (*Apium graveolens* L.) (Zidorn et al. 2005). Other derivatives of falcarindiol (compounds 30–38) were obtained from *Notopterygium incisium* Ting (Liu et al. 2014). A C_{19}-triyne called yuccifolol (22) was isolated from the above-ground organs of *Eryngium yuccifolium* Michaux (Ayoub et al. 2006). The tetraol (23) and its mono-acetate derivatives (24, 25) were isolated from *Hydrocotyle leucocephala* Cham. & Schlecht. (Ramos et al. 2006). From the methanolic extract of *Centella asiatica* Urb., polyacetylene cadiyenol (27) was isolated and described (Govindan et al. 2007).

17.3.3 Bioactive Properties of Polyacetylenes from Carrot and Other Apiaceae

A wide range of biological activities have been reported for falcarinol-type polyacetylenes, including antibacterial and antifungal activity as well as anti-inflammatory, anti-platelet aggregatory, allergenic, neuritogenic, serotonergic and anticancer effects. This section briefly reviews the main bioactivities of polyacetylenes from carrot and other Apiaceae. For more comprehensive reviews on this topic, see Christensen and Brandt (2006) and Christensen (2011).

17.3.3.1 Allergenic Effects

It has been demonstrated that falcarinol (1) is a potent contact allergen at high concentrations as can be found in some ornamental plants like *Hedera helix* (Araliaceae) (Hausen et al. 1987). However, allergic contact dermatitis by carrot and other Apiaceae vegetables is rare (Murdoch and Dempster 2000), presumably due to the relatively low concentrations of falcarinol in food plants compared to ornamental and wild plant species (Christensen 2011). On the other hand, polyacetylenes such as falcarindiol and falcarinone (compounds 2 and 3, respectively) do not seem to be allergenic (Hansen and Boll 1986).

17.3.3.2 Antimicrobial Activity

Falcarinol (1) and falcarindiol (2) have been shown to inhibit spore germination of various phytopathogenic fungi in concentrations ranging from 20 to 200 µg/mL. These polyacetylenes also displayed antibacterial and antimycobacterial effects in vitro against a number of species, including *Mycobacterium* sp. and gram-positive bacteria *Staphylococcus aureus*, at concentrations nontoxic for humans, suggesting that falcarinol-type polyacetylenes may be potentially useful as an antibiotic for treating a number of human illnesses caused by these microbes.

17.3.3.3 Neurotoxic and Neuritogenic Effects

Polyacetylenes are the most important toxic metabolites of various poisonous Apiaceae plants. For example, cicutoxin and oenanthotoxin are C_{17}-polyacetylenes and the most characteristic constituents of *Cicuta virosa* and *Oenanthe crocata*, two historical poisonous plants (Pohl 1894). These compounds act directly on the central nervous system, causing convulsions and respiratory paralysis and hence are extremely poisonous (Wittstock et al. 1997). The mode of action of these neurotoxins was further discussed by Christensen (2011). Contrary to the neurotoxic effects of cicutoxin and oenanthotoxin, falcarinol has been proved to have neuritogenesis effects on paraneurons such as PC12 and Neuro2a cells (Yamazaki et al. 2001), as well as neuroprotective effects on induced neuronal apoptosis (Nie et al. 2008). According to Christensen (2011), if the neuritogenic and neuroprotective effects of falcarinol are further confirmed in vivo, this polyacetylene may have potential for treating neurodegenerative diseases such as Alzheimer's disease.

17.3.3.4 Anti-inflammatory and Anti-platelet Effects

The three main C_{17}-polyactilenes of carrot (1-3) demonstrated anti-inflammatory activity by decreasing lipopolysaccharide (LPS)-induced production of pro-inflammatory cytokines interleukin-6 (IL-6), TNF-α and NO (Metzger et al. 2008). Falcarinol and falcarindiol are also strong inhibitors of lipoxygenases that are involved in the synthesis of inflammatory mediators such as prostaglandin E_2 (Schneider and Bucar 2005a, b). In addition, platelets' anti-aggregatory effects have been reported for falcarinol and falcarindiol (Kuo et al. 1990; Appendino et al. 1993), and possible mechanisms of action, likely associated with the anti-inflammatory properties of these polyacetylenes, were discussed (Christensen 2011).

17.3.3.5 Cytotoxic and Anticancer Effects

Falcarinol (1), panaxydiol (12) and panaxytriol were highly cytotoxic to various cancer cell lines, including leukemia (L-1210), human gastric adenocarcinoma (MK-1), mouse melanoma (B-16) and mouse fibroblast-derived tumor cells (L-929), showing comparatively low toxicity against normal healthy cells (Matsunaga et al. 1989, 1990). The selective in vitro cytotoxicity of falcarinol and related polyacetylenes against cancer cells compared to normal cells appears to depend on the tested cell line, as comparable toxicities were reported for these compounds against normal and cancerous human intestinal epithelial cells (Purup et al. 2009). Falcarindiol (2) also displays cytotoxic and anti-mutagenic effects in vitro (reviewed by Christensen 2011), although displaying less potent antiproliferative effects than falcarinol. The cytotoxic and anticancer modes of action of these polyacetylenes were further discussed by Christensen (2011).

17.3.3.6 Bioactivities of Other Polyacetylenes from Apiaceae Species

A polyacetylene from *Bupleurum longiradiatum* (compound 15) showed antiangiogenic activity in human umbilical vein endothelial cells (You et al. 2002). A methyl ether derivative of falcarindiol (29) isolated from celery (*Apium graveolens* L.) displayed in vitro cytotoxicity against leukemia, lymphoma and myeloma cells, comparable with that of falcarindiol (the most active compound against leukemia cell lines) and falcarinol (Zidorn et al. 2005).

Polyacetylenes isolated from the roots and rhizomes of *Notopterygium incisium* (compounds 30–32 and 38), a Chinese medicinal plant, displayed agonistic effects on the 'peroxisome proliferator-activated receptor gamma' (PPARγ), which is involved in the regulation of various metabolic and inflammatory processes. In addition, one of these compounds, namely notoincisol A (38), also exhibited inhibitory activity of nitric oxide (NO) production of stimulated RAW 264.7 macrophages (Liu et al. 2014). The protective effect of notoincisol A by reducing NO levels relies in the fact that high concentration of NO promotes inflammation and NO is involved in the pathophysiology of many diseases.

The tetraol and its mono-acetate derivative polyacetylenes (23–25) isolated from the Brazilian aquatic plant *Hydrocotyle leucocephala* showed immunosuppressive activity in a lipopolysaccharide (LPS)-induced cytokine induction assay, suggesting that these polyacetylene compounds from *H. leucocephala* may be effective therapeutic agents for Th2 (T Helper Type 2)-type diseases.

The polyacetylene cadiyenol (27), isolated from the methanolic extract of *Centella asiatica*, revealed in vitro induction of apoptosis in mouse lymphoma cells independent of cell cycle regimen. It also reduced NO production in lipopolysaccharide-activated mouse macrophages (Govindan et al. 2007).

17.3.4 Discovery of Key Genes Involved in Carrot Polyacetylene Biosynthesis

Until very recently, no carrot genes involved in polyacetylene biosynthesis had been described. In plants in general, very little is known on the genetic regulation and the enzymes involved in the production of these compounds. Previous work in parsley (*Petroselinum crispum*, Apiaceae), a close relative of carrot, identified a divergent form of FAD2 that (i) was upregulated in response to pathogen attack and (ii) when expressed in soybean embryos resulted in the production of crepenynic acid and, by the action of an unassigned enzyme, dehydrocrepenynic acid (Kirsch et al. 1997; Cahoon et al. 2003). These results, indicating a pathogen-responsive divergent FAD2-mediated pathway leading to the accumulation of acetylenic fatty acids, suggest that a similar pathway may exist in carrot.

Very recently, a study was conducted to identify and functionally characterize key enzymes and candidate genes involved in carrot polyacetylene biosynthesis, in particular those that can divert pathway flux from linolenic acid synthesis into the production of dehydrocrepenynic acid and, ultimately, C_{17}-falcarinol-type polyacetylenes (Busta et al. 2018). To this end, thin layer chromatography (TLC) was combined with gas chromatography-mass spectrometry (GC-MS) and gas chromatography-flame ionization detection (GC-FID) to identify and quantify polyacetylenic metabolites in five different carrot tissues for which transcriptomic (RNA-Seq) data were publicly available from the carrot genome sequencing project (Iorizzo et al. 2016). The sequences and tissue expression profiles of potential FAD2 and FAD2-like genes annotated in the carrot genome were compared with the metabolite data to identify candidate genes, followed by their functional characterization using yeast and *Arabidopsis* as heterologous expression systems.

GC-MS analytical data revealed the highest concentration of total polyacetylenes and falcarindiol in the root periderm (Fig. 17.3). Two previously unreported polyacetylenes were revealed, falcarintriol-8-acetate and falcarintriol-9-acetate (Fig. 17.1, compounds 13 and 14), presenting the highest content in leaves and petioles. In order to search for candidate FAD2-like genes associated with this distribution pattern of polyacetylenes in the carrot plant, first, the carrot genome was BLASTed with FAD2 sequences from parsley and other species, revealing a total of 24 carrot FAD2s—the largest number of members for this gene family found in a plant genome to date—distributed throughout seven of the nine carrot chromosomes. Sequence comparisons and clustering analysis of the carrot FAD2s with previously characterized canonical FAD2 desaturases and known divergent FAD2s from other species revealed four carrot genes associated with canonical FAD2s and 20 carrot genes associated with divergent FAD2s. Having identified which carrot FAD2s were likely canonical and likely divergent, transcriptomic analysis of these 24 genes in the carrot tissues previously characterized for polyacetylene content and composition (Fig. 17.3), along with their analysis in additional RNA-Seq data from carrot calli treated and untreated (control) with a fungal elicitor, was performed in order to identify candidate carrot FAD2-like genes. One of the putative canonical FAD2 genes (DCAR_013547) was upregulated both in whole root tissue (containing the periderm) and in response to fungal elicitation, thereby becoming the best candidate *Δ12 desaturase* associated with the carrot polyacetylene pathway (on the basis of previous results reported in parsley). Heterologous expression of DCAR_013547 in yeast confirmed Δ12 desaturase activity for this gene (by the appearance of linoleic acid in a culture medium containing only oleic acid). Using the same approach, Δ12 desaturase activity was also demonstrated for two other putative carrot FAD2s (DCAR_019786 and DCAR_019845).

Having identified a canonical FAD2 with periderm-enhanced expression (DCAR_013547), analysis of co-expression between the latter and the 20 putative divergent carrot FAD2s was performed to identify candidate Δ12 acetylenase genes. Three DCARs (DCAR_013552, DCAR_017011 and DCAR_013548) were strongly co-expressed with the canonical FAD2, becoming the best candidates for carrot Δ12 acetylenases. Functional characterization of these three DCARs was performed using transgenic *Arabidopsis* as a heterologous expression system. To this end, each DCAR was expressed under the control of a seed-specific promoter in the *Arabidopsis* double mutant *fad3fae1*, which accumulates large amounts of linoleic acid in its seeds. GC analysis revealed the accumulation of crepenynic acid in the transgenic lines but not in the controls (transformed with an empty vector), which only contained linoleic acid, confirming Δ12 acetylenase activity for the enzymes encoded by DCAR_013552, DCAR_017011 and DCAR_013548. Additional heterologous expression analysis of these genes in yeast using a high oleic acid medium revealed that none of these genes catalyze desaturation of oleic acid in the Δ12 position in addition to acetylation of linoleic acid in the Δ12 position.

Biosynthesis of polyacetylenes requires further desaturation of crepenynic acid in the Δ14 position to produce dehydrocrepenynic acid, a required precursor of C_{17}-falcarinol-type polyacetylenes (Fig. 17.2). Thus, in order to test whether the same enzymes catalyzing Δ12 desaturation of oleic acid (encoded by DCAR_013547, DCAR_019786 and DCAR_019845) can catalyze desaturation in the Δ14 position, thereby converting crepenynic acid to dehydrocrepenynic acid, DCAR_013547 and DCAR_019786 were expressed together with DCAR_013552 (i.e., in a background containing crepenynic acid). It was found that both enzymes, encoded by DCAR_013547 and DCAR_019786, were able to catalyze Δ14 crepenynic acid desaturation in addition to Δ12 oleic acid desaturation (Fig. 17.2).

In conclusion, this study identified 24 FAD2-like sequences in the carrot genome, several of which exhibit enhanced expression in the taproot periderm. At least three of these genes encode Δ12 oleic acid desaturases, at least three encode Δ12 linoleic acid acetylenases, and at least two of the Δ12 oleic acid desaturases catalyze Δ14 crepenynate desaturation. These enzymes and their corresponding genes are major candidates for the regulation of falcarinol-type polyacetylenes in carrot. Based on the structural comparisons between dehydrocrepenynic acid and falcarinol, it seems likely that a Δ14 acetylenase, an ω-3 hydroxylase, as well as machinery that catalyzes ω-desaturation and head group removal, still remains to be identified in carrot (Fig. 17.2).

Interestingly, the six carrot FAD2 genes functionally characterized in this study reside close to one another in a small region (29.28–29.39 Mb) of chromosome four. This region appears to have experienced local tandem duplications, as data from microsyntenic and phylogenomic analyses suggest (Busta et al. 2018). Similar arrangements in clusters and tandem duplications for other multi-gene family members, such as the R2R3-MYB transcription factor family, were recently reported in carrot (Iorizzo et al. 2018) and other species (discussed further in Chap. 15).

References

Alabran DM, Mabrouk AF (1973) Carrot flavor, sugars and free nitrogenous compounds in fresh carrots. J Agric Food Chem 21:205–208

Alasalvar C, Grigor JM, Zhang D, Quantick PC, Shahidi F (2001) Comparison of volatiles, phenolics, sugars, antioxidant vitamins, and sensory quality of different colored carrot varieties. J Agric Food Chem 49:1410–1416

Appendino G, Tagliapietra S, Nano GM (1993) An anti-platelet acetylene from the leaves of *Ferula communis*. Fitoterapia 64:179

Ayoub N, Al-Azizi M, König W, Kubeczka KH (2006) Essential oils and a novel polyacetylene from *Eryngium yuccifolium* Michaux. (Apiaceae). Flavour Fragrance J 21:864–868

Badami RC, Patil KB (1980) Structure and occurrence of unusual fatty acids in minor seed oils. Prog Lipid Res 19:119–153

Bajaj KL, Kaur G, Sukhija BS (1980) Chemical composition and some plant characteristics in relation to quality of some promising cultivars of carrot (*Daucus carota* L.). Qual Plant Food Human Nutr 30:97–107

Baranska M, Schulz H (2005) Spatial tissue distribution of polyacetylenes in carrot root. Analyst 130:855–859

Baranska M, Schulz H, Baranski R, Nothnagel T, Christensen LP (2005) In situ simultaneous analysis of polyacetylenes, carotenoids and polysaccharides in carrot roots. J Agric Food Chem 53:6565–6571

Baranski R, Allender C, Klimek-Chodacka M (2012) Towards better tasting and more nutritious carrots: carotenoid and sugar content variation in carrot genetic resources. Food Res Int 47:182–187

Barley GC, Ewart J, Thaller V (1988) Crepenynate as a precursor of falcarinol in carrot tissue culture. Bioact Mol 7:85–91

Barrero AF, Herrador MM, Akssira M, Arteaga P, Romera JL (1999) Lignans and polyacetylenes from *Bupleurum acutifolium*. J Nat Prod 62:946–948

Bohlman F (1988) Naturally-occurring acetylenes. In: Lam J, Breteler H, Hansen L (eds) Chemistry and biology of naturally-occurring acetylenes and related compounds (NOARC). Elsevier, Amsterdam, pp 1–19

Busta L, Yim W, LaBrant EW, Wang P, Grimes L, Malyszka K, Cushman JC, Santos P, Kosma DK, Cahoon EB (2018) Identification of genes encoding enzymes catalyzing the early steps of carrot polyacetylene biosynthesis. Plant Physiol. https://doi.org/10.1104/pp.18.01195

Cahoon EB, Schnurr JA, Huffman EA, Minto RE (2003) Fungal responsive fatty acid acetylenases occur widely in evolutionarily distant plant families. Plant J 34: 671–683

Carlton BC, Peterson CE (1963) Breeding carrots for sugar and dry matter content. Proc Am Soc Hort Sci 82:332–340

Chauhan RS, Nautiyal MC, Tava A, Mella M (2012) Chemical composition of the volatile oil from the roots of *Selinum tenuifolium* Wall. Helv Chim Acta 95:780–787

Christensen LP (2011) Aliphatic C_{17}-polyacetylenes of the falcarinol type as potential health promoting compounds in food plants of the Apiaceae family. Recent Pat Food Nutr Agric 3:64–77

Christensen LP, Brandt K (2006) Bioactive polyacetylenes in food plants of the Apiaceae family: occurrence, bioactivity and analysis. J Pharm Biomed Anal 41:683–693

Czepa A, Hofmann T (2003) Structural and sensory characterization of compounds contributing to the bitter off-taste of carrots (*Daucus carota* L.) and carrot puree. J Agric Food Chem 51:3865–3873

Czepa A, Hofmann T (2004) Quantitative studies and sensory analyses on the influence of cultivar, spatial tissue distribution and industrial processing on the bitter off-taste of carrots (*Daucus carota* L.) and carrot products. J Agric Food Chem 52:4508–4514

Copeland L (1990) Enzymes of sucrose metabolism. In: Lea PJ (ed) Methods in plant biochemistry: enzymes of primary metabolism, vol 3. Academic Press, New York, pp 73–84

Dawid C, Dunemann F, Schwab W, Nothnagel T, Hofmann T (2015) Bioactive C_{17}-polyacetylenes in carrots (*Daucus carota* L.): current knowledge and future perspectives. J Agric Food Chem 63:9211–9222

Freeman RE, Simon PW (1983) Evidence for simple genetic control of sugar type in carrot (*Daucus carota* L.). J Am Soc Hort Sci 108:50–54

Garrod B, Lewis BG (1979) Location of the antifungal compound falcarindiol in carrot root tissue. Trans Br Mycol Soc 72:515–517

Govindan G, Sambandan TG, Govindan M, Sinskey A, Vanessendelft J, Adenan I, Rha CK (2007) A bioactive polyacetylene compound isolated from *Centella asiatica*. Planta Med 73:597–599

Grzebelus D, Yau YY, Simon PW (2006) *Master*: a novel family of *PIF/Harbinger*-like transposable elements identified in carrot (*Daucus carota* L.). Mol Genet Genomics 275:450–459

Hasselbring H (1927) Carbohydrate transformation in carrots during storage. Plant Physiol 2:225–243

Hansen L, Boll PM (1986) Polyacetylenes in Araliaceae: Their chemistry, biosynthesis and biological significance. Phytochem 25:285–293

Hausen BM, Bröhan J, König WA, Faasch H, Hahn H, Bruhn G (1987) Allergic and irritant contact dermatitis from falcarinol and didehydrofalcarinol in common ivy (*Hedera helix* L.). Contact Dermatitis 17:1–9

Huang HQ, Zhang X, Shen YH, Su J, Liu XH, Tian JM, Lin J, Shan L, Zhang WD (2009) Polyacetylenes from *Bupleurum longiradiatum*. J Nat Prod 72:2153–2157

Huber SC, Huber JL (1996) Role and regulation of sucrose-phosphate synthase in higher plants. Annu Rev Plant Physiol Plant Mol Biol 47:199–222

Iorizzo M, Ellison S, Senalik D et al (2016) A high-quality carrot genome assembly provides new insights into carotenoid accumulation and asterid genome evolution. Nature Genet 48:657–666

Iorizzo M, Cavagnaro PF, Bostan H et al (2018) A cluster of MYB transcription factors regulates anthocyanin biosynthesis in carrot (*Daucus carota* L.) root and petiole. Front Plant Sci. Accepted

Jones ERH, Safe S, Thaller V (1966) Natural acetylene. Part XXIII. A C18 polyacetylenic keto aldehyde related to falcarinone from an umbellifer (*Pastinaca sativa* L.). J Chem Soc C Org

Kawazu K, Noguchi H, Fujishita K, Iwasa J, Egawa H (1973) Two new antifungal compounds from *Dendropanax trifidus*. Tetrahedron Lett 14:3131–3132

Klann EM, Chetelat RT, Bennett AB (1993) Expression of acid invertase gene controls sugar composition in tomato (*Lycopersicon*) fruit. Plant Physiol 103:863–870

Kidmose U, Hansen SL, Christensen LP, Edelenbos M, Larsen E, Norbaek R (2004) Effect of genotype, root size, storage, and processing on bioactive compounds in organically grown carrots (*Daucus carota*). J Food Sci 69:388–394

Killeen DP, Sansom CE, Lill RE, Eason JR, Gordon KC, Perry NB (2013) Quantitative Raman spectroscopy for the analysis of carrot bioactivities. J Agric Food Chem 61:2701–2708

Kirsch C, Hahlbrock K, Somssich IE (1997) Rapid and transient induction of a parsley microsomal A12 fatty acid desaturase mRNA by fungal elicitor. Plant Physiol 115:283–937

Kjellenberg L, Johansson E, Gustavsson KE, Olsson ME (2010) Effects of harvesting date and storage on the amounts of polyacetylenes in carrots, *Daucus carota*. J Agric Food Chem 58:11703–11708

Kurimoto SI, Okasaka M, Kashiwada Y, Kodzhimatov OK, Takaishi Y (2010) A C_{14}-polyacetylenic glucoside with an α-pyrone moiety and four C_{10}-polyacetylenic glucosides from *Mediasia macrophylla*. Phytochemistry 71:688–692

Kuo S-C, Teng C-M, Lee J-C, Ko F-N, Chen S-C, Wu T-S (1990) Antiplatelet components in *Panax ginseng*. Planta Med 56:164–167

Laurière C, Laurière M, Sturm A, Faye L, Chrispeels MJ (1988) Characterization of β-fructosidase, an extracellular glycoprotein of carrot cells. Biochimie 70:1483–1491

Lee HS, Sturm A (1996) Purification and characterization of neutral and alkaline invertase from carrot. Plant Physiol 112:1513–1522

Liu C, Kunert O, Blunder M, Fakhrudin N, Noha SM, Malainer C, Schinkovitz A, Heiss EH, Atanosov AG, Kollroser M, Schuster D, Dirsch VM, Bauer R (2014) Polyyne hybrid compounds from *Notopterygium incisum* with peroxisome proliferator activated receptor gamma agonistic effects. J Nat Prod 77:2513–2521

Liu YJ, Wang GL, Ma J, Xu ZS, Wang F, Xiong AS (2018) Transcript profiling of sucrose synthase genes involved in sucrose metabolism among four carrot (*Daucus carota* L.) cultivars reveals distinct patterns. BMC Plant Biol 18:8

Lund ED (1992) Polyacetylenic carbonyl compounds in carrots. Phytochemistry 31:3621–3623

Lund DL, White JM (1990) Polyacetylenes in normal and water-stressed "Orlando-gold" carrots (*Daucus carota*). J Sci Food Agric 51:507–516

Lorenz K, Lienhard S, Sturm A (1995) Structural organization and differential expression of carrot, β-fructofuranosidase genes: identification of a gene coding for a flower bud-specific isozyme. Plant Mol Biol 28:189–194

Matsunaga H, Katano M, Yamamoto H, Mori M, Takata K (1989) Studies on the panaxytriol of *Panax ginseng* C. A. Meyer. Isolation, determination and antitumor activity. Chem Pharm Bull 37:1279–1281

Matsunaga H, Katano M, Yamamoto H, Fujito H, Mori M, Takata K (1990) Cytotoxic activity of polyacetylene compounds in Panax ginseng C. A. Meyer. Chem Pharm Bull 38:3480–3482

Metzger BT, Barnes DM (2009) Polyacetylene diversity and bioactivity in orange market and locally grown colored carrots (*Daucus carota* L.). J Agric Food Chem 57:11134–11139

Metzger BT, Barnes DM, Reed JD (2008) Purple carrot (*Daucus carota* L.) polyacetylenes decrease lipopolysaccharide-induced expression of inflammatory proteins in macrophage and endothelial cells. J Agric Food Chem 56:3554–3560

Minto RE, Blacklock BJ (2008) Biosynthesis and function of polyacetylenes and allied natural products. Prog Lipid Res 47:233–306

Murdoch SR, Dempster J (2000) Allergic contact dermatitis from carrot. Contact Dermatitis 42:236

Negri R (2015) Polyacetylenes from terrestrial plants and fungi: recent phytochemical and biological advances. Fitoterapia 106:92–109

Nie B-M, Jiang X-Y, Cai J-X, Fu S-L, Yang L-M, Lin L et al (2008) Panaxydol and panaxynol protect cultured cortical neurons against Aβ25–35-induced toxicity. Neuropharmacology 54:845–853

Pferschy-Wenzig EM, Getzinger V, Kunert O, Woelkart K, Zahrl J, Bauer R (2009) Determination of falcarinol in carrot (*Daucus carota* L.) genotypes using liquid chromatography/mass spectrometry. Food Chem 114:1083–1090

Phan CT, Hsu H (1973) Physical and chemical changes occurring in the carrot root during growth. Can J Plant Sci 53:629–634

Phan CT, Hsu H, Sarkar SK (1973) Physical and chemical changes occurring in the carrot root during storage. Can J Plant Sci 53:635–641

Pohl J (1894) Poisonous constituents of *Oenanthe crocata* and of *Cicuta virosa*. Arch Exp Pathol Pharmakol 34:258–267

Purup S, Larsen E, Christensen LP (2009) Differential effects of falcarinol and related aliphatic C_{17}-polyacetylenes on intestinal cell proliferation. J Agric Food Chem 57:8290–8296

Platenius H (1934) Physiological and chemical changes occurring in the carrots during growth and storage. Cornell Univ Agric Expt Sta 161:1–19

Ramloch-Lorenz K, Knudsen S, Sturm A (1993) Molecular characterization of the gene for carrot cell wall β-fructosidase. Plant J 4:545–554

Ramos F, Takaishi Y, Kawazoe K, Osorio C, Duque C, Acuña R, Fujimoto Y, Sato M, Okamoto M, Oshikawa T, Ahmed SU (2006) Immunosuppressive diacetylenes, ceramides and cerebrosides from *Hydrocotyle leucocephala*. Phytochemistry 67:1143–1150

Roman M, Dobrowolski JC, Baranska M, Baranski R (2011) Spectroscopic studies on bioactive polyacetylenes and other plant components in wild carrot root. J Nat Prod 74:1757–1763

Rygg GL (1945) Sugars in the root of the carrot. Plant Physiol 20:47–50

Sebkova V, Unger C, Hardegger Sturm A (1995) Biochemical, physiological, and molecular characterization of sucrose synthase from *Daucus carota*. Plant Physiol 108:75–83

Shallenberger RS, Birch GG (1975) Sugar chemistry. Avi Publishing Co., Westport, Connecticut, p 221

Schmiech L, Alayrac L, Witulski B, Hofmann T (2009) Structure determination of bisacetylenic oxylipins in carrots (*Daucus carota* L.) and enantioselective synthesis of falcarindiol. J Agric Food Chem 57:11030–11040

Schneider I, Bucar F (2005a) Lipoxygenase inhibitors from natural plant sources. Part 1. Medicinal plants with inhibitory activity on arachidonate 5-lipoxygenase and 5-lipoxygenase/cyclooxygenase. Phytother Res 19:81–102

Schneider I, Bucar F (2005b) Lipoxygenase inhibitors from natural plant sources. Part 2. Medicinal plants with inhibitory activity on arachidonate 12-lipoxygenase, 15-lipoxygenase and leukotriene receptor antagonists. Phytother Res 19:263–272

Simon PW (1996) Inheritance and expression of purple and yellow storage root color in carrot. J Hered 87:63–66

Simon PW, Peterson CE, Lindsay RC (1980a) Correlations between sensory and objective parameters of carrot flavor. J Agric Food Chem 28:559–562

Simon PW, Peterson CE, Lindsay RC (1980b) Genetic and environmental influences on carrot flavor. J Amer Soc Hort Sci 105:416–420

Simon PW, Peterson CE, Lindsay RC (1980c) Genotype, soil, and climate effects on sensory and objective components of carrot flavor. J Amer Soc Hort Sci 107:644–648

Stommel JR, Simon PW (1989) Phenotypic recurrent selection and heritability estimates for total dissolved solids and sugar type in carrot. J Amer Soc Hort Sci 114:695–699

Stommel JR, Simon PW (1990) Multiple forms of invertase from *Daucus carota* cell cultures. Phytochemistry 29:2087–2089

Sturm A (1996) Molecular characterization and functional analysis of sucrose-cleaving enzymes in carrot (*Daucus carota* L.). J Exp Bot 47:1187–1192

Sturm A, Chrispeels MJ (1990) cDNA cloning of carrot extracellular β-fructosidase and its expression in response to wounding and bacterial infection. Plant Cell 2:1107–1119

Sturm A, Šebková V, Lorenz K, Hardegger M, Lienhard S, Unger C (1995) Development- and organ-specific expression of the genes for sucrose synthase and three isozymes of acid β-fructofuranosidase in carrot. Planta 195:601–610

Sturm A, Tang GQ (1999) The sucrose-cleaving enzymes of plants are crucial for development, growth and carbon partitioning. Trends Plant Sci 4:401–407

Sturm A, Lienhard S, Schatt S, Hardegger M (1999a) Tissue-specific expression of two genes for sucrose synthase in carrot (*Daucus carota* L.). Plant Mol Biol 39:349–360

Sturm A, Hess D, Lee H-S, Lienhard S (1999b) Neutral invertase is a novel type of sucrose-cleaving enzyme. Physiol Plant 107:159–165

Tang GQ, Lüscher M, Sturm A (1999) Antisense repression of vacuolar and cell wall invertase in transgenic carrot alters early plant development and sucrose partitioning. Plant Cell 11:177–189

Tang GQ, Sturm A (1999) Antisense repression of sucrose synthase in carrot (*Daucus carota* L.) affects growth rather than sucrose partitioning. Plant Mol Biol 41:465–479

Unger C, Hardegger M, Liehard S, Sturm A (1994) cDNA cloning of carrot (*Daucus carota*) soluble acid β-furctofuranosidases and comparison with the cell wall isoenzyme. Plant Physiol 104:1351–1357

Unger C, Hofsteenge J, Sturm A (1992) Purification and characterization of a soluble β-fructofuranosidase from *Daucus carota*. Eur J Biochem 204:915–921

Vivek BS, Simon PW (1999) Linkage relationships among molecular markers and storage root traits of carrot (*Daucus carota* L. ssp. *sativus*). Theor Appl Genet 99:58–64

Werner HO (1941) Dry matter, sugar, and carotene content of morphological portions of carrots through the growing and storage season. Proc Amer Soc Hort Sci 38:267–272

Wittstock U, Lichtnow KH, Teuscher E (1997) Effects of cicutoxin and related polyacetylenes from *Cicuta virosa* on neuronal action potentials: a comparative study on the mechanism of the convulsive action. Planta Med 63:120–124

Xu ZS, Tan HW, Wang F, Hou XL, Xiong AS (2014) CarrotDB: a genomic and transcriptomic database for carrot. Database J Biol Databases Curation 40:1–8

Yau YY, Simon PW (2003) A 2.5-kb Insert eliminates acid soluble invertase isozyme II transcript in a carrot (*Daucus carota* L.) roots, causing high sucrose accumulation. Plant Mol Biol 53:151–162

Yau YY, Santos K, Simon PW (2005) Molecular tagging and selection for sugar type in carrot roots using co-dominant, PCR-based markers. Mol Breed 16:1–10

Yamazaki M, Hirakura K, Miyaichi Y, Imakura K, Kita M, Chiba K et al (2001) Effect of polyacetylenes on the neurite outgrowth of neuronal culture cells and scopolamine-induced memory impairment in mice. Biol Pharm Bull 24:1434–1436

You YJ, Lee IS, Kim Y, Bae KH, Ahn BZ (2002) Antiangiogenic activity of *Bupleurum longiradiatum* on human umbilical venous endothelial cells. Arch Pharm Res 25:640–642

Zidorn C, Hrer KJ, Ganzera M, Schubert B, Sigmund EM, Mader J, Greil R, Ellmerer EP, Stuppner H (2005) Polyacetylenes from the Apiaceae vegetables carrot, celery, fennel, parsley, and parsnip and their cytotoxic activities. J Agric Food Chem 53:2518–2523

18 Genetics and Genomics of Carrot Biotic Stress

Lindsey J. du Toit, Valérie Le Clerc and Mathilde Briard

Abstract

Carrot (*Daucus carota* ssp. *sativus*) production can be affected by a wide range of pests and pathogens. At least five diseases of carrot are caused by bacterial pathogens, 36 by fungal and oomycete pathogens, two by phytoplasmas, and 13 by viruses; and seven genera of nematodes and two genera of parasitic plants affect carrot. In addition, numerous insect and mite pests can cause losses. There have been extensive efforts to select carrot cultivars with partial or complete resistance to many of these pathogens and pests, and to identify wild species with resistance to specific biotic stresses for introgression into breeding populations and commercial cultivars. For some pathogens and pests, significant advances have been made at identifying resistance and mapping that resistance to the carrot genome. For others, resistance has been identified, but the genetic basis is yet to be determined. For a majority of these diverse stresses, however, there has been little success at identifying highly effective resistance and understanding the genetic basis of resistance. The diversity of stresses as well as interactions among these pests and pathogens can complicate efforts to develop cultivars with resistance to all key biotic stresses in a region that also meet market and consumer expectations. New approaches to identifying resistant material and speeding traditional breeding are being developed with molecular breeding tools, including simple sequence repeat markers and deep-coverage libraries of the carrot genome. These valuable genomic resources will enhance efforts to identify and breed for resistance to carrot pests and pathogens.

18.1 Introduction

Diseases and insect or mite pests limit carrot production to various degrees in most regions of carrot production in the world (Rubatzky et al. 1999). The foliar diseases of primary concern tend to be Alternaria leaf blight (caused by *Alternaria dauci*), Cercospora leaf spot (*Cercospora carotae*), bacterial blight (*Xanthomonas hortorum* pv. *carotae*), and powdery mildew (*Erysiphe heraclei*) (Davis and Raid 2002). The most widespread soilborne root pathogens of carrot are cavity spot (caused by several species of *Pythium*), white mold (*Sclerotinia sclerotiorum*), and root-knot nematodes (various species of *Meloidogyne*) (Davis and Raid 2002).

L. J. du Toit (✉)
Department of Plant Pathology, Washington State University Mount Vernon NWREC, Mount Vernon, WA 98273, USA
e-mail: dutoit@wsu.edu

V. Le Clerc · M. Briard
IRHS, Agrocampus Ouest, INRA, Université d'Angers, SFR QuaSaV, Angers 49045, France

P. Simon et al. (eds.), *The Carrot Genome*, Compendium of Plant Genomes,
https://doi.org/10.1007/978-3-030-03389-7_18

Widespread pests of concern to carrot producers include the carrot rust fly (*Psila rosae*), aphids (e.g., the willow-carrot aphid, *Cavariella aegopodii*), and the two-spotted spider mite (*Tetranychus urticae*) (Simon et al. 2008). Other carrot pathogens and pests cause losses of regional significance (Davis and Raid 2002), such as violet root rot, bacterial soft rots, and Fusarium dry rot. For most of these biotic stresses, breeders have relied on natural infection in areas where the pathogen or pest is well established to facilitate identifying and selecting genetic resistance. Highly susceptible cultivars or breeding lines sometimes are planted at intervals among carrot entries to promote the development of the biotic stress. For some of these pests and pathogen, these screening and breeding efforts include plants inoculated with the pathogen or infested with the pest. Screening for resistance to soilborne pests and diseases can be particularly complex because of the difficulty of establishing uniform soilborne disease pressure, especially for screening large numbers of lines and for stresses caused by multiple species or races of a pathogen (e.g., cavity spot and root-knot nematodes) or a pest for which there could be different types or sources of resistance (e.g., root-knot nematodes and aphids). This chapter describes efforts to identify resistance to specific pests and pathogens of carrot, phenotypic screening methods evaluated, and what is known about the genetic basis of resistance, including inheritance of resistance and annotation of resistance genes on the carrot genome. Unfortunately, for a majority of the diverse biotic stresses of carrot, resistance genes have not been identified and/or little is known about the genetic basis of resistance that has been identified. The numerous gaps in understanding of the genetics of carrot germplasm reactions to these biotic stresses, as detailed in this chapter, highlight the need for additional research.

This chapter is not a comprehensive review of the literature on resistance to all known pathogens and pests of carrot. The chapter focuses on some key pathogens and pests for which there have been efforts to screen for resistance and to evaluate the genetic basis and genomics of resistance. Some pathogen and pest names used in older literature cited in this chapter have changed. The effort was made to use current scientific nomenclature. Synonyms of these pests and pathogens are noted. The carrot diseases reviewed in this chapter are divided into those caused by soilborne pathogens and those caused by foliar pathogens, followed by a section on nematode and insect pests.

18.2 Carrot Diseases

18.2.1 Soilborne Diseases

18.2.1.1 Cavity Spot (*Pythium* spp.)

Cavity spot has been documented in almost all regions of carrot production in the world (McDonald 2002). The disease is caused by several species of *Pythium*, the most common being *P. violae* and *P. sulcatum*, two slow-growing species that typically are the most virulent on carrot roots (McDonald 2002). Other species associated with cavity spot include *P. ultimum*, *P. intermedium*, *P. irregulare*, and *P. sylvaticum*. The disease rarely causes a reduction in root yield but can have significant economic impact because the shallow, surface lesions render roots unsuitable for fresh and processing markets (McDonald 2002) (Fig. 18.1a). *Pythium* spp. typically infect carrot roots within the first four to six weeks after seeding and probably throughout the growth of the carrot (McDonald 1994b). Cavity spot will continue to develop on roots in storage. Root lesions can be invaded by secondary microorganisms, including bacteria, which can lead to discoloration around the cavities, particularly during heating/blanching (Fig. 18.1b). Severity of cavity spot generally increases the longer the roots are in soil (Montfort and Rouxel 1988; Vivoda et al. 1991).

Resistance screening *Partial resistance to cavity spot*. Differences in susceptibility of carrot cultivars to cavity spot have been identified, but no commercially available carrot cultivars are completely resistant (Groom and Perry 1985; McDonald 1994b; McDonald 2002; Soroker et al. 1984; Sweet et al. 1986; Vivoda et al. 1991;

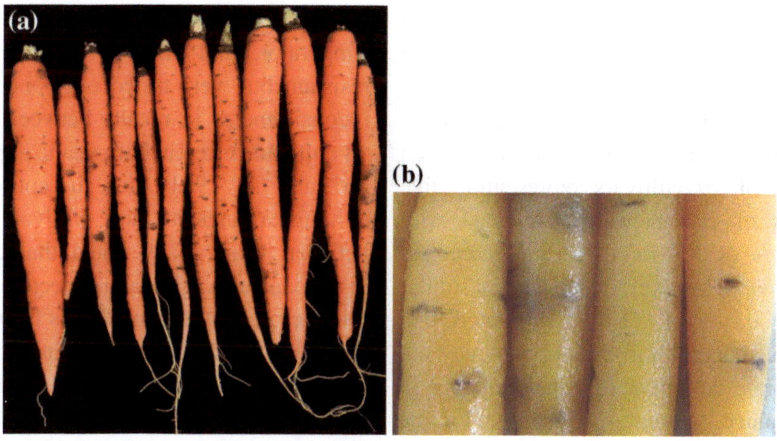

Fig. 18.1 **a** Severe symptoms of cavity spot on carrot roots caused by *Pythium sulcatum* (Alex Batson, Washington State University). **b** Lesions and discoloration of peeled and blanched carrot roots resulting from cavity spot (Lindsey du Toit, Washington State University)

White 1988). Guba et al. (1961) first reported differences in cultivar susceptibility to cavity spot. 'Hutchinson' roots developed less cavity spot than those of 'Waltham Hicolor,' but the differences were greater among lots of 'Waltham Hicolor' than among cultivars. The National Institute of Agricultural Botany (1991) in the UK reported variation in susceptibility among groups of carrots (Sweet et al. 1989). 'Redca' was a more resistant Chantenay cultivar than 'Supreme,' and 'Nandor' was a more resistant Nantes cultivar than 'Tino.' That study also documented increased severity of cavity spot on later maturing types or when cultivars were harvested later in the autumn; e.g., the Autumn King type Vita Long was more susceptible at late harvest compared to early harvest (National Institute of Agricultural Botany 1991).

In growth chamber trials, six Imperator cultivars commonly grown in California were planted in a potting medium inoculated with *P. violae* and *P. ultimum* and maintained at 20 °C (Vivoda et al. 1991). All six cultivars were susceptible to cavity spot caused by the two species, but *P. violae* isolates were more virulent than *P. ultimum* isolates. 'Topak' was the most susceptible cultivar to both species. The other five cultivars varied in response to *P. violae* but were similarly susceptible to *P. ultimum*. 'Caropak' and 'Pakmor' were the next most susceptible after 'Topak,' followed by 'Sierra' and 'Dominator.' The cultivars originated from a few parental lines which might account for the limited variation in response to *P. violae* and *P. ultimum* (Vivoda et al. 1991).

White et al. (1987) screened 19 carrot cultivars representing five main types of carrots (Amsterdam Forcing, Nantes, Chantenay, Berlicum, and Autumn King) for resistance to cavity spot caused by each of *P. violae*, *P. sulcatum*, and *P. intermedium*. Roots of each cultivar were grown in a greenhouse, washed, and inoculated with colonized agar plugs of each *Pythium* species (10 plugs per root, with 5 roots tested for each of two replicate trays per cultivar). For *P. violae*, there were no significant differences among the five carrot types or the 19 cultivars. For *P. sulcatum*, differences were detected among types of carrots but not among cultivars, and then only for one of the three measures of cavity spot (percentage of agar disks that resulted in lesions on roots two days after inoculation). For *P. intermedium*, White et al. (1987) only detected significant differences among cultivars, not types of carrots, and only for one of the three measures of cavity spot (percentage of disks causing lesions four days after inoculation). They concluded there was no 'useful' genetic resistance to the three *Pythium* spp. among the cultivars tested.

White et al. (1988) found no differences in frequency of recovery of *Pythium* spp. from the periderm of asymptomatic carrots of 'Chantenay New Supreme,' 'Fingo,' and 'Sweetheart.' They also found no useful genetic resistance when mature roots of 19 cultivars representing five main groups of carrots were inoculated with mycelial plugs of *P. violae*, *P. sulcatum*, and *P. intermedium*. However, results of a similar trial by Vivoda et al. (1991) suggested that inoculation of roots with mycelial plugs may not provide accurate determination of differences in cultivar resistance. In contrast, Benard and Punja (1995) indicated that inoculation of 36 carrot cultivars with *P. violae* in a laboratory study demonstrated differences in susceptibility that correlated significantly with field results. Various private breeding programs have made effective progress at developing cultivars with enhanced resistance to cavity spot using combinations of field nurseries and greenhouse screening protocols with inoculated soil or potting medium, as well as laboratory assays that entail root inoculation.

McDonald (1994b) found the partially resistant cultivar 'Six Pak' effectively suppressed cavity spot in field trials in Ontario, Canada. 'SR-481,' 'Eagle,' and 'Red Core Chantenay' had intermediate levels of resistance, and 'Chanton' and 'Huron' were the most susceptible. Similarly, 'Six Pak' was more resistant than 'Cellobunch' and 'Chancellor.' Interestingly, 'Eagle' was as resistant as 'Six Pak' in non-irrigated plots but was more susceptible under irrigated conditions. The cultivars had little effect on early season development of cavity spot, with differences in susceptibility only becoming evident as roots matured. This was the first study to demonstrate that older carrots are not necessarily more susceptible to cavity spot than younger carrots, based on seeding carrots on different dates in the same plots. Also, a marked decrease in cavity spot incidence was observed late in the season (McDonald 1994b).

Using an in vitro mature carrot root inoculation protocol, Benard and Punja (1995) screened 37 carrot cultivars for reaction to cavity spot. 'Panther,' 'E0792,' 'Caropride,' 'Fannia,' and 'Navajo' were the most resistant. 'Six Pak,' 'Imperator,' and 'XPH 3507' also appeared very resistant but were tested only once. There were inconsistencies in results between years among 18 cultivars tested in both years; e.g., 'Eagle' was resistant in 1991 but susceptible in 1992, although ratings of most cultivars were similar between years. They noted that cultivars with discrepancies might have reflected differences in rootage or growing conditions between years.

Cooper et al. (2004) examined morphological and biochemical responses of commercial carrot cultivars Bertan, Narbonne, and Bolero as well as the 'Eastern' carrot genebank variety 'Purple Turkey' to inoculation with *P. violae* in a greenhouse bioassay and field trials. 'Purple Turkey' was less susceptible than all commercial cultivars. The small cell size in the roots and higher constitutive levels of enzymes in the roots of 'Purple Turkey' was hypothesized as the basis for resistance of this line to cavity spot. Of the commercial cultivars tested, 'Bolero' was the least susceptible, 'Narbonne' was intermediate, and 'Bertan' was the most susceptible.

Cooper et al. (2006) screened seed progeny from 19 tissue culture-derived carrot somaclone families for resistance to cavity spot caused by *P. violae* in greenhouse and field trials, along with 'Bertran,' 'Nando,' 'Bolero,' and 'Vita Longa' as commercial control cultivars. There was little relationship between greenhouse and field trial results although the results suggested there might be genetic variation in susceptibility to cavity spot in some of the somaclones.

For several years, McDonald et al. (2017) have screened experimental carrot breeding lines from the United States Department of Agriculture (USDA) carrot breeding program alongside commercial carrot cultivars in a field site at the Muck Crops Research Station of the University of Guelph in the Holland Marsh of Ontario, Canada. The site has a high level of natural infestation with the cavity spot pathogen. Each year, a wide range in incidence and severity of cavity spot has been observed in breeding lines and cultivars. Lines with low cavity spot incidence and severity displayed a consistent response among years, including crosses with these more resistant lines, e.g., the orange lines CS736 (pedigree 1137A), and CS732 (1137B-F_2M_5), and the USDA parent lines 1137, 5367, and 6526. A similar consistent response was observed for lines that had the greatest cavity spot incidence and severity ratings, and many crosses with those lines, e.g., 2205, 5494, and CS 724 (2205B). However, despite the relatively uniform disease pressure in this nursery, some lines did not respond consistently among years, illustrating the difficulty of screening for resistance to cavity spot (McDonald et al. 2017). Forking of carrot roots, which has been attributed in some studies to be caused by the same *Pythium* spp. that cause cavity spot, was not correlated with cavity spot incidence and severity ratings in the muck nursery trials (McDonald et al. 2017).

Phenotypic screening methods. Uneven distribution of inoculum in fields and the very sporadic nature of the disease within and among fields make screening for resistance to cavity spot very difficult. Wide variation in responses among roots of the same cultivar necessitates evaluating large numbers of roots of each carrot line in replicated and randomized plots over multiple seasons for robust differentiation of responses among cultivars. The occurrence and severity of cavity spot in phenotypic screening methods can be influenced significantly by soil temperature, soil moisture, other soil properties (including soil microflora), the species of *Pythium*, age of carrot roots, etc. (Benard and Punja 1995; McDonald 1994b, 2002). Higher soil moisture, particularly flooding, and cool soil temperatures ($\sim$15 °C) tend to be optimal for cavity spot development.

Severity of cavity spot generally increases with the length of time roots remain in the soil (Montfort and Rouxel 1988; Sweet et al. 1989; Vivoda et al. 1991). This could reflect increased susceptibility of roots as they mature, accumulation of lesions over the season, expansion of lesions as the root diameter increases, or increased change of infection as the root surface increases (Vivoda et al. 1991; Wagenvoort et al. 1989). However, McDonald (1994b) demonstrated in field trials in Ontario, Canada, that older carrots are not necessarily more susceptible to cavity spot than younger carrots when carrots were planted on different dates in the same plots. McDonald (1994b) observed that cultivars had little effect on early season development of cavity spot, but severity and incidence of the disease differed among cultivars as the season progressed, despite similar inoculum levels and environmental conditions. She stated that an increase in severity of cavity spot during the growing season did not indicate roots become more susceptible as they aged, only that the disease continued to develop. Benard and Punja (1995) also found that carrot age (1–3 months) did not affect cavity spot development. Vivoda et al. (1991) found that the incidence of cavity spot did not increase during the season, but the number of lesions per root increased with plant age from three to five months after planting. McDonald (1994b) demonstrated that changes in cavity spot during the season appear more closely related to environmental factors than plant age. These temporal characteristics of cavity spot development illustrate the potential influence of timing of cavity spot evaluations on efforts to screen for resistance to the disease.

A number of breeding programs have used inoculation of mature carrot roots with agar plugs colonized by *Pythium* spp. that cause cavity spot to facilitate testing large numbers of roots per carrot line and large numbers of lines to counter the variability associated with efforts to screen for resistance to cavity spot. However, lesions induced by root inoculation with colonized agar plugs typically are more superficial, discolored, and have indistinct margins compared to cavity spot lesions that develop when roots are grown in infested soil or planting media (Vivoda et al. 1991). Vivoda et al. (1991) suggested that screening for resistance to cavity spot using colonized agar plugs may not reflect accurately the response of cultivars or breeding lines in soil conditions. However, others have demonstrated that inoculating carrot roots with agar plugs colonized by *P. violae* only works if roots are inoculated within 24 h of harvest because rapid suberization of the epidermis following harvest limits infection of the roots by that species. In contrast, inoculation of roots with *P. sulcatum*-inoculated plugs can be done as late as a week after harvest as long as the roots are stored in cool conditions to limit the extent of root suberization and desiccation. To avoid these issues, some bioassays that entail root inoculations entailed removing the tops after harvest of the roots to prevent excessive dehydration, and submerging the roots in water until they were inoculated to improve the reliability of the bioassays (e.g., Cooper et al. 2004). Other recommendations for increasing consistency in results with root inoculations include not surface-sterilizing the roots prior to inoculation (just washing the roots gently to avoid damaging the periderm), incubating the roots in the dark at cool temperatures (15–20 °C), and incubating the roots at high relative humidity (e.g., with regular misting or in dew chambers) for a 7- to 10-day duration of incubation before rating severity of cavity spot. Variability in lesion size among inoculation sites on the same root and among roots of the same entry necessitates inoculating and rating large numbers of roots (e.g., inoculating and rating 40–50 roots/entry/replication). The tedious nature of root agar plug inoculation protocols has limited the use in carrot breeding programs. Others have dipped roots of entries into a slurry of inoculum prepared by blending colonized agar plates in water, and then incubating the inoculated roots at high relative humidity before rating the roots for severity of cavity spot. Suffert and Montfort (2007) developed a soil infestation method in which an inoculated and infected carrot root was planted in close proximity to healthy roots to induce typical symptoms of cavity spot. Cavity spot lesions were induced more efficiently with this method than inoculating soil with *P. violae*.

Rating carrot roots for cavity spot. Various ways of assessing cavity spot have been reported. Some have been based on the incidence of roots with lesions, severity of lesions (e.g., number of lesions per root or the size of the lesions using horizontal and/or vertical length of each lesion), combinations of the two lesion dimensions (McDonald 1994b), length of the largest lesion/root, or categorizing lesions as small, medium, and large to facilitate rating large numbers of roots. The use of different assessment methods can make it difficult to compare results among studies. Assessment of cavity spot incidence or severity on a single harvest date can give variable results because cavity spot levels can increase or even decrease during the season. For this reason, McDonald (1994b) found the area under disease progress curve (AUDPC) to be more informative than incidence ratings for differentiating treatment effects in field trials in Canada. However, calculation AUDPC necessitates multiple assessments. Comparing slopes and elevations of disease progress curves also provided useful information for assessing the resistance of cultivars to cavity spot in field trials (McDonald 1994b).

Genetics of resistance. Several studies have suggested that cavity spot lesions represent a hypersensitive response of carrot root tissues to abort *Pythium* infections (Endo and Colt 1974; Klisiewicz 1968). Others have demonstrated that resistance generally is quantitative based on relatively minor differences in severity among cultivars (e.g., Johnston and Palmer 1985; White 1991). To date, there appear to be no published (publicly accessible) studies on the genetics of resistance to cavity spot.

Pectate lyase and cellulose produced by *Pythium* spp. are involved in the development of cavity spot lesions (Cooper et al. 2004). Induction of cell wall-degrading enzymes occurs after extensive penetration of root tissue by these pathogens, with enzyme production localized near the area of hyphal penetration (Campion et al. 1988; Guérin et al. 1994). Benard and Punja (1995) showed that highly virulent isolates of *Pythium* spp. produced significantly greater concentrations of pectolytic enzymes compared to moderately or weakly virulent isolates. The pathogens caused disintegration of host cells and the development of hyphae beneath the epidermis, followed by collapse of the infected area to form a cavity. Carrot roots respond to infection with synthesis and deposition of material around the site of infection, including oxidized phenolics and phenylalanine-ammonia lyase. The latter is thought to be associated with deposition of lignin around the lesion, providing physical protection against the pathogen. Impeding internal spread of *Pythium* has been proposed as one component of resistance to cavity spot (Endo and Colt 1974). Root defense mechanisms are thought to be activated after cell collapse to impede infection, as protease, peroxidase, and polyphenol oxidase activity were elevated in cavity tissue compared to healthy surrounding root tissue (Perry and Harrison 1979; Soroker et al. 1984). The phenol content of cavity tissue increased proportionally to the severity of cavity spot (Soroker et al. 1984). Suberin and lignin were deposited in the cell walls of periderm surrounding the lesions and accumulated in the phloem parenchyma cells near the wound surface (Perry and Harrison 1979). Garrod et al. (1982) indicated the development of these structural barriers was less important than accumulation of antifungal compounds in resistance to *Pythium*. Some antifungal compounds have been detected in non-inoculated root tissue, e.g., falcarindiol (Garrod et al. 1978), and others are elicited in response to injury or infection by pathogens, e.g., the phytoalexin 6-methoxymellein (Kurosaki et al. 1985). This was confirmed by structural analysis by Guérin et al. (1998) of susceptible and partially resistant cultivars who suggested that cell walls of the more resistant cultivars were better preserved, possibly as a result of fungitoxic phenolic compounds synthesized in response to infection. As noted above, Cooper et al. (2004) suggested the small cell size in roots, and higher constitutive levels of enzymes in the roots of 'Purple Turkey' might account for the resistance of this line to cavity spot compared to commercial cultivars they screened for morphological and biochemical responses of roots to *P. violae*.

White et al. (1988) suggested the speed at which a carrot root responds to infection with these defense mechanisms might determine the degree of susceptibility to cavity spot. *Pythium* spp. were recovered more frequently from juvenile tissue approximately eight weeks after seeding compared to isolations as plants matured. Fast-growing *Pythium* spp. were recovered readily from asymptomatic periderm but not from symptomatic tissues, which suggested that carrot defense mechanisms prevent infection by these fast-growing species or the roots do not react to these species (McDonald 1994b). In contrast, the cavity reaction is elicited by slow-growing species such as *P. violae* and *P. sulcatum* (White et al. 1988). This supported observations by Zamski and Peretz (1995) that fast-growing species did not cause lesions, only slow-growing species which penetrated root tissue for 3–4 days, releasing small amounts of wall-degrading enzymes before a host response occurred. They observed a lag of about 5 days before lignin deposition increased linearly.

Severity of cavity spot typically increases while carrot roots are in cold storage (McDonald 1994b). The increase in susceptibility with storage may be associated with changes in the carrot root that occur with the onset of bolting, a vernalization-induced physiological shift from vegetative to reproductive growth. An increase in the number of lesions per root also can occur in storage, which suggests latent infections present on the roots at harvest can progress to active infections in storage. However, wound healing during storage may heal some smaller cavity spot lesions (McDonald 1994b).

18.2.1.2 Phytophthora Root Rot or Rubbery Brown Rot (*Phytophthora* spp.)

Phytophthora root rot of carrot, also called rubbery brown rot, is generally a minor disease but can cause significant losses in waterlogged soils and usually occurs after periods of excessive rain or irrigation (Browne 2002). The disease can be caused by several species of *Phytophthora*, including *P. cactorum*, *P. cryptogea*, *P. megasperma*, and *P. porri*. Phytophthora root rot has been documented in Canada, France, Norway, Australia, and the USA (Browne 2002; Ho 1983; Rader 1952; Saude et al. 2007; Stelfox and Henry 1978; White 1945). Symptoms can develop in the field, usually close to harvest, and in storage, with roots becoming dark brown to black and rubbery (Fig. 18.2). However, often symptoms only become visible after roots have been in storage for some time. The firm, water-soaked lesions usually develop in the

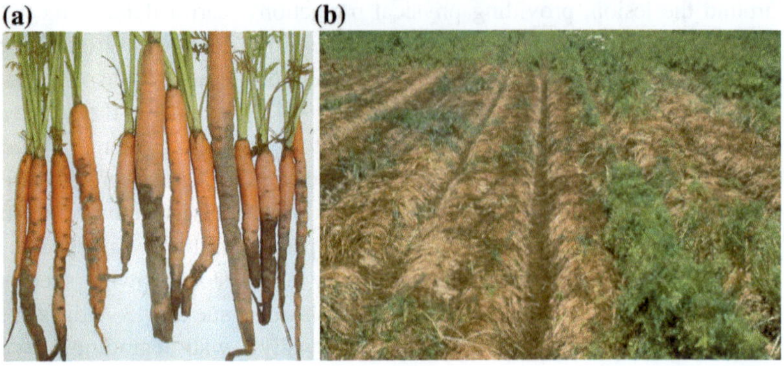

Fig. 18.2 Phytophthora root rot (rubbery brown rot) symptoms on individual carrot roots (**a**) and in a low-lying, saturated area of a carrot crop (**b**) (R. Michael Davis)

middle and crown areas of roots (Saude et al. 2007). Losses have been documented in fields in France in the winters. The pathogen can produce white mycelium on root lesions. Secondary invasion of the lesions by bacteria and fungi can lead to a soft rot. Periods of prolonged saturation during carrot growth, processing, or storage favor production and release of swimming spores called zoospores. Cool to moderate conditions favor inoculum production and spread of the pathogen. The pathogen spreads readily in storage.

Resistance screening: To date, there appear to be no published studies on screening for resistance to Phytophthora root rot of carrot. Stelfox and Henry (1978) noted the disease was documented in Alberta, Canada, in stored carrots of 'Imperator II' in the winter of 1969–1970. Symptoms were not observed on roots at harvest or during the washing operation. Saude et al. (2007) reported this disease in processing carrot fields in multiple counties in Michigan, but no information was provided on specific cultivars affected or differences in severity of the disease among cultivars. Based on the protocols used for testing pathogenicity of isolates of *Phytophthora* on carrot roots (e.g., Saude et al. 2007; Stelfox and Henry 1978), it should be possible to screen carrot cultivars or breeding lines for resistance to rubbery root rot using a protocol similar to the root agar plug inoculation method described for cavity spot. Colonized agar plugs of the pathogen, taken from 7- to 14-day-old cultures, could be placed on washed carrot roots, the surface of the roots moistened or misted regularly with sterilized water, and the inoculated roots incubated at high relative humidity for up to seven days at cool to moderate temperature. Various studies have incubated inoculated roots at temperatures ranging from 20 to 25 °C, but the optimal temperature might depend on the particular *Phytophthora* species being used to screen for resistance. Symptoms developed within a week when roots were stored at 20 °C, but only after seven weeks when stored at 0 °C (McDonald 1994d). Saude et al. (2007) demonstrated that wounding was not necessary to get typical symptoms of rubbery root rot with this method of inoculation. In fact, wounding resulted in slightly different symptoms. Key features of a phenotypic resistance screening protocol are likely to include the equivalent of saturated soil conditions that are necessary for the development of Phytophthora root rot, and storing carrots at 20 °C with high relative humidity (>95%).

18.2.1.3 Diseases Caused by *Rhizoctonia* spp.

Three species of *Rhizoctonia* have been demonstrated to cause diseases of carrot: *R. carotae* (= *Fibularhizoctonia carotae*, sexual stage *Athelia arachnoidea*), *R. crocorum* (sexual stage *Helicobasidium brebissonii* = *H. purpureum*), and *R. solani* (sexual stage *Thanatephorus cucumeris*) (Davis and Raid 2002). All three species are soilborne. *R. solani* is found in most soils.

Rhizoctonia solani is one of the multiple soilborne pathogens that can cause damping-off of carrot seedlings (Nuñez and Westphal 2002), and the fungus also causes crown rot of mature carrots (Punja 2002b) (Fig. 18.3). Isolates of this

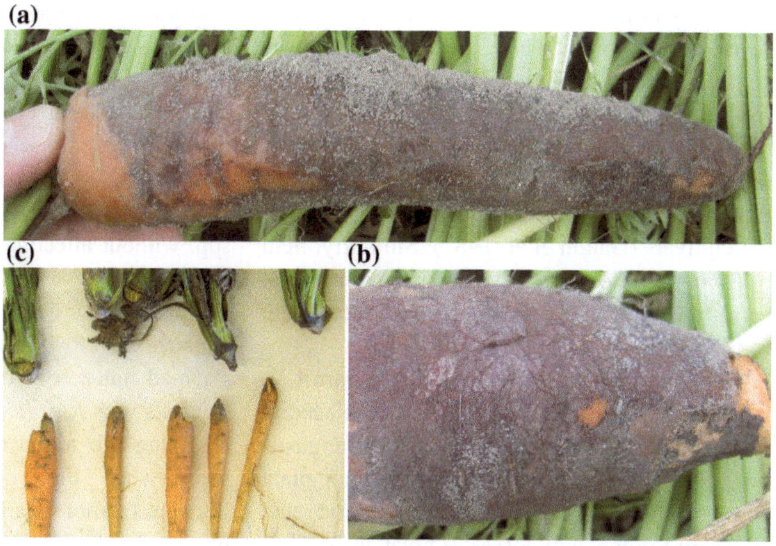

Fig. 18.3 Severe symptoms of violet root rot of carrot (**a** and **b**) caused by *Rhizoctonia crocorum*, and detached crowns (**c**) as a result of crown rot caused by *Rhizoctonia solani* (Lindsey du Toit, Washington State University). *Note* the web-like growth of purple-brown hyphae on the surface of the root infected with violet root rot (**b**)

pathogen that cause damping-off of carrot tend to belong to one of three anastomosis groups (AGs), AG-2 (primarily) and, to a lesser extent, AG-1 and AG-4 (Grisham and Anderson 1983; Nuñez and Westphal 2002). Conditions that delay seed germination and emergence, e.g., cool, and wet soils, favor damping-off. Damping-off results in poor seed germination (seed rot), root dieback as a result of loss of the apical meristem, death of seedlings pre-emergence or post-emergence, and seedlings with extremely poor vigor and stunting (Nuñez and Westphal 2002). Crown rot tends to be problematic in muck soils with high levels of organic matter, particularly in regions with warm, wet conditions close to harvest (Howard and Williams 1976; Punja 2002b). The disease may only become evident late in the season when the foliage senesces prematurely, sometimes in patches. The petioles and crowns rot, and dark brown, sunken lesions develop near the crown and sometimes further down the root (Punja 2002b). Crown rot lesions can be similar to those associated with cavity spot. Lesions on the crown or taproot render roots unmarketable, and secondary invasion of lesions by bacteria can initiate soft rot. Web-like mycelium can develop in lesions under very moist conditions. Lesions continue to expand when roots are placed in storage. Although empirical observations of carrot cultivars in growers' fields indicated all cultivars were susceptible to crown rot, partial resistance to crown rot has been suggested (Howard and Williams 1976) based on cultivar responses in fields with different amounts of inoculum and favorability of conditions for this disease.

R. crocorum causes violet root rot of carrot, celery, fennel, parsley, and parsnip as well as many other vegetable crops (e.g., table beets and potato) and weeds (Cheah and Page 1999; McDonald 1994e; Punja and McDonald 2002). Violet root rot of carrot occurs in many regions of carrot production but has caused greater losses in Europe, New Zealand, and Australia than in North America. Patches of dead or dying plants are usually the first evidence of this disease, with soil adhering to roots pulled out of the ground. Dark purple-brown, firm lesions develop on roots, on the surface of which a dense mat of mycelium of the fungus forms that can become violet to dark brown and leathery. The fungus can grow between plants as a thick, brown, mycelial mat on the soil surface (McDonald 1994b). Soft rotting of the root develops beneath the lesions. Symptoms tend to appear later in the season, and violet root rot can continue to develop in storage. The pathogen can infect carrot roots at soil temperatures ranging from 5 to 30 °C, with an optimum of 20 °C. This disease can occur in all soil types but tends to be more severe at high soil moisture, low soil pH, and low soil nitrogen (Cheah and Page 1999; Garrett 1949). An effort to screen 54 carrot cultivars for resistance to violet root rot in each of three sites naturally infested with *R. crocorum* in the UK did not reveal differences in susceptibility among the cultivars (Dalton et al. 1981). Similarly, field trials in New Zealand revealed all commercial carrot cultivars tested to be susceptible to violet root rot (Cheah and Page 1999).

R. carotae causes crater rot of carrot, a postharvest disease observed on carrots placed in longer-term storage (Punja 2002a). The fungus is not known to cause disease on any other plant species. Crater rot occurs in North America and Northern Europe, with losses as great as 50–70% recorded in Denmark (McDonald 1994a). Dry, sunken craters or pits form on the surface of roots under very humid, cool conditions in storage, with white mycelium lining the lesions and appressed to the root surface, and in which dark brown sclerotia may form (McDonald 1994c; Punja 2002a). The fungus spreads readily in storage. Crater rot is a dry rot, but infected roots can become colonized by bacteria, leading to soft rot. Latent infections of roots occur in the field, and roots with senescent foliage attached to the crown may harbor greater inoculum. The fungus can develop on roots at temperatures as low as 2–3 °C, with infection favored when roots are held at high relative humidity or a film of water develops on the roots (Punja 2002a). *R. carotae* can even grow at −1 °C (Punja 1987). Delayed harvest of carrots to late autumn appears to exacerbate disease pressure.

Resistance screening: Damping-off: Since cool, wet soil conditions delay seed germination and seedling emergence, and favor damping-off, planting carrot seed into cool, wet, poorly drained soils or providing excessive irrigation after seeding can favor damping-off and enhance efforts to screen for resistance to damping-off (Nuñez and Westphal 2002). Raised beds increase soil draining, so planting into flat fields can enhance damping-off in screening trials. However, these conditions favor all damping-off pathogens, not just *R. solani*, so it may be difficult to separate responses of carrot germplasm to different causal agents of damping-off, including *Pythium* spp., unless carrots are screened in sterilized or pasteurized soil or other planting media to which the target pathogen has been added, or seed is treated with a fungicide such as mefenoxam which can control *Pythium* spp. without affecting *Rhizoctonia* spp.

Crown rot: Howard and Williams (1976) attempted to screen carrot lines for resistance to Rhizoctonia crown rot by planting carrot seed in steamed muck soil, inoculating the carrots after 4 weeks with *R. solani* infested corn kernels, and assessing the number of normal and abnormal roots when the roots were harvested 16–20 weeks after planting. Seven-day-old agar plugs of a highly virulent isolate of *R. solani* grown on cornmeal agar were added to flasks containing autoclaved corn kernels and incubated at 20–24 °C for two weeks with the flasks shaken every 2–3 days to facilitate uniform colonization of the corn kernels by the fungus. Inoculum age (2–16 weeks after preparation) did not affect the reaction of 'Royal Chantenay' and 'Scarlet Nantes,' but they recommended using 'fresh' inoculum for each test. They observed no difference in disease incidence/severity when inoculum was added to the soil 2, 3, 4, or 6 weeks after planting, but found it most convenient to thin carrots three weeks after seeding and add inoculum a week later, similar to the protocol used by Mildenhall and Williams (1970). Howard and Williams (1976) also recommended maintaining soil moisture at approximately −0.1 bars and growing carrots at 20, 24, or 28 °C for optimal carrot plant and crown rot development. Crown rot pressure can be enhanced by placing infested soil or infested carrot debris in contact with the crown and petioles, by close spacing of carrots to promote a humid microclimate once the canopy closes, and if temperatures are warm (>18 °C) (Gurkin and Jenkins 1985; Punja 2002b). Planting carrot crops into infested debris, following perennial crops such as alfalfa, and adding inoculum (e.g., colonized grain kernels) to soil or other potting media can increase disease pressure in resistance screening trials (e.g., Breton et al. 2003).

Violet root rot: Since violet root rot is favored by high soil moisture, low soil pH, and low soil nitrogen levels, screening for resistance could be enhanced by using field sites with acid soils or using acid planting media, maintaining high soil moisture, and keeping roots in infested soil or planting medium as long as possible as disease incidence and severity increase the longer roots are in infested soil (Cheah and Page 1999; Garrett 1949; McDonald 1994e; Punja and McDonald 2002). However, Dalton et al. (1981) were not able to detect any differences in susceptibility among 54 commercial carrot cultivars (9 Amsterdam Forcing selections, 11 Nantes, 10 Chantenay, 9 Autumn King, 5 Danvers, 3 Berlicum, 6 Feonia or Imperator, and 1 unknown type) tested in three field sites in the UK that were naturally infested with *R. crocorum*. Similarly, Cheah and Page (1999) did not observe differences in susceptibility among commercial carrot cultivars. Slight differences in severity of violet root rot at one site were not significant statistically because of inadequate disease pressure, and severe disease pressure at another site still did not enable differentiation of cultivar reactions to violet root rot. Dalton et al. (1981) hypothesized the lack of differences may reflect the fact that western carrots have been developed from closely related types—Late Half Long, Early Half Long, and Early Scarlet Horn, all of which were derived from Long Orange by selection or intercrossing. They recommended searching for resistance in pre-cursor types to western types, namely anthocyanin and yellow types. There remains a need for an effective and efficient protocol to screen for resistance to violet root rot.

Crater rot: Hyphae of *R. carotae* can grow over a carrot root within a few days, penetrating the root surface without forming appressoria or other infection structures, and killing root cells in advance of hyphal penetration (McDonald 1994a). Roots placed in storage can be rendered unmarketable within three weeks of infection. There does not appear to have been any effort to screen for resistance to crater rot of carrot, but a root screening protocol should be feasible given crater rot is a postharvest disease and the pathogen is highly virulent in cool, moist storage conditions. Wounding of roots increases the severity of crater rot, so wounding could be incorporated into a screening protocol. It may be difficult to develop a soil inoculation protocol that mimics field infection, given the latent nature of field infections.

18.2.1.4 Bacterial Soft Rots

Several bacteria can cause soft rots of carrot, including *Pectobacterium carotovorum* subsp. *carotovorum* (formerly *Erwinia carotovora* subsp. *carotovora*), *Dickeya dadantii* (formerly *D. chrysanthemi* = *Pectobacterium chrysanthemi* = *Erwinia chrysanthemi*), and *Pectobacterium atrosepticum* (formerly *P. carotovorum* subsp. *atrosepticum* = *E. carotovora* subsp. *atroseptica*) (Farrar 2002; McDonald 1994a; Nuñez and Davis 2016). These bacteria tend to be ubiquitous in soils and can infect a wide range of plant species, including most vegetables. Bacterial soft rot of carrot is a problem primarily in storage, where the pathogens can cause major losses as secondary invaders of roots that were wounded or infected with other pathogens. Soft rot symptoms usually only develop in the field in low-lying areas or other areas that become saturated (e.g., near broken irrigation pipes). Sporadic reports of severe outbreaks in fields are associated with extended periods of saturated soil conditions and warm temperatures as these bacteria are thermophilic, facultative anaerobes (Farrar 2002). The pathogens also can be found in sources of water used for irrigation or water used to wash carrot roots after harvest (Segall and Dow 1973), and can be disseminated by insects (Phillips and Kelman 1982). The bacteria infect carrot roots through wounds or natural openings, causing small, water-soaked lesions that enlarge rapidly (Fig. 18.4). The pathogens degrade roots most rapidly under warm conditions (20–25 °C for *P. carotovorum* subsp. *carotovorum* and 30–35 °C for *D. dadantii*), and infected roots become mushy and soft. The outer surface of infected roots may remain intact over a softened interior, and macerated interior tissue may ooze through cracks that form on the root surface (McDonald 1994a).

Fig. 18.4 Severe root pitting symptoms caused by bacterial soft rot (**a** and **b**), and infection of the base of seed stalks of a bolted carrot plant by *Pectobacterium carotovorum* subsp. *carotovorum* (**c**) (Lindsey du Toit, Washington State University)

Resistance screening: There have been various attempts to screen for resistance to soft rot of carrot (Bedlan 1984; Lebeda 1985; Michalik et al. 1992; Michalik and Sleczek 1997; Skadow 1978). Although Segall and Dow (1973) did not focus on screening for resistance to soft rot, they demonstrated that holding carrot roots naturally infected with *P. carotovorum* subsp. *carotovorum* at 21 °C for four days resulted in more severe soft rot than holding roots at 2 °C for three days and then at 21 °C for four days. They suggested that less severe soft rot associated with refrigerated storage before and after inoculation may reflect the development of phenolic or related compounds in refrigerated carrots. For example, the antimicrobial compound 3-methyl-6-methoxy-8-hyrdoxy-3, 4-dihydroxoisocoumarin was found in carrot roots stored for 4–8 weeks at 0 °C but not in freshly harvested carrot roots. Segall and Dow (1973) suggested this may contribute to resistance to bacterial soft rot after cold storage of carrots.

In an effort to develop an efficient, reliable method of screening carrot breeding material for resistance to soft rot caused by *P. carotovorum* subsp. *carotovorum* and *P. atrosepticum*, Michalik et al. (1992) evaluated four methods of inoculation of roots. They used roots harvested from an organic soil in Wisconsin that had been stored for 1–3 weeks at 0–4 °C, washed with tap water, surface-sterilized with 0.05% NaOCl for 40 min and 70% ethanol for 5 min, rinsed in sterilized water, and air-dried. The inoculation methods included: (1) injecting a 10 μl aliquot of bacterial suspension into each of two holes (1 mm diameter × 2 cm deep) in the cambium region through the cut surface of the crown portion of each root (top third of the root, including 2 cm of trimmed petioles attached); (2) the same injection method using the middle third of the root; (3) 5-mm-thick, cross-sectional slices of the root on each of which a 5-mm-diameter filter paper disk was placed on the cambial region of the proximal cut surface after the disk had been soaked in a bacterial suspension for 30 min; and (4) root slices each inoculated with a 10 μl aliquot of bacterial suspension placed on the cambial region of the freshly cut surface without additional wounding (Michalik et al. 1992). Inoculated root samples were incubated at 22 °C for 48–96 h in boxes lined with wet paper towel, sealed with plastic wrap, and enclosed in plastic bags. They observed differences in responses of inbred lines and open-pollinated cultivars to the two pathogens, with *P. carotovorum* subsp. *carotovorum* causing more severe soft rot than the isolate of *P. atrosepticum*, although the severity of soft rot increased with increasing inoculum concentration for both bacteria. They did not detect bacterial strain × carrot line or strain × inoculum concentration interactions. The two root cross-section inoculation methods resulted in more severe soft rot and less variability in reactions than methods using larger root sections. The most consistent responses were achieved with bacterial-soaked filter disks as the disks reduced evaporative drying of the inoculum. The use of root slices also enabled replicated screening from individual roots, and the crown of the root could be maintained for seed production. The duration of postharvest storage of carrot roots (2, 6, or 12 weeks) did not influence the severity of soft rot. Results were similar for stored intact versus cut roots, although the root tip tended to be more susceptible than the crown or middle of the root (Michalik et al. 1992). The variation observed among carrot lines suggested advances could be made in breeding for resistance to soft rot, as observed by others.

In a follow-up study, Michalik and Ślęczek (1997) evaluated progenies from crosses of orange carrot cultivars with five wild *D. carota* subspecies and four local Mirzoe cultivars from Uzbekistan to identify a source of resistance to soft rot caused by *P. carotovorum* subsp. *carotovorum*. Although genetic variation in sensitivity to soft rot had been observed in orange carrot cultivars, they considered the variation too limited for breeding purposes. They inoculated carrot root disks with filter disks soaked in a bacterial suspension (5×10^6 CFU/ml) for 30 min, placed on the cambial region of each disk as described above. Inoculations of roots from the progeny of crosses of *D. carota* subsp. *sativus* with *D. carota* subsp. *commutatus*, *D. c. gummifer*, *D. c. drepanensis*, *D. c. maritimus*, and *D. c. gadecaei* did not indicate the presence of resistance genes to soft rot in these wild species since all the F2 and BC1 generations had more severe soft rot than the original orange cultivars. However, one of the four local Mirzoe cultivars showed some promise as a source of partial resistance to soft rot, although soft rot severity increased in the F2 generation. Stein and Nothnagel (1995) noted that pronounced differences in susceptibility to bacterial soft rot have been detected among lines, F1 hybrids, and open-pollinated carrot lines in Germany. For any laboratory screening method, results need to be correlated with storage and field evaluations, and it is important to use roots grown, harvested, and stored together to avoid confounding the various factors that can influence severity of soft rot (Lebeda 1985; Michalik et al. 1992; Michalik and Sleczek 1997; Skadow 1978).

The genetic basis of resistance of carrot lines to bacterial soft rot pathogens has not been determined, but variation in responses to soft rot bacteria among carrot lines suggests the potential for molecular screening methods to enhance the identification of QTLs associated with resistance to these pathogens.

18.2.1.5 Black Rot (*Alternaria radicina*)

Black rot of carrot is caused by *Alternaria radicina* (formerly *Stemphylium radicinum*). Black rot has been documented mainly as a postharvest disease during root storage, as a disease affecting seedlings as a result of planting infected seed, and as a disease affecting carrot seed crops. In the field, *A. radicina* can cause a black decay on the foliage, petioles, and umbels (Meier et al. 1922). First described in New York, black rot has now been reported all over the world. The pathogen is seedborne and seed transmitted, causing seed rot, poor seedling establishment, and/or damping-off. *A. radicina* can persist in the soil for long periods (as long as eight years) and cause disease in subsequent carrot crops (Farrar et al. 2004; Maude 1966; Pryor et al. 1998; Scott and Wenham 1972). Black rot is characterized by black, sunken necrotic lesions on the taproots and crowns (Fig. 18.5). Under wet conditions, infection of the crown can result in rotting of petioles and leaf blight symptoms similar to those caused by *Alternaria dauci*, leading to significant crop losses because of the tops breaking during mechanical harvest with harvesters that pull roots out of the ground by their tops (Farrar et al. 2004; Grogan and Snyder 1952; Pryor et al. 1998). Once roots are infected, the pathogen spreads readily between roots in storage. Infection on umbels can reduce seed yield and seed germination.

Seed treatment with fungicides like azoxystrobin, fludioxonil, iprodione, or thiram, or with hot water or disinfectants like sodium

Fig. 18.5 Severe symptoms of black rot caused by *Alternaria radicina* on carrot stecklings (vernalized roots used for a carrot seed crop)

hypochlorite can be effective methods to control seedborne inoculum of this pathogen (Biniek and Tylkowska 1987; Maude 1966; Pryor et al. 1994; Soteros 1979). Chen and Wu (1999) documented significant effects of two biological control agents, *Burkholderia cepacia* No. 229 and *Bacillus amyloliquefasciens* No. 224 against *A. radicina*. Kordowska-Wiater et al. (2012) showed that application of the yeast *Candida melibiosica* to carrot roots before inoculation with *A. radicina* partially reduced progress in the development of black rot.

Resistance screening: Pryor et al. (2000) evaluated 46 carrot cultivars under field conditions using a toothpick inoculation method and observed significant differences in lesion size among cultivars. Relatively resistant cultivars included 'Panther' and 'Caropak,' and susceptible cultivars included 'Royal Chantenay' and 'Nogales.' Lesion development was greater in cold storage conditions than in the field, but the relative ranking of cultivars in terms of resistance to *A. radicina* was similar. In 2008–2009, Karkleliene et al. (2012) completed a black rot field experiment with organic production conditions under which they also observed varietal differences. The cultivar Magi was the most susceptible to *A. radicina* among 13 cultivars screened. Cwalina-Ambroziak et al. (2014) reported more severe symptoms on 'Koral' than on 'Bolero.'

Phenotypic screening methods: Pryor et al. (2000) developed a toothpick inoculation method for black rot resistance evaluation. After sterilization, toothpicks were incubated with 2 ml of a suspension of *A. radicina* conidia (1×10^4 conidia/ml) and incubated for five days at 28 °C in the dark. The colonized end of each toothpick was inserted into the shoulder of a 10- to 12-week-old carrot root, and the area of the lesion measured 9–10 weeks later. Grzebelus et al. (2013) set up an in vitro selection protocol for plants with superior phenotypic performance against *A. radicina*. They isolated somaclonal variants within protoplast cultures that were challenged by fungal culture filtrates and obtained regenerated plants with greater tolerance to the pathogen. Cwalina-Ambroziak et al. (2014) inoculated petioles and seedlings directly with agar disks (each 5 mm in diameter) taken from 5-day-old cultures of *A. radicina*.

Genetics of resistance: As detailed for gray mold caused by *Botrytis cinerea*, Baranski et al. (2008) confirmed the positive impact of chitinase on *A. radicina* when using transgenic plants expressing CHIT36, one of the chitinase lytic enzymes secreted by *T. harzianum* that exhibits antifungal activity in vitro. Infection by *A. radicina* was reduced by 50%. When looking for a modification in systemic acquired resistance (SAR), Wally et al. (2009a) achieved a significant reduction in the severity of taproot symptoms caused by *A. radicina* (diameter of lesions reduced by 50%) and in the number of foliar necrotic spots (greater than 33% reduction in foliar disease severity index) by inoculating transgenic plants expressing the *NPR1* gene of *Arabidopsis*. Wally and Punja (2010) examined the mechanisms of resistance in a transgenic carrot line, P23, which constitutively overexpressed the rice cationic peroxidase OsPrx114 and which exhibited enhanced resistance to necrotrophic foliar pathogens. OsPrx114 overexpression led to a slight enhancement of constitutive transcript levels of pathogenesis-related (PR) genes, and taproots had increased lignin formation in the outer periderm tissues, particularly after inoculation with *A. radicina.*

18.2.1.6 Fusarium Dry Rot (*Fusarium* spp.)

Fusarium dry rot has been reported in the USA, Canada, France, Japan, and China (Rubatzky et al. 1999; Sherf and MacNab 1986; Villeneuve 2014; Zhang et al. 2014) and can be of great economic importance in some regions. Zhang et al. (2014) reported losses of up to 80% in Tuo Ke Tuo County, China. Symptoms include round, 3–4 cm in diameter, black lesions on root surfaces. The lesions evolve into a soft rot or brown canker, resulting in unmarketable roots. Symptoms may also include black spots on the crown that reduce nutrient translocation between the root and foliage and, therefore, impact root quality and yield. The disease also can cause significant losses during storage. Four species have been described as causing this disease, *Fusarium solani*, *F. avenaceum*, *F. culmorum* and, more recently, *F. caeruleum*. Zhang et al. (2014) described two ways to reproduce typical symptoms that could be used as a screening tool for breeders to evaluate varietal differences. The first was with mature carrot roots inoculated with colonized agar plugs (5 mm in diameter) cut from the margin of actively growing colonies on potato dextrose agar plates. One mycelial plug was placed on each carrot root, with the mycelial side facing the root. The inoculated roots were then incubated in a humid chamber (90% relative humidity) at 25 °C. Four days after incubation, mycelium had covered most of the surface of the root, and brown lesions were observed on the root. The second protocol entailed a potting trial in which carrot seeds were sown in sterilized soil in pots (30 cm × 25 cm) with 15 seeds per pot. The soil was infested by adding a spore suspension at a final concentration of 1×10^4 CFU/g soil. Plants grown in non-infested soil served as the control treatment. All the plants in inoculated pots were placed in a field. After 13 weeks, symptoms of dry rot were evident. Even though there are no known sources of resistance or published varietal screening trials, genetic transformation has been reported by Sidorova and Miroshnichenko (2013). They reported that 'Nantskaya 4' transgenic carrot plants overexpressed a single-gene coding for a thaumatin II protein and showed enhanced tolerance to infection by *F. avenaceum.*

18.2.1.7 Gray Mold (*Botrytis cinerea*)

Gray mold is caused by the fungus *Botrytis cinerea* and can result in considerable losses in temperate regions of Europe, North America, and Asia (Rubatzky et al. 1999). Primary infections occur in fields, principally from airborne spores. The development of symptoms mainly occurs in cold storage. The fungus generally spreads into carrot roots at the base of petioles or on the crown. Watery brown lesions expand rapidly to become water-soaked, dark brown lesions covered with gray mycelium and, as the lesions age, small sclerotia. Resistance tests based on root inoculation were developed for screening the susceptibility of carrot cultivars to *B. cinerea* in storage and to study the process of induced resistance (Bowen and Heale 1987; Goodliffe and Heale 1975). Baranski et al. (2006) set up a leaf assay using colonized agar plugs to get rapid assessment of carrot leaf susceptibility to gray mold for a non-destructive, preliminary evaluation of precious and limited carrot source materials. Mercier et al. (2000) reported that heat-killed conidia of *B. cinerea* induced systemic resistance to *B. cinerea* in carrot slices through enhanced suberization and local accumulation of the phytoalexin 6-methoxymellein.

They questioned the role of a 24-kDa chitinase in the induced resistance response. Baranski et al. (2008) confirmed the impact of chitinase on gray mold when using transgenic plants expressing CHIT36, one of the chitinase lytic enzymes secreted by the biological control fungus *Trichoderma harzianum* that exhibits antifungal activity in vitro. Transgenic plants reduced *B. cinerea* attack by as much as 50%.

18.2.1.8 Sclerotinia Soft Rot or White Mold (*Sclerotinia sclerotiorum*)

Carrot foliage and roots in fields may be destroyed by Sclerotinia soft rot or white mold, but it is mainly in cold storage conditions and long distance transportation that damage to this disease is significant. Small translucent spots on roots are covered rapidly by white, flocculent mycelium, which develops into melanized, black structures called sclerotia (Fig. 18.6). Sclerotia can survive up to 10 years in soils. Three species are cited as causal agents: *Sclerotinia sclerotiorum*, *S. minor*, and *S. subarctica* (Leyronas et al. 2018). *Sclerotium rolfsii*, which causes southern blight of carrot, is a basidiomycete that is not related to the white mold fungi. The white mold pathogens are ascomycetes. White mold is distributed worldwide (Kora et al. 2003; Rubatzky et al. 1999) and has a host range of more than 500 species, including weeds.

No resistance sources have been identified in carrot germplasm, but a phenotyping test with *S. sclerotiorum* was described by Ojaghian et al. (2016). Freshly harvested carrot roots were disinfected with 2% sodium hypochlorite for three minutes, and then washed with sterilized tap water and dried on sterilized filter paper. Roots were inoculated using fungal isolates grown on carrot dextrose agar. A 5-mm diameter colonized agar plug taken from the leading edge of a 3-day-old culture was then placed centrally on the root with the colonized agar surface facing the root. The carrots were placed in plastic boxes (12 per box) and covered with three thin layers of plastic to provide a moist chamber. To increase humidity, moist cottonwool pieces were placed in the boxes, and the roots were stored at 21–23 °C. Disease severity was determined six days after inoculation on a scale of y1 to y4, where: y1 = no lesion,

Fig. 18.6 Matted foliage in a carrot crop as a result of white mold caused by *Sclerotinia sclerotiorum* (**a**), dieback of bolted carrot plants in a seed crop following root infection by *S. sclerotiorum* (**b**), and severe rot of a carrot root on which black sclerotia of *S. sclerotiorum* had formed (**c**) (Lindsey du Toit, Washington State University)

y2 = 1–4 cm long lesions on roots without sclerotium formation, y3 = 4–8 cm long lesions with 1–4 mature or immature sclerotia, y4 = ≥ 8 cm long lesions on the roots with more than 4 mature or immature sclerotia. A disease index was then calculated using the formula: Disease index = $[(1.25 \times y2) + (2.5 \times y3) + (3.75 \times y4)]$/total number of carrots $\times$ 1/0.05. In this formula, 0.05 is a constant coefficient (Ojaghian et al. 2016). Punja and Chen (2004) reported that transgenic carrot plants expressing a thaumatin-like protein from rice showed significantly enhanced tolerance to *S. sclerotiorum* when detached petioles and leaflets were inoculated under controlled environmental conditions. Wally et al. (2009b) showed that carrot lines overexpressing OsPrx114 peroxidase were highly resistant to *S. sclerotiorum* without showing any visible phenotypic abnormalities of the roots. The resistance was associated with increased transcript levels of pathogenesis-related (PR) genes when tissues were treated with cell wall fragments of *S. sclerotiorum* (Wally and Punja 2010).

18.2.1.9 Common Scab (*Streptomyces scabies*)

Common scab of carrot, caused by *Streptomyces scabies*, occurs in many areas of carrot production but is particularly problematic in Canada and Europe, especially France and the Netherlands (Janse 1988; Villeneuve 2014). Infections occur through wounds or lateral secondary roots, and death of infected epidermal cells occurs during dry periods. After a few months, a corky bulge expands horizontally through the root surface, particularly toward the top of the root (Fig. 18.7). *Streptomyces scabies* can survive in soils for several years as a saprophyte. Schoneveld (1994) demonstrated that 4–5 weeks after spring sowing was the most susceptible stage of growth for infection by *S. scabies*. A phenotyping test was described by Janse (1988). A 60-ml aliquot of a bacterial suspension (10^7 spores/ml) prepared from a 4-week-old culture on yeast malt agar was added to 20 L of a steam-sterilized loamy soil at pH 5.9 prior to sowing carrot seed. Plants were grown at 18 °C and 80% RH with 10 000 lx of light and at 50% soil saturation. Plant roots were harvested four months after sowing and examined for symptoms. There are no reports of resistance to common scab in carrot germplasm.

18.2.2 Foliar Diseases

18.2.2.1 Alternaria Leaf Blight (*Alternaria dauci*), Cercospora Leaf Spot (*Cercospora carotae*), and Bacterial Leaf Blight (*Xanthomonas hortorum* pv. *carotae*)

Carrot leaf blights can be caused by two fungal pathogens, *Alternaria dauci* and *Cercospora carotae*, and the bacterial pathogen *Xanthomonas hortorum* pv. *carotae* (formerly *Xanthomonas campestris* pv. *carotae*) (Fig. 18.8). Alternaria leaf blight caused by *A. dauci* is the

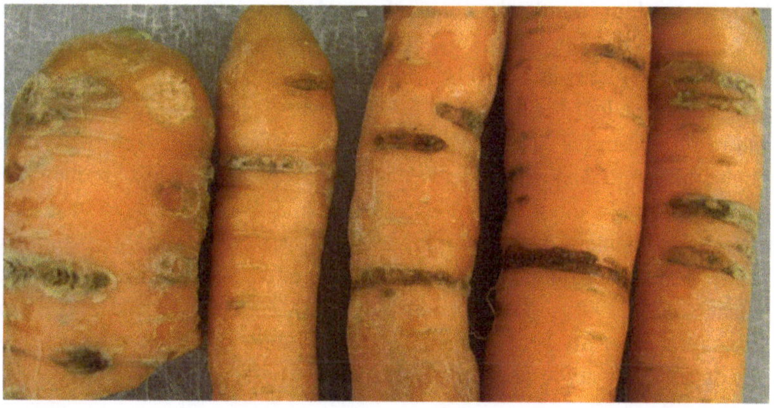

Fig. 18.7 Common scab on carrot roots caused by *Streptomyces scabies* (Lindsey du Toit, Washington State University)

Fig. 18.8 Symptoms of Alternaria leaf blight caused by *Alternaria dauci* (**a**), bacterial leaf blight caused by *Xanthomonas hortorum* pv. *carotae* (**b**), and Cercospora leaf spot caused by *Cercospora carotae* (**c**) (Lindsey du Toit, Washington State University)

major foliage disease of carrots in most areas of production. First described in 1855 in Germany and 1890 in the USA, *A. dauci* causes severe defoliation in carrot crops all over the world where there are conditions of high moisture and temperature (Farrar et al. 2004). The relatively large, dematiaceous spores (Maude 1966) can be produced and dispersed aerially over a wide range of temperatures (8–28 °C) and moisture conditions throughout the growing season (Strandberg 1977), although periods of dark and leaf wetness durations of at least 10 h also favor sporulation (Langenberg et al. 1977). Foliar symptoms appear 8–10 days after infection as small, green-brown lesions. The lesions enlarge and infected tissue becomes dark brown to black, sometimes surrounded by a chlorotic halo (Farrar et al. 2004). While foliar symptoms are the most common, *A. dauci* can also infect the inflorescences and seeds developing in umbels. Seedborne inoculum can lead to seed transmission and infection of seedlings, resulting in damping-off (Farrar et al. 2004; Maude 1966).

Cercospora leaf spot, caused by *C. carotae*, can result in similar symptoms to Alternaria leaf blight, although lesions on leaves and petioles tend to be more circular, and each lesion may develop a distinct, dark margin with a lighter brown center (Bourgeois et al. 1998; Carisse and Kushalappa 1990; Gugino et al. 2007; Milosavljević et al. 2014; Raid 2002). The fungus only infects aerial parts of carrot plants. The optimum temperature range for infection is 20–28 °C with short periods of leaf wetness (<6 h) followed by high relative humidity sufficient to result in infection (Carisse and Kushalappa 1992).

As highlighted by Pfleger et al. (1974), bacterial leaf blight is caused by the seedborne pathogen *X. hortorum* pv. *carotae*. This disease can be indistinguishable from the fungal blights caused by *A. dauci* and *C. carotae* based on foliar symptoms alone. However, bacterial leaf blight lesions sometimes are accompanied by production of a gummy bacterial exudate, particularly lesions on petioles, umbels, and seed stalks (du Toit et al. 2005). Bacterial leaf blight was first reported in California in 1934. The disease can be found worldwide wherever carrots are grown. The pathogen can infect the foliage, stems, umbels, and seed (Pfleger et al. 1974; du Toit et al. 2005). Some studies have suggested roots can become infected, but these probably reflect infection limited to the crown where petioles attach to the root. Seeds may be infected internally or contaminated on the surface. Infection of seed by *X. hortorum* pv. *carotae* does not necessarily reduce seed germination or vigor, but the need to use hot water seed treatment to eradicate the pathogen or to reduce seed infection levels can impact seed germination or vigor.

Where severe outbreaks of foliar diseases of carrot occur, regardless of the causal agent, harvest operations can be hindered as lesions coalesce and weaken the tops which may break during mechanical harvest. A reduction in green leaf area for photosynthesis may also reduce yields. Planting seed lots that have been tested for the pathogen, or infected seed lots that have been treated (e.g., with hot water for the bacterial blight pathogen or with fungicides for the fungal pathogens), and the use of cultural and chemical control methods in carrot fields are important management strategies for these foliar diseases. Using healthy seed lots or treated seed lots remains particularly important for bacterial leaf blight as copper bactericides or antibiotics are not permitted in some countries, and copper bactericides are purely protectant, so the disease cannot be controlled once infection is established in a crop. *A. dauci* and *C. carotae* can be managed with timely applications of relevant fungicides to crops during the growing season. Although chemical control is a primary means used by some growers for management of the fungal foliar diseases, the use of resistant cultivars, when available, together with chemical and cultural control measures, remains the most effective integrated approach to managing these diseases.

Resistance screening: Traditional carrot breeding methods have relied on phenotypic evaluations of large numbers of carrot accessions. Most of the literature on phenotypic evaluations have been on characterizing the resistance mechanisms of carrot against *A. dauci*, with very few published studies on *C. carotae* and *X. hortorum* pv. *carotae*.

Numerous carrot cultivars have been evaluated for resistance to *A. dauci*, and several sources have been identified. However, none of the cultivars evaluated to date has been completely resistant to Alternaria leaf blight. Strandberg et al. (1972) screened 90 breeding lines and 241 Plant Introduction (PI) lines from 31 countries. They identified nine lines with high levels of resistance. Significant differences in resistance to *A. dauci* were observed among four cultivars evaluated under natural infection in Brazil, with 'Brasilia' being the most resistant (Boiteux et al. 1993). Field evaluations of tropical germplasm were also reported by Pereira et al. (2012), Silva et al. (2009), and Carvalho et al. (2015). Amirov et al. (2014) evaluated 86 accessions in Kazakhstan. Information on the stability of resistance evaluated in different environments has been important for breeders to utilize the documented resistance. Using 21 isolates of *A. dauci* collected from commercial carrot fields in northeastern North America, Rogers and Stevenson (2010) detected variety-by-isolate interactions with three commercial carrot cultivars. Conversely, testing 11 *A. dauci* isolates from different parts of the world on eight varieties or inbred lines and one segregating population for *A. dauci* resistance, Le Clerc et al. (2015a) did not find a significant interaction between isolates and varieties. The different conclusions may reflect differences in carrot varieties, fungal isolates, and environmental conditions in the two studies. As suggested by Le Clerc et al. (2015b), it is possible that some isolate-specific resistance factors in some varieties may confer varying degrees of resistance in different environments. The disease phenotyping by Rogers and Stevenson (2010) was done 8 and 16 days after inoculation, while this was done 20 and 35 days after inoculation in greenhouse trials and 30 days after inoculation (and every 15 days thereafter) in tunnel trials by Le Clerc et al. (2015b). For a given isolate, disease development varies among carrot genotypes, which may reflect activation of different resistance mechanisms at different time periods after inoculation.

Simon and Strandberg (1998) confirmed that evaluations for resistance to *A. dauci* in field conditions generally correlate well with resistance ratings in greenhouse trials. Although widely used, field testing can be time-consuming, expensive, potentially affected by uncontrollable environmental conditions, and generally can only be done once a year. To address these issues, tests have been developed under controlled conditions such as greenhouses, tunnels, or growth chambers. While field evaluations generally can include assessment of a large number of plants and entries, assays in controlled conditions tend to necessitate using fewer plants, even one plant per variety or detached parts of plants. Baranski et al. (2007) evaluated the resistance of transgenic plants using a laboratory-based assay with detached leaflets and petioles that were inoculated with fungal pathogens. Pawelec et al. (2006) reported effective ranking of carrot cultivars using plants grown in pots in a greenhouse and inoculated with the pathogens, whereas detached leaf and hypocotyl tests failed to discriminate responses among cultivars. For detached leaf and petiole tests, a drop inoculation method was developed to facilitate more rapid screening than field evaluations and to use less plant material (Boedo et al. 2010). The responses of carrot lines to *A. dauci* have also been evaluated in vitro (Dugdale et al. 2000). Regenerant somaclone plants initiated from seedling hypocotyls were evaluated for susceptibility to a pathogen by measuring the loss of chlorophyll of infected, detached leaves. Lecomte et al. (2014) and Courtial et al. (2018) challenged embryogenic cell cultures with fungal extracts to evaluate the resistance of carrot genotypes to *A. dauci*. However, these tests were mainly dedicated to comprehension of resistance mechanisms than to high-throughput phenotyping and would need to be automated to be of value for more extensive phenotyping.

There is very little information on screening for resistance of carrot cultivars to *C. carotae*. Lebeda et al. (1988) evaluated a worldwide collection of 142 carrot cultivars for resistance to *C. carotae*. A large proportion of the cultivars was highly susceptible, with only ~30% expressing resistance under field conditions. Field trials also were done by Gugino et al. (2007). None of the cultivars displayed complete resistance although there was variability in reaction among cultivars. Data on the genetics of resistance to Cercospora leaf spot are incomplete, and there do not appear to have been any public efforts to breed for resistance to Cercospora leaf spot.

Similarly, genetic resistance to *X. hortorum* pv. *carotae* is not well documented and there has been very little public research on screening for resistance. No commercial cultivars currently are marketed as resistant to bacterial blight (Christianson et al. 2015). Pfleger et al. (1974) indicated varietal differences in response to bacterial blight among six cultivars and breeding lines. Christianson et al. (2015) screened 66 PI lines, two public inbred lines, and 17 commercial carrot cultivars and carrot wild relatives for response to *X. hortorum* pv. *carotae* in a greenhouse by rating severity of bacterial blight and quantifying the amount of *X. hortorum* pv. *carotae* that developed on the leaves of each inoculated line. Eight putative resistant PI lines and five highly susceptible PI lines identified in the first screening were tested again with an additional two PI lines, 12 cultivars, two inbred lines, and 12 carrot wild relatives. PI lines 418967, 432905, and 432906 were identified as partially resistant to bacterial blight, with potential value in breeding more resistant cultivars. None of the accessions had complete resistance. Of the 12 carrot wild relatives, only Ames 7674 and SS10 OR had relatively limited bacterial blight. Overall, the severity of symptoms and the amount of *X. hortorum* pv. *carotae* recovered from the foliage differed significantly among the accessions tested. Christianson et al. (2015) showed that using visual foliar disease severity ratings instead of the highly resource- and labor-intensive *X. hortorum* pv. *carotae* quantification protocol was effective, subject to using adequate number of replications for accurate assessment, as foliar severity ratings were positively correlated with *X. hortorum* pv. *carotae* quantification in both trials ($r = 0.52–0.62$ at $P < 0.0001$). This study illustrated that the *Daucus* germplasm in the United States Department of Agriculture's National Plant Germplasm System represents a valuable public source of potential resistance for breeders. As suggested by Christianson et al. (2015), the susceptible and resistant PI lines identified in that study could be used to study the inheritance of *X. hortorum* pv. *carotae* resistance in carrot.

Genetics of resistance: In order to develop hybrid carrot cultivars with high level of resistance, knowledge of the heritability and combining abilities of sources of resistance is needed in breeding programs. Studying resistance to *A. dauci*, 40% narrow-sense heritability (h^2) was calculated for the open-pollinated cultivar Brasilia (Boiteux et al. 1993). Vieira et al. (1991)

found broad-sense heritabilities ranging from 45 to 82% when evaluating foliar leaf blight resistance (without distinguishing the potential causal agent(s) as *A. dauci*, *C. carotae*, or *X. hortorum* pv. *carotae*), with the higher heritability observed for a hybrid population of Kuroda and Nantes cultivars. Evaluating different hybrid combinations, Simon and Strandberg (1998) suggested that a preponderance of additive variation with some dominant gene action and epistasis may contribute to resistance to Alternaria leaf blight. In 2009, Le Clerc et al. (2009) confirmed the polygenic nature of resistance to this disease, with identification of three quantitative trait loci (QTL) regions in a population of $F_{2:3}$ progeny. The phenotypic variation explained by each QTL ranged from 10 to 23%. Some QTLs were only detected in the tunnel trial or field trial, and only at one screening date, suggesting that expression of these QTLs might be influenced by the environment, with a delay in expression after infection. Two other populations with different genetic backgrounds were evaluated under field conditions over two years, from which 11 QTLs were identified (Le Clerc et al. 2015b). Complementarity between the parental origins of the favorable alleles at each QTL provides potential opportunities for breeders to combine resistance in one genotype in an effort to achieve higher levels of resistance. It would be valuable to understand the mechanisms underlying these QTLs in order to select those with complementary actions; e.g., some QTLs may delay penetration of carrot foliar tissue by the pathogen, while some QTLs may confine the pathogen after penetration into the leaf.

Little information is available on the genetic nature of resistance in carrot lines to *C. carotae* and *X. hortorum* pv. *carotae*. Lebeda et al. (1988) suggested that heredity of resistance to *C. carotae* could be oligogenic, with different degrees of phenotypic expression. Using glasshouse experiments, Angel and Gabelman (1968) found that a single dominant gene determined resistance of inbred line WCR 1.

Comprehension of the mechanisms underlying resistance to foliar diseases of carrot is important to develop durable and highly resistant cultivars, i.e., by combining resistance mechanisms. Boedo et al. (2008) characterized the different stages of fungal infection and development in carrot leaves using a resistant and a susceptible cultivar to *A. dauci*. Based on scanning electron microscopy, differences in *A. dauci* development between the two cultivars were only obvious 21 days post-inoculation (dpi). In contrast, the fungus was able to invade the leaf tissues of the susceptible cultivar rapidly. This was supported by significantly greater fungal biomass detected in leaves of the susceptible cultivar than leaves of the resistant cultivar, measured by quantitative, real-time PCR assay. The result was confirmed by Boedo et al. (2010) with two partially resistant cultivars in which significantly less fungal biomass was detected 15 dpi than in a susceptible cultivar. Investigating the germination of *A. dauci* conidia on carrot leaves *in vivo*, they also highlighted that a greater mean number of germ tubes per conidium (up to 3.42 ± 0.35) was observed on the two partially resistant cultivars than on the susceptible cultivar (1.26 ± 0.18). Penetration of the fungus into the epidermis of the susceptible cultivar was rapid. Several germ tubes per conidium were produced by spores infecting the resistant cultivar, which might reflect multiple attempts by the fungus to penetrate the epidermis.

Since plant resistance to pathogens can be due, in part, to production of secondary metabolites, Lecomte et al. (2012) investigated the role of 6-methoxymellein (6-MM) and falcarindiol in the resistance of carrot lines to *A. dauci*. After inoculation with *A. dauci*, significantly faster production of 6-MM in the resistant cultivar Bolero than the susceptible cultivar Presto suggested that this phytoalexin was involved in the resistance reaction by slowing pathogen spread, but with relatively limited effect. Falcarindiol was more active than 6-MM at reducing in vitro development of the fungus and was able to induce permeabilization of the *A. dauci* plasma membrane. Greater accumulation of this metabolite in leaves of Bolero than Presto suggested this metabolite contributes to resistance to the fungus. Lecomte et al. (2014) demonstrated that carrot tolerance to toxins produced by *A. dauci* could be another component of partial resistance. Embryogenic cellular cultures from carrot genotypes with different level of resistance were challenged with fungal extracts. A differential response between resistant and susceptible cultivars was demonstrated with a positive correlation between resistance to the fungus on the whole plant and resistance at the cellular level to fungal exudates. This suggested that phytotoxic compounds were present in the exudates and should be characterized. In 2018, Courtial et al. identified a new phytotoxin named aldaulactone that was shown to mimic the effect of fungal extracts on carrot embryogenic cell cultures, but at lower toxicity. Aldaulactone could be a major toxin produced by the fungus. Cellular targets of aldaulactone should be investigated. In order to identify compounds potentially associated with resistance to *A. dauci*, Koutouan et al. (2018) compared the leaf metabolome of four carrot genotypes with different levels of resistance, and the metabolomes of a set of resistant and susceptible progenies, by bulk segregant analysis. Based on results for the two parental lines, luteolin 7-*O*-glucuronide and feruloylquinic acid were expressed and accumulated differentially to camphene, α-pinene, apigenin 4′-*O*-glucoside, luteolin 4′-*O*-glucoside, caryophyllene, and β-bisabolene in the susceptible and resistant bulk populations. Analyses are in progress to characterize the involvement and role of those secondary metabolites in resistance to *A. dauci*, and their association with QTLs previously identified by developing a metabolite QTL approach as well as microarray analysis to characterize gene expression in the metabolic pathways.

Mercier and Kuć (1996) studied the induction of systemic resistance in carrot to Cercospora leaf spot by inoculation with *C. carotae*. Interestingly, newly emerged leaves of carrot plants previously inoculated with *C. carotae* developed significantly fewer lesions than comparable leaves of the control plants after challenge with the pathogen, suggesting that the foliar pathogen induced defense mechanisms in carrot leaves.

Instead of classical screening of wild and cultivated carrot accessions to identify new sources of resistance, Arbizu et al. (2017) suggested relying on prediction approaches by investigating the association between *Daucus* clades and severity ratings for Alternaria leaf blight. Using a phylogenetic linear regression model, they identified plant height as the best explanatory variable to predict resistance to this disease based on a study of 106 accessions of wild and cultivated *Daucus* spp. and related genera. The authors concluded that *D. carota* subsp. *capillifolius* and *D. c.* subsp. *maximus*, as well as *D. crinitus*, may provide new sources of resistance. They established the feasibility of hybridization of these species and subspecies with the cultivated carrot.

Strategies to develop transgenic plants exhibiting enhanced resistance to foliar fungal and bacterial pathogens of carrot have been evaluated. Lysozymes of plant origin have been suggested to be protective and defensive against bacterial or fungal pathogens. The human lysozyme cleaves peptidoglycan in the bacterial cell wall or chitin in the fungal cell wall. The human lysozyme gene was used to transform carrot via *Agrobacterium tumefaciens*, resulting in transgenic plants with enhanced resistance to *A. dauci* (Takaichi and Oeda 2000). Punja (2005) obtained two transgenic carrot lines via transformation with *A. tumefaciens* to express a thaumatin-like protein. Both lines had significantly less severe disease caused by *A. dauci*, *A. radicina*, *B. cinerea*, and *S. sclerotiorum*. Carrot transgenic plants expressing a microbial factor 3 (MF3) gene from a plant-growth promoting rhizobacterium, *Pseudomonas fluorescence*, were evaluated for resistance to *A. dauci*, *A. radicina*, and *B. cinerea* (Baranski et al. 2007). The homology of MF3 to FKB proteins suggested that MF3 is involved in the signaling pathway affecting induced systemic resistance. The transgenic plants had significantly enhanced resistance to the three pathogens (by 20–40%) in comparison to non-transformed plants. Less convincing results were obtained from the transformation of polyethylene glycol carrot protoplast with a chitinase gene. While slower disease progress caused by *A. dauci* was observed for two of the clones, a third clone was more susceptible than the control plants (Baranski et al. 2008). Wally et al. (2009a) considered it more efficient to manipulate the plant innate defense signaling pathways by controlling a large number of induced genes instead of using traditional gene expression strategies (i.e., relying on the expression of a single pathogenesis-related protein) to develop transgenic plants highly resistant to a range of pathogens. Modification of systemic acquired resistance was obtained through overexpression of the controlling gene NPR1 introduced into a carrot cultivar. The transgenic lines challenged with *X. hortorum* pv. *carotae* exhibited an 80% reduction in disease severity and 35–50% reduction in disease severity when challenged with *B. cinerea*, *A. radicina*, and *S. sclerotiorum*, confirming that this strategy offered the ability to control a range of pathogens. Klimek-Chodacka et al. (2018) reported the first use of the CRISPR/Cas9 system for efficient, site-targeted mutagenesis of the carrot genome, which offers new perspectives for improving carrot resistance to diseases such as those caused by foliar fungal and bacterial pathogens.

18.2.2.2 Powdery Mildew (*Erysiphe heraclei*, *Leveillula lanuginosa*, *L. taurica*)

Two types of powdery mildew can occur on carrot, the common *Oidium* type caused by *Erysiphe heraclei* (synonyms *E. polygoni* and *E. umbelliferarum*), and the less common *Oidiopsis* type caused by *Leveillula lanuginosa* and *L. taurica* (Aegerter 2002). *Erysiphe heraclei* has been documented wherever carrots are grown, but tends to be most severe in warm, semiarid regions or seasons. The severity of powdery mildew is influenced by weather conditions, growth stage of the crop at the onset of disease, production practices, and cultivar (Abercrombie and Finch 1976; Aegerter 2002; du Toit and Derie 2008; Palti 1975). Powdery mildew tends to be most severe on susceptible cultivars or parent lines grown with drip or furrow irrigation in warm and semiarid regions. Severe leaf infections can impede mechanical harvesters that pull roots out of the ground by the tops. The disease can be severe in greenhouses (Geary and Wall 1976). *Leveillula* spp. are favored by very warm, very dry climates and have been documented infecting carrot primarily in the Middle East, Armenia, India, Kazakhstan, and other countries in central Asia and Mediterranean regions of Europe and Africa. In Israel, these species only occur in the driest parts of that country (Palti 1975).

Erysiphe heraclei is ectotypic on carrot; i.e., the fungus grows externally on the plant surface and produces haustoria that penetrate the host epidermal cells (Aegerter 2002). The fungus produces white mycelium and sporulation on any aboveground part of carrot plants, including leaves, petioles, flower stalks, bracts, and umbels (Aegerter 2002) (Fig. 18.9). Severely infected foliage can become chlorotic and leaves may senesce prematurely. In contrast, *L. lanuginosa* and *L. taurica* produce mycelium that is both endophytic and ectotopic. These two fungi produce conidia at the ends of long conidiophores that protrude through stomata. Powdery mildew caused by *Leveillula* spp. usually appears as pale yellow lesions on the upper leaf surface, with white sporulation on the lower leaf surface (Aegerter 2002). Lesions can appear angular because the leaf veins limit expansion of infections. As the disease progresses, sporulation can develop on the upper leaf surface and chlorotic areas become necrotic. The white fungal growth is not nearly as conspicuous as that caused by *E. heraclei*.

Conidia produced by powdery mildew fungi are carried long distances by air movement (Aegerter 2002). Unlike most fungal plant pathogens, the spores do not require free water for germination and infection of plants, but require high humidity and moderate temperatures. Powdery mildews are most severe in shaded areas as sunlight damages conidia and mycelium. Therefore, powdery mildew tends to start on the older foliage and spread to newer leaves, reflecting greater humidity and shading lower in the canopy. Carrot plants become more susceptible to powdery mildew as they age

Fig. 18.9 Severe powdery mildew on a carrot plant infected with *Erysiphe heraclei* and transplanted into a field trial to increase disease pressure (**a**) and colonization of an umbel by *E. heraclei* in a carrot seed crop (**b**) (Lindsey du Toit, Washington State University)

(Aegerter 2002). Sporulation can occur within 7–14 days of infection, with symptoms developing rapidly under favorable conditions for this highly polycyclic disease. The disease can be particularly severe in carrot seed crops because the tall, dense canopy results in highly favorable conditions after canopy closure (du Toit and Derie 2008; du Toit et al. 2009). There is no evidence the powdery mildew pathogens infect carrot seed, but seed lots can be contaminated with cleistothecia (sexual fruiting bodies) (Boerema et al. 1963; Vasudeva 1963).

Erysiphe heraclei can infect at least 86 plant species in Apiaceae, although cross-inoculation studies suggest a degree of host specialization (Aegerter 2002; Braun 1987; Cunnington et al. 2008; Glawe et al. 2005; Hammarlund 1925; Marras 1962). Therefore, a powdery mildew on one host species might not serve as a source of inoculum for another plant species. Nonetheless, some isolates have been demonstrated to infect multiple plant species and genera, including differences in degree of virulence on different Apiaceae genera and species (e.g., Cunnington et al. 2008; Koike and Saenz 1994, 1997). Similarly, *L. lanuginosa* can infect different Apiaceae genera and species, with evidence of host specialization among isolates (e.g., Cirulli 1975). *Leveillula taurica* has an even broader host range, infecting many plant families with varying degrees of host specialization having been reported (Aegerter 2002; Braun 1987; Palti 1975).

Resistance screening: Partial resistance to powdery mildew was identified in four subspecies of *Daucus* (Bonnet 1977; Umiel et al. 1975) used to initiate breeding for resistance. Single-gene, dominant resistance to powdery mildew was identified by Bonnet (1983) in *D. c.* subsp. *dentatus*. A backcross study with the susceptible cultivar Touchon demonstrated that resistance was controlled by a single dominant gene, *Eh*. Resistant lines were selected based on an orange root color. Bonnet (1983) also documented *D. siculus* as well as the *D. carota* cultivar Bauers Kieler Rote as potential sources of resistance to powdery mildew. Lebeda and Coufal (1987) screened 111 cultivars of *D. c.* subsp. *sativus* under field conditions in Czechoslovakia for resistance to natural infection by *E. heraclei*. One cultivar, 'Gavrilovskaya' from the former USSR, remained asymptomatic, and 13 developed very limited powdery mildew. They categorized about half of the 111 cultivars as 'moderately vulnerable' and interpreted the results to indicate possible incomplete dominance and quantitative resistance to powdery mildew. Takaichi and Oeda (2000) developed transgenic versions of the carrot cultivars Kurodagosun and Nantes Scarlet using *A. tumefaciens* to transfer a plasmid containing the human lysozyme under control of the constitutive

CaMV 35S promoter. As detailed above, the enzyme has lytic activity against plant pathogenic fungi and bacteria. Two of the transgenic plants of 'Nantes Scarlet' displayed partial resistance to powdery mildew, and one was also partially resistant to Alternaria leaf blight. The increase in resistance in these lines was correlated with an increase in the production of the human lysozyme. Wally et al. (2009a) developed transgenic lines of the carrot cultivar Nantes Coreless containing the *Arabidopsis* (*At*) *NPR1* gene (non-expressor of PR genes) for enhancing the plant innate defense system. Two transgenic lines, NPR1-I and NPR1-XI, displayed increased duration and intensity of expression of DcPR-1, -2, and -5 genes when the lines were treated with purified cell wall fragments of the white mold pathogen, *S. sclerotiorum* or with 2,6-dichloroisonicotinic acid. Leaves of these lines had 90% less severe powdery mildew when inoculated with *E. heraclei* compared to non-transgenic lines of the cultivar. Overexpression of the systemic acquired resistance (SAR) master switch, *NPR1*, resulted in increased resistance to powdery mildew as well as necrotrophic pathogens such as *B. cinerea*, *A. radicina*, *S. sclerotiorum*, and *X. hortorum* pv. *carotae* (Wally et al. 2009a).

Although Lebeda and Coufal (1987) relied on natural infection to screen cultivars for resistance in field trials in Czechoslovakia, adequate disease pressure only occurred in one of three field seasons to assess cultivars for differences in susceptibility to powdery mildew. If conditions remain relatively warm and dry, fairly uniform powdery mildew pressure can be generated readily under field conditions or in greenhouses using plants of highly susceptible cultivars infected with powdery mildew as 'spreader' plants. Inoculum can be maintained by growing infected plants in a greenhouse and adding new plants alongside infected plants at regular intervals to keep propagating the pathogen. Placing plants close together increases relative humidity for promoting powdery mildew. Infected plants can be transplanted into field sites to establish powdery mildew nurseries. Using this protocol, du Toit and Derie (2008) and du Toit et al. (2009) were able to get very severe powdery mildew pressure in carrot seed crops to assess the impact of this disease on carrot seed yields.

18.2.2.3 Virus Diseases

Approximately 14 virus diseases of carrot are recognized (Moran et al. 2002; Nuñez and Davis

Table 18.1 Carrot diseases caused by viruses (Nuñez and Davis 2016)

Disease	Virus name	Virus genus
Alfalfa mosaic or calico	*Alfalfa mosaic virus* (AMV)	Alfamovirus
Carrot latent	*Carrot latent virus* (CLtV)	Nucleorhabdovirus
Carrot mottle	*Carrot mottle virus* (CMoV)	Umbravirus
Carrot red leaf	*Carrot red leaf virus* (CRLV)	Polerovirus
Carrot thin leaf	*Carrot thin leaf virus* (CTLV)	Potyvirus
Carrot yellow leaf	Coinfection with *Carrot yellow leaf* (CYLV) and	Closterovirus
	Carrot virus Y (CarVY)	Potyvirus
Celery mosaic	*Celery mosaic virus* (CeMV)	Potyvirus
Cucumber mosaic	*Cucumber mosaic virus* (CMV)	Cucumovirus
Curly top	*Beet curly top virus* (BCTV)	Curtovirus
Parsnip yellow fleck	*Parsnip yellow fleck virus* (PYFV)	Sequivirus
Tobacco necrosis	*Tobacco necrosis virus* (TNV)	Necrovirus
Tomato spotted wilt	*Tomato spotted wilt virus* (TSWV)	Tospovirus
Carrot motley dwarf	Coinfection with CRLV and CMoV	Polerovirus and Umbravirus

2016) (Table 18.1). Some of these diseases are of minor importance economically (e.g., AMV, CTLV, and TSWV), while others can cause significant decreases in root yields or seed production, including total loss (Lebeda and Coufal 1985; Nuñez and Davis 2016; Stein and Nothnagel 1995). Motley dwarf (CRLV and CMoV) is probably the most important and persistent virus disease of carrot (Waterhouse 1985; Watson and Sarjeant 1964). The majority of viruses infecting carrot are transmitted by aphids, and early efforts to screen for resistance to viruses and/or the vectors indicated a lack of highly effective resistance (Van Dijk and Bos 1985; Elnagar and Murant 1978). However, efforts to screen for variation in susceptibility to some of these viruses have demonstrated differences among carrot breeding lines, with the potential to improve the resistance of commercial cultivars.

Motley dwarf: Motley dwarf was first detected in Australia in 1948, and now occurs in most regions of carrot production in the world where the climate is relatively cool (Koike et al. 2002). The two viruses that cause motley dwarf also infect cilantro, dill, and parsley. Each virus can infect carrot plants alone, but motley dwarf only occurs when plants are infected with both CRLV and CMoV. The viruses are both transmitted by the willow-carrot aphid, *Cavariella aegopodii*, which transmits the viruses in a circulative, non-propagative manner. If plants are infected with CRLV alone, CRLV can be transmitted by aphids but not transmitted mechanically. The opposite is true for carrots infected only with CMoV, which can be transmitted mechanically but not by aphids in the absence of CRLV (Koike et al. 2002; Waterhouse and Murant 1983). A third virus-like RNA was identified in motley dwarf-infected carrots in California (Watson et al. 1998). This small, CRLV-associated RNA (CRLVaRNA) is transmitted by the carrot-willow aphid with CRLV and CMoV, although it is not known if this RNA affects symptom expression. The viruses have a relatively narrow host range within Apiaceae, as does the aphid vector, which preferentially feeds on carrot. As a result, older carrot plantings or overwintered carrots infected with motley dwarf are the most common inoculum sources (Howell and Mink 1977; Watson and Falk 1994). Losses to motley dwarf can be severe if infection occurs early in the season and carrots are grown when conditions are relatively cool (15–20 °C) with low light conditions (e.g., late winter and early spring plantings in California, and plantings in the mild, maritime region of the Pacific Northwest USA). Symptoms vary depending on the age of plants at the time of infection, but typically include reddening and yellowing of leaves along with stunting if plants are infected at an early growth stage. Symptoms can resemble those caused by nutrient deficiency. Roots can be affected severely, and plants may die. Symptoms on older plants tend to be less severe, and plants may be symptomless at temperatures >24 °C.

CarVY: CarVY was first described on carrot in Australia in 2002 (Moran et al. 2002) and is now well established in that country (Jones 2005). The virus has not been reported in other countries. CarVY is transmitted non-persistently by at least 14 aphid species, with infected carrot crops and volunteer carrot plants serving as the main reservoir for newly planted crops. This is particularly problematic in Australia where carrots can be grown year-round, typically using sequential plantings. CarVY causes chlorotic mottle of the foliage, a feathery appearance of foliage from increased subdivision of leaflets, marginal necrosis or reddening of leaves, and stunting (Latham and Jones 2004). Roots become stubby and knobby if plants are infected at an early growth stage, and severely symptomatic roots are unmarketable. There is no evidence of CarVY being seedborne or seed transmitted in carrot (Jones 2005). Carrot was thought to be the only known host until Jones (2005) demonstrated that CarVY can infect *Chenopodium amaranticolor* and *C. quinoa*, although neither became infected systemically. Carrot appears to be the primary source of inoculum for spread of CarVY in Australia.

PYFV: PYFV has caused significant economic losses in carrot root production in the UK and Germany, and carrot seed production in the Netherlands (Murant and Spence 2002). The virus infects carrot, celery, chervil, coriander, dill, and parsnip, as well as wild chervil

(*Anthriscus sylvestris*) and cow parsley (*Heracleum sphondylium*). There are two serotypes of PYRV, the parsnip serotype which infects celery, cow parsley, and parsnip; and the anthriscus serotype which infects carrot, chervil, coriander, cow parsley, and dill. Infection of young carrot plants by the anthriscus serotype can cause severe stunting, necrosis, and death of plants. Infection of older plants leads to mottled foliage with yellow flecks. Infected plants can develop misshapen rots and/or secondary roots. PYFV causes premature dieback in carrot seed crops in the Netherlands (Van Dijk and Bos 1985). The virus is transmitted semi-persistently by the carrot-willow aphid, *C. aegopodii*, as well as *C. pastinacae* but not by *C. theobaldi* or the green peach aphid, *Myzus persicae*. A helper virus, *Anthriscus yellows virus* (AYV), must be present in source plants of anthriscus for transmission of PYFV. AYV is phloem limited, unlike PYFV, is not transmissible mechanically, and is in the family Sequiviridae. PYFV is not seed transmitted. Aphids fed on plants infected with AYV alone, and then on plants infected with PYFV alone, were able to transmit both viruses, but not if fed in the reverse order. Infective aphid adults can transmit the viruses for up to four days, but nymphs cease to transmit the viruses after molting. AYV and PYFV host ranges overlap, and only plant species that are a host to both serve as a source for aphid transmission (Murant and Spence 2002).

Resistance screening: Motley dwarf: A wide range in responses of carrot cultivars to motley dwarf has been documented (Fig. 18.10), and there is genetic resistance to motley dwarf (Koike et al. 2002). In California, the cultivar Danvers was very susceptible while CVC-14 was partially resistant (Watson and Falk 1994). Differentiating resistance of cultivars to motley dwarf versus resistance or tolerance to the carrot-willow aphid vector complicates efforts to screen for resistance (Dunn 1970). Dunn (1970) noted that the cultivar Autumn was highly susceptible to the aphid but

Fig. 18.10 Carrot cultivars can vary widely response to motley dwarf, caused by coinfection with *Carrot red leaf virus* (CRLV) and *Carrot mottle virus* (CMoV), as illustrated in these small breeding plots in western Washington State in 2017 and 2018, with dramatic differences observed in severity of symptoms of breeding lines in adjacent plots (Lindsey du Toit, Washington State University)

displayed partial resistance to motley dwarf in multiple trials, whereas Nantes was not as tolerant of the aphid as the other cultivars screened and was highly susceptible to motley dwarf. Tomlinson (1965) reported 'Kurnella Strongtop' and 'Western Red' to be tolerant of motley dwarf, and Kinsella (1966) reported similar results for 'Western Red,' despite these cultivars being well colonized by the vector *C. aegopodii*. Tomlinson (1965) noted severe motley dwarf symptoms on 'Nantes' and the Chantenay cultivars Early Market and Cluseed Stumprooted. Dunn (1970) also demonstrated both to be highly intolerant of *C. aegopodii*, whereas 'Berlikum' appeared the least susceptible to the aphid of seven cultivars evaluated.

CarVY: All carrot cultivars commonly grown in Australia were reported to be susceptible to CarVY, but symptoms varied in severity among cultivars (Latham and Jones 2004). Jones (2005) inoculated plants of 22 Apiaceae species in a glasshouse using viruliferous green peach aphids (*M. persicae*). Aphids were maintained on canola inside cages at 15–20 °C. For inoculation, the aphids were starved for two hours and then fed on infected leaves of the carrot cultivar Stefano for 10 min and transferred to healthy plants of the target species using a paint brush. The aphids fed for one hour before they were killed with insecticide. Jones (2005) observed systemic infection by CarVY on carrot, four other *Daucus* species (*D. bicolor*, *D. hispidifolius*, *D. muricatus*, and *D. littoralis*), five Apiaceae herbs (anise, chervil, coriander, cumin, and dill), a naturalized weed (Bishop's weed, *Ammi majus*), and two Australian native plants in Apiaceae (Australian carrot, *D. glochidiatus*, and native parsnip, *Trachymene pilosa*). Seven of the 22 host species became infected in field sites, with wide variation in type and severity of symptoms among host plants. An additional screen of 34 wild carrot germplasm accessions and 16 other *Daucus* spp. using viruliferous green peach aphids in a greenhouse again revealed wide variation in severity of symptoms and no complete or extreme resistance. Systemic CarVY infection was observed in plants of all of the following wild germplasm accessions: 21 accessions from the Polish germplasm collection, including 7 wild carrot accessions, 6 of *D. muricatus*, 2 of *D. bicolor*, and 6 of an unidentified *Daucus* sp.; and 29 from a UK collection, including 27 wild carrot accessions and 1 each of *D. hispidifolius* and *D. littoralis*. Some accessions had to be inoculated several times to establish infection, and some did not develop infection, suggesting these lines may have resistance to CarVY. The accessions also were screened in field trials in Australia, where symptoms were less diverse compared to the greenhouse tests.

PYFV: There do not appear to be any reports on screening for resistance to PYFV in carrot.

The genetic nature of resistance of carrot lines to virus diseases such as motley dwarf, CarVY, and other viruses remain to be determined. The variation in severity of symptoms observed among carrot lines infected with specific viruses suggests much potential for using molecular screening methods to identify resistance genes, including QTLs, associated with resistance to these viruses.

18.2.2.4 Diseases Caused by Phytoplasmas and Spiroplasmas

A number of phloem-limited mollicutes (phytoplasmas and spiroplasmas) can cause diseases of carrot. Aster yellows and beet leafhopper-transmitted virescence agent (BLTVA) yellows are phytoplasma diseases that affect a wide variety of wild and cultivated plant species, including carrot and more than 300 other vegetables, weeds, and ornamentals (Blomquist 2002). They are both vectored by leafhoppers. Although losses to these phytoplasmas tend to be sporadic in carrot, aster yellows has been found in carrot production regions worldwide, whereas BLTVA yellows occurs only in the western USA. The aster yellows phytoplasma is in the aster yellows group, 16SrI, and the BLTVA yellows phytoplasma is in subgroup A of the clover proliferation group, 16SrVI (Lee et al. 2006). Symptoms caused by these phytoplasmas are similar. Aster yellows symptoms start with chlorosis of leaf veins and progress to chlorosis of entire leaves. Infected leaves may be narrower than leaves of healthy plants. Dormant buds in the crown sprout, leading to upright, adventitious shoots. Older leaves become bronze, red, or purple and break readily, interfering with mechanical harvest using the tops (Blomquist 2002). The main root tends to stay small if plants are infected at an early stage, and the taproot of infected plants may develop prolific secondary roots (Fig. 18.11). In carrot seed crops, phyllody (development of leaf-like petals on the flowers) and virescence (greening of flowers) occur after bolting (Fig. 18.11). Similar symptoms develop on plants infected with the BLTVA yellows phytoplasma, but plants infected with this phytoplasma also tend to bolt (flower) prematurely, unlike plants infected with aster yellows, and the taproot may be thin, woody, and develop a proliferation of secondary roots (Fig. 18.11). Seed is not viable in umbels that develop virescence and phyllody.

Lee et al. (2006) documented natural infection of carrot plants in Washington State, USA by *Spiroplasma citri*. The symptomatic plants displayed chlorosis, purpling, and bronzing of foliage, rosette formation of the crown, stunting of shoots and roots, proliferation of fibrous secondary roots, and even development of multiple taproots. Infected plants were found in multiple processing carrot crops in central Washington. Some plants were co-infected with *S. citri* and the aster yellows phytoplasma or BLTVA yellows phytoplasma. *S. citri* also causes citrus greening of citrus in Florida and California.

Phytoplasmas and spiroplasmas are prokaryotes that infect the phloem sieve cells of host plants where they reproduce by budding or division (Blomquist 2002). They also reproduce in the bodies of their leafhopper vectors. These obligate pathogens cannot be cultured on agar media, so infection is confirmed using enzyme-linked immunosorbent assays (ELISAs) or polymerase chain reaction (PCR) assays with primers specific to the group of phytoplasma or spiroplasma. Aster yellows can be transmitted by many species of leafhoppers, but the aster leafhopper, *Macrosteles fascifrons*, is the most important vector (Blomquist 2002; Boivon 1994). In contrast, the BLTVA yellows phytoplasma and *S. citri* are acquired and transmitted by the beet leafhopper, *Circulifer tennellus*. Once infected, leafhoppers transmit these phytoplasmas and spiroplasma persistently and remain infective until they die.

Significant losses to aster yellows in carrot production occur periodically in the Midwestern USA, where aster leafhoppers migrate from southern states each spring, and where they acquire the aster yellows phytoplasma from

Fig. 18.11 Carrot plants infected with aster yellows or beet leafhopper-transmitted virescence agent (BLTVA) yellows phytoplasmas. Symptoms include yellow, purple, and/or bronze foliage (**a**), excessive secondary roots (**a**), stunting, and virescence and phyllody of umbels (**b**, healthy umbels on the left and infected umbels on the right) (Lindsey du Toit, Washington State University)

infected weeds and other crops. In the western and eastern USA, aster leafhoppers do not go through an annual migration, but local populations that survive the winter acquire phytoplasmas from infected crops and weeds. In the western USA, beet leafhoppers acquire the BLTVA yellows phytoplasma from infected wild vegetation in hills bordering cropland, and the vectors migrate into irrigated fields when wild vegetation dries during the dry season each summer. Neither of the phytoplasmas nor *S. citri* is seed transmitted in carrot. Female leafhoppers cannot transmit the pathogens to their offspring (Blomquist 2002).

Resistance screening and breeding: Gabelman et al. (1994) initiated a breeding program in 1982 to develop aster yellows resistance in carrot. They developed an aster yellows synthetic (AYSYN) population comprising four open-pollinated cultivars and five inbred lines by screening 200 carrot entries (open-pollinated cultivars and inbred lines) in the field. Since aster leafhoppers feed preferentially on lettuce, each 4-row bed of carrot lines was bordered by rows of lettuce to ensure adequate leafhopper pressure. Leafhoppers infected with phytoplasmas were reared in cages in a greenhouse and distributed evenly throughout the field in June and July. They rated plots visually in October for symptoms of aster yellows to calculate the incidence (percentage) of infection. The top 10% of the 200 lines was selected, and 189 roots from the lines were planted in a greenhouse after vernalization for pollination. Roots of 20 plants that flowered were selected, including five inbred lines (derived from crosses with W33, W263, W266, and W93), six inbreds from a double-cross of OSU260 with an F1 of an unnamed Russian line and W33, and four open-pollinated cultivars (Royal Chantenay, Scarlet Nantes, Gold King, and Nanco). Seed harvested from the crosses formed the AYSYN population, from which inbred lines were extracted using several schemes. Inbred lines also were developed from the Wisconsin carrot breeding program (WBP) using a third method detailed by Gabelman et al. (1994). The inbred W1-1 was developed by selecting four roots from inbred lines in the WBP that had good resistance to aster yellows, and intermating and inbreeding these lines for eight generations. Inbred lines derived from the AYSYN population were developed with three methods, from inbred progeny of the population, from AYSYN-derived inbred lines intermated with inbred selections from that population, and from intermating the AYSYN population with high-color inbred lines. AYSYN lines were inbred for at least five generations and then used to produce AYSYN hybrids. Field trials in 1990, 1991, and 1993 were used to assess the relative resistance to aster yellows of 26 selected lines compared to six commercial carrot cultivars, based on the visual rating of symptoms. Gabelman et al. (1994) detected significant differences among carrot genotypes with the resistant lines ranging from 2.5 to 35.3% infection/plot versus 12 to 42% for the standard cultivars. Many of the selected resistant lines displayed significantly less incidence of aster yellows. 'Scarlet Nantes,' 'Royal Chantenay,' and 'Gold King' exhibited the least aster yellows (average 15.3% infection), whereas 'Danvers 126,' 'Py-60,' and 'Spartan Bonus 60' had greater incidences of infected plants (average 33.3%). Leafhopper populations were comparable across the diverse genotypes, suggesting that resistance was not associated with reduced feeding of the vector. Feeding preferences on specific carrot genotypes were not observed. The most successful breeding scheme for resistance was using the AYSYN population as a source of inbreds, suggesting that combining a synthetic population with established inbred lines was highly effective.

The genetic basis of aster yellows resistance selected by Gabelman et al. (1994) has not been determined, but the inbreeding in that study may have revealed recessive alleles that contribute to resistance. Selection in a naturally infested and inoculated field site increased the ability to screen for resistance by maintaining strong selection pressure. The results suggest resistance to aster yellows is quantitative based on the phenotypic distribution and the influence of environmental conditions on the disease reactions observed.

18.3 Carrot Nematode and Insect Pests

18.3.1 Root-Knot Nematodes (*Meloidogyne* spp.)

Among the root-knot nematode (RKN) species (*Meloidogyne* spp.) attacking carrots, *Meloidogyne hapla* Chitwood, *M. javanica* (Treub) Chitwood, and *M. incognita* (Kofoid and White) Chitwood are of worldwide economic importance for carrot cultivation. Losses up to 100% may occur with yield reduction and shape deformation, i.e., taproot forking and galling that render carrots unmarketable (Roberts and Mullens 2002) (Fig. 18.12). *M. hapla* is the predominant species in fields in cooler production areas (temperate areas or at higher altitudes in subtropical and tropical areas), while *M. javanica* and *M. incognita* are major pests in warmer areas (Bridge and Starr 2007; Parsons et al. 2015). Even less widespread are *M. chitwoodi* and *M. fallax*, which can cause considerable losses by deforming the carrot taproot. *M. chitwoodi* causes severe galling near the lenticels, resulting in a rough surface to the carrot taproot (Wesemael and Moens 2008). Control of RKN is dependent on various methods, including soil applications of nematicides, crop rotation to non-host species, and flooding. However, genetic resistance appears to be the most effective and environmentally sound method to reduce damage caused by RKN.

Significant genetic variability for resistance to nematodes has been identified in carrot germplasm (Fig. 18.12). Twenty-one cultivars and breeding lines with various root types were evaluated for their responses to *M. hapla* under controlled greenhouse conditions or in field conditions by Yarger and Baker (1981). Gall formation and nematode tolerance appeared to be associated with root type; e.g., Nantes and Long Chantenay root types exhibited tolerance in general while Danvers root types were mostly susceptible. While some cultivars are tolerant or partially resistant to *M. hapla*, based on the taproots being less susceptible to deformation, others exhibit tolerance because roots become parasitized but reproduction of the nematode is inhibited (Wang and Goldman 1996).

Huang et al. (1986) established a reliable method for evaluating varietal responses to *M. javanica* in the greenhouse, highlighting that primary root galling in the carrot seedling stage was an efficient parameter for resistance evaluation to this nematode species. While it appeared that ‘Brasilia’ and other tropical carrot cultivars had the same resistance level to *M. incognita* and *M. javanica*, Nantes and Kuroda groups showed more severe symptoms, suggesting different resistance mechanisms to these two nematode species compared with *M. hapla*. The resistance

(a) (b)

Fig. 18.12 Symptoms on carrot roots infected with the root-knot nematode, *Meloidogyne incognita* (**a**), and comparison of reactions of a resistant breeding line (left) versus a highly susceptible carrot line (right) (**b**) (Lindsey du Toit, Washington State University)

in the cultivar Brasilia was associated with retarded nematode penetration, development, and egg production, and fast plant growth that culminated in a low nematode population density (Huang 1986). Yunhee et al. (2014) screened 170 Korean carrot lines for resistance to *M. incognita* race 1. Among them, 61 resistant lines represented potential genetic resources for breeders. While susceptible root tissues infected with *M. incognita* seven weeks after inoculation showed the formation of well-developed giant cells surrounding the nematodes, root tissues resistant to this RKN did not show giant cells but, instead, relatively small modified cells around the infecting nematode (Yunhee et al. 2014). The expression of resistance genes against RKN could be responsible for the formation of necrotic layers around the modified cells formed in resistant carrot root tissues. 'DR-333', a cultivar commonly grown in north India, was found to be tolerant to *M. incognita* (Siddiqui et al. 2011).

The diversity of nematodes should be considered in screening for resistance, as this can affect responses of cultivars. This is especially true for *M. chitwoodi* as three races have been identified in the USA (Wesemael and Moens 2008). Fifteen carrot cultivars, including Imperator, Nantes, Chantenay and hybrid carrot types, were identified as good hosts for *M. hapla* while their reaction to *M. chitwoodi* varied depending on the inoculated race (Santo et al. 1988). Among the 15 cultivars, 13 were moderate to good hosts of *M. chitwoodi* race 1. On the contrary, all of the cultivars, except Orlando Gold (moderate host), were non-hosts or poor hosts of *M. chitwoodi* race 2. Wesemael and Moens (2008) identified great variation in host status of 19 carrot cultivars to *M. chitwoodi* in glasshouse conditions, with some cultivars (Berlanda, Bolero, Chantenay, Nantucket, and Parmex) observed not to have egg masses, which indicated potential resistance. In 2009, Charchar et al. (2009) identified *M. polycephannulata*, a new species of RKN able to parasitize two important vegetables crops (i.e., carrot and tomato) that are cultivated extensively in Brazil. The identification of resistant cultivars of carrot for use in crop rotations is becoming a necessity for control of this RKN.

Knowledge of the genetic nature of resistance is essential to develop RKN-resistant carrot cultivars. A major part of this type of work has been done with *M. javanica* and *M. incognita*. Huang et al. (1986) found relatively low narrow-sense heritabilities of resistance to *M. javanica*, from 0.16 to 0.48 for root galling and from 0.31 to 0.35 for egg mass production, depending on the cultivar. Simon et al. (2000) identified simply inherited dominant resistance to *M. javanica* (one or two duplicated dominant genes at a single locus, *Mj-1*) in a selection of the carrot variety Brasilia, which was also effective against *M. incognita* based on field tests. Using bulk segregant analysis, Boiteux et al. (2004) identified RAPD markers linked to the *Mj-1* locus that are, therefore, useful for marker-assisted selection to develop hybrids resistant to *M. javanica*. Investigating the association between expression of resistance to this nematode and locus dosage, Boiteux et al. (2004) suggested that phenotypic resistance is affected by *Mj-1* locus dosage, and that the *Mj-1* region may contain a quantitative resistance locus. Screening for additional sources of resistance, Ali et al. (2014) identified a new source of resistance to *M. javanica* in a segregating population bred for resistance to both *M. javanica* and *M. incognita*. A single major gene, designated *Mj-2* and identified on the same chromosome as *Mj-1* but not at the same locus, contributes to this resistance. Working with resistance to *M. incognita*, Parsons et al. (2015) identified five QTLs from the analysis of three segregating populations, with the QTLs located on carrot chromosomes 1, 2, 4, 8, and 9. One QTL, common to all three populations, is on chromosome 8 and co-localized with *Mj-1*, which confers resistance to *M. javanica*. Broad-sense heritability for resistance to *M. incognita* was calculated to be 0.33 and 0.25 in two carrot populations derived from a cross among three sources of resistance from Syria, South America, and Europe (Parsons et al. 2015).

Wang and Goldman (1996) identified two homozygous recessive genes with epistatic control of *M. hapla* resistance, suggesting that this resistance may be relatively simply introgressed into inbred lines via backcrossing. However, previous studies have reported both quantitative and qualitative resistance to nematode infection. More recently, Yunhee et al. (2014) reported that resistance to *M. incognita* may be governed by one or a few genes.

Commercial cultivars have been released with resistance genes to *Meloidogyne* species from germplasm of *Daucus* spp. and lines derived from the cross 'Brasília 1252' × 'B6274' (Simon et al. 2000), or a population derived from the resistant cultivar Brasília (Vieira et al. 2003) with resistance to both *M. javanica* and *M. incognita*. While 'Brasilia' was developed in 1981, Brasília-derived germplasm is still considered one of the most promising sources of stable, wide-spectrum field resistance to RKN species in carrot (Vieira et al. 2003). In 2009, Embrapa Vegetables in Brazil (Pinheiro et al. 2011) released the cultivar BRS Planalto, which has a high level of tolerance to RKN.

As highlighted by Ali et al. (2014), conventional breeding protocols for developing RKN resistance often include greenhouse and extensive field evaluations for phenotyping, which are time-consuming and labor-intensive. A promising strategy could be the application of RNA interference (RNAi) to confer resistance to host plants engineered to express dsRNA and small interfering RNAs (siRNA) to target and silence specific nematode genes (Roderick et al. 2018). In this manner, Tan et al. (2013) were able to reduce the expression level of two genes involved in structural integrity and proper muscle contraction in *Pratylenchus thornei* and *P. zeae*, by soaking nematodes in dsRNA.

18.3.2 Carrot Fly or Carrot Rust Fly (*Psila rosae*)

The carrot fly, *Psila rosae* (F.), is one of the most significant insect pests of carrot and other Apiaceae crops, with the host range restricted to Apiaceae species (Hardman and Ellis 1982). Females search out carrot plants to lay eggs around the crown. The larvae migrate down to feed on carrot roots, making the root unmarketable due to larval damage (Ellis 1999) (Fig. 18.13). Losses are mostly due to a reduction in quality rather than yield (Dufault and Coaker 1987). While antixenosis reduces the initial infestation of flies and contributes more to resistance than antibiosis against larvae in some Apiaceae species, it was shown to be the opposite for carrot varieties (Degen et al. 1999a, b, c).

Carrot lines have been screened for susceptibility to carrot fly. Ellis et al. (1978) reported that many evaluations of carrot cultivars for resistance to carrot fly have not led to consistent results among experiments. When comparing 11 cultivars for the effects of different rooting systems on insecticide efficiency against the carroty fly, the authors identified 'Royal Chantenay' and Speed's 'Norfolk Giant' as having the extremes of a range of resistance to susceptibility among the 11 cultivars. Compared to other methods of assessment, the damage index based on recording weights and numbers of roots in four damage categories discriminated among the cultivars efficiently, especially when carrot fly attack was severe. Michalik and Wiech (2000) divided carrot genotypes into four groups according to the percentage of damaged roots, and identified five breeding lines as resistant to this pest. The greatest resistance found in cultivated carrot resulted in a 50% reduction in damage by *P. rosae*. Several *Daucus* spp. evaluated for reaction to carrot fly also had potential for hybridizing with cultivated carrots to develop cultivars with greater level of resistance (Ellis 1999). Since the first substantial source of resistance discovered in the Nantes cultivar Sytan (partial resistance based on antibiosis) by Ellis and Hardman (1981), breeding programs have been initiated and resistant F3 and F4 lines resulting from the crosses between carrot cultivars and *D. capillifolius* have been released. Nine inbred lines derived from a cross between two carrot cultivars, Sytan and Long Chantenay, with moderate level of resistance to the carrot fly were also released in 1991 with the aim of developing new cultivars (Ellis et al. 1991). Cultivars and wild accessions have been used in breeding programs as source of resistance, resulting in release of the partially resistant cultivar Flyaway

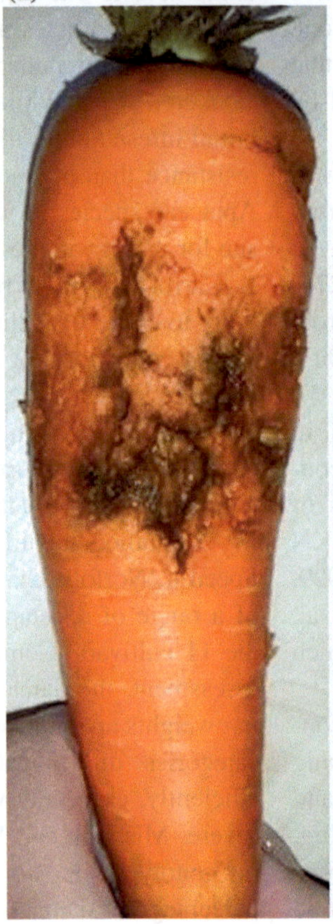

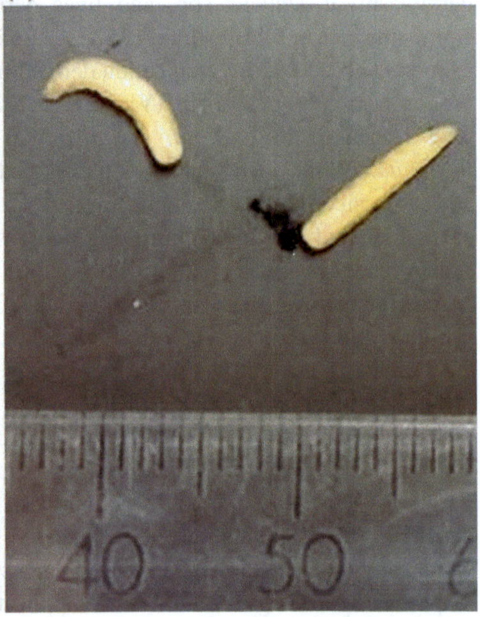

Fig. 18.13 Damage to a carrot root by the rust fly, *Psila rosae* (**a**), and larvae (maggots) of the rust fly (**b**) (Anita Sequeira, Greenway Farms, South Africa—the ruler is demarcated in millimeters [short lines] and centimeters [numbered lines])

and lines with significantly greater resistance than Sytan (Ellis 1999; Simlat et al. 2013).

Understanding the biological, biochemical, and genetic mechanisms of resistance to the carrot fly could help breeders identify appropriate crosses to make between valuable genotypes. The chemical signature of the leaf surface is probably a complex mixture of more than one key compound (Guerin et al. 1983; Städler and Buser 1984). However, many volatile oviposition stimulants for the carrot fly have been identified from carrot leaves (Guerin and Visser 1980; Städler and Buser 1984). Most of these compounds (propenylbenzenes, coumarins, and polacetylene) are powerful defensive compounds against bacteria, fungi, plants, and herbivores (Städler and Buser 1984). Some studies have attempted to characterize the underlying mechanisms involved in resistance to the carrot fly. 'Oviposition non-preference' has been suggested as one mechanism. Guerin and Stadler (1984) evaluated the impact of the quantity of foliar chemostimulants produced by four cultivars on this parameter. The results suggested olfactory and contact chemostimuli may be involved in preference shown by the carrot fly for specific cultivars, but other factors such as leaf color and morphological characteristics were also involved in host selection and oviposition. While some varieties had antixenosis resistance, resulting in fewer eggs being laid, the principal basis of resistance was found in the root of plants (Guerin and Ryan 1983). The concentration of chlorogenic acid in carrot roots was also correlated with susceptibility to damage by carrot fly larvae (Cole 1985). However, when selected lines of 'Sytan' were screened for levels of this compound, discrimination among the lines was not a reliable indicator of resistance, suggesting that this compound was not the chemical basis of resistance (Ellis 1999). More recently, Simlat et al. (2013) demonstrated a positive correlation between the ratio of certain phenolic compounds and resistant carrot phenotypes. Greater levels of expression of *PAL1* and *PAL3* were also observed in the most resistant carrot lines.

In conclusion, many sources of resistance to the carrot fly have been identified in wild and cultivated materials, and potentially could be combined to develop even better resistance. Ellis (1999) suggested that resistance to this pest is inherited polygenically, but few studies have investigated the genetic basis of resistance to the carrot fly. This information would be of great help in developing cultivars with greater resistant to this important pest of carrot.

18.3.3 Aphids

Numerous aphid species can feed on carrot, with plants damaged by the mechanical feeding action as well as toxic saliva from the aphids (Rubatzky et al. 1999). Aphid feeding weakens plants, and the aphids deposit a sticky, sugary waste product on the foliage called honeydew. Honeydew is a highly suitable medium for growth of saprophytic fungi, which can cover the photosynthetic surface area of foliage. In addition, some aphid species are vectors of viruses that infect carrot, e.g., the carrot-willow aphid, *C. aegopodii*, vectors CMoV and CRLV, which together cause motley dwarf (see virus diseases above). The green peach aphid, *M. persicae*, is considered the most widespread and damaging aphid species on carrot and is a vector of several viruses of carrot, such as CarVY and others. Other aphids that feed on carrot include the carrot-willow aphid, *C. aegopodii*, violet aphid (*Myzus ornatus*), pea aphid (*Acyrothosiphon pisum*), bean aphid (*Aphis fabae*), melon aphid (*Aphis gossypii*), hawthorn-carrot aphid (*Dysaphis crataegi*), honeysuckle aphids (*Rhopalosiphum conii*, *Hyadaphis foeniculi*, and *H. coriandri*), and potato aphid (*Macrosteles fascifrons*), among others.

Resistance screening: Lamb (1953) noted that the Australian carrot cultivar Osborne Park might be partially resistant to the carrot-willow aphid, partly as a result of the strong vigor of this cultivar. In Britain, the vigorous cultivar Autumn King was thought to be resistant to this aphid compared to cultivars with smaller tops, as this cultivar showed less severe symptoms of motley dwarf. Therefore, Dunn (1970) investigated whether this was related to Autumn King being partially resistant to motley dwarf or partially resistant to the aphid. Cultivars representing Chantenay, Nantes, Berlikum, and Autumn King, as well as three Australian cultivars were tested for susceptibility to the aphid at different temperatures in cages, and by evaluating the entries in field trials over three years. Dunn (1970) noted differences in susceptibility to aphid attack but large numbers of aphids were counted on all the cultivars and the differences were too minor to be of practical value in breeding for resistance. In contrast to the observation by Lamb (1953), 'Osborne Park' had intermediate susceptibility to the carrot-willow aphid. In addition, 'Autumn King' was the most susceptible variety but was identified as 'tolerant' to motley dwarf. The Nantes cultivar was rated as susceptible to both the aphid and virus disease, whereas the Berlikum cultivar seemed the least susceptible to the aphid. Dunn (1970) noted that temperature had less effect on cultivar susceptibility than it did on aphid fecundity. Painter (1951) suggested three components of resistance to aphids feeding on plants: preference or non-preference, antibiosis, and tolerance. In the field trials, Dunn (1970) noticed that 20–30% fewer immigrant alate (winged) aphids settled on 'Berlikum', which might infer host non-preference, even though there was no evidence of preference among cultivars in the cage trials. However, 'Berlikum' was the shortest cultivar in these trials, so the fewer number of incoming alates observed on this cultivar may have reflected a form of escape rather than resistance.

Painter (1951) categorized plant resistance to aphids and aphid feeding into three categories: antibiosis, antixenosis, and tolerance. The terminology was used for many years, with modifications, until the advent of plant and arthropod genomics. Smith and Chuang (2014) reviewed the extensive literature on plant resistance to aphid feeding. They summarized the literature related to advances in understanding of behavioral, biochemical, and physiological aspects of aphid selection of plant hosts for feeding, putative biophysical and chemical resistance factors involved in plant defenses to aphid herbivory, and the genetics and genomics of aphid-resistant cultivars that have been developed for various plant species. They described the relationship between constitutively expressed and aphid-induced plant resistance genes, as well as aphid virulence and the deployment of aphid resistance genes into crops for management of these pests. They noted that aphid resistance often is inherited as a dominant trait but can be polygenic and inherited as recessive or incompletely dominant traits. However, virulence to aphid resistance genes in plants has been documented in at least 17 aphid species, illustrating the need to identify new and diverse sources of resistance. For some plant species, genetic linkage maps and plant microarrays have provided invaluable tools for understanding the identity and location of aphid and aphid-vectored virus resistance genes in plants. However, currently there do not appear to be published studies documenting resistance of carrot cultivars to aphid feeding.

18.3.4 Thrips

Thrips are not typically considered a major pest of carrot, but feeding with their rasping mouthparts leads to scarring or silvering of carrot leaves and petioles (Rubatzky et al. 1999). Species that can cause some losses to carrot include the western flower thrips, *Frankliniella occidentalis*, flower thrips, *F. tritici*, and onion thrips, *Thrips tabaci*. In addition to causing damage to carrot by feeding, the western flower thrips also vectors a virus that can infect carrot, *Tomato spotted wilt virus* (TSWV), although this is not an important virus disease of carrot.

Leiss et al. (2013) investigated quantitative resistance to the western flower thrips, *F. occidentalis*, using cultivated carrots, wild carrots, and biofortified carrots with increased concentrations of the antioxidant chlorogenic acid. They evaluated six commercial cultivars (Chantenay, Sugarsnax, Paris Market, Ingot, and Nantes), four wild carrots (S1, D2, D2, and D3 derived from individual plants collected in the Netherlands), and four biofortified accessions (two accessions with high chlorogenic acid levels, purple-yellow 309-2 and purple-orange B7262 from the University of Wisconsin carrot breeding program, and a purple and an orange accession from a seed company). Carrot leaves varied in thrips resistance, with a ten-fold difference observed in the degree of silvering (feeding damage) between the most resistant and most susceptible carrot lines. Comparison of the carrot morphological traits and leaf metabolic profiles of the three most resistant and susceptible carrots using nuclear magnetic resonance microscopy (NMR) revealed the wild carrots were not more resistant to thrips than the cultivated carrots. The most resistant cultivar was Ingot, which is also resistant to the carrot fly, *P. rosae*. The biofortified carrots were not resistant to thrips. In fact, three of the four biofortified carrots were the most susceptible to thrips feeding. There were no significant differences in plant size, leaf area, and number of leaf hairs between the thrips resistant and susceptible carrots, but the metabolic profiles of the leaves of the resistant cultivars differed from that of leaves of the susceptible cultivars. The flavanoid luteolin, the phenylpropanoid sinapic acid, and the amino acid β-alanine were present in greater amounts in leaves of resistant cultivars than those of susceptible cultivars. In vitro bioassays confirmed the inhibitory effects of these compounds on thrips. Leiss et al. (2013) suggested the results could be utilized to improve resistance to thrips based on the natural variation in these metabolites in cultivated carrots. The compounds also function as antioxidants, providing an additional benefit to efforts at breeding for resistance to thrips. They also suggested more sensitive metabolomics might detect additional metabolites associated with host resistance.

18.3.5 Other Insect and Mite Pests

Many other insect and mite pests can feed on carrot roots and/or foliage, causing losses in root and/or seed production (Rubatzky et al. 1999). This includes the carrot psyllid (*Trioza apicalis*), leafhoppers (of which the aster leafhopper and beet leafhopper are important vectors of phytoplasmas—see the section on phytoplasmas and spiroplasmas above), lygus bugs (several *Lygus* species), carrot weevils (*Lisonotus oregonensis* and *L. latiusculas*), spider mites (particularly the two-spotted spider mite, *Tetranychus urticae*), carrot leafminer (*Napomyza carotae*), whiteflies (*Bemisia* spp. and *Trialeurodes* spp.), a diversity of lepidopteran pests such as cutworms (e.g., the common cutworm, *Agrostis segetum*, black cutworm, *A. ipsilon*, variegated cutworm, *Peridroma saucia*, and others), armyworm (*Spodoptera exigua*), cabbage looper (*Trichoplusia ni*), webworm (*Depressaria heracliana*), loopers (e.g., *Autographa californica* and *A. falcifera*), corn earworm (*Helicoverpa zea*), and European corn borer (*Ostinia nubilalis*), and other types of insect pests such as symphylans (*Scutigerella immaculata*), wireworms (*Limonius* and other genera), springtails (Collembola family), carrot beetle (*bothynus biggosus*), ground beetles (Tenebrionidae), flea beetles (e.g., the potato flea beetle, *Epitric cucumeris*, and the palestriped flea beetle, *Systena blanda*), and other beetles. For most of these pests, very little is known about variation in susceptibility among carrot cultivars or accessions aside from anecdotal evidence from breeders and growers, and even less is known about potential sources of resistance and genetic mechanisms of resistance. For those pests that also serve as vectors of viruses, phytoplasmas, and spiroplasmas of carrot, differentiating resistance to the vector versus the pathogen further complicates efforts to screen for resistance and develop an understanding of resistance to the insect pest versus the pathogen.

Although not considered a major pest of carrot root crops, lygus bugs can cause significant losses in carrot seed production. The insects feed preferentially on flowers and developing seed, destroying the embryos of carrot seed and resulting in non-viable seed. Scott (1970) reported observing differences in resistance of 'Nantes,' 'Imperator' (which he named 'Imperida'), and 'Royal Chantenay' to feeding by two species of lygus bugs, *L. hesperus* and *L. elisus*. He caged lygus bugs on the umbels of carrot plants in flower, counted the percent mortality of the insects in the cages, and considered percent mortality of the insects on the umbels to represent resistance of the cultivar to feeding by this pest. Scott (1977) used the same approach to attempt to select for resistance to lygus bugs in field trials in Idaho. There was no evidence in either study that Scott (1970, 1977) assessed the umbels for severity of damage to the developing seed by lygus bugs. The insects could have died from a diversity of causes not related to feeding on carrot umbels or seed, and not related to potential differences in resistance of the cultivars to this pest. There were many factors confounding these attempt to compare cultivars for reaction to lygus bugs (e.g., each cultivar was evaluated in a different season). Some umbels may not have provided adequate nutrition for the lygus bugs to persist, which also could have impacted the development of the seeds. He recorded as much variation in lygus bug mortality among plants of the same cultivar as he did among cultivars. The very poor rigor of those studies and the fact that losses to lygus bugs continue to be a major constraint in some areas of carrot seed production, raise doubt about the conclusions reached in those studies.

Kainulainen et al. (2002) examined the essential oil composition in the leaves of seven carrot cultivars in relation to oviposition acceptance by sucking insects with different feeding strategies. They evaluated the carrot psyllid, *T. apicalis*, as a carrot specialist, *T. anthrisci* as an Apiaceae specialist psyllid, and the lygus bug, *L. rugulipennis*, as a generalist in growth chamber, greenhouse, and field trials. The carrot psyllid is a significant pest of carrot in northern Europe, where feeding by nymphs reduces root growth, and saliva injected as the adults feed causes curling of the leaves. As noted above, lygus bugs can be particularly damaging in carrot seed production as they puncture seed to feed on the developing embryo, resulting in non-viable seed (Scott 1977). Kainulainen et al. (2002) observed that oil composition differed significantly among leaves of the cultivars Flakkeer 2, Nantura, Parano, Napoli, Panther, Splendid, and Nantes 3 Express. The mean number of eggs laid by the carrot psyllid and *T. anthrisci* did not differ significantly among cultivars. However, lygus bug females laid more eggs on Nantes Express 3 than on Panther, even though this preference was not observed in an olfactometer test, which suggests that physical contact with the plant is more important in host selection for this generalist insect than olfactory orientation. There was no evidence of a correlation between the composition of essential oils and egg-laying preference of the lygus bug. A negative linear relationship was detected between concentration of the oil limonene and the number of eggs laid by the carrot psyllid, indicating that this oil is a repellant to the carrot psyllid. The carrot psyllid was attracted to high concentrations of sabinene. Other studies have also demonstrated attraction of the carrot psyllid to Apiaceae species that produce large amounts of α-pinene and sabinene (Nehlin et al. 1996; Valterova et al. 1997). A positive correlation was detected between myrcene concentration and number of eggs laid by the Apiaceae psyllid, *T. anthrisci*. Interestingly, this is the main compound in leaves of wild chervil (*Anthriscus sylvestris*), the main host of *T. anthrisci*. The study illustrated that some compounds influence the egg-laying preference of this psyllid, but these may not be the primary components of essential oils in carrot leaves. Overall, this study indicated that the host preference of the Apiaceae specialized pest, *T. anthrisci*, was affected more by the essential oil composition of carrot cultivars than lygus bugs, a generalist pest. Carrot cultivars with high concentrations of limonene might be more resistant to the carrot psyllid.

18.4 Conclusions

As detailed in this chapter, there is wide variation in the degree of understanding of carrot germplasm reactions to the numerous pathogens and pests of carrot. For a few of these biotic stresses, significant phenotypic resistance screening has been completed, candidate genes have been identified, and the resistance incorporated into commercially acceptable cultivars. For others, almost nothing is known about potential sources of resistance, and phenotypic screening methods have not yet been developed. Significant resources are needed to extend this effort to many more carrot pathogens and pests of concern regionally and internationally. Resistance to biotic stresses, combined with chemical (foliar fungicide sprays and seed treatments) and cultural control measures, has contributed significantly to disease and pest suppression compared with individual control measures (Ben-Noon et al. 2003), but far more can be achieved in screening for resistance and incorporating resistance into breeding programs. The absence of crossing barriers between wild carrot species and cultivated carrot greatly facilitates the introduction of resistance genes into cultivated carrot. Furthermore, the development of molecular markers and related tools has facilitated much more rapid progress at identifying resistance genes and developing cultivars with improved resistance (Stein and Nothnagel 1995). A novel set of 300 simple sequence repeat markers, combined with a deep-coverage, highly redundant carrot genome library with >17X coverage (Cavagnaro et al. 2009, 2011) have provided invaluable genomic resources for carrot breeding and genetics. Recent analysis of ~1.74 Mb of BAC-end sequences gave the first overview of the composition and organization of the carrot nuclear genome. Iorizzo et al. (2016) reported a high-quality genome assembly accounting for ~90% of the estimated carrot genome. They predicted 634 putative pest and disease resistance genes that will be very helpful for identifying candidate genes underlying biotic and abiotic stresses, and other important traits. More recently, Wang et al. (2018) sequenced the genome of 'Kurodagosun,' a major carrot variety in Japan and China, accounting for 78.5% of the estimated genome (473 Mb). These genomic resources enhance fundamental and applied research on carrot, including breeding for resistance to pests and diseases such as leaf blights. Klimek-Chodacka et al. (2018) reported the first use of the CRISPR/Cas9 system for efficient, site-targeted mutagenesis of the carrot genome, which offers new prospectives for improving carrot resistance to diseases and pests. The ultimate challenge is to combine resistance to multiple pests and pathogens with other traits important for carrot root production, seed production, storage, flavor, nutritional qualities, and processing.

References

Abercrombie K, Finch HC (1976) Powdery mildew of carrot in California. Plant Dis Report 60:780–781

Aegerter BJ (2002) Powdery mildew. In: Davis RM, Raid RN (eds) Compendium of umbelliferous crop diseases. American Phytopathological Society, St. Paul, pp 23–24

Ali A, Matthews WC, Cavagnaro PF, Iorizzo M, Roberts PA, Simon PW (2014) Inheritance and mapping of *Mj-2*, a new source of root-knot nematode (*Meloidogyne javanica*) resistance in carrot. J Hered 105:288–291

Amirov BM, Amirova ZS, Manabaeva UA, Zhasybaeva KR (2014) Carrot breeding for Alternaria leaf blight resistance. Acta Hortic 1053:223–226

Angel FF, Gabelman WH (1968) Inheritance of resistance in carrot *Daucus carota* var *sativa* to leaf spot fungus *Cercospora carotae*. Am Soc Hortic Sci 93:434

Arbizu CI, Tas PM, Simon PW, Spooner DM (2017) Phylogenetic prediction of Alternaria leaf blight resistance in wild and cultivated species of carrots. Crop Sci 57:2645–2653

Baranski R, Krämer R, Klocke E (2006) A laboratory leaf assay of carrot susceptibility to *Botrytis cinerea*. J Phytopathol 154:637–640

Baranski R, Klocke E, Nothnagel T (2007) Enhancing resistance of transgenic carrot to fungal pathogens by the expression of *Pseudomonas fluorescence* microbial factor 3 (*MF3*) gene. Physiol Mol Plant Pathol 71:88–95

Baranski R, Klocke E, Nothnagel T (2008) Chitinase CHIT36 from *Trichoderma harzianum* enhances resistance of transgenic carrot to fungal pathogens. J Phytopathol 156:513–521

Bedlan G (1984) Wichtige krankheiten der karotten. Pflanzenarzt 37:140–142

Benard D, Punja ZK (1995) Role of *Pythium* spp. in the development of cavity spot on carrots in British Columbia. Can J Plant Pathol 17:31–45

Ben-Noon E, Shtienberg D, Shlevin E, Dinoor A (2003) Joint action of disease control measures: a case study of Alternaria leaf blight of carrot. Phytopathology 93:1320–1328

Biniek A, Tylkowska K (1987) Germination and mycoflora of carrot seeds treated with thiram and conditioned in polyethylene glycol (PEG 6000). Acta Hortic 215:225–230

Blomquist CL (2002) Aster yellows and beet leafhopper-transmitted virescence agent yellows. In: Davis RM, Raid RN (eds) Compendium of umbelliferous crop diseases. American Phytopathological Society, St. Paul, pp 58–59

Boedo C, Le Clerc V, Briard M et al (2008) Impact of carrot resistance on development of the Alternaria leaf blight pathogen (*Alternaria dauci*). Eur J Plant Pathol 121:55–66

Boedo C, Berruyer R, Lecomte M et al (2010) Evaluation of different methods for the characterization of carrot resistance to the Alternaria leaf blight pathogen (*Alternaria dauci*) revealed two qualitatively different resistances. Plant Pathol 59:368–375

Boerema GH, Dorenbosch MMJ, Van Kesteren HA (1963) Some notable fungus infections. II. Jaarboek 1962. Versl Plziekt Dienst Wageningen (Rev Appl Mycol 43:2185c, 1964)

Boiteux LS, Della Vecchia PT, Reifschneider FJB (1993) Heritability estimate for resistance to *Alternaria dauci* in carrot. Plant Breeding 110:165–167

Boiteux LS, Hyman JR, Bach IC et al (2004) Employment of flanking codominant STS markers to estimate allelic substitution effects of a nematode resistance locus in carrot. Euphytica 136:37–44

Boivon G (1994) Aster leafhopper. In: Howard RJ, Garland JA, Seaman WL (eds) Diseases and pests of vegetable crops in Canada. Canadian Phytopathological Society, Guelph, pp 76–77

Bonnet A (1977) Resistance a l'Oidium. INRA Stat. d'Amelior. Plantes Maraicheres, pp 33–35

Bonnet A (1983) *Daucus carota* L. subsp. *dentatus* Bertol., géniteur de résistance à l'oïdium pour l'amélioration de la carotte cultivée. Agronomie 3:33–38

Bourgeois G, Brodeur C, Kushalappa A (1998) Effet de la brûlure cercosporéenne, causée par le *Cercospora carotae*, sur le développement, la croissance et le rendement de la carotte. Phytoprotection 79:9–19

Bowen RM, Heale JB (1987) Resistance in carrot root tissue. Physiol Mol Plant Pathol 30:55–66

Braun U (1987) A monograph of the Erysiphales (powdery mildews). Beih Nova Hedwigia 89:1–700

Breton D, Béasse C, Montfort F, Villeneuve F (2003) Focus on the recent evolution of soil-borne diseases of carrot in France. In: Proceedings of the 30th international carrot conference, Muskegon, 7–10 Sept 2003

Bridge J, Starr JL (2007) Plant nematodes of agricultural importance: a colour handbook. CRC Press, London, p 152
Browne GT (2002) Phytophthora root rot. In: Davis RM, Raid RN (eds) Compendium of umbelliferous crop diseases. American Phytopathological Society, St. Paul, pp 37–38
Campion C, Vian B, Nicole M, Rouxel FA (1988) A comparative study of carrot root tissue colonization and cell wall degradation by *Pythium violae* and *Pythium ultimum*, two pathogens responsible for cavity spot. Can J Microbiol 44:221–230
Carisse O, Kushalappa AC (1990) Development of an infection model for *Cercospora carotae* on carrot based on temperature and leaf wetness duration. Phytopathology 80:1233–1238
Carisse O, Kushalappa AC (1992) Influence of interrupted wet periods, relative humidity, and temperature on infection of carrots by *Cercospora carotae*. Phytopathology 82:602–606
Carvalho AD, Silva GO, Pereira RB, Pinheiro JB (2015) Productivity and tolerance to the leaf blight disease of hybrid and open-pollinated carrot cultivars. Hortic Bras 33:299–304
Cavagnaro PF, Chung SM, Szklarczyk M et al (2009) Characterization of a deep-coverage carrot (*Daucus carota* L.) BAC library and initial analysis of BAC-end sequences. Mol Genet Genomics 281:273–288
Cavagnaro PF, Chung SM, Manin S et al (2011) Microsatellite isolation and marker development in carrot—genomic distribution, linkage mapping, genetic diversity analysis and marker transferability across Apiaceae. BMC Genom 12:386
Charchar JM, Eisenback JD, Vieira JV, De N, Fonseca-Boiteux ME, Boiteux LS (2009) *Meloidogyne polycephannulata* n. sp. (Nematoda: Meloidogynidae), a root-knot nematode parasitizing carrot in Brazil. J Nematol 41:174–186
Cheah L-H, Page BBC (1999) Epidemiology and control of violet root rot of carrots. In: Proceedings of the 52nd N.Z. plant protection conference. New Zealand Institute for Crop and Food Research Limited, Palmerston North, pp 157–161
Chen TW, Wu WS (1999) Biological control of carrot black rot. J Phytopathol 147:99–104
Christianson CE, Jones SS, du Toit LJ (2015) Screening carrot germplasm for resistance to *Xanthomonas hortorum* pv. *carotae*. HortScience 50:341–350
Cirulli M (1975) The powdery mildew of parsley caused by *Leveillula lanuginosa* (Fuck.) Golovin. Phytopathol Mediterr 14:94–99
Cole RA (1985) Relationship between the concentration of chlorogenic acid in carrot roots and the incidence of carrot fly larval damage. Ann Appl Biol 106:211–217
Cooper C, Isaac S, Jones MG, Crowther T, Smith BM, Collin HA (2004) Morphological and biochemical response of carrots to *Pythium violae*, causative agent of cavity spot. Physiol Mol Plant Pathol 64:27–35
Cooper C, Crowther T, Smith BM, Isaac S, Collin HA (2006) Assessment of the response of carrot somaclones to *Pythium violae*, causal agent of cavity spot. Plant Pathol 55:427–432
Courtial J, Hamama L, Helesbeux JJ et al (2018) Aldaulactone—an original phytotoxic secondary metabolite involved in the aggressiveness of *Alternaria dauci* on carrot. Front Plant Sci 9:1–29
Cunnington JH, Watson A, Liberato JR, Jones RH (2008) First record of powdery mildew on carrots in Australia. Australas Plant Dis Notes 3:38–41
Cwalina-Ambroziak B, Amarowicz R, Glosek M, Janiak M (2014) Changes in the concentrations of phenolic acids in carrot plants inoculated with *Alternaria radicina* Meier, Drechsler & Eddy. Acta Sci Pol Hortorum Cultus 13:97–108
Dalton LP, Epton HAS, Bradshaw NJ (1981) The susceptibility of modern carrot cultivars to violet root rot caused by *Helicobasidium purpureum*. J Hortic Sci 56:95–96
Davis RM, Raid RN (eds) (2002) Compendium of umbelliferous crop diseases. American Phytopathological Society, St. Paul, p 75
Degen T, Städler E, Ellis PR (1999a) Host-plant susceptibility to the carrot fly, *Psila rosae*. I. Acceptability of various host species to ovipositing females. Ann Appl Biol 134:1–11
Degen T, Städler E, Ellis PR (1999b) Host-plant susceptibility to carrot fly. II. Suitability of various hosts. Ann Appl Biol 134:27–34
Degen T, Städler E, Ellis PR (1999c) Host-plant susceptibility to the carrot fly. III. The role of oviposition preferences. Ann Appl Biol 134:13–26
du Toit LJ, Derie ML (2008) Effect of powdery mildew on seed yield and quality in a carrot seed crop, 2006–2007. Plant Dis Manage Rep 2:V007
du Toit LJ, Crowe FJ, Derie ML, Simmons RB, Pelter GQ (2005) Bacterial blight in carrot seed crops in the Pacific Northwest. Plant Dis 89:896–907
du Toit LJ, Derie ML, Wohleb CH (2009) Effect of powdery mildew on seed yield and quality in a carrot seed crop, 2007–2008. Plant Dis Manage Rep 3:V136
Dufault CP, Coaker TH (1987) Biology and control of the farrot fly, *Psila rosae* (F.). Agric Zool Rev 2:97–134
Dugdale LJ, Mortimer AM, Isaac S, Collin HA (2000) Disease response of carrot and carrot somaclones to *Alternaria dauci*. Plant Pathol 49:57–67
Dunn JA (1970) The susceptibility of varieties of carrot to attack by the aphid, *Cavariella aegopodii* (Scop.). Ann Appl Biol 66:301–312
Ellis PR (1999) The identification and exploitation of resistance in carrots and wild Umbelliferae to the carrot fly, *Psila rosae* (F.). Integr Pest Manage Rev 4:259–268
Ellis PR, Hardman JA (1981) The consistency of the resistance of eight carrot cultivars to carrot fly attack at several centres in Europe. Ann Appl Biol 98:491–497
Ellis PR, Wheatley GA, Hardman JA (1978) Preliminary studies of carrot susceptibility to carrot fly attack. Ann Appl Biol 88:159–170
Ellis PR, Saw PL, Crowther TC (1991) Development of carrot inbreds with resistance to carrot fly using a

single seed descent programme. Ann Appl Biol 119:349–357

Elnagar S, Murant AF (1978) Relations of carrot red leaf and carrot mottle viruses with their aphid vector, *Cavariella aegopodil*. Ann Appl Biol 89:237–244

Endo RM, Colt WM (1974) Anatomy, cytology and physiology of infection by *Pythium*. Proc Am Phytopathol Soc 1:215–223

Farrar JJ (2002) Soft rot. In: Davis RM, Raid RN (eds) Compendium of umbelliferous crop diseases. American Phytopathological Society, St. Paul, pp 14–15

Farrar JJ, Pryor BM, Davis RM (2004) Alternaria diseases of carrot. Plant Dis 88:776–784

Gabelman WH, Goldman IL, Breitbach DW (1994) Evaluation and selection for resistance to aster yellows in carrot (*Daucus carota* L.). J Am Soc Hort Sci 119:1293–1297

Garrett SD (1949) A study of violet root rot, 2. Effect of substratum on survival of *Helicobasidium purpureum* colonies in the soil. Trans Br Mycol Soc 32:217–223

Garrod B, Lewis BG, Coxon DT (1978) Cis-heptadeca-1,9-diene-4,6-diyne-3,8-diol, an antifungal polyacetylene from carrot root tissue. Physiol Plant Pathol 13:241–246

Garrod B, Lewis BG, Brittain MM, Davies WP (1982) Studies on the contribution of lignin and suberin to the impedece of wounded carrot root tissue to fungal invasion. New Phytol 90:99–108

Geary JR, Wall CJ (1976) New or uncommon plant diseases and pests. Plant Pathol 25:165

Glawe DA, Pelter GQ, du Toit LJ (2005) First report of powdery mildew of carrot and parsley caused by *Erysiphe heraclei* in Washington State. Online. Plant Health Prog. https://doi.org/10.1094/php-2005-0114-01-hn

Goodliffe JP, Heale JB (1975) Incipient infections caused by *Botrytis cinerea* in carrots entering storage. Ann App Biol 8:243–246

Grisham MP, Anderson NA (1983) Pathogenicity and host specificity of *Rhizoctonia solani* isolated from carrots. Phypathology 73:1564–1569

Grogan RG, Snyder WC (1952) The occurrence and phytopathological effects of *Stemphylium radicinum* on carrots in California. Phytopathology 42:215–218

Groom MR, Perry DA (1985) Induction of "cavity spot-like" lesions on roots of *Daucus carota* by *Pythium violae*. Trans Br Mycol Soc 84:755–757

Grzebelus E, Kruk M, Macko-Podgorni A, Grzebelus D (2013) Response of carrot protoplasts and protoplast-derived aggregates to selection using a fungal culture filtrate of *Alternaria radicina*. Plant Cell Tissue Organ Cult 115:209–222

Guba EF, Young RE, Ci T (1961) Cavity spot disease of carrot and parsnip roots. Plant Dis Report 45:102–105

Guerin PM, Ryan MF (1983) Relationship between root volatiles of some carrot cultivars and their resistance to the carrot fly, *Psila rosae*. Field experiments assessment of larval damage to carrots. Entomol Exp Appl 36:217–224

Guerin PM, Stadler E (1984) Carrot fly cultivar preferences: some influencing factors. Ecol Entomol 9:413–420

Guerin PM, Visser JH (1980) Electroantennogram responses of the carrot fly, *Psila rosae*, to volatile plant components. Physiol Entomol 5:111–119

Guerin PM, Städler E, Buser HR (1983) Identication of host plant attractants for the carrot fly, *Psila rosae*. J Chem Ecol 9:843–861

Guérin L, Briard M, Rouxel F (1994) Biochemical characterization of *Pythium* spp. involved in cavity spot of carrots in France. Ann Appl Biol 125:255–265

Guérin L, Benhamou N, Rouxel F (1998) Ultrastructural and cytochemical investigation of pathogen development and host reaction in susceptible and partially resistant carrot roots infected by *Pythium violae*, the major causal agent of cavity spot. Eur J Plant Pathol 104:653–655

Gugino BK, Carroll JE, Widmer TL, Chen P, Abawi GS (2007) Field evaluation of carrot cultivars for susceptibility to fungal leaf blight diseases in New York. Crop Prot 26:709–714

Gurkin RS, Jenkins SF (1985) Influence of cultural practices, fungicides, and inoculum placement on southern blight and Rhizoctonia crown rot of carrot. Plant Dis 69:477–481

Hammarlund C (1925) Zur genetik, biologie und physiologie einiger Erysiphaceen. Hereditas 6:1–126

Hardman JA, Ellis PR (1982) An investigation of the host range of the carrot fly. Ann Appl Biol 100:1–9

Ho HH (1983) *Phytophthora porri* from stored carrots in Alberta. Mycologia 75:747–751

Howard RJ, Williams PH (1976) Methods for detecting resistance to *Pythium* and *Rhizoctonia* root diseases in seedling carrots. Plant Dis Report 60:151–156

Howell WE, Mink GI (1977) The role of weed hosts, volunteer carrots, and overlapping growing seasons in the epidemiology of carrot thin leaf and carrot motley dwarf viruses in central Washington. Plant Dis Report 61:217–222

Huang SP (1986) Penetration, development, reproduction, and sex ratio of *Meloidogyne javanica* in three carrot cultivars. J Nematol 18:408–412

Huang SP, Vecchia PT, Ferreira PE (1986) Varietal response and estimates of heritability of resistance to *Meloidogyne javanica* in carrots. J Nematol 18:496–501

Iorizzo M, Ellison S, Senalik D et al (2016) A high quality carrot genome assembly provides new insights into carotenoid accumulation and asterid genome evolution. Nat Genet 48:657–666

Janse JD (1988) *Streptomyces* species identified as the cause of carrot scab. Neth J Plant Pathol 94:303–306

Johnston LF, Palmer GK (1985) Symptom variability and selection for reduced severity of cotton seedling disease caused by *Pythium ultimum*. Plant Dis Report 52:209–212

Jones RAC (2005) Further studies on *Carrot virus Y*: hosts, symptomatology, search for resistance, and tests for seed transmissibility. Austral J Agric Res 56:859–868

Kainulainen P, Nissinen A, Piirainen A, Tiilikkala K, Holopainen JK (2002) Essential oil composition in leaves of carrot varieties and preference of specialist and generalist sucking insect herbivores. Agric Forest Entomol 4:211–216

Karkleliene R, Radzevicius A, Dambrauskiene E, Surviliene E, Bobinas C, Duchovskiene L, Kavaliauskaite D, Bundiniene O (2012) Root yield, quality and disease resistance of organically grown carrot (*Daucus sativus* Röhl.) hybrids and cultivars. Žemdirbystė Agric 99:393–398

Kinsella MN (1966) Vegetable patch. J Agric Vict Dept Agric 64:468

Klimek-Chodacka M, Oleszkiewicz T, Lowder LG, Qi Y, Baranski R (2018) Efficient CRISPR/Cas9-based genome editing in carrot cells. Plant Cell Rep 37:575–586

Klisiewicz JM (1968) Relation of *Pythium* spp. to root rot and damping-off of safflower. Phytopathology 58:1384–1386

Koike ST, Saenz GS (1994) Occurrence of powdery mildew on parsley in California. Plant Dis 78:1219

Koike ST, Saenz GS (1997) First report of powdery mildew caused by *Erysiphe heraclei* on celery in North America. Plant Dis 81:231

Koike ST, Nuñez JJ, Falk BW (2002) Carrot motley dwarf. In: Davis RM, Raid RN (eds) Compendium of umbelliferous crop diseases. American Phytopathological Society, St. Paul, pp 51–52

Kora C, McDonald MR, Boland GJ (2003) Sclerotinia rot of carrot. Plant Dis 87:456–470

Kordowska-Wiater M, Wagner A, Hetman B (2012) Efficacy of *Candida melibiosica* for control of post-harvest fungal diseases of carrot (*Daucus carota* L.). Acta Sci Pol Hortorum Cultus 11:55–65

Koutouan C, Le Clerc V, Baltenweck R, Claudel P, Walter D et al (2018) Link between carrot leaf secondary metabolites and resistance to *Alternaria dauci*. Sci Rep 8:13746

Kurosaki F, Tsurusawa Y, Nishi A (1985) Partial purification and characterization of elicitors for 6-methoxymellein production in cultures carrot cells. Physiol Plant Pathol 27:209–217

Lamb KP (1953) Observations on yield and varietal susceptibility of some carrot varieties to insect attack in the field. New Zeal J Sci Technol Sect A 34:351

Langenberg WJ, Sutton JC, Gillespie TJ (1977) Relation of weather variables and periodicities of airborne spores of *Alternaria dauci*. Phytopathology 67:879–883

Latham LJ, Jones RAC (2004) *Carrot virus Y*: symptoms, losses, incidence, epidemiology and control. Virus Res 100:89–99

Le Clerc V, Pawelec A, Birolleau-Touchard C, Suel A, Briard M (2009) Genetic architecture of factors underlying partial resistance to Alternaria leaf blight in carrot. Theor Appl Genet 118:1251–1259

Le Clerc V, Suel A, Pawelec A, Marques S, Huet S, Lecomte M, Poupard P, Briard M (2015a) Is there variety × isolate interaction in the polygenic quantitative resistance of carrot to *Alternaria dauci*? Euphytica 202:235–243

Le Clerc V, Marques S, Suel A, Huet S, Hamama L, Voisine L, Auperpin E, Jourdan M, Barrot L, Prieur R, Briard M (2015b) QTL mapping of carrot resistance to leaf blight with connected populations: stability across years and consequences for breeding. Theor Appl Genet 128:2177–2187

Lebeda A (1985). Response of certain carrot species (*Daucus carota*) to artificial inoculation by *Erwinia carotovora* spp. *carotovora*. In: Eucarpia meeting on breeding of root vegetables, Olomouc, 6–9 Sept 1985, pp 82–88

Lebeda A, Coufal J (1985). Relationship between virus infection symptoms, carrot (*Daucus carota*) root quality and seed yield. In: Eucarpia meeting on breeding of root vegetables, Olomouc, 6–9 Sept 1985, pp 107–117

Lebeda A, Coufal J (1987) Evaluation of susceptibility of *Daucus carota* varieties to natural infection with *Erysiphe heraclei*. Arch Züchtungsforsch Berlin 17:73–76

Lebeda A, Coufal J, Kvasnička P (1988) Evaluation of field resistance of *Daucus carota* cultivars to *Cercospora carotae* (carrot leaf spot). Euphytica 39:285–288

Lecomte M, Berruyer R, Hamama L, Boedo C, Hudhomme P, Bersihand S, Arul J, N'Guyen G, Gatto J, Guilet D, Richomme P, Simoneau P, Briard M, Le Clerc V, Poupard P (2012) Inhibitory effects of the carrot metabolites 6-methoxymellein and falcarindiol on development of the fungal leaf blight pathogen *Alternaria dauci*. Physiol Mol Plant Pathol 80:58–67

Lecomte M, Hamama L, Voisine L, Gatto J, Hélesbeux JJ, Séraphin D, Pena-Rodriguez L, Richomme P, Boedo C, Yovanopoulos C, Gyomlai M, Briard M, Simoneau P, Poupard P, Berruyer R (2014) Partial resistance of carrot to *Alternaria dauci* correlates with in vitro cultured carrot cell resistance to fungal exudates. PLoS One 9(7)

Lee I-M, Bottner KD, Munyaneza JE, Davis RE, Crosslin JM, du Toit LJ, Crosby T (2006) Carrot purple leaf: a new spiroplasmal disease associated with carrots in Washington State. Plant Dis 90:989–993

Leiss KA, Cristorfori G, van Steenis R, Verpoorte R, Klinkhamer PGL (2013) An eco-metabolomic study of host plant resistance to western flower thrips in cultivated, biofortified and wild carrots. Phytochemistry 93:63–70

Leyronas C, Troulet C, Duffaud M et al (2018) First report of *Sclerotinia subarctica* in France detected with a rapid PCR-based test. Can J Plant Pathol 40:248–253

Marras F (1962) Intorno ad *Erysiphe umbelliferarum* de Bary parasitta della carota, del finocchio e del prezzemolo in Sardegna. Studi Sassaressi Sez III 9, 12 pp. (Rev Appl Mycol 43:635, 1964)

Maude RB (1966) Studies on the etiology of black rot, *Stemphylium radicinum* (Meier, Drechsl. & Eddy) Neerg., and leaf blight, *Alternaria dauci* (KuHn) Groves & Skolko, on carrot crops; and on fungicide control of their seed borne infection phases. Ann Appl Biol 57:83–93

McDonald MR (1994a) Bacterial soft rot. In: Howard RJ, Garland JA, Seaman WL (eds) Diseases and pests of vegetable crops in Canada. Canadian Phytopathological Society, Guelph, p 62

McDonald MR (1994b) Cavity spot of carrot (*Pythium* spp.): etiology, epidemiology and control. PhD dissertation, University of Guelph, Ontario, 314 pp

McDonald MR (1994c) Crater rot. In: Howard RJ, Garland JA, Seaman WL (eds) Diseases and pests of vegetable crops in Canada. Canadian Phytopathological Society, Guelph, p 68

McDonald MR (1994d) Rubbery brown rot. In: Howard RJ, Garland JA, Seaman WL (eds) Diseases and pests of vegetable crops in Canada. Canadian Phytopathological Society, Guelph, pp 71–72

McDonald MR (1994e) Violet root rot. In: Howard RJ, Garland JA, Seaman WL (eds) Diseases and pests of vegetable crops in Canada. Canadian Phytopathological Society, Guelph, p 73

McDonald MR (2002) Cavity spot. In: Davis RM, Raid RN (eds) Compendium of umbelliferous crop diseases. American Phytopathological Society, St. Paul, pp 27–29

McDonald MR, van der Kooi K, Simon P (2017) Evaluation of carrot breeding lines for susceptibility to cavity spot, 2017. In: Muck crops research station annual report, University of Guelph, Ontario, 6 pp. https://www.uoguelph.ca/muckcrop/pdfs/Muck%20Vegetable%20Cultivar%20Trial%20&%20Research%20Report%202017%20Complete.pdf. Accessed Aug 2018

Meier FC, Drechsler C, Eddy ED (1922) Black rot of carrots caused by *Alternaria radicina* n. sp. Phytopathology 12:157–168

Mercier J, Kuć J (1996) Induced systemic resistance to Cercospora leaf spot of carrot by inoculation with *Cercospora carotae*. J Phytopathol 144:75–77

Mercier J, Roussel D, Charles MT, Arul J (2000) Systemic and local responses associated with UV- and pathogen-induced resistance to *Botrytis cinerea* in stored carrot. Phytopathology 90:3–8

Michalik B, Ślęczek S (1997) Evaluation of *Daucus carota* germplasm for tolerance to *Erwinia carotovora*. J Appl Genet 38A:86–90

Michalik B, Wiech K (2000) Differences in the resistance of carrot lines and cultivars to carrot fly [*Psila rosae* (Fabr.)] attack. Folia Hortic 12:43–51

Michalik B, Simon P, Gabelman WH (1992) Assessing susceptibility of carrot roots to bacterial soft rot. HortScience 27:1020–1022

Mildenhall JP, Williams PH (1970) Rhizoctonia crown rot and cavity spot of muck-grown carrots. Phytopathology 60:887–890

Milosavljević A, Pfaf-Dolovac E, Mitrović M, Jović J, Toševski I, Duduk N, Trkulja N (2014) First rport of *Cercospora apii*, causal agent of Cercospora early blight of celery, in Serbia. Plant Dis 98:1157

Montfort F, Rouxel F (1988) La maladie de la "tache" de la carotte due a *Pythium violae* Chesters et Hickman: donnees symptomatoloiques et etiologiques. Agronomie 8:701–706

Moran J, van Rijswijk B, Traicevski V, Kitajima E, Mackenzie AM, Gibbs AJ (2002) Potyviruses, novel and known, in cultivated and wild species of the family Apiaceae in Australia. Arch Virol 147:1855–1867

Murant AF, Spence N (2002) Parsnip yellow fleck. In: Davis RM, Raid RN (eds) Compendium of umbelliferous crop diseases. American Phytopathological Society, St. Paul, pp 56–57

National Institute of Agricultural Botany (1991) Vegetable growers leaflet no. 5. N.T.A.B., Cambridge

Nehlin G, Valterova I, Borg-Karlson A-K (1996) Monoterpenes released from Apiaceae and the egg-laying preferences of the carrot psyllid, *Trioza apicalis*. Entomol Exp Appl 80:83–86

Nuñez JJ, Davis RM (2016) Diseases of carrot (*Daucus carota* L. subsp. *sativus* (Hoffm.) Arcang.). Common names of plant diseases. American Phytopathological Society, St. Paul. http://www.apsnet.org/publications/commonnames/Pages/Carrot.aspx. Accessed Oct 2018

Nuñez JJ, Westphal A (2002) Damping-off. In: Davis RM, Raid RN (eds) Compendium of umbelliferous crop diseases. American Phytopathological Society, St. Paul, pp 31–33

Ojaghian MR, Wang Q, Li X et al (2016) Inhibitory effect and enzymatic analysis of e-cinnamaldehyde against Sclerotinia carrot rot. Pestic Biochem Physiol 127:8–14

Painter R (1951) Insect resistance in crop plants. Macmillan Company, New York, p 520

Palti J (1975) Erysiphaceae affecting umbelliferous crops, with special reference to carrot, in Israel. Phytopathol Mediterr 14:87–93

Parsons J, Matthews W, Iorizzo M, Roberts P, Simon PW (2015) *Meloidogyne incognita* nematode resistance QTL in carrot. Mol Breeding 35:114

Pawelec A, Dubourg C, Briard M (2006) Evaluation of carrot resistance to Alternaria leaf blight in controlled environments. Plant Pathol 55:68–72

Pereira RB, Carvalho ADF, Pinheiro JB, Silva GO, Vieira JV (2012) Resistência de populações de cenoura à queima-das-folhas com diferentes níveis de germoplasma tropical. Hortic Bras 30:489–493

Perry DA, Harrison JG (1979) Cavity spot of carrots. I. Symptomology and calcium involvement. Ann Appl Biol 93:101–108

Pfleger FL, Harman GE, Marx GA (1974) Bacterial blight of carrots: interaction of temperature, light, and inoculation procedures on disease development of various carrot cultivars. Phytopathology 64:746–749

Phillips JA, Kelman A (1982) Direct fluorescent antibody stain procedure applied to insect transmission of *Erwinia carotovora*. Phytopathology 72:898–901

Pinheiro JB, De Mendonça JL, De Santana JP (2011) Reaction of wild Solanaceae to *Meloidogyne incognita* race 1 and *M. javanica*. Acta Hortic 917:237–241

Pryor B, Davis RM, Gilbertson RL (1994) Detection and eradication of *Alternaria radicina* on carrot seed. Plant Dis 78:452–456

Pryor B, Davis RM, Gilbertson RL (1998) Detection of soilborne *Alternaria radicina* and its occurrence in California carrot fields. Plant Dis 82:891–895

Pryor B, Davis RM, Gilbertson RL (2000) A toothpick inoculation method for evaluating carrot cultivars for resistance to *Alternaria radicina*. HortScience 35:1099–1102

Punja JK (1987) Mycelial growth and pathogenesis by *Rhizotonia carotae* on carrot. Can J Plant Pathol 9:24–31

Punja JK (2002a) Crater rot. In: Davis RM, Raid RN (eds) Compendium of umbelliferous crop diseases. American Phytopathological Society, St. Paul, p 42

Punja ZK (2002b) Crown rot of carrot. In: Davis RM, Raid RN (eds) Compendium of umbelliferous crop diseases. American Phytopathological Society, St. Paul, p 31

Punja ZK (2005) Transgenic carrots expressing a thaumatin-like protein display enhanced resistance to several fungal pathogens. Can J Plant Pathol 27:291–296

Punja ZK, Chen WP (2004) Transgenic carrots expressing enhanced tolerance to herbicide and fungal pathogen infection. Acta Hortic 637:295–302

Punja ZK, McDonald MR (2002) Violet root rot. In: Davis RM, Raid RN (eds) Compendium of umbelliferous crop diseases. American Phytopathological Society, St. Paul, pp 40–41

Rader WE (1952) Diseases of stored carrots in New York State. N Y (Cornell) Agric Exp Stn Bull 889:35–38

Raid RN (2002) Cercospora leaf blight of carrot. In: Davis RM, Raid RN (eds) Compendium of umbelliferous crop diseases. American Phytopathological Society, St. Paul, p 18

Roberts PA, Mullens TR (2002) Diseases caused by nematodes. In: Davis RM, Raid RN (eds) Compendium of umbelliferous crop diseases. American Phytopathological Society, St. Paul, pp 45–46

Roderick H, Urwin PE, Atkinson HJ (2018) Rational design of biosafe crop resistance to a range of nematodes using RNA interference. Plant Biotechnol J 16:520–529

Rogers PM, Stevenson WR (2010) Aggressiveness and fungicide sensitivity of *Alternaria dauci* from cultivated carrot. Plant Dis 94:405–412

Rubatzky VE, Quiros CF, Simon PW (1999) Diseases, disorders, insects and other pests. In: Carrots and related vegetables umbelliferae. CABI Publishing, New York, pp 173–220

Santo GS, Mojtahedi H, Wilson JH (1988) Host-parasite relationship of carrot cultivars and *Meloidogyne chitwoodi* races and *M. hapla*. J Nematol 20:555–564

Saude C, Hausbeck MK, Hurtado-Gonzales O, Rippetoe C, Lamour KH (2007) First report of *Phytophthora cactorum* causing root rot of processing carrots (*Daucus carota*) in Michigan. Plant Dis 91:459

Schoneveld JA (1994) Effect of irrigation on the prevention of scab in carrots. Acta Hortic 354:135–144

Scott DR (1970) Lygus bugs feeding on developing carrot seed: plant resistanc to that feeding. J Econ Entomol 63:959–961

Scott DR (1977) Selection for lygus bug resistance in carrot. HortScience 12:452

Scott DJ, Wenham HT (1972) Occurrence of two seed-borne pathogens, *Alternaria radicina* and *Alternaria dauci*, on imported carrot seed in New Zealand. New Zeal J Agr Res 16:247–250

Segall RH, Dow AT (1973) Effects of bacterial contamination and refrigerated storage on bacterial soft rot of carrots. Plant Dis Rep 57:896–899

Sherf AF, MacNab AA (1986) Carrot. In: Vegetable diseases and their control. In: MacNab AA, Sherf AF (eds) Wiley Interscience Publication, Wiley, New York, pp 138–139

Siddiqui ZA, Nesha R, Varshney A (2011) Response of carrot cultivars to *Meloidogyne incognita* and *Pectobacterium carotovorum* subsp. *carotovorum*. J Plant Pathol 93:503–506

Sidorova T, Miroshnichenko D (2013) Transgenic carrot expressing thaumatin II gene has enhanced resistance against *Fusarium avenaceum*. Acta Agric Scand 974:123–130

Silva GO, Vieira JV, Vilela MS, Reis A, Boiteux LS (2009) Genetic parameters of the resistance to the leaf blight disease complex in carrot populations. Hortic Bras 27:354–356

Simlat M, Stobiecki M, Szklarczyk M (2013) Accumulation of selected phenolics and expression of PAL genes in carrots differing in their susceptibility to carrot fly (*Psila rosae* F.). Euphytica 190:253–266

Simon PW, Strandberg JO (1998) Diallel analysis of resistance in carrot to Alternaria leaf blight. J Am Soc Hortic Sci 123:412–415

Simon PW, Matthews WC, Roberts PA (2000) Evidence for simply inherited dominant resistance to *Meloidogyne javanica* in carrot. Theor Appl Genet 100:735–742

Simon PW, Freeman RE, Vieira JV, Boiteux LS, Briard M, Nothnagel T, Michalik B, Kwon Y-S (2008) Carrot. In: Prohens J, Nuez F (eds) Handbook of plant breeding. Vegetables II. Fabaceae, Liliaceae, Solanaceae, and Umbelliferae. Springer, New York, pp 327–357

Skadow K (1978) Eine objective, rationelle methode der resistenz-prüfing von möhren gegen *Erwinia carotovora* (Jones) Bergey et al. var. *carotovora* dye. Arch Phytopathol Pflanzenschutz 14:27–31

Smith CM, Chuang W-P (2014) Plant resistance to aphid feeding: behavioral, physiological, genetic and molecular cues regulate aphid host selection and feeding. Pest Manage Sci 70:528–540

Soroker E, Bashan Y, Okon Y (1984) Reproducible induction of cavity spot in carrots and physiological and microbial changes occurring during cavity formation. Soil Biol Biochem 16:541–548

Soteros IJ (1979) Pathogenicity and control of *Alternaria radicina* and *A. dauci* in carrots. New Zeal J Agric Res 22:191–196

Städler E, Buser HR (1984) Defense chemicals in leaf surface wax synergistically stimulate oviposition by a phytophagous insect. Experientia 40:1157–1159

Stein M, Nothnagel T (1995) Some remarks on carrot breeding (*Daucus carota sativus* Hoffm.). Plant Breeding 114:1–11

Stelfox D, Henry AW (1978) Occurrence of rubbery brown rot of stored carrots in Alberta. Can Plant Dis Survey 58:87–91

Strandberg JO (1977) Spore production and dispersal of *Alternaria dauci*. Phytopathology 77:1262

Strandberg JO, Bassett MJ, Peterson CE, Berger RD (1972) Sources of resistance to *Alternaria dauci*. HortScience 7:345

Suffert F, Montfort F (2007) Demonstration of secondary infection by *Pythium violae* in epidemics of carrot cavity spot using root transplantation as a method of soil infestation. Plant Pathol 56:588–594

Sweet JB, Lake SE, Wright IR, Priestley RH (1986) Resistance of carrot varieties to cavity spot disease. Aspects Appl Biol 12:235–245

Sweet JB, Beale RE, Wright IR (1989). Cavity spot disease in six carrot cultivars treated with a metalaxyl and thiram fungicide. Tests of agrochemicals and cultivars 10. Ann Appl Biol 114:38–39 (Supplement)

Takaichi M, Oeda K (2000) Transgenic carrots with enhanced resistance against two major pathogens, *Erysiphe heraclei* and *Alternaria dauci*. Plant Sci 153:135–144

Tan JAC, Jones MG, Fosu-Nyarko J (2013) Gene silencing in root lesion nematodes (*Pratylenchus* spp.) significantly reduces reproduction in a plant host. Exp Parasitol 133:166–178

Tomlinson JA (1965) Rep Natn Veg Res Stn for 1964, p 70

Umiel N, Jacobson R, Globerson D (1975) Pollination of the cultivated carrot (*Daucus carota* L.) by the wild carrot (*D. carota* var. *maximus*) and its implication on commercial seed production. Hassadeh 56:478–480

Valterova I, Nehlin G, Borg-Karlson AK (1997) Host plant chemistry and preferences in egg laying Trioza apicalis (Homeoptera, Psylloidea). Biochem Syst Ecol 25:477–491

Van Dijk P, Bos L (1985) Viral dieback of carrot and other Umbelliferae caused by the Anthriscus strain of parsnip yellow fleck virus, and its distinction from carrot motley dwarf. Neth J Plant Pathol 91:169–187

Vasudeva RS (1963) Report of division of mycology and plant pathology. Sci Rep Agric Res Inst New Delhi 1961:87–100

Vieira JV, Dias Casali VW, Milagres JC, Cardoso AA, Regazzi AJ (1991) Heritability and genetic gain for resistance to leaf blight in carrot (*Daucus carota* L.) populations evaluated at different times after sowing. Rev Bras Genét 14:501–508

Vieira JV, Charchar JM, Aragão FAS, Boiteux LS (2003) Heritability and gain from selection for field resistance against multiple root-knot nematode species (*Meloidogyne incognita* race 1 and *M. javanica*) in carrot. Euphytica 130:11–16

Villeneuve F (2014) Les Maladies: Symptomes et biologie. In: Villeneuve F (ed) La carotte: maladies, ravageurs et protection. CTIFL, Paris, pp 71–131

Vivoda E, Davis RM, Nunez JJ, Guerard JP (1991) Factors affecting the development of cavity spot of carrot. Plant Dis 75:519–522

Wagenvoort WA, Blok I, Monbarg HFM, Velhuinzen T (1989) Cavity spot of carrot in relation to a *Pythium* sp. Gartenbauwissenschaft 54:70–73

Wally ZK, Punja O (2010) Enhanced disease resistance in transgenic carrot (*Daucus carota* L.) plants over-expressing a rice cationic peroxidase. Planta 232:1229–1239

Wally O, Jayaraj J, Punja ZK (2009a) Broad-spectrum disease resistance to necrotrophic and biotrophic pathogens in transgenic carrots (*Daucus carota* L.) expressing an Arabidopsis NPR1 gene. Planta 231:131–141

Wally O, Jayaraj J, Punja ZK (2009b) Comparative resistance to foliar fungal pathogens in transgenic carrot plants expressing genes encoding for chitinase, β-1,3-glucanase and peroxidise. Planta 231:131–142

Wang M, Goldman IL (1996) Resistance to root knot nematode (*Meloidogyne hapla* Chitwood) in carrot is controlled by two recessive genes. J Hered 87:119–123

Wang F, Long G, Xi W et al (2018) The genome sequence of 'Kurodagosun', a major carrot variety in Japan and China, reveals insights into biological research and carrot breeding. Mol Genet Genomics 293:861–871

Waterhouse PM (1985) Isolation and identification of carrot red leaf virus from carrot and dill growing in the Australian Capital Territory. Austral Plant Pathol 14:32–34

Waterhouse PM, Murant AF (1983) Further evidence on the nature of the dependence of carrot mottle virus on carrot red leaf virus for transmission by aphids. Ann Appl Biol 103:455–464

Watson MT, Falk BW (1994) Ecological and epidemiological factors affecting carrot motley dwarf development in carrots grown in the Salinas Valley of California. Plant Dis 78:477–481

Watson MT, Sarjeant EP (1964) The effect of motley dwarf virus on yield of carrots and its transmission in the field by *Cavariella aegopodii* Scop. Ann Appl Biol 53:77–83

Watson MT, Tian T, Estabrook E, Falk BW (1998) A small RNA resembling the beet western yellows luetovirus ST9-associated RNA is a component of the California carrot motley dwarf complex. Phytopathology 88:164–170

Wesemael W, Moens M (2008) Quality damage on carrots (*Daucus carota* L.) caused by the root-knot nematode *Meloidogyne chitwoodi*. Nematology 10:261–270

White NH (1945) Fungal soft-rot of carrots. Tasman J Agric 16:59–60

White JG (1988) Studies on the biology and control of cavity spot of carrots. Ann Appl Biol 113:259–268

White JG (1991) Curing spotty carrots. Grower 9–10

White JG, Dowker BD, Crowther TC (1987) Screening carrot cultivars against *Pythium* spp. associated with cavity spot. Tests of agrochemicals and cultivars 8. Ann Appl Biol 110 (Supplement)

White JG, Dowker BD, Crowther TC, Wakeham AJ (1988) Laboratory screening of carrot cultivars with reported differential field performance for cavity spot to three *Pythium* spp. Tests of agrochemicals and cultivars 9. Ann Appl Biol 112 (Supplement)

Yarger LW, Baker LR (1981) Tolerance of carrot to *Meloidogyne hapla*. Plant Dis 65:337–339

Yunhee S, Park J, Kim YS, Park Y, Kim YH (2014) Screening and histopathological characterization of Korean carrot lines for resistance to the root-knot nematode *Meloidogyne incognita*. Plant Pathol J 30:75–81

Zamski E, Peretz I (1995) Cavity spot of carrots: interactions between the host and pathogen, related to the cell wall. Ann Appl Biol 127:23–32

Zhang XY, Hu J, Zhou HY et al (2014) First report of *Fusarium oxysporum* and *F. solani* causing Fusarium dry rot of carrot in China. Plant Dis 98:1273

19 Genetics and Genomics of Carrot Abiotic Stress

Dariusz Grzebelus

Abstract

Plant reaction to abiotic stresses leading to stress tolerance is a complex and multi-level process comprising several inter-dependent mechanisms. While it has been extensively studied in model plant species, it has not been a subject of systematic investigation in carrot. Only few reports pointing at the importance of particular proteins in response to stressors have been published. No attempt has been made to describe regulatory mechanisms governing tolerance to heat, cold, drought, salinity and other abiotic stresses in carrot. Nevertheless, the issue seems vital, as agriculture is coping with global climate changes. Also, the area of carrot cultivation worldwide is growing and its adaptation to environmental conditions outside the temperate climatic zone would provide health benefits to human populations suffering from malnutrition. In the present chapter, we review the existing knowledge on the reaction of carrot to abiotic stresses, with particular emphasis on molecular or genetic mechanisms governing stress tolerance.

D. Grzebelus (✉)
Faculty of Biotechnology and Horticulture, Institute of Plant Biology and Biotechnology, University of Agriculture in Krakow, Al. 29 Listopada 54, 31-425 Krakow, Poland
e-mail: dariusz.grzebelus@urk.edu.pl

19.1 Introduction

The response of plants to abiotic stresses essentially requires complex reprogramming of an array of genes, initiated by stress perception and resulting in physiological changes allowing plants to survive the period of stress. Stress sensing and the ability of plants to withstand unfavorable conditions rely upon cross-talk of several key components, including abscisic acid (ABA) (Qin et al. 2011), redox homeostasis (Foyer and Noctor 2005; Gill and Tuteja 2010), soluble sugars (Rosa et al. 2009), membrane-localized receptor-like kinases (RLKs) (Osakabe et al. 2013), and calcium-regulated effector proteins (Knight and Knight 2001). Stress perception and signal transduction affect the activity of transcription factors which in turn alter expression levels of genes directly involved in the physiological response (Chinnusamy et al. 2004). Moreover, reaction to abiotic stresses may also be controlled on the epigenetic level. Histone modifications and DNA (de)methylation can be induced upon stress resulting in altered chromatin status and promoting changes in gene expression (Chinnusamy and Zhu 2009). Epigenetic reprogramming may extend beyond the period of stress leading to the formation of stable epiallels and improving long-term transgenerational adaptation to suboptimal environmental conditions (Mirouze and Paszkowski 2011). Small RNAs are involved in both methylation-dependent silencing and posttranscriptional

P. Simon et al. (eds.), *The Carrot Genome*, Compendium of Plant Genomes,
https://doi.org/10.1007/978-3-030-03389-7_19

silencing of plant genes in response to stressors (Khraiwesh et al. 2012; Sunkar et al. 2012).

Existence of the above-mentioned interdependent mechanisms makes studies on molecular mechanisms of reactions to abiotic stresses notoriously difficult and requires a great amount of basic knowledge on the genome biology of the investigated species. It is why most of the key mechanisms have been reported for model plant species, for which robust genomic resources have been available since the beginning of the 'post-genome era' in plant science. In carrot, the availability of genomic resources has been very limited until recently, resulting in fragmentary knowledge on possible genetic factors affecting tolerance to abiotic stresses. In the present chapter, we will summarize reported results on the effects of abiotic stresses on carrot and present postulated genetic mechanisms of tolerance.

19.2 Mechanisms of Tolerance to Abiotic Stress in Carrot

19.2.1 Temperature

Stress-induced cell reprogramming in plants is required for adaptation to unfavorable conditions. One of the possible mechanisms driving such reprogramming, the activity of alternative oxidases (*AOX*) required for an alternative respiratory pathway, has been thoroughly studied in carrot (Table 19.1). *AOX* genes might be responsible for relieving environmentally induced oxidative stress by limiting the formation of reactive oxygen species (ROS) in the mitochondria (Nogales et al. 2016). Unlike many other plant species, carrot carries four genes representing each of the two *AOX* families, *AOX1* (two paralogs) and *AOX2* (two paralogs), which makes it especially suited to study their importance in stress response (Costa et al. 2009). With respect to carrot reaction to abiotic stresses, *AOX* genes were reported as highly responsive to temperature changes. The expression of *DcAOX1* in the callus tissue increased fivefold when the temperature was raised from 21 to 28 °C. Exposure of plants to chilling also induced immediate and substantial changes in the expression levels of both *DcAOX1* and *DcAOX2a* (Campos et al. 2016). As *DcAOX1* was shown to be more responsive to environmental regulation, it was postulated that the existing allelic variability within that gene could have an effect on the host tolerance to abiotic stresses, thus the gene could be a target for marker-assisted selection in carrot breeding

Table 19.1 A list of genes involved in abiotic stress tolerance in carrot

Stress type	Proposed gene name	Function	Reference
General abiotic	*DcLac1, DcLac2*	Laccases	Ma et al. (2015)
	DcHsp17.7	Small heat shock protein	Malik et al. (1999)
			Song and Ahn (2010)
			Song and Ahn (2011)
			Ahn and Song (2012)
Temperature	*AFP*	Antifreeze protein	Meyer et al. (1999) Smallwood et al. (1999)
	Dc cyclin	Cyclin	Kumar et al. (2013)
	Dc profilin	Profilin	Kumar et al. (2013)
	Dc WD	Transcription factor	Kumar et al. (2013)
	DcAOX1, DcAOX2a	Alternative oxidases	Campos et al. (2016)
Drought	*dcTLP*	Thaumatin-like protein	Jung et al. (2005)
Wounding	EDGP	Extracellular dermal glycoprotein	Satoh et al. (1992)
UV irradiation	*DcMYB1*	Transcription factor	Maeda et al. (2005)
Hypoxia	*DcADH1–3*	Alcohol dehydrogenases	Que et al. (2018)

(Nogales et al. 2016). Potential functional markers have also been developed for *DcAOX2a* (Cardoso et al. 2009).

Heat shock proteins are important components of the plant reaction to a range of abiotic stresses. They function as molecular chaperones providing protection against temperature changes to other proteins. DcHsp17.7 is a heat shock protein identified in carrot and thoroughly studied with respect to its protective function (Table 19.1). It was first described as playing a critical role in tolerance of carrot to high (42 °C) temperatures. Manipulation of the *DcHsp17.7* transcript levels resulted in variable responses to heat stress (Malik et al. 1999). Park et al. (2013) showed that it was rapidly produced in response to exposure to heat, remained steady up to two days following heat treatment and decayed afterward. Night exposure to heat resulted in higher accumulation of DcHsp17.7 than the same stress imposed during daytime. Interestingly, DcHsp17.7 also accumulated upon low (2 °C) temperature treatment and was also shown to enhance cold tolerance by preventing cold-induced protein degradation (Song and Ahn 2010). A more general analysis of heat shock factors in carrot and their expression under abiotic stress conditions has been reported by Huang et al. (2015). Most *DcHsfs* were up-regulated by heat stress, while their reaction to cold stress was more variable. Interestingly, *DcHsf09* appeared to be down-regulated by heat and up-regulated by cold.

Kumar et al. (2013) attempted to identify and clone genes induced upon cold stress using suppression subtractive hybridization (SSH). Genes encoding proteins involved in signal transduction, regulation of transcription and translation, protein folding and osmolyte synthesis were revealed as up-regulated upon stress (Table 19.1).

Two carrot laccase-encoding genes *DcLac1* and *DcLac2* were shown to be differentially regulated upon heat and cold treatment (Table 19.1). Laccases are blue copper oxidases that might be involved in plant defense against stresses. The reported changes in their expression levels in carrots treated with temperature, salt, and metal stresses may suggest their importance in the plant reaction (Ma et al. 2015).

A prominent feature of carrot is to withstand freezing. This mechanism has been well described and the main component of freezing tolerance has been identified (Smallwood et al. 1999). It is based on the synthesis of the antifreeze protein (AFP) (Table 19.1) which is a member of the polygalacturonase inhibitor protein (PGIP) gene family. PGIP proteins, including AFP, are secreted to the extracellular space. However, AFP acquired a new function possibly allowing it to interact with ice through an LRR (leucine-rich repeat) domain by modifying the shape of crystals and inhibiting ice recrystallization in the apoplast (Meyer et al. 1999). Carrot AFP has been shown to confer freezing tolerance when transferred to other plant species via genetic transformation, e.g., in tomato (Kumar et al. 2014).

19.2.2 Drought

Drought is among the abiotic stresses of high importance for carrot production and its significance will likely be even more pronounced as global warming continues. In carrot, a *dcTLP* gene encoding a thaumatin-like protein (TLP) was shown to be specifically up-regulated upon dehydration, independent from the developmental stage of the plant. It was not regulated by abscisic acid (ABA), salicylic acid (SA), or jasmonic acid (JA) (Jung et al. 2005). TLPs are classified as pathogenesis-related (PR) proteins that are involved in reaction to a number of stresses including abiotic stresses. DcTLP may therefore be one of the key components governing physiological adaptation of carrot plants to the drought stress (Table 19.1). Other drought-induced genes identified by Jung et al. (2005) in the carrot embryogenic callus encoded putative lipid transfer proteins, peroxidases, metallothionins, pyruvate decarboxylase, damage-specific DNA-binding protein, elongation factor, GTP-binding protein, mitochondrial half-ABC transporter, and myosin PfM-B-like protein. Also, a small heat shock protein

DcHsp17.7 has been reported as being accumulated upon osmotic stress (Ahn and Song 2012), pointing at the general importance of this protein to protection against abiotic stresses in carrot.

19.2.3 Salinity

Carrot is generally considered as sensitive to increased levels of salinity in soil, which negatively affects seed germination and plant growth (Schmidhalter and Oertli 1991). Reduced rates of photosynthesis and stomatal conductance were observed in salt-stressed carrot plants (Gibberd et al. 2002). Other physiological responses to salinity stress in carrot comprised changes in the antioxidative defense system, resulting in increased amounts of glycinebetaine (GB), malondialdehyde (MDA) and ascorbic acid, and decreased activity of key antioxidative enzymes; peroxidases, catalases and superoxide dismutases (Bano et al. 2014). One of the possible mechanisms mediating salt tolerance is the induction of phytoene synthase 2 (*DcPSY2*) gene by abscisic acid (ABA). *DcPSY2* is required to produce carotenoids in carrot roots (Fuentes et al. 2012; Wang et al. 2014), while carotenoids are precursors of ABA. Simpson et al. (2018) showed that salt stress and ABA boost expression of *DcPSY2* through binding of DcAREB3 transcription factor to ABA-responsive elements (ABREs) located in the promoter of *DcPSY2*. DcHsp17.7 may also be a component of the plant response to high salinity (Song and Ahn 2011).

Some wild *Daucus carota* subspecies are well adapted to grow in saline areas. However, it may be difficult to introduce salt tolerance from the wild crop relatives to cultivars. Carrot landraces of Iranian origin highly tolerant to salinity stress have been described (Kasiri et al. 2013). A higher salt tolerance of Iranian carrot populations has been confirmed in the experiment combining in vitro selection for salt tolerance and subsequent evaluation of regenerants for reaction to high salinity. Plants derived from one of the two Iranian populations showed markedly higher survival rate, as compared to the control western carrot cultivar 'Dolanka' of the Flakkee type. At the same time, 'Dolanka' showed a much higher range of diversity in response to selection for salinity tolerance in protoplast cultures than the eastern carrot populations, resulting in the appearance of more tolerant regenerants. Increased tolerance to salinity was correlated with more intensive anthocyanin pigmentation of petioles and increased leaf and petiole hairiness, pointing at possible genetic and epigenetic mechanisms (Kiełkowska et al. 2019). F2 populations produced by crossing a western-type carrot breeding line and an eastern-type salt-tolerant accession segregated with respect to their ability to germinate on a medium supplemented with 150 mM NaCl. Only ca. 20% seeds germinated on the medium supplemented with NaCl, while 98% germination rate was observed for the control. Also, morphology of salt-stressed seedlings was different than those grown on the salt-free medium, the former being thicker, yellowish, and swollen (Fig. 19.1).

Boron (B) is frequently present in large amounts in saline soils imposing additional abiotic stress to crops. Eraslan et al. (2007) reported on morphological and physiological effects of salinity and boron stress on carrots. Salinity stress alone caused increase in root diameter and decrease in dry weight, while combined salt and boron stress somewhat reversed the tendency. However, adverse physiological changes were more severe under combined salt and boron stress. Application of salicylic acid (SA) to some extent alleviated boron stress. The effect of boron excess on carrot cells was also investigated using in vitro cultures. Six proteins were identified that were accumulated in stressed cells, mostly involved in redox homeostasis, including glyoxalase 1 (a detoxification protein), glutathione peroxidase, isocitrate dehydrogenase, adenosyl homocysteinase, but interestingly also carrot major allergen Dau c1 and extracellular dermal glycoprotein (EDPG) precursor (Demiray et al. 2011). The latter was earlier to be rapidly produced in response to wounding (Satoh et al. 1992).

Fig. 19.1 Differences in the morphology of carrot seedlings obtained from a population segregating with respect to reaction to salt stress. Left panel (**a** and **c**): seedlings grown on the salt-free medium, right panel (**b** and **d**): seedlings grown on the medium supplemented with 150 mM NaCl. *Photograph* M. Klimek-Chodacka

19.2.4 Heavy Metals

Accumulation of heavy metals in carrot roots has been investigated, as they are toxic to consumers. Safety limits of heavy metal content have been defined in most countries, with a special focus on cadmium and lead. The thresholds are even more stringent for carrots used for baby food production. Heavy metals, especially cadmium, can be readily stored in carrot roots at levels largely exceeding those thresholds when grown in polluted soils. Carrot cultivars accumulate large amounts of cadmium and show little variability regarding the tendency for cadmium uptake. Thus, carrot cultivation should be avoided on cadmium-contaminated soils (Zheng et al. 2008). The uptake of heavy metals may be affected by agricultural practices, e.g., fertilization (Smoleń and Sady 2006, 2007).

Besides being harmful to humans, excess of heavy metals may impose severe stress to the plant. Sanità di Toppi et al. (2012) demonstrated in the in vitro experiment that prolonged exposition of carrots to high cadmium concentrations resulted in the induction of adaptive response likely aiming at restricting cadmium movement in the root, followed by gradual root deterioration. However, no genetic mechanisms leading to increased tolerance against heavy metals in carrot have been proposed. Carrot heat shock proteins may be involved in the protection against heavy metal stress, as they accumulated in plants exposed to lead and arsenic (Park et al. 2013).

19.2.5 Hypoxia

Deficiency of oxygen in the root zone or hypoxia is an abiotic stress which may affect plant growth and yield. In carrot, owing to the common cultivation practice, it seems to have little importance. However, Que et al. (2018) investigated the effect of hypoxia on carrot root development using hydroponics. They showed that carrot roots grown under the low oxygen regime became more lignified. Three genes encoding alcohol dehydrogenases (*ADH1–3*) were found to be highly up-regulated in roots suffering from hypoxia, in comparison with those grown in aerated cultures.

19.3 Carrot Transcription Factors Regulating Response to Abiotic Stresses

Several gene families encoding transcription factors (TFs) have been identified and described in carrot. Of all annotated carrot genes, ca. 10% (3267 genes) were attributed to provide regulatory functions. Of those, six TF families (ZF-GFR, JmjC, TCP, GeBP, B3, and response regulators) expanded in the carrot genome, as compared to other plant species (Iorizzo et al. 2016). Obviously, their modes of operation are varying and they regulate a range of processes, from developmental to adaptive. Few TF families have been further characterized with respect to their possible involvement in the plant reaction to abiotic stresses. The APETALA2/Ethylene Responsive Factor (AP2/ERF) TF family comprises 267 genes in the carrot genome. Eight randomly selected *AP2/ERF* genes representing different subfamilies showed altered expression and different expression patterns under abiotic stress conditions (cold, heat salt, and drought). Their expression depended on the type of stress, but also on the accession (Li et al. 2015). The basic helix-loop-helix (bHLH) TFs were also shown to be involved in plant reaction to abiotic stresses. In carrot, 109 genes encoding bHLH proteins were revealed. Again, eight *bHLH* genes (classified to subfamily 15) showed differential expression, both accession- and tissue-specific, as well as depending on the type of abiotic stress. They were primarily up-regulated upon heat, cold, and drought, while the reaction to the salt stress depended on the variety (Chen et al. 2015). The DNA-binding one zinc finger (Dof) family in carrot was shown to include 46 genes. It was revealed that representatives of this TF family showed very diverse and irregular expression patterns in response to abiotic stresses, possibly reflecting their specific reaction to a particular stressor (Huang et al. 2016). WRKY is another prominent family of plant transcription factors regulating a range of processes, including plant reaction against abiotic stresses. In carrot, 95 *WRKY* genes were characterized, of which 71 were expressed, mostly in a developmentally coordinated manner. Twelve carrot *WRKY* genes were selected on the basis of their presumed involvement in the reaction to abiotic stresses, as shown for their respective orthologs in *Arabidopsis*. Their expression patterns were investigated in cold, heat salt, and drought-stressed plants. Changes in expression depended on the gene and the type of stress, sometimes peaking at 80-fold increase (*DcWRKY27*/salinity stress) (Li et al. 2016).

The above examples reflect significance of the research on the molecular regulation of carrot responses to abiotic stresses. The general characteristics of TF families involved in the reaction to stress are only the first step leading to elucidation of key regulatory mechanisms. To date, results on TF expression are only fragmentary, and more systematic efforts are needed.

19.4 Perspectives

Systematic investigations on the genetics of abiotic stress tolerance in carrot are of high significance, as they are essential for the development of new cultivars better adapted to the changing conditions, e.g., those imposed by global warming. It can be obtained by exploring the existing genetic diversity both in the cultivated gene pool and in the wild crop relatives. Genetic modifications might be another method of choice, depending on the public acceptance of genetic transformation and novel, more precise techniques of gene editing. Abiotic stresses can be applied post-harvest, in order to increase synthesis of valuable biologically active secondary metabolites.

19.4.1 Exploration of the Existing Variability in the Cultivated Carrot and its Wild Relatives

While carrot has been widely cultivated in temperate climatic zones, efforts have been undertaken to breed for varieties that could be cultivated in warmer regions. In Brazil, breeding of carrot cultivars suitable for production in the subtropical climate using well-adapted local landraces of European origin was successful. The open-pollinated cultivar 'Brasilia' and its derivatives constitute the major fraction of carrot production (Simon et al. 2008). Elucidation of major genetic determinants of adaptation to abiotic stresses and incorporation of molecular tools in breeding would certainly shorten the time required for develop and select plant materials showing desired characteristics, that could subsequently be introduced for production in regions suffering from malnutrition and vitamin A deficiency, supporting previous efforts implementing conventional selection methods (Tabor et al. 2016). Application of molecular techniques (e.g., marker-assisted backcrossing) might also support more efficient transfer of abiotic stress tolerances present in the wild *Daucus carota* gene pool.

19.4.2 Genetic Engineering of Resistance to Abiotic Stresses

Carrot is highly amenable for genome engineering, using both transgenesis (Baranski 2008) and CRISPR/Cas9 genome editing (Klimek-Chodacka et al. 2018) (see also Chap. 10). The latter technology has appeared very recently as a new possibility, and it has not been implemented as a tool to modify the reaction of plants to abiotic stresses. However, genetic transformation has been used to improve carrot tolerance and several reports on the expression of heterologous stress-related genes in carrot have been published. Transgenic carrot plants carrying a gene coding for betaine-aldehyde dehydrogenase (BADH) showed highly increased betaine content and significantly improved tolerance to salt stress (Kumar et al. 2004). Carrot transformation with mammalian 6-phosphofructo-2-kinase/fructose 2,6-bisphosphatase (*6-PF-2-K/Fru 2,6-P2ase*) gene resulted in highly increased levels of fructose 2,6-bisphosphate (Fru-2,6-P_2) in roots of the transgenic plants. Under drought and cold stress, it allowed to mobilize energy reserves by gluconeogenesis (Kovács et al. 2006). A tobacco gene encoding osmotin also conferred increased drought tolerance in genetically modified carrots (Annon et al. 2014).

19.4.3 Abiotic Stresses as Elicitors of the Production of Bioactive Compounds

Post-harvest abiotic stresses can be applied purposely, in order to increase the synthesis of secondary metabolites involved in the protective reaction. These compounds can be subsequently recovered and used industrially, as they show a range of activities beneficial to human health. Heat shock and UV-C irradiation were shown to increase the content of phenolic compounds and carotenoids in stored carrot roots (Alegria et al. 2012). UV-B treatment and the resulting over-expression of *DcPAL1*, a key gene in the phenylpropanoid biosynthesis pathway encoding phenylalanine ammonia-lyase, resulted in high accumulation of phenolics three days after treatment of shredded carrots (Formica-Oliveira et al. 2017). Ultrasound treatment also increased *DcPAL1* expression and stimulated accumulation of 3-*O*-caffeoylquinic acid and isocoumarin, as well as ethylene production, in stored carrot roots (del Rosario Cuéllar-Villarreal et al. 2016). Maeda et al. (2005) reported on a possible regulatory mechanism leading to the up-regulation of *DcPAL1* expression following UV irradiation. UV treatment induced the expression of *DcMYB1* encoding a MYB transcription factor that in turn increased the expression of *DcPAL1*.

Wounding combined with ethylene and methyl jasmonate treatment also induced *DcPAL1* and altered the composition of main phenolic compounds, depending on the wounding intensity (Heredia and Cisneros-Zevallos 2009). A combination of wounding and water loss resulted in overexpression of phenylalanine ammonia-lyase and 3-deoxy-D-arabino-heptulosanate synthase, but subsequent lignification resulted in lower accumulation of phenolics (Becerra-Moreno et al. 2015). Reactive oxygen species (ROS) are likely an important factor in the accumulation of phenolic compounds in wounded carrots, as they upregulate genes encoding key enzymes in the phenylpropanoid pathway (Han et al. 2017). Sánchez-Rangel et al. (2014) developed a technology for extraction of bioactive compounds, mainly chlorogenic acid and its derivatives, produced in carrot roots following wounding.

19.5 Conclusions

Genetics of carrot tolerance to abiotic stresses is a largely unexplored area of research. As shown above, the present knowledge is very limited, most presented evidence points at possible significance of particular proteins, of which small heat shock protein Hsp17.7 and antifreeze protein AFP have been more thoroughly characterized. No systematic information is present on the regulation of reaction to stress in carrot, except some very preliminary data on the expression changes in reaction to abiotic stressors of some representatives of few TF families. Well-designed studies integrating the existing knowledge coming from investigations on model plant species and high-throughput capabilities of modern genomic technologies are urgently needed. Efficient phenotyping of plant materials showing contrasting reactions to stresses combined with omics analyses comprising genomics (e.g., genome-wide association studies, GWAS), transcriptomics (changes in gene expression), methylome and small RNA sequencing should provide a good basis for elucidating mechanisms governing abiotic stress tolerance.

Acknowledgements This study was supported by the Polish Ministry of Science and Higher Education fund for statutory activities for the University of Agriculture in Krakow.

References

Ahn Y-J, Song N-H (2012) A cytosolic heat shock protein expressed in carrot (*Daucus carota* L.) enhances cell viability under oxidative and osmotic stress conditions. HortScience 47:143–148

Alegria C, Pinheiro P, Duthoit M, Gonçalves EM, Moldão-Martins M, Abreu M (2012) Fresh-cut carrot (cv. Nantes) quality as affected by abiotic stress (heat shock and UV-C irradiation) pre-treatments. LWT Food Sci Technol 48:197–203

Annon A, Rathore K, Crosby K (2014) Overexpression of a tobacco osmotin gene in carrot (*Daucus carota* L.) enhances drought tolerance. In Vitro Cell Dev Biol Plant 50:299–306

Bano S, Ashraf M, Akram NA (2014) Salt stress regulates enzymatic and nonenzymatic antioxidative defense system in the edible part of carrot (*Daucus carota* L.). J Plant Interact 9:324–329

Baranski R (2008) Genetic transformation of carrot (*Daucus carota*) and other *Apiaceae* species. Transgenic Plant J 2:18–38

Becerra-Moreno A, Redondo-Gil M, Benavides J, Nair V, Cisneros-Zevallos L, Jacobo-Velázquez DA (2015) Combined effect of water loss and wounding stress on gene activation of metabolic pathways associated with phenolic biosynthesis in carrot. Front Plant Sci 6:837

Campos MD, Nogales A, Cardoso HG, Kumar SR, Nobre T, Sathishkumar R, Arnholdt-Schmitt B (2016) Stress-induced accumulation of *DcAOX1* and *DcAOX2a* transcripts coincides with critical time point for structural biomass prediction in carrot primary cultures (*Daucus carota* L.). Front Genet 7:1

Cardoso HG, Campos MD, Costa AR, Campos MC, Nothnagel T, Arnholdt-Schmitt B (2009) Carrot alternative oxidase gene *AOX2a* demonstrates allelic and genotypic polymorphisms in intron 3. Physiol Plantarum 137:592–608

Chen Y-Y, Li M-Y, Wu X-J, Huang Y, Ma J, Xiong A-S (2015) Genome-wide analysis of basic helix-loop-helix family transcription factors and their role in responses to abiotic stress in carrot. Mol Breeding 35:125

Chinnusamy V, Zhu J-K (2009) Epigenetic regulation of stress responses in plants. Curr Opin Plant Biol 12:133–139

Chinnusamy V, Schumaker K, Zhu J-K (2004) Molecular genetic perspectives on cross-talk and specificity in abiotic stress signalling in plants. J Exp Bot 55:225–236

Costa JH, Cardoso HG, Campos MD, Zavattieri A, Frederico AM, Fernandes de Melo D, Arnholdt-Schmitt B (2009) *Daucus carota* L.—an old model for cell reprogramming gains new importance through a novel expansion pattern of alternative oxidase (AOX) genes. Plant Physiol Biochem 47: 753–759

del Rosario Cuéllar-Villarreal M, Ortega-Hernández E, Becerra-Moreno A, Welti-Chanes J, Cisneros-Zevallos L, Jacobo-Velázquez DA (2016) Effects of ultrasound treatment and storage time on the extractability and biosynthesis of nutraceuticals in carrot (*Daucus carota*). Postharvest Biol Technol 119:18–26

Demiray H, Dereboylu AE, Altan F, Zeytünlüoğlu A (2011) Identification of proteins involved in excess boron stress in roots of carrot (*Daucus carota* L.) and role of niacin in the protein profiles. Afr J Biotechnol 10:15545–15551

Eraslan F, Inal A, Gunes A, Alpaslan M (2007) Impact of exogenous salicylic acid on the growth, antioxidant activity and physiology of carrot plants subjected to combined salinity and boron toxicity. Sci Hortic 113:120–128

Formica-Oliveira AC, Martínez-Hernández GB, Díaz-López V, Artés F, Artés-Hernández F (2017) Effects of UV-B and UV-C combination on phenolic compounds biosynthesis in fresh-cut carrots. Postharvest Biol Technol 127:99–104

Foyer H, Noctor G (2005) Redox homeostasis and antioxidant signaling: a metabolic interface between stress perception and physiological responses. Plant Cell 17:1866–1875

Fuentes P, Pizarro L, Moreno JC, Handford M, Rodriguez-Concepcion M, Stange C (2012) Light-dependent changes in plastid differentiation influence carotenoid gene expression and accumulation in carrot roots. Plant Mol Biol 79:47–59

Gibberd MR, Turner NC, Storey R (2002) Influence of saline irrigation on growth, ion accumulation and partitioning and leaf gas exchange of carrot (*Daucus carota* L.). Ann Bot 90:715–724

Gill SS, Tuteja N (2010) Reactive oxygen species and antioxidant machinery in abiotic stress tolerance in crop plants. Plant Physiol Biochem 48:909–930

Han C, Li J, Jin P, Li X, Wang L, Zheng Y (2017) The effect of temperature on phenolic content in wounded carrots. Food Chem 215:116–123

Heredia JB, Cisneros-Zevallos L (2009) The effect of exogenous ethylene and methyl jasmonate on pal activity, phenolic profiles and antioxidant capacity of carrots (*Daucus carota*) under different wounding intensities. Postharvest Biol Technol 51:242–249

Huang Y, Li M-Y, Wang F, Xu Z-S, Huang W, Wang G-L, Ma J, Xiong A-S (2015) Heat shock factors in carrot: genome-wide identification, classification, and expression profiles response to abiotic stress. Mol Biol Rep 42:893–905

Huang W, Huang Y, Li M-Y, Wang F, Xu Z-S, Xiong A-S (2016) Dof transcription factors in carrot: genome-wide analysis and their response to abiotic stress. Biotechnol Lett 38:145–155

Iorizzo I, Ellison E, Senalik D, Zeng P, Satapoomin P, Huang J, Bowman M, Iovene M, Sanseverino W, Cavagnaro P, Yildiz M, Macko-Podgórni A, Moranska E, Grzebelus E, Grzebelus D, Ashrafi H, Zheng Z, Cheng S, Spooner D, Van Deynze A, Simon P (2016) A high-quality carrot genome assembly provides new insights into carotenoid accumulation and asterid genome evolution. Nat Genet 48:657–666

Jung YC, Lee HJ, Yum SS, Soh WY, Cho DY, Auh CK, Lee TK, Soh HC, Kim YS, Lee SC (2005) Drought-inducible—but ABA-independent—thaumatin-like protein from carrot (*Daucus carota* L.). Plant Cell Rep 24:366–373

Kasiri MR, Hassandokht MR, Kashi A, Shahi-Gharahlar A (2013) Evaluation of genetic diversity in Iranian yellow carrot accessions (*Daucus carota* var. *sativus*), an exposed to extinction rooty vegetable, using morphological characters. Int J Agric Crop Sci 6:151–156

Khraiwesh B, Zhu J-K, Zhu J (2012) Role of miRNAs and siRNAs in biotic and abiotic stress responses of plants. Biochim Biophys Acta 1819:137–148

Kiełkowska A, Grzebelus E, Lis-Krzyścin A, Maćkowska K (2019) Application of the salt stress to the protoplast cultures of the carrot (*Daucus carota* L.) and evaluation of the response of regenerants to soil salinity. Plant Cell Tiss Organ Cult. https://doi.org/10.1007/s11240-019-01578-7

Klimek-Chodacka M, Oleszkiewicz T, Lowder LG, Qi Y, Baranski R (2018) Efficient CRISPR/Cas9-based genome editing in carrot cells. Plant Cell Rep 37:575–586

Knight H, Knight MR (2001) Abiotic stress signalling pathways: specificity and cross-talk. Trends Plant Sci 6:262–267

Kovács G, Sorvari S, Scott P, Toldi O (2006) Pyrophosphate: fructose 6-phosphate 1-phosphotransferase operates in net gluconeogenic direction in taproots of cold and drought stressed carrot plants. Acta Biol Szeged 50:25–30

Kumar S, Dhingra A, Daniell H (2004) Plastid-expressed betaine aldehyde dehydrogenase gene in carrot cultured cells, roots, and leaves confers enhanced salt tolerance. Plant Physiol 136:2843–2854

Kumar SR, Anandhan S, Dhivya S, Zakwan A, Sathishkumar R (2013) Isolation and characterization of cold inducible genes in carrot by suppression subtractive hybridization. Biol Plantarum 57:97–104

Kumar SR, Kiruba R, Balamurugan S, Cardoso HG, Arnholdt-Schmitt B, Zakwan A, Sathishkumar R (2014) Carrot antifreeze protein enhances chilling tolerance in transgenic tomato. Acta Physiol Plant 36:21–27

Li M-Y, Xu Z-S, Huang Y, Tian C, Wang F, Xiong A-S (2015) Genome wide analysis of AP2/ERF transcription factors in carrot (*Daucus carota* L.) reveals evolution and expression profiles under abiotic stress. Mol Genet Genomics 290:2049–2061

Li M-Y, Xu Z-S, Tian C, Huang Y, Wang F, Xiong A-S (2016) Genomic identification of WRKY transcription factors in carrot (*Daucus carota*) and analysis of evolution and homologous groups for plants. Sci Rep 6:23101

Ma J, Xu Z-S, Wang F, Xiong A-S (2015) Isolation, purification and characterization of two laccases from carrot (*Daucus carota* L.) and their response to abiotic and metal ions stresses. Protein J 34:444–452

Maeda K, Kimura S, Demura T, Takeda J, Ozeki Y (2005) DcMYB1 acts as a transcriptional activator of the carrot phenylalanine ammonia-lyase gene (*DcPAL1*) in response to elicitor treatment, UV-B irradiation and the dilution effect. Plant Mol Biol 59:739–752

Malik MK, Slovin JP, Hwang CH, Zimmerman JL (1999) Modified expression of a carrot small heat shock protein gene *Hsp17.7*, results in increased or decreased thermotolerance. Plant J 20:89–99

Meyer K, Keil M, Naldrett MJ (1999) A leucine-rich repeat protein of carrot that exhibits antifreeze activity. FEBS Lett 447:171–178

Mirouze M, Paszkowski J (2011) Epigenetic contribution to stress adaptation in plants. Curr Opin Plant Biol 14:267–274

Nogales A, Nobre T, Cardoso HG, Muñoz-Sanhueza L, Valadas V, Campos MD, Arnholdt-Schmitt B (2016) Allelic variation on *DcAOX1* gene in carrot (*Daucus carota* L.): an interesting simple sequence repeat in a highly variable intron. Plant Gene 5:49–55

Osakabe Y, Yamaguchi-Shinozaki K, Shinozaki K, Phan Tran L-S (2013) Sensing the environment: key roles of membrane-localized kinases in plant perception and response to abiotic stress. J Exp Bot 64:445–458

Park H, Ko E, Jang E, Park S, Lee J, Ahn Y-J (2013) Expression of *DcHsp17.7*, a small heat shock protein gene in carrot (*Daucus carota* L.). Hortic Environ Biotechnol 54:121–127

Qin F, Shinozaki K, Yamaguchi-Shinozaki K (2011) Achievements and challenges in understanding plant abiotic stress responses and tolerance. Plant Cell Physiol 52:1569–1582

Que Q, Wang G-L, Feng K, Xu Z-S, Wang F, Xiong A-S (2018) Hypoxia enhances lignification and affects the anatomical structure in hydroponic cultivation of carrot taproot. Plant Cell Rep. https://doi.org/10.1007/s00299-018-2288-3

Rosa M, Prado C, Podazza G, Interdonato R, Gonzalez JA, Hilal M, Prado FE (2009) Soluble sugars—metabolism, sensing and abiotic stress. A complex network in the life of plants. Plant Signal Behav 4–5:388–393

Sánchez-Rangel JC, Jacobo-Velázquez DA, Cisneros-Zevallos L, Benavides J (2014) Primary recovery of bioactive compounds from stressed carrot tissue using aqueous two-phase systems strategies. J Chem Technol Biotechnol 91:144–154

Sanità di Toppi L, Vurro E, De Benedictis M, Falasca G, Zanella L, Musetti R, Lenucci MS, Dalessandro G, Altamura MM (2012) A bifasic response to cadmium stress in carrot: early acclimatory mechanisms give way to root collapse further to prolonged metal exposure. Plant Physiol Biochem 58:269–279

Satoh S, Sturm A, Fujii T, Chrispeels MJ (1992) cDNA cloning of an extracellular dermal glycoprotein of carrot and its expression in response to wounding. Planta 188:432–438

Schmidhalter U, Oertli JJ (1991) Germination and seedling growth of carrots under salinity and moisture stress. Plant Soil 132:243–251

Simon PW, Freeman R, Vieira JV, Boiteux LS, Briard M, Nothnagel T, Michalik B, Kwon Y-S (2008) Carrot. In: Prohens J, Nuez F (eds) Handbook of plant breeding, vegetables II. Fabaceae, Liliaceae, Solanaceae, and Umbelliferae. Springer, New York, pp 327–357

Simpson K, Fuentes P, Quiroz-Iturra LF, Flores-Ortiz C, Contreras R, Handford M, Stange C (2018) Unraveling the induction of phytoene synthase 2 expression by salt stress and abscisic acid in *Daucus carota*. J Exp Bot

Smallwood M, Worrall D, Byass L, Elias L, Ashford D, Doucet CJ, Holt C, Telford J, Lillford P, Bowles DJ (1999) Isolation and characterization of a novel antifreeze protein from carrot (*Daucus carota*). Biochem J 340:385–391

Smoleń A, Sady W (2006) The content of Cd, Cu and Zn in carrot storage roots as related to differentiated nitrogen fertilization and foliar nutrition. Pol J Environ Stud 15:503–509

Smoleń A, Sady W (2007) The effect of nitrogen fertilizer form and foliar application on Cd, Cu and Zn concentrations in carrot. Folia Hortic 19:87–96

Song N-H, Ahn Y-J (2010) DcHsp17.7, a small heat shock protein from carrot, is upregulated under cold stress and enhances cold tolerance by functioning as a molecular chaperone. HortScience 45:469–474

Song N-H, Ahn Y-J (2011) DcHsp17.7, a small heat shock protein in carrot, is tissue-specifically expressed under salt stress and confers tolerance to salinity. New Biotechnol 28:698–704

Sunkar R, Li Y-F, Jagadeeswaran G (2012) Functions of microRNAs in plant stress responses. Trends Plant Sci 17:196–203

Tabor G, Yesuf M, Haile M, Kebede G, Tilahun S (2016) Performance of some Asian carrot (*Daucus carota* L. ssp. *sativa* Hoffm.) cultivars under Ethiopian conditions: carrot and seed yields. Sci Hortic 207:176–182

Wang H, Ou CG, Zhuang FY, Ma ZG (2014) The dual role of phytoene synthase genes in carotenogenesis in carrot roots and leaves. Mol Breeding 34:2065–2079

Zheng R-L, Li H-F, Jiang R-F, Zhang F-S (2008) Cadmium accumulation in the edible parts of different cultivars of radish, *Raphanus sativus* L., and carrot, *Daucus carota* var. *sativa*, grown in a Cd-contaminated soil. Bull Environ Contam Toxicol 81:75–79

MIX
Papier aus verantwortungsvollen Quellen
Paper from responsible sources
FSC® C105338

If you have any concerns about our products,
you can contact us on
ProductSafety@springernature.com

In case Publisher is established outside the EU,
the EU authorized representative is:
Springer Nature Customer Service Center GmbH
Europaplatz 3, 69115 Heidelberg, Germany

Printed by Libri Plureos GmbH
in Hamburg, Germany